Practical Seismic Data Analysis

This modern introduction to seismic data processing in both exploration and global geophysics demonstrates practical applications through real data and tutorial examples. The underlying physics and mathematics of the various seismic analysis methods are presented, giving students an appreciation of their limitations and potential for creating models of the subsurface. Designed for a one-semester course, this textbook discusses key techniques within the context of the world's ever-increasing need for petroleum and mineral resources – equipping upper undergraduate and graduate students with the tools they need for a career in industry.

Key features

- Examples throughout the texts allow students to compare different data analysis methods and can be demonstrated using the instructor's software of choice.
- Exercises at the end of sections allow the students to check their understanding and put the theory into practice.
- Further reading lists encourage exploration of more advanced and detailed literature relating to topics covered in the book.

Hua-Wei Zhou is Professor and Robert Sheriff Endowed Chair of Applied Seismology at the University of Houston, and has held the position of "Lv-Ka" Scholar at Ocean University of China in Qingdao since 2010. He is one of the few experts in seismic imaging to have done extensive research in both exploration geophysics and solid Earth geophysics, having worked for Exxon Production Research Company as well as having 20 years' academic teaching experience. He has taught the materials for this book in graduate classes as well as in industry short courses, given in the USA, South Africa, and China.

"This book is a valuable reference for senior undergraduates and graduates in exploration geophysics and seismology. It covers all the common methods and steps of seismic data processing, using clearly presented mathematics. In particular, helpful Boxes in the text enable readers to better understand both basic and crucial information, supported well by the Exercises and Further Reading lists also provided."
– Jingyi Chen, *Decker Dawson Assistant Professor of Geophysics, University of Tulsa*

Practical Seismic Data Analysis

HUA-WEI ZHOU

University of Houston

CAMBRIDGE
UNIVERSITY PRESS

CAMBRIDGE
UNIVERSITY PRESS

University Printing House, Cambridge CB2 8BS, United Kingdom

Published in the United States of America by Cambridge University Press, New York

Cambridge University Press is part of the University of Cambridge.

It furthers the University's mission by disseminating knowledge in the pursuit of education, learning, and research at the highest international levels of excellence.

www.cambridge.org
Information on this title: www.cambridge.org/9780521199100

© Hua-Wei Zhou 2014

First published 2014

Printed in the United Kingdom by TJ International, Padstow, Cornwall

A catalog record for this publication is available from the British Library

Library of Congress Cataloging in Publication data
Zhou, Hua-Wei, 1957–
Practical seismic data analysis / Hua-Wei Zhou.
    pages   cm
Includes bibliographical references and index.
ISBN 978-0-521-19910-0 (hardback)
1. Seismology – Textbooks.   I. Title.
QE534.3.Z56   2013
551.22 – dc23   2013035758

ISBN 978-0-521-19910-0 Hardback

Additional resources for this publication at www.cambridge.org/zhou

Cambridge University Press has no responsibility for the persistence or accuracy of
URLs for external or third-party Internet websites referred to in this publication,
and does not guarantee that any content on such websites is, or will remain,
accurate or appropriate.

CONTENTS

PREFACE

Seismic data analysis transfers seismic records measured at the surface or along wellbores into imagery, estimates, and models of subsurface structures and properties. It covers the topics of digital seismic data processing, seismic migration, and subsurface model building that are useful in both exploration geophysics and solid Earth geophysics. Although several excellent books have covered these topics either from the viewpoint of exploration geophysics or that of solid Earth geophysics, I was motivated to write this book to deal with common seismic analysis methods for both aspects of geophysics. This book is intended as an introductory text on common and practical methods in seismic data analysis.

Most of the materials for this book originated as lecture notes for graduate courses in geophysics at University of Houston and Texas Tech University. Students on these courses usually have a variety of backgrounds: many are recent graduates from geophysics, geology, engineering, computer sciences, or other physical science disciplines, and others are employees in the petroleum industry. They intend to apply seismic data analysis skills to problems in exploration geophysics, solid Earth geophysics, and engineering and environmental sciences. Although they may have access to some commercial or free software in seismic processing, most of these students have not gone through a systematic review of common approaches to seismic data analysis and the practical limitations of each method. Hence, an effort has been made in this book to emphasize the concepts and practicality of common seismic analysis methods using tutorial and case examples or schematic plots.

The first six chapters of the book prepare the background and deal mostly with time processing. Chapter 1 introduces seismic data and issues of sampling, amplitude, and phase. Chapter 2 addresses pre-processing of reflection seismic data using examples on normal moveout (NMO) analysis, noise suppression, and near-surface statics. The topics of discrete Fourier transform and wavelet transfer are both discussed in Chapter 3 in terms of the law of decomposition and superposition. Chapter 4 is devoted to the meaning and assessment of seismic resolution and fidelity. Chapter 5 discusses filtering of time series using z-transform and Fourier transform methods. Chapter 6 covers several common deconvolution methods.

Each of the final four chapters may be studied independently: Chapters 7 to 9 are on three main branches of seismic data analysis, and Chapter 10 covers several special topics. Chapter 7 introduces several seismic migration methods that have served as the main subsurface seismic imaging tools in exploration geophysics. Chapter 8 is on seismic velocity analysis using semblance, migration, and tomography. Chapter 9 discusses the basic issues and relationship between seismic modeling and inversion. Chapter 10 addresses processing issues in topics of seismic data acquisition, suppressing of multiple reflections, seismic velocity anisotropy, multi-component seismic data, and seismic attributes.

Each chapter starts with an overview paragraph describing the sections to follow. Terms defined are indicated by bold font. For students, it is especially important to comprehend the meaning of common terms and concepts in the field because this often reflects the depth of their understanding. A large number of figures are given that illustrate concepts or applications. Several boxes are provided in each chapter to examine specific case studies or ideas. There is an exercise at the end of each main section. Each chapter ends with a summary of key concepts, and a list of further reading. All serious learners should read several technical papers from the suggested reading lists, to draw connections between the issues covered by the chapter and the reference papers.

The mathematical content has been kept to a minimum, although I assume that readers are comfortable with basic calculus and linear algebra including matrices. Most parts of the book should be readable by those with an undergraduate degree in physical science or engineering. Readers without much mathematical training should focus on the main concepts and physical meanings.

This book could not have been completed without the encouragement of Dr. Robert E. Sheriff, my colleague and mentor. I would like to thank my fellow geophysicists for granting permission to reproduce figures from their publications. I acknowledge the assistance of many people in the preparation of this book, especially those students who provided feedback. I particularly thank Kurt Marfurt, Oong Youn, Mike Thornton, Zhihui Zou, Fang Yuan, and Wendy Zhang. This book is dedicated to my parents.

Hua-Wei Zhou

1 Introduction to seismic data and processing

The discipline of subsurface seismic imaging, or mapping the subsurface using seismic waves, takes a remote sensing approach to probe the Earth's interior. It measures ground motion along the surface and in wellbores, then puts the recorded data through a series of data processing steps to produce seismic images of the Earth's interior in terms of variations in seismic velocity and density. The ground movements recorded by seismic sensors (such as geophones and seismometers onshore, or hydrophones and ocean bottom seismometers offshore) contain information on the media's response to the seismic wave energy that traverses them. Hence the first topic of this chapter is on seismic data and their acquisition, processing, and interpretation processes. Because nearly all modern seismic data are in digital form in order to be stored and analyzed in computers, we need to learn several important concepts about sampled time series such as sampling rate and aliasing; the latter is an artifact due to under-sampling. In exploration seismology, many useful and quantifiable properties of seismic data are called seismic attributes. Two of the most common seismic attributes are the amplitude and phase of seismic wiggles. They are introduced here together with relevant processing issues such as gain control, phase properties of wavelets, and the Hilbert transform, which enables many time-domain seismic attributes to be extracted. To process real seismic data, we also need to know the basic issues of data formats, the rules of storing seismic data in computers. To assure that the data processing works, we need to conduct many quality control checks. These two topics are discussed together because in practice some simple quality control measures need to be applied at the beginning stage of a processing project.

A newcomer to the field of seismic data processing needs to know the fundamental principles as well as common technical terms in their new field. In this book, phrases in **boldface** denote where special terms or concepts are defined or discussed. To comprehend each new term or concept, a reader should try to define the term in his or her own words. The subject of seismic data processing often uses mathematical formulas to quantify the physical concepts and logic behind the processing sequences. The reader should try to learn the relevant mathematics as much as possible, and, at the very least, try to understand the physical basis and potential applications for each formula. Although it is impossible for this book to endorse particular seismic processing software, readers are encouraged to use any commercially or openly accessible seismic processing software while learning seismic data processing procedures and exercises. An advanced learner should try to write computer code for important processing steps to allow an in-depth comprehension of the practical issues and limitations.

1.1 Seismic data and their acquisition, processing, and interpretation

As a newcomer, you first want to know the big picture: the current and future objectives and practices of seismic data processing, and the relationship of this field to other related disciplines. You will need to comprehend the meanings of the most fundamental concepts in this field. This section defines seismic data and a suite of related concepts such as signal-to-noise ratio (SNR or S/N), various seismic gathers, common midpoint (CMP) binning and fold, stacking, pre-stack versus post-stack data, and pre-processing versus advanced processing. The relationship between acquisition, processing, and interpretation of seismic data is discussed here, since these three processes interrelate and complement each other to constitute the discipline of subsurface seismic imaging.

1.1.1 Digital seismic data

Seismic data are physical observations, measurements, or estimates about seismic sources, seismic waves, and their propagating media. They are components of the wider field of geophysical data, which includes information on seismic, magnetic, gravitational, geothermal, electromagnetic, rock physics, tectonophysics, geodynamics, oceanography, and atmospheric sciences. The form of seismic data varies, and can include analog graphs, digital time series, maps, text, or even ideas in some cases. This book treats the processing of a subset of seismic data, those in digital forms. We focus on the analysis of data on body

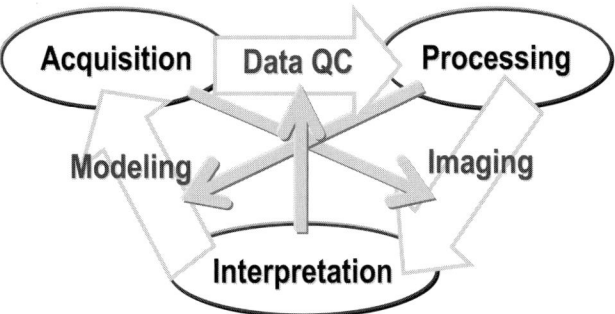

Figure 1.1 Relationship between data acquisition, processing, and interpretation.

waves, mostly P-waves, in their transmission, reflection, diffraction, refraction, and turning processes. The processing of other seismic data and many non-seismic data often follows similar principles.

The purpose of acquiring and processing seismic data is to learn something about the Earth's interior. To understand certain aspects of the Earth, we initially need to figure out some specific relations between the intended targets and measurable parameters. Then our first step is to conduct **data acquisition** designed for the problem, our second step to use **data processing** to identify and enhance the desired signal, and our third step to conduct **data interpretations** based on the processed data. In reality, the processes of data acquisition, processing and interpretation are interconnected and complement each other; their relationship may be viewed as shown in Figure 1.1.

After data acquisition and before data processing, we need to conduct the process of data **quality control**, or **QC**. This involves checking the survey geometry, data format, and consistency between different components of the dataset, and assuring ourselves that the quality and quantity of the dataset are satisfactory for our study objectives. The data QC process is typically part of the **pre-processing**. After pre-processing to suppress various kinds of noise in the data, **seismic imaging** is conducted to produce various forms of imagery for the interpretation process. The seismic imaging methods include seismic migration, seismic tomography, and many other methods of extracting various seismic attributes. Some people call seismic imaging methods the advanced processing. The scope of this book covers the entire procedure from pre-processing to seismic imaging.

After data interpretation, we often conduct **seismic modeling** using the interpreted model and the real data geometry to generate predictions to compare with the real measurements, and hence further verify the interpretation. The three inner arrows shown in Figure 1.1 show how the interactions between each pair of components (namely the data QC, imaging, or modeling processes) are influenced by the third component.

1.1.2 Geometry of seismic data gathers

Seismic data acquisition in the energy industry employs a variety of acquisition geometries. In cross-section views, Figure 1.2 shows two **seismic acquisition spreads**, the arrangements of shots and receivers in seismic surveys. Panel (a) shows a **split spread**, using a shot located in the middle and many receivers spread around it. This spread is typical of onshore

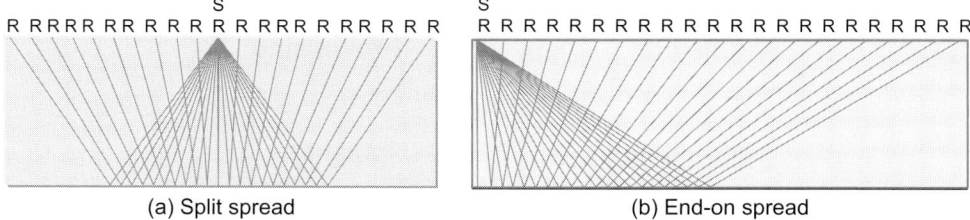

(a) Split spread   (b) End-on spread

Figure 1.2 Cross-section views of two seismic data acquisition spreads and raypaths.

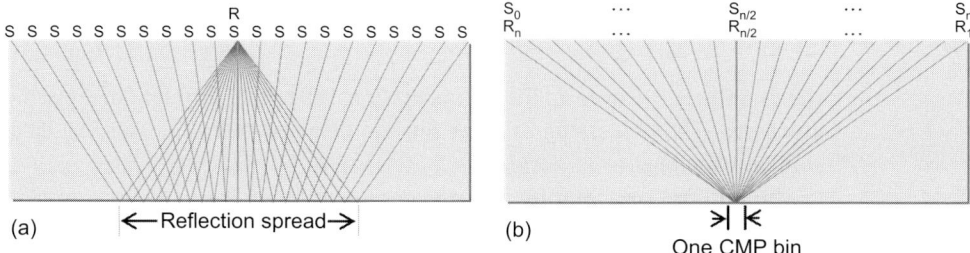

(a)   ←Reflection spread→   (b)   →| |←
                                             One CMP bin

Figure 1.3 Cross-section views of (a) a common receiver gather and (b) a common midpoint (CMP) gather.

acquisition geometry using dynamite or Vibroseis technology as sources and geophones as receivers. The real-world situation is much more complicated, with topographic variations, irregular source and receiver locations in 3D, and curving raypaths. Panel (b) shows an **end-on spread**, with a shot located at one end and all receivers located on one side of the shot. This spread is the case for most offshore seismic surveys using airgun or other controlled sources near the boat and one or more streamers of hydrophones as receivers. In comparison with onshore seismic data, offshore seismic data usually have much higher quality because of a number of favorable conditions offshore, including consistent and repeatable sources, good coupling conditions at sources and receivers, and the uniform property of water as the medium. However, offshore seismic data may have particular noise sources, especially multiple reflections, and at present most 3D offshore seismic surveys have much narrower azimuthal coverage than their onshore counterparts.

The seismic data traces collected from many receivers that have recorded the same shot, such as that shown in Figure 1.2, produce a **common shot gather (CSG)**. A **seismic gather** refers to a group of pre-stack seismic traces linked by a common threading point. The phrase "pre-stack traces" refers to data traces retaining the original source and receiver locations; they are in contrast to the "post-stack" or "stacked traces" that result from stacking or summing many traces together.

A **common receiver gather (CRG)** as shown in Figure 1.3a is a collection of traces recorded by the same receiver from many shots, and a **common midpoint (CMP) gather** (Figure 1.3b) is a collection of traces with their source-to-receiver midpoint falling within the same small area, called a **CMP bin**. Among the three common types of seismic gathers, the **reflection spread**, or the lateral extent of reflection points from a seismic gather across a reflector, is zero for the CMP gather in the case of a flat reflector beneath a constant

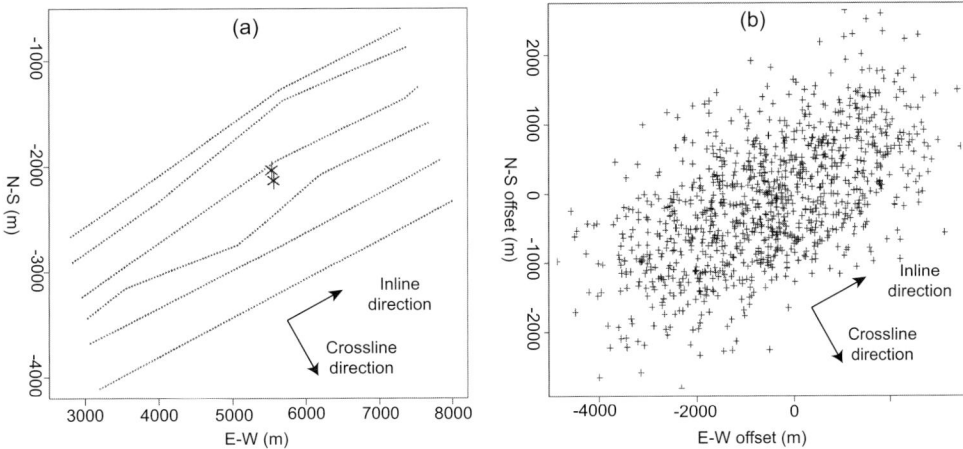

Figure 1.4 Map views of an acquisition geometry from the Canadian Rockies (Biondi, 2004).
(a) Locations of shots (asterisks) and receivers (dots) for two consecutive shot gathers.
(b) Offsets of 1000 traces, randomly selected.

velocity medium (Figure 1.3b). There are other gathers, such as a **common image-point (CIG) gather**, which is a collection of migrated traces at the same image bin location. Some people call a collection of traces with the same amount of source-to-receiver offset as a common offset gather, though it is logically a **common offset section**.

1.1.3   CMP binning and seismic illumination

Owing to the minimum spread of reflection points, traces of each CMP gather can be summed or stacked together to form a single **stacked trace**, A stacked trace is often used to approximate a zero-offset trace, which can be acquired by placing a shot and a receiver at the same position. The stacked trace has good signal content because the stacking process allows it to take all the common features of the original traces in the gather. Consequently, the CMP gathers are preferred to other gathers in many seismic data processing procedures. However, because the CSG or CRG data are actually collected in the field, a process of re-sorting has to be done to reorganize the field data into the CMP arrangement. This is done through a process called **binning**, by dividing the 2D line range or the 3D survey area into a number of equal-sized **CMP bins** and, for each bin, collecting those traces whose **midpoints** fall within the bin as the CMP gather of this bin. The number of traces, or midpoints, within each CMP bin is called the **fold**. As an important seismic survey parameter, the fold represents the multiplicity of CMP data (Sheriff, 1991).

Figures 1.4 and 1.5, respectively, show the geometries of two 3D surveys onshore and offshore. In each of these figures the left panel shows the locations of the shots and receivers, and the right panel shows the midpoint locations of 1000 traces randomly selected from the corresponding survey. To maintain a good **seismic illumination**, the fold should be high enough and distributed as evenly as possible over the survey area. In practice, the desire for good seismic illumination has to be balanced against the desire to make the survey as efficient as possible to reduce the cost in money and time. In 3D onshore seismic surveys,

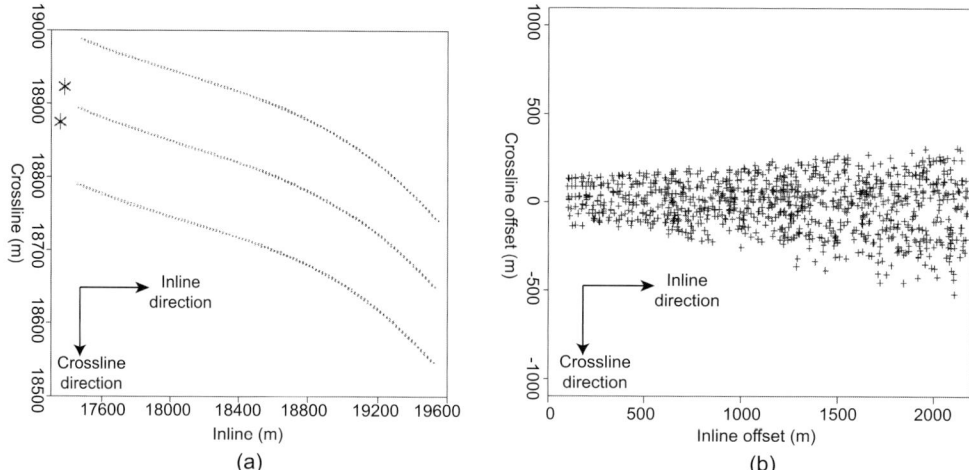

Figure 1.5 Map views of a three-streamer acquisition from the North Sea (Biondi, 2004).
(a) Locations of shots (asterisks) and receivers (dots) for two consecutive shot gathers.
(b) Offsets of 1000 traces, randomly selected.

the orientations of the shot lines are often perpendicular to the orientations of the receiver lines in order to maximize the azimuthal coverage of each **swath**, which is a patch of area recorded by an array of sensors at one time. Typically there is an **inline direction** along which the spatial sampling is denser than the perpendicular **crossline direction**. The inline is often along the receiver line direction, like that shown in Figure 1.4a, because the spacing between receivers is typically denser than the spacing between shots. In the case of irregular distributions of shots and receiver lines, however, the inline direction may be decided based on the distribution of midpoints of data, like that shown in Figure 1.5b.

Sometimes special layouts of shot and receivers are taken to optimize the seismic illumination. Figure 1.6 shows an example of a special 3D seismic survey geometry over the Vinton salt dome in southwest Louisiana. The survey placed receivers along radial lines and shots in circular geometry centered right over the subsurface salt diapir. In most applied sciences, quality and cost are the two main objectives that often conflict with each other, and the cost is in terms of both money and time. Because geophones today are connected by cables, they are most effectively deployed in linear geometry, such as along the radial lines in this example. The sources here were Vibroseis trucks which can easily be run along the circular paths. Similarly, in carrying out seismic data processing projects, we need to satisfy both the quality and cost objectives.

1.1.4   SNR and CMP stacking

With respect to the objectives of each project, geophysical data may contain relevant information – the **signal** – and irrelevant components – **noise**. A common goal for digital data processing in general and for seismic data processing in particular is to improve the **signal-to-noise ratio** or **SNR**. In seismology the SNR is often expressed as the ratio between the amplitude of the signal portion and the amplitude of the noise portion of seismic traces.

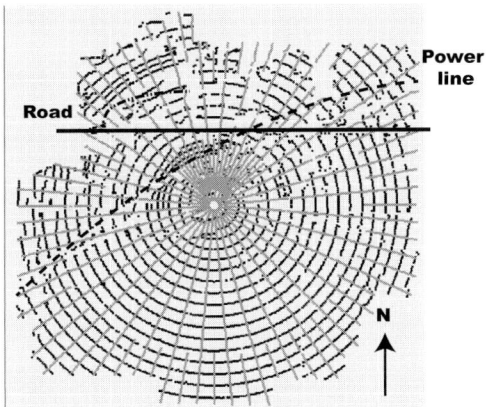

Figure 1.6 Map view of a 3D seismic survey over the Vinton salt dome in west Louisiana. The straight radial lines denote receiver positions, and the circular lines denote shot positions. The geometry of the shot and receiver layout is designed to optimize the coverage of reflection waves from the boundary of the underlying salt dome.

Box 1.1 Why use CMP stacking and what are the assumptions?

The main reason is to improve the SNR and focus the processing on the most coherent events in the CMP gather. CMP stacking is also a necessary step for post-stack migration where each stacked trace is regarded as a zero-offset trace. The assumption is there is a layer-cake depth velocity model, at least locally within each CMP gather.

In practice the meaning of signal versus noise is relative to the objectives of the study and the chosen data processing strategy. Similarly, the meanings of **raw data** versus **processed data** may refer to the input and output of each specific processing project. The existence of noise often demands that we treat seismic data from a statistical point of view.

Common midpoint (**CMP**) **stacking** (see Box 1.1) refers to summing up those seismic traces whose reflections are expected to occur at the same time span or comparable reflection depths. The main motivation for such stacking is to improve the SNR. In fact, stacking is the most effective way to improve the SNR in many observational sciences. A **midpoint** for a source and receiver pair is simply the middle position between the source and receiver. In a **layer-cake model** of the subsurface, the reflection points on all reflectors for a pair of source and receiver will be located vertically beneath the midpoint (Figure 1.7). Since the layer-cake model is viewed as statistically the most representative situation, it is commonly taken as the default model, and the lateral positions of real reflectors usually occur quite close to the midpoint. Consequently on cross-sections we usually plot seismic traces at their midpoints. Clearly, many traces share the same midpoint. In the configuration of CMP binning, the number of traces in each CMP bin is the fold.

It is a common practice in seismic data processing to conduct CMP stacking to produce stacked sections. Thus, reflection seismic data can be divided into pre-stack data and

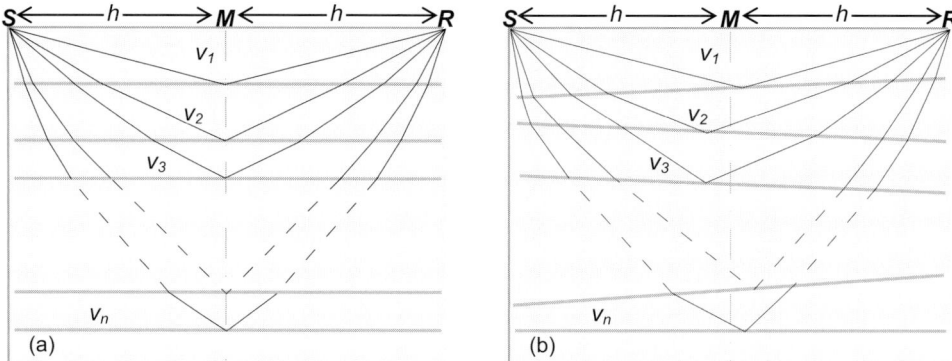

Figure 1.7 Reflection rays (black lines) from a source S to a receiver R in (a) a layer-cake model; and (b) a model of dipping layers. All reflection points are located vertically beneath the midpoint M in the layer-cake model.

post-stack data, and processing can be divided into pre-stack processing and post-stack processing. The traditional time sections are obtained through the process of stacking and then post-stack migration. Modern processing often involves pre-stack processing and migration to derive depth sections that have accounted for lateral velocity variations and therefore supposedly have less error in reflector geometry and amplitude. One can also conduct depth conversion from time section to depth section using a velocity–depth function. Post-stack seismic processing is cheaper and more stable but less accurate than pre-stack seismic processing. In contrast, pre-stack seismic processing is more costly, often unstable, but potentially more accurate than post-stack seismic processing.

1.1.5   Data processing sequence

The primary objective of this book is to allow the reader to gain a comprehensive under-standing of the principles and procedures of common seismic data processing and anal-ysis techniques. The sequence of processing from raw seismic data all the way to final forms ready for interpretation has evolved over the years, and many general aspects of the sequence have become more-or-less conventional. It is a non-trivial matter to design a proper sequence of seismic data processing, called a **processing flow**. Figure 1.8 shows an example of a processing flow for reflection seismic data more than 30 years ago (W. A. Schneider, unpublished class notes, 1977). The general procedure shown in this figure still holds true for today's processing flow for making post-stack sections.

    The goal of seismic data processing is to help interpretation, the process of deciphering the useful information contained in the data. The task is to transfer the raw data into a form that is optimal for extracting the signal. The word "optimal" implies making the best choice after considering all factors. Hence we need to make decisions in the process of seismic data analysis. All methods of seismic data analysis rely on physical and geological theories that tie the seismic data and the geological problem together. For instance, a problem of inferring aligned fractures may involve the theory of seismic anisotropy. The subsequent data processing will attempt to utilize this theory to extract the signal of the fracture

a. Data Conditioning

Input: Field tapes
1. Gain removal $G\text{-}1(t)$
2. Source array stack
3. Source correction (Vibroseis, etc.)
4. True amplitude recovery
5. Trace editing
6. Wavelet deconvolution
7. Reverberation deconvolution
8. CDP sorting
Output: CDP file

b. Parameter Analysis

Input: CDP file
1. Statics correction
2. Stacking velocity analysis
3. Residual statics analysis / Velocity interpretation
4. QC stack
Output: CDP, statics, & velocity files

c. Data Enhancement

Input: CDP, statics, & velocity files
1. NMO & statics corrections
2. CDP stack
3. Earth absorption compensation
4. Time variant band-pass filtering
5. Display of time section
Output: Time section

d. Migration / Depth Conversion

Input: CDP & velocity files
1. Time migration
2. Migration velocity analysis
3. Time migration & depth conversion
4. Depth migration
Output: Migrated volumes

e. Modeling & Interpretation

Produce reservoir models based on seismic, geology, & well data

f. Exploration Decision Making

1. Where, when, & how to drill?
2. Analysis risks & economics

Figure 1.8 A general processing flow, after Schneider (unpublished class notes from 1977). Steps **c**, **d**, and **e** are usually iterated to test different hypotheses. Pre-stack processing is often conducted after a post-stack processing to help the velocity model building process. There are also reports of pre-stack processing using limited offsets to increase the efficiency.

orientation according to the angular variation of traveling speed, and to suppress the noise that may hamper the signal extraction process.

Exercise 1.1

1. How would you estimate the fold, the number of the source-to-receiver midpoints in each CMP bin, from a survey map like that shown in Figure 1.6? Describe your procedure and assumptions.

2. As shown in Figure 1.7, the shapes of reflection raypaths tend to resemble the letter "U" rather than the letter "V". Explain the reason behind this phenomenon.

3. Update the processing flow shown in Figure 1.8 by finding and reading at least two papers published within the past 10 years. What happened to those processing steps in Figure 1.8 that are missing from your updated processing flow?

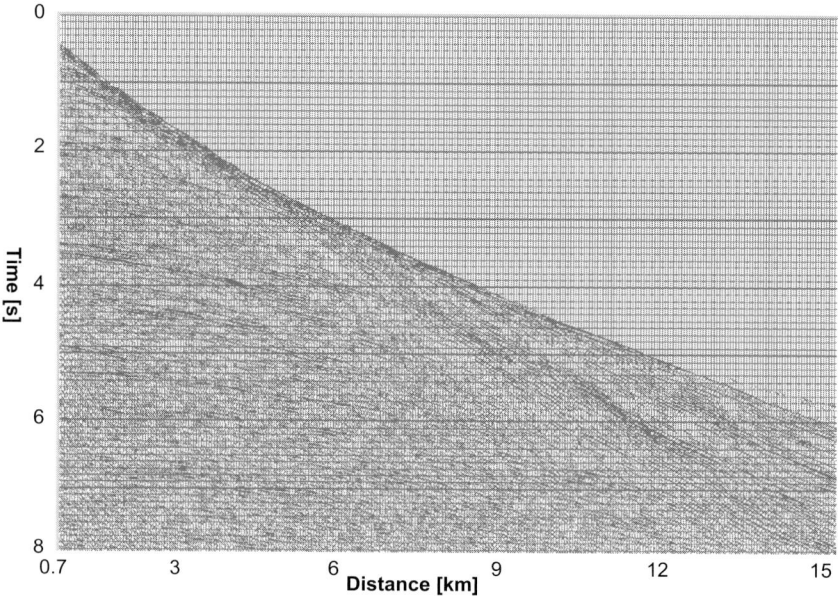

Figure 1.9 A common shot gather from an offshore seismic survey.

1.2   Sampled time series, sampling rate, and aliasing

Through their propagation history, seismic waves vary in a continuous manner in both temporal and spatial dimensions. However, measurements of seismic data need to be sampled into digital form in order to be stored and processed using computers. At the acquisition stage each trace of seismic wiggles has been digitized at a constant sample interval, such as 2 ms (milliseconds). The resulted string of numbers is known as a time series, where the number represents the amplitude of the trace at the corresponding sample points. In the following, some basic properties of the sampled time series are introduced.

1.2.1   Sampled time series

Figure 1.9 shows an example of offshore seismic data for which the streamer of hydrophones is nearly 20 km long. We treat each recorded seismic trace as a **time series**, which is conceptualized as an ordered string of values, and each value represents the magnitude of a certain property of a physical process. The word "time" here implies sequencing or connecting points in an orderly fashion. A continuous geological process may be sampled into a discrete sequence called a **sampled time series**. Although the length of the sample is usually finite, it may be extrapolated to infinity when necessary. All the data processing techniques discussed in this book deal with sampled time series. A 1D time series is usually taken to simplify the discussion. However, we should not restrict the use of time series to just the 1D case, because there are many higher-dimensional applications.

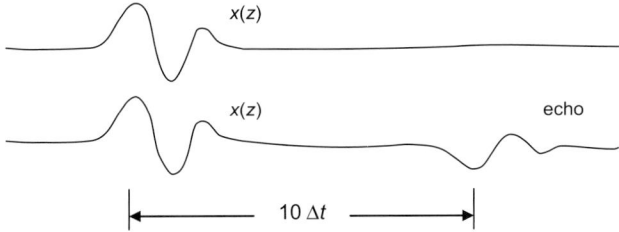

Figure 1.10 A signal $x(z)$ and its echo.

The z-transform

Perhaps the easiest way to represent a 1D time series is using the **z-transform**, a polynomial in z in which the coefficients are filled with the values of the time series (Claerbout, 1985a). For instance, a time series

$$\ldots, \; a_{-2}, \; a_{-1}, \; a_0, \; a_1, \; a_2, \; a_3, \ldots \tag{1-1}$$

is represented by

$$x(z) = \cdots + a_{-2}z^{-2} + a_{-1}z^{-1} + a_0z^0 + a_1z^1 + a_2z^2 + a_3z^3 + \cdots \tag{1-2}$$

So in the z-transform, the coefficients are the value of the time series, and the exponents denote the corresponding positions in the time series. The operator z can be interpreted as the **unit-delay operator** (or a unit-delay filter, to be described later). For instance, multiplying $x(z)$ by z will shift the whole time series by one sample point:

$$zx(z) = \cdots + a_{-2}z^{-1} + a_{-1}z^0 + a_0z^1 + a_1z^2 + a_2z^3 + a_3z^4 + \cdots \tag{1-3}$$

Comparing (1–2) and (1–3), we note that the term with coefficient a_k corresponds to z^k or time step k in (1–2), but z^{k+1} or time step $k + 1$ in (1–3). In the opposite direction, z^{-1} is the **unit-advance operator**. Therefore, the z-transform offers a convenient algebraic way to represent discrete geophysical data or time series.

The z-transform notation eases our understanding about processing digital data. An important fact is that using a complex variable z, it transforms data from a discrete time domain into a continuous frequency domain. This transform can also be used to describe more complicated signals. For instance, if **linearity** (the legitimacy of simple linear addition of two time series) holds true, a primary wave $x(z)$ plus an echoed wave of half the strength arriving 10 sample points later will be

$$y(z) = x(z) - 0.5\,x(z)\,z^{10} \tag{1-4}$$

Because linearity is a property of all low-amplitude waves, the equation can be seen as a seismogram of a primary phase $x(z)$ with its reflection echo, as shown in Figure 1.10.

Sampling rate, aliasing, and Nyquist condition

Let us now turn to **sampling rate** (or sample rate), the rate at which a continuous process is sampled into a time series. If the sampling rate is too slow, the sampled series may differ from the original continuous process. The distortion of the true frequency content due to

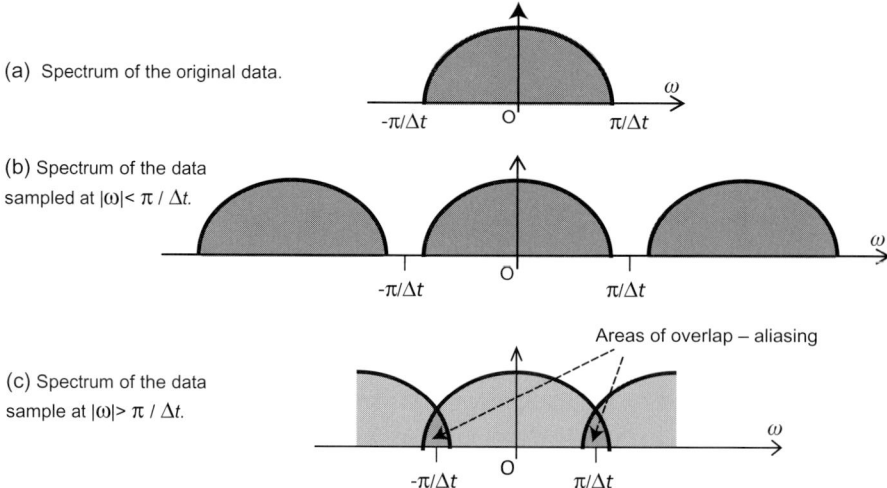

(a) Spectrum of the original data.

(b) Spectrum of the data sampled at $|\omega| < \pi / \Delta t$.

(c) Spectrum of the data sample at $|\omega| > \pi / \Delta t$.

Areas of overlap – aliasing

Figure 1.11 Aliasing as seen in the frequency domain.

under-sampling is called **aliasing**, which is a harmful artifact. On the other hand, if the sampling rate is too high, we may waste extra processing time and data storage. In the case that we know the frequency content of a continuous signal prior to sampling, we can use a sampling rate that is just high enough to prevent aliasing. This brings in the idea of the Nyquist condition to prevent aliasing.

The Nyquist condition can be examined using the z-transform by inserting the complex variable z in terms of

$$z = \exp(i\omega\Delta t)$$
$$= \cos(\omega\Delta t) + i \sin(\omega\Delta t) \tag{1–5}$$

where ω is the angular frequency and Δt is the sampling interval in the time domain. The above expression simply says that z is a complex variable with a phase angle $\omega\Delta t$.

As shown in Figure 1.11, we suppose that a time domain function $b(t)$ and its frequency domain counterpart $B(\omega)$ are both continuous. We want to compare the true spectrum $B(\omega)$ with $\underline{B}(\omega)$, the spectrum corresponding to the sampled time series $\{b_n\}$. Note that the spectrum here merely means the transformed function in the frequency domain. The z-transform of the sampled time series is

$$\underline{B}(\omega) = \sum b_n z^n \tag{1–6}$$

$\underline{B}(\omega)$ is already in the continuous frequency domain if we use (1–5). To see the spectrum using $\{b_n\}$, we may evaluate it along the **unit circle** as a function of the phase angle within $[-\pi, \pi]$. This is

$$\underline{B}(\omega) = \sum b_n \exp(i\omega n \Delta t) \tag{1–7}$$

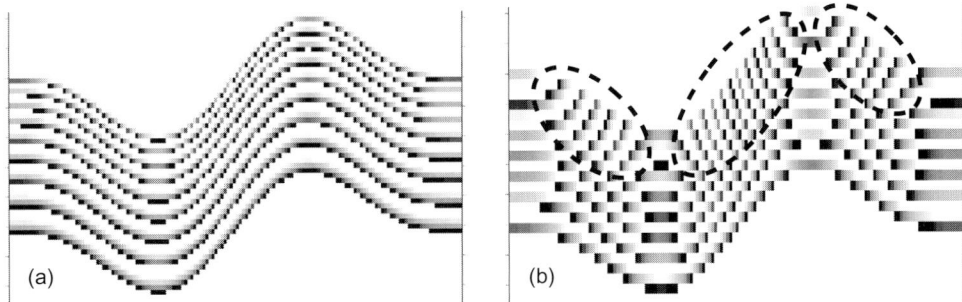

Figure 1.12 Examples of spatial aliasing. (a) A finely sampled section of folded strata.
(b) The same folded strata sampled coarsely. The poorly sampled steeply dipping thin layers,
as highlighted by three dashed ellipses, show a spatial aliasing artifact.

$\underline{B}(\omega)$ will be equal to the true $B(\omega)$ provided that the phase is restricted to within one cycle, i.e.,

$$|\omega| \leq \pi/\Delta t \qquad (1\text{–}8a)$$

Or using $f = \omega/2\pi$, we get

$$f \leq 1/(2\Delta t) \qquad (1\text{–}8b)$$

Hence, **a time series must be sampled for at least two points per wavelength cycle to avoid aliasing**; this is the **Nyquist condition**. The frequency $\omega_N = \pi/\Delta t$ is called the **Nyquist frequency**. The energy at frequencies higher than ω_N folds back into the principal region $(-\omega_N, \omega_N)$, known as the aliasing or edge folding phenomenon. In practice, the sampling rate may be set at 5 to 10 points per cycle, with the considerations that the signal frequency may be higher than anticipated and that a slight redundancy may help in constraining the noise without too much waste in over-sampling.

1.2.4   Spatial aliasing

In two or higher dimensions, under-sampling of dipping events may produce a particularly harmful imaging artifact called **spatial aliasing**. Such an artifact consists of false events in multi-dimensional data due to alignment of spatially under-sampled dipping events of high frequencies. In Figure 1.12, panel (a) shows a series of folded strata whose thickness increases with depth. There is not much evidence of spatial aliasing in this panel as it has a sufficient sampling rate. In panel (b), however, a coarse sampling rate is used for the same section as in (a). Spatial aliasing appears in the shallow and steeply dipping portions, as highlighted by the three dashed ellipses. The shallow layers are thinner (higher frequency) than the deep layers. By comparing the two panels in this figure, we see that the dipping direction of the spatial aliased artifact is usually symmetrically opposite to the correct dipping direction of the targets.

With a fixed sampling rate, the chance of spatial aliasing increases if we increase either the frequency or the dipping angle of the sampled dipping events. Hence, for those data processing steps that will increase the signal frequency or dipping angle, special measures are necessary to reduce the risk of spatial aliasing. For instance, anti-aliasing filtering is commonly applied after many seismic migration methods.

Exercise 1.2

1. A given seismic trace is sampled at 4 milliseconds. If the signal frequency is known to be up to 60 Hz, find a way to reduce the total number of sampling points without losing the signal.

2. In Figure 1.9 try to identify as many of the seismic events as you can. What are the main primary reflection events? What are the factors affecting the amplitudes of seismic events in this figure?

3. In your own words, define the following terms: signal, time series, z-transform, processing artifact, and spatial aliasing.

1.3   Seismic amplitude and gain control

1.3.1   Seismic amplitude

Seismic amplitude refers to the magnitude of the wiggles of seismic records and quantifies the energy level of seismic waves. This amplitude is one of the most important attributes of seismic data because it represents the distribution of the energy of seismic waves as a function of the propagating space, recording time, and frequency. Throughout seismic data processing, those processes that alter the amplitude of seismic data are called **gain controls**.

To examine the distribution of amplitude over frequency, we can use Fourier theory to decompose each seismic trace into a suite of frequencies, which will be discussed later in Chapter 3. Figure 1.13a gives an example from Yilmaz (1987), where the input trace on the left is a time series which is decomposed into 128 frequencies, shown as the other monochromatic time traces in this figure. Applying a time stack or horizontal summation of these 128 single-frequency time traces will result in the original input trace. The amplitude spectrum shown in Figure 1.13b is a plot of the amplitude of the monochromatic time traces against the frequency.

1.3.2   Source radiation pattern and media attenuation

The amplitude of seismic data is a function of three factors: the source, the receiver, and the media. The source factor may be quantified by the source **radiation pattern**. An explosion in fluids may have an "expanding ball" radiation pattern, while shear faulting may have a double-couple, or DC, "beach ball" radiation pattern. The radiation pattern of a real source is often complicated, and varies with time, frequency, and spatial angle from the source. Even for airguns, which are among the simplest seismic sources, the radiation pattern is

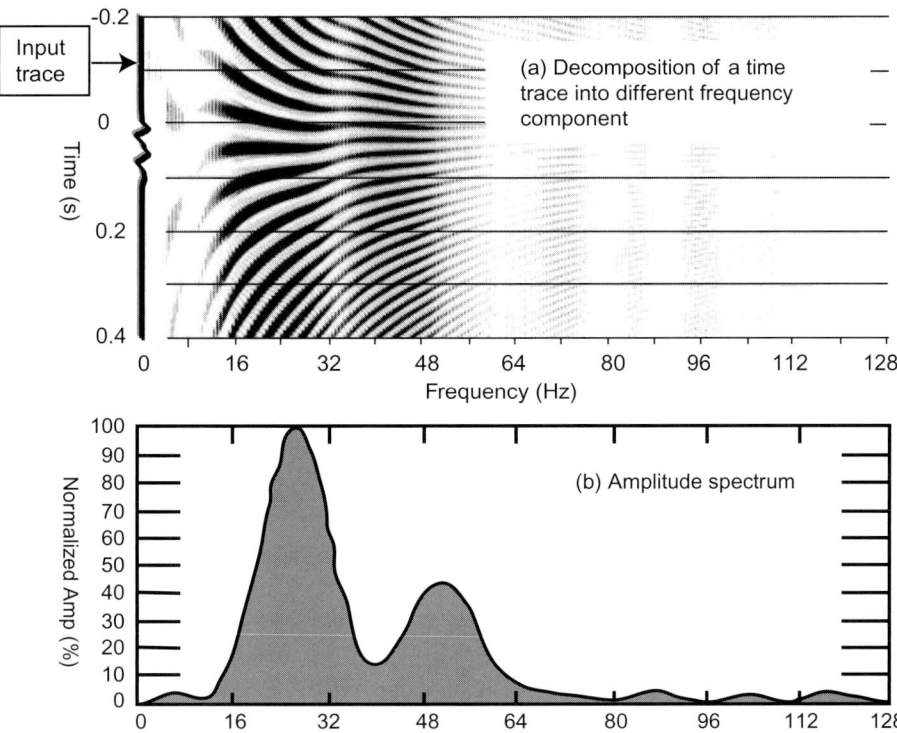

Figure 1.13 (a) The Fourier decomposition of a time series, the input trace on the left side, into a discrete number of frequency components. (b) The amplitudes of different frequencies form the amplitude spectrum. (Modified from Yilmaz, 1987.)

a function of the sample frequency. Figure 1.14 shows the radiation pattern of an airgun array using four different central frequencies. Each of the panels shows a lower-hemisphere projection of the source amplitude on a map view. The rings denote different **take-off angles** from the source, with the center of the rings pointing vertically downwards. As shown in this figure, as the frequency increases, the source radiation pattern of an airgun array worsens and becomes less "**omni-directional**", meaning invariant with respect to the azimuth and take-off angles.

The second factor, the characteristics of the receiver, will certainly affect the amplitude and other properties of seismic data. However, the effect of the receiver is usually known or measurable and therefore accounted for. In most applications we want to minimize the difference between different receivers and also to minimize the drifting of each receiver's response over time. An effective source includes the physical source plus the portion of the medium within several wavelengths from the source, and an effective receiver includes its neighboring media. This is the main reason that offshore seismic data, with nearly homogeneous and well coupled media surrounding the sources and receivers, have far better quality than onshore seismic data (see Box 1.2).

The final factor, the effect of the media, is the most interesting because many seismic studies are geared towards finding the properties of the media. We need to look at three aspects of **seismic attenuation** due to media properties: intrinsic attenuation, geometric

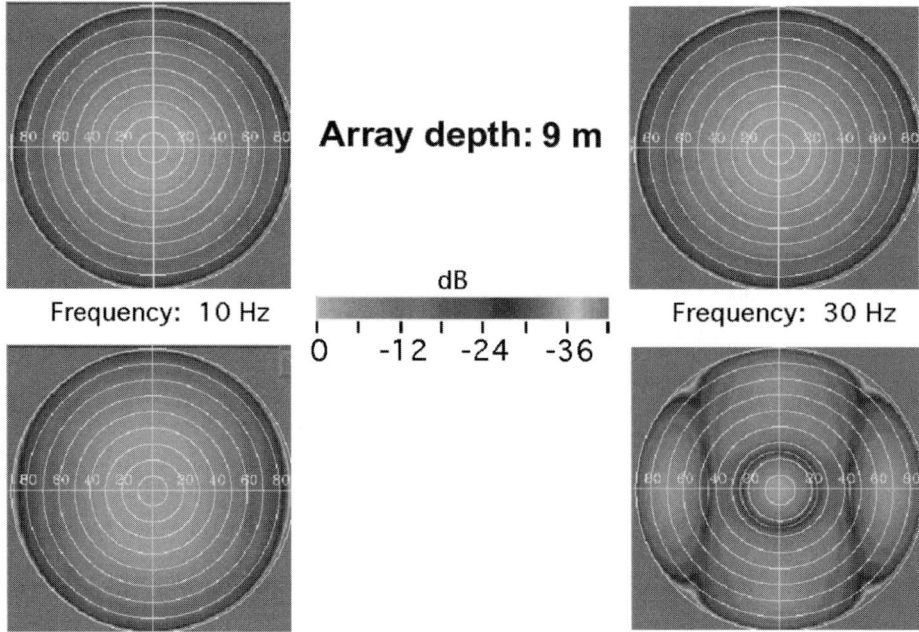

Figure 1.14 Radiation pattern of an airgun array at a tow depth of 9 m. Each panel is a lower-hemisphere projection of the wave amplitude at a particular frequency as a function of the azimuth and dip angles (Caldwell & Dragoset, 2000). For color version see plate section.

spreading, and structural properties. **Intrinsic attenuation** is due to the anelastic behavior of Earth material, and it is quantified by the **quality (Q) factor** which is inversely proportional to attenuation. The Q factor is usually assumed to be independent of frequency. Thus, for a given rock there will be a certain amount of loss of elastic energy per wave cycle. Consequently, at a given distance, the higher-frequency components have endured more intrinsic attenuation than the lower-frequency components. Therefore the **frequency content** of seismic data usually shifts toward lower frequency with increasing time or distance from the source.

Geometrical spreading refers to the systematic decay of the wave amplitude in response to the expansion of the propagating wavefront. Seismic amplitude is proportional to the square root of the energy density, which is the seismic energy in a unit volume in the seismic wave train. In a homogeneous space, the geometric spreading of a line source will be cylindrical and inversely proportional to the square root of the distance from the source. In contrast, the geometric spreading of a point source in a homogenous space will be inversely proportional to the distance from the source as the wavefront expands like a sphere. Furthermore, in a layer-cake model of the Earth, the amplitude decay from a point source may be described approximately by $1/[t v^2(t)]$, where t is the two-way traveltime and $v(t)$ is the **root-mean-square (rms) velocity** of the primary reflection (Newman, 1973). In inhomogeneous media, wavefront varies according to the variation of velocity gradient. As shown in Figure 1.15, the amplitude decay of a typical seismic trace is somewhere between that of spherical spreading and cylindrical spreading.

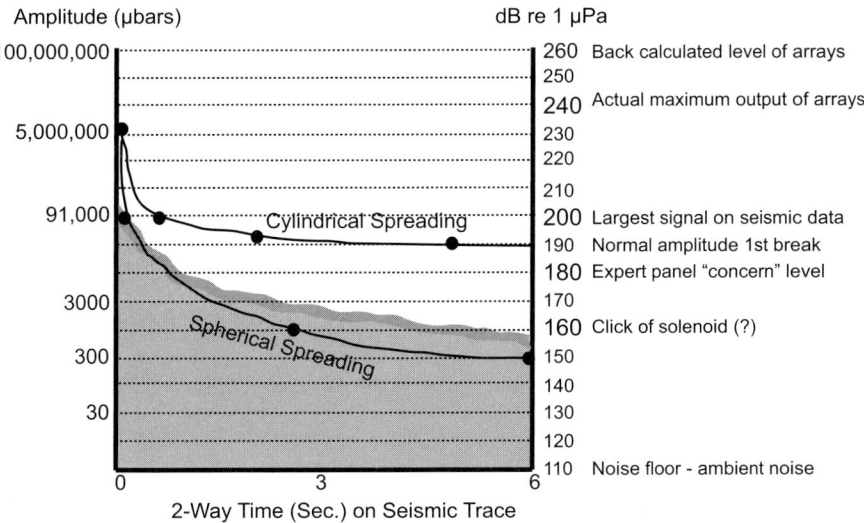

Figure 1.15 Amplitude level for a typical seismic trace as a function of recording time denoted by the gray curve (Caldwell & Dragoset, 2000).

Box 1.2 Comparison between land and marine seismic data

Land and marine seismic data have many similarities and differences due to their different environments and survey conditions. Can you make a table to compare and contrast them? As a start, the table below is an example.

Box 1.2 Table 1.1 Comparing land and marine seismic data

Aspect	Land seismic data	Marine seismic data
SNR	Poor	Usually moderate to high
Shot and receiver coupling	Usually poor and variable	Usually high and repeatable
Shot and receiver coverage	Possibly wide azimuth coverage	Narrow azimuth coverage
Surface multiples	N/A	Usually strong
Internal multiples	Yes, but often disorganized	Yes, often coherent
Near-surface statics	Yes, usually strong	Less apparent, usually of long wavelength
Ground rolls	Usually strong and dispersive	Weak, but can be strong in shallow water
Other 1?		
Other 2?		
Other 3?		

The third aspect of the seismic properties of a medium is its **structural properties**, which are the target of most seismologic studies. The structural properties include variations in **elastic impedance** (the product of density and seismic velocity) at all scales and in all directions, known as **seismic inhomogeneity** and **seismic anisotropy**, respectively. Two end-member descriptions of the structural properties are the **layered model** and the **gradient model**. Because the partition of wave energy is about amplitude, it is easier to consider in a layered model than in more realistic models. A layered model allows us to focus on the interfaces between layers of varying thickness, such as the partition of wave energy across a layer interface. In contrast, a gradient model allows us to study the gradual evolution of seismic waves through long-wavelength variation of the velocity and density fields in the medium.

1.3.3   Gain control

To pursue our interest in inferring structural properties from seismic data, we want to remove the influence of other factors such as source radiation pattern, receiver response, geometric spreading, and the attenuating effects of the medium. One practical way is to apply **gain control**, which balances time-variant amplitude variations. Although gain control is often applied to improve the display of seismic data, appropriate gain control can effectively enhance many processing tools. Gain control may be based on our understanding of a physical process, such as intrinsic attenuation and geometrical spreading. It may also be based on simple statistical statements to balance the amplitude of a section, such as **automatic gain control or correction (AGC)**. However, AGC as a statistics-based gain control is very harmful to those data processing and interpretation projects that rely on the amplitude variation of seismic data. Consequently, we need to be careful in using AGC, and record the details of the AGC operator if we have to use it.

The **rms amplitude** is the square root of the mean squared amplitude of all samples within a time gate. An rms amplitude AGC is based on the rms amplitude within a specified time gate on an input trace. The **gate length**, the length span that the operator is applied to, can either be constant or increase with time or depth. The ratio of desired rms amplitude to the input rms value is assigned as the value of the gain function at the center of the gate. There are also **instantaneous AGC**, in which we assign the ratio to any desired time sample of the time gate rather than to the sample at the center of the gate, and the time gate slides along the time axis one sample a time. An example of gain control is shown in Figure 1.16, for a shot gather obtained from a physical modeling experiment. There are surface-consistent gain controls that associate attenuation factors with each source and geophone location. **Surface-consistent** processing means that it accounts for all the near-surface effects such as the locations of the shots and receivers, topography, and possibly near-surface velocities.

1.3.4   Amplitude versus offset (AVO)

Although its effect is ubiquitous, the use of AGC in practice requires great care because it is a statistical approach that will harm the characteristics and integrity of real reflections. The AGC must not be used if the amplitude and phase of the seismic data are at the core of the study, as in the case of **amplitude versus offset (AVO)** studies such as that shown in Figure 1.17. AVO studies aim to reveal the presence of fluids such as gas, oil, or brine,

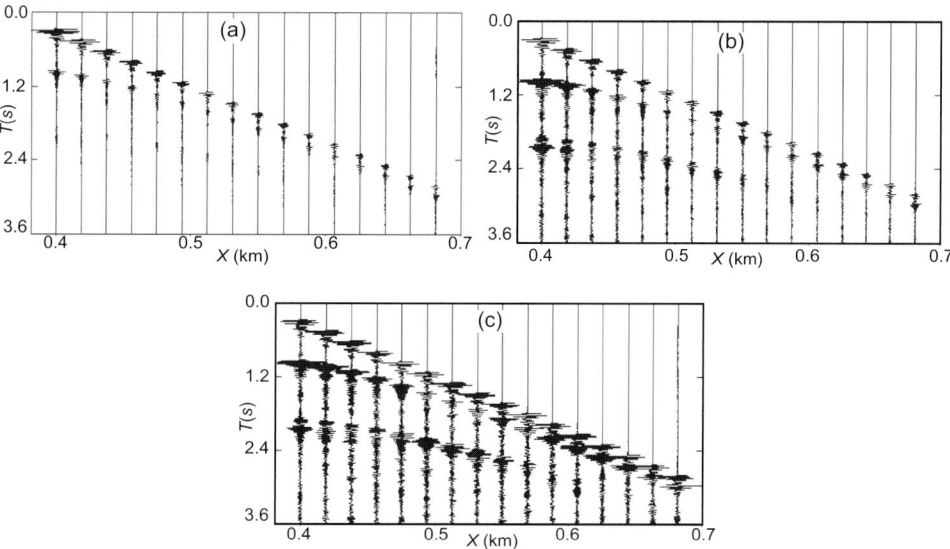

Figure 1.16 (a) A 16-trace common shot gather from a physical modeling experiment. (b) After applying automatic gain correction for each trace. (c) After trace balancing.

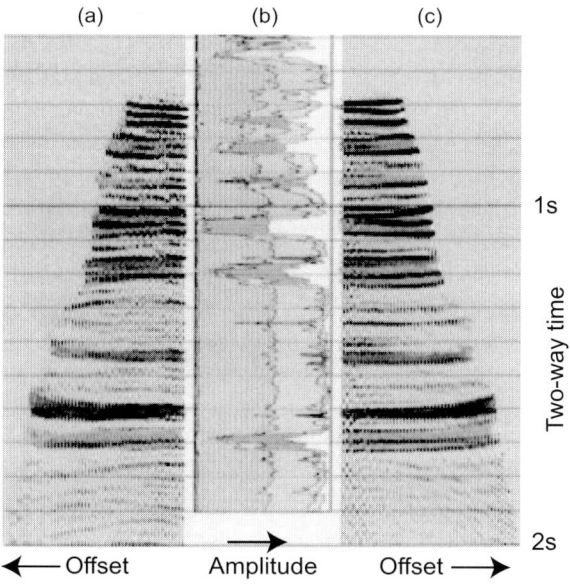

Figure 1.17 (a) A migrated CMP gather before stack at well location. (b) Well logs at the location. (c) Elastic AVO synthetic derived from well-log rock properties (Hilterman, 1990).

and/or type of lithology of a particular formation by analyzing the reflectivity viewed from different reflection angles. With high-quality data and assistance from understanding the structural properties of the formation and removing the effect of over-burden strata, the AVO may serve as a **direct hydrocarbon indicator** (**DHI**), particularly for the exploration of natural gas, because its elastic impedance differs so much from the ambient rocks and other types of fluids. Further details on AVO will be given in Section 10.5.1.1.

1.3.4 Exercise 1.3

1. Find an example of AGC in seismic data processing from the literature. What was the gain function used and what was the reason for it?

2. In Figure 1.14 why do the panels with lower central frequencies have smoother radiation patterns? What was the likely orientation of the airgun array and why?

3. Search the literature to find the basic physical principles for the use of AVO to detect fluid properties in subsurface. Why is AGC not allowed in AVO studies?

1.4   Phase and Hilbert transforms

1.4.1   Phase and phase spectrum

In seismology, **seismic phases** refer to distinguishable groups of body waves and surface waves that have particular propagation paths and particle motions, such as P, S, Pg, Pn, PmP, PmS, LQ, and LR waves (e.g., Storchak *et al.*, 2003). In the context of seismic data processing, **phase** quantifies the angular difference between the amplitude peak of a seismic wiggle and a reference point that is usually at time $t = 0$. As an important seismic attribute, phase usually means **phase lag**, the angular difference between two phase angles. The concept arose from the general harmonic expression of a seismic signal in the form of $A(\omega) \exp[i\phi(\omega)]$, where $A(\omega)$ is the amplitude component, $\phi(\omega)$ is the phase component, and ω is angular frequency. In seismology, propagation-induced alternation in the amplitude component is called **attenuation**, and propagation-induced alternation of the phase component is called **dispersion**. For monochromatic waves like that shown in Figure 1.18, the phase angle measures the angular difference between where the time zero is defined and the nearest peak. Thus, the peaks of such waves are always at a phase angle of zero degrees, in parallel with the definition of the cosine function.

We may understand the concept of the **phase spectrum** by applying the above definition of phase for a monochromatic wave to the result of Fourier decomposition of a time series,

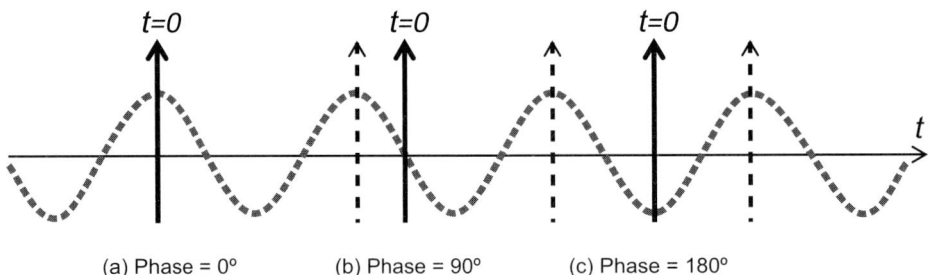

(a) Phase = 0°   (b) Phase = 90°   (c) Phase = 180°

Figure 1.18 Phase angles of a monochromatic wave for three different definitions for the location of time $= 0$.

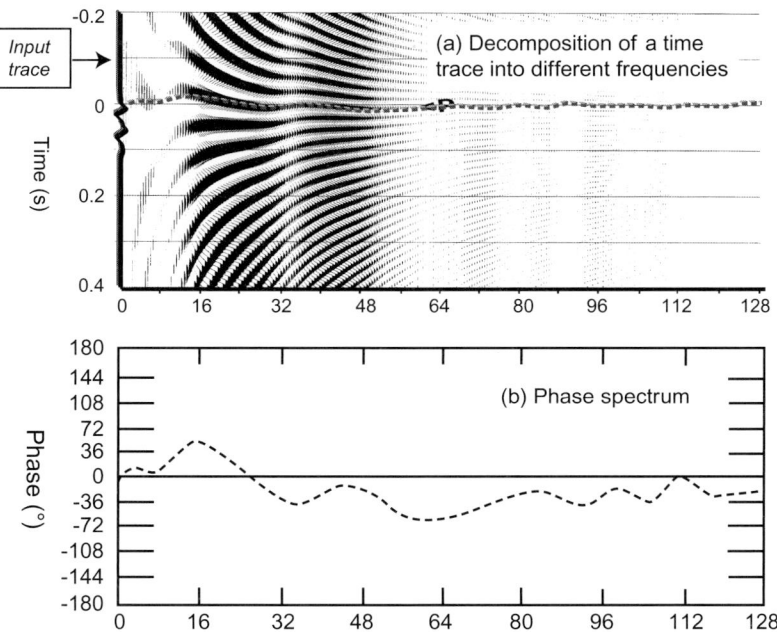

Figure 1.19 (a) The Fourier decomposition of an input trace as shown in Figure 1.13. (b) Its phase spectrum (dotted curves). When plotting the phase spectrum curve along the zero-time line in the Fourier decomposition, the curve follows the peaks of the monochromatic time series; this is because the phase angle at each frequency is the angular distance from time zero to the nearest peak. (From Yilmaz, 1987.)

as shown in Figure 1.19. Figure 1.19a is the same as Figure 1.13a, a decomposition of an input trace into 128 monochromatic time series. The dotted curve connects the peaks of these time series that are the nearest to the zero-time line. This curve is the phase spectrum as shown in Figure 1.19b, where the vertical scale unit is the phase angle in degrees. Because the phase angle is confined within a range of $\pm 180°$, a "**wrap-around**" may occur along the phase spectrum when the phase angle goes beyond the $\pm 180°$ window:

$$\text{Phase} \pm 2\pi n = \text{Phase} \tag{1–9}$$

1.4.2 Phase of a wavelet

A **wavelet** is defined as a time series that is confined within a finite time window and a finite amplitude range. The finite time window means that a wavelet may consist of different frequencies. Then the phase of a wavelet means the collective value of the phases of all frequency components with respect to time zero. In practice, the phase of a wavelet depends on its shape. As shown in Figure 1.20, a **zero-phase wavelet** is symmetric with respect to the origin, with its maximum peak located at the zero-time position. This zero-time position is taken to be special time position, such as the two-way reflection time of the

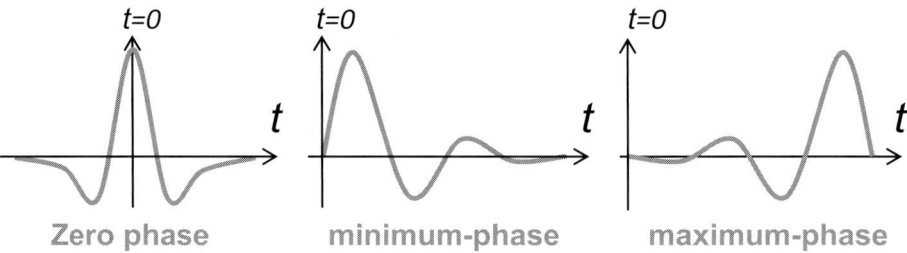

Figure 1.20 The phase of a wavelet depends on its shape.

upper interface of a geologic formation, rather than a moment on the clock. The zero-time position is at the beginning of both the minimum-phase and maximum-phase wavelet. A **minimum-phase wavelet** has most of its energy concentrated near its front edge, so it is **front-loaded**. Conversely, a **maximum-phase wavelet** has most of its energy concentrated near its end, so it is **tail-loaded**.

Why are the minimum-phase and maximum-phase wavelets front-loaded and tail-loaded, respectively? The physical cause is the interference between different frequency components within a wavelet. Among all wavelets of the same amplitude spectrum, the minimum-phase is the one that minimizes the phase lags between different frequencies. This minimization means there is constructive inference of the frequency components near time zero, which is defined to be at the beginning of the wavelet. In contrast, a maximum-phase wavelet is one that maximizes the phase lags between its frequencies; hence there is destructive interference near the zero-time position. In practice most wavelets, including zero-phase wavelets, have mixed phase. Most interpreters prefer zero-phase wavelets because it is easy to pick out the peaks. However, many time processing steps require the use of minimum-phase wavelets.

1.4.3   Analytical signal and Hilbert transform

Some seismic attributes such as the envelope and instantaneous phase are called **instantaneous attributes** because they are functions of a particular moment in time. The derivation of these attributes requires the use of a **complex trace** or **analytical signal**. For a real time series $x(t)$, its analytic signal $\mathbf{x}(t)$ is defined as

$$\mathbf{x}(t) = x(t) - i H[x(t)] \tag{1–10}$$

As shown in Figure 1.21, the analytic signal is a complex time series: its real part is the original time series, and its imaginary part is the negative of the **Hilbert transform** of the original time series (e.g., Choy & Richards, 1975; Clayton et al., 1976). The Hilbert transform $H[]$ advances the phase of all frequency components by $\pi/2$. For instance, it converts a sine into a cosine. It is therefore called the **quadrature filter**, since $\pi/2$ is one-quarter of a full cycle.

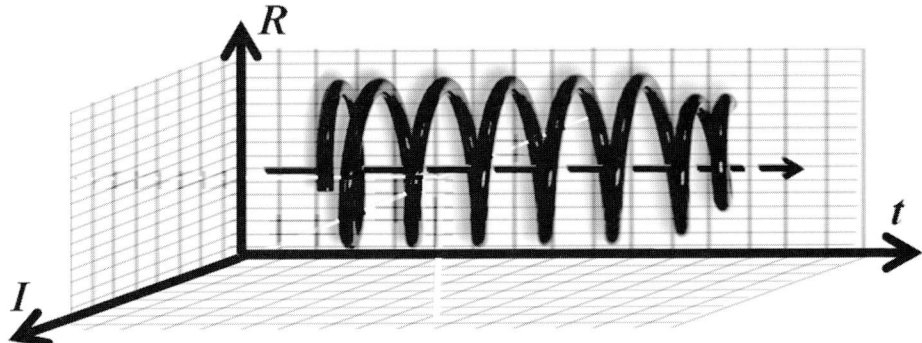

Figure 1.21 A schematic view of an analytical signal in the form of a coiled spring. Its projection on the R–t plane is the original time series. Its projection on the I–t plane is the negative of the Hilbert transform of the original time series.

In the time domain, the transform can be expressed as a convolution of the original signal with $(-1/\pi t)$

$$H\left[x\left(t\right)\right] = x\left(t\right) * \left(-1/\pi t\right) \tag{1-11}$$

Since the Fourier transform of $(-1/\pi t)$ is equal to $i\omega/|\omega| = i\,\text{sgn}\,\omega$, where sgn( ) is the **sign function**, we have

$$H\left[x\left(t\right)\right] = -\int X\left(\omega\right) i\,\text{sgn}\left(\omega\right) e^{i\omega t}\, d\omega$$

$$= \int X\left(\omega\right) e^{-i(\pi/2)\text{sgn}(\omega)} e^{i\omega t}\, d\omega$$

$$= \int X\left(\omega\right) e^{-i[\omega t - (\pi/2)\text{sgn}(\omega)]}\, d\omega \tag{1-12}$$

The above equation verifies that the Hilbert transform advances the phase of the original signal by $\pi/2$.

In practice the time-domain convolution approach for the Hilbert transform is very costly to compute because $(-1/\pi t)$ is a slowly decaying function with respect to time. In other words, many terms are needed by the convolution to make a good approximation. In contrast, it is much more efficient to conduct the Hilbert transform in the frequency domain in which the analytical signal is defined (e.g., Clayton *et al.*, 1976) as

$$X\left(\omega\right) = X\left(\omega\right)\left[1 + \text{sgn}\left(\omega\right)\right] \tag{1-13}$$

This leads to the following procedure to derive the analytical signal:

- Transform the input signal $x(t)$ to frequency domain $X(\omega)$;
- Double the values of the positive frequency terms, and let the negative frequency terms be zero (this enforces causality in the frequency domain);
- The inverse Fourier transform of the terms from te previous step is the analytical signal.

1.4.4   Instantaneous attributes

The expression of the analytical signal (1–10) allows the computation of many instantaneous attributes. For example, the **envelope** of the original time series is

$$e(t) = |\mathbf{x}(t)| \tag{1–14}$$

Since the envelope is the absolute value of the analytical signal, we may call it the **instantaneous amplitude**. Next, the **instantaneous phase** of the original time series $x(t)$ is

$$\phi(t) = \tan^{-1}\{-H[x(t)]/x(t)\} \tag{1–15}$$

The **instantaneous frequency** is just the time derivative of the instantaneous phase

$$\omega_{\text{ins}}(t) = d\phi(t)/dt \tag{1–16}$$

The instantaneous properties of a simple trace are given in Figure 1.22.

Box 1.3 Generalized Hilbert transform and an application

Attribute extraction and feature detections are among common usages of seismic data volumes today. An example is the detection of fluvial channels from 3D volumes of seismic imageries (Luo *et al.*, 2003). Box 1.3 Figure 1 shows cross-sections of three tests for detecting the edges of a synthetic channel model as the input. The generalized Hilbert transform (GHT) detected both edges of the channel with good resolution. Box 1.3 Figure 2 shows an example of applying the GHT to channel detection in field data.

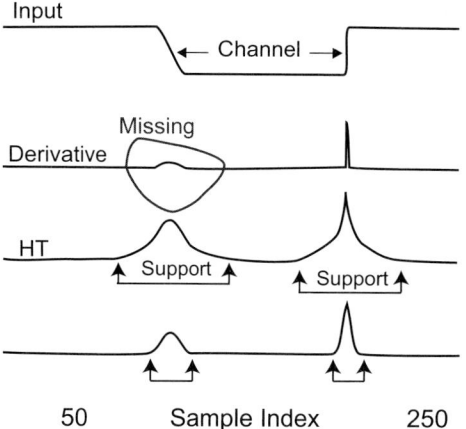

Box 1.3 Figure 1 Channel detection in cross-section views. The synthetic input trace has a channel with a sloping edge to its left and a sharp edge to its right. The derivative operator detected the sharp edge but not the sloping edge of the channel. The Hilbert transform (HT) detected both edges, but the horizontal resolution, or support, is too wide. The generalized Hilbert transform (GHT) detected both edges with higher resolution. (After Luo *et al.*, 2003.)

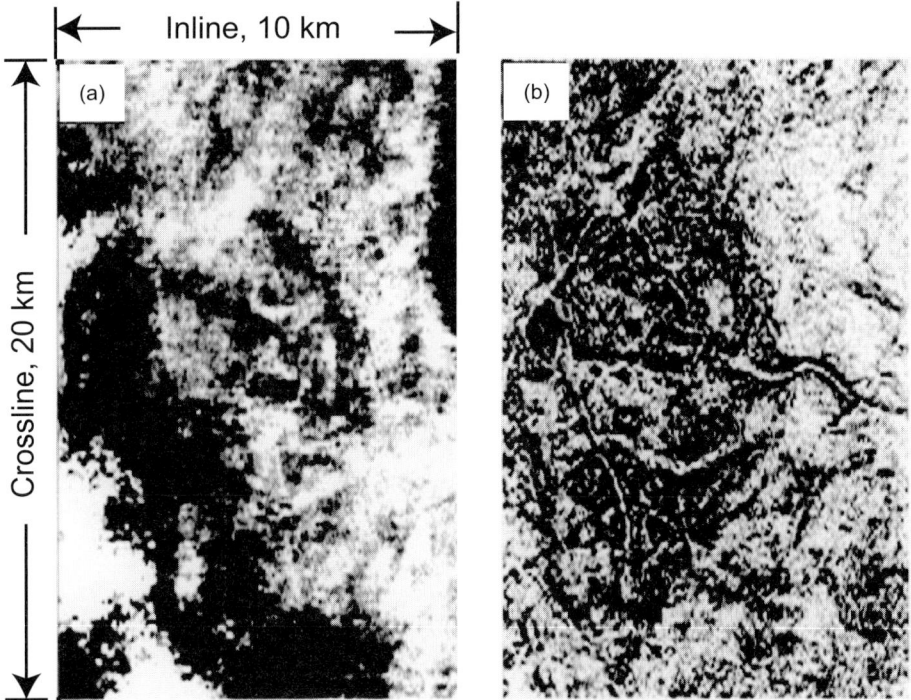

Box 1.3 Figure 2 Time slices of a seismic imagery volume showing amplitude in gray tone. (a) Input data. (b) Result of channel detection using GHT. Many channels are clearly seen in (b). (After Luo *et al.*, 2003.)

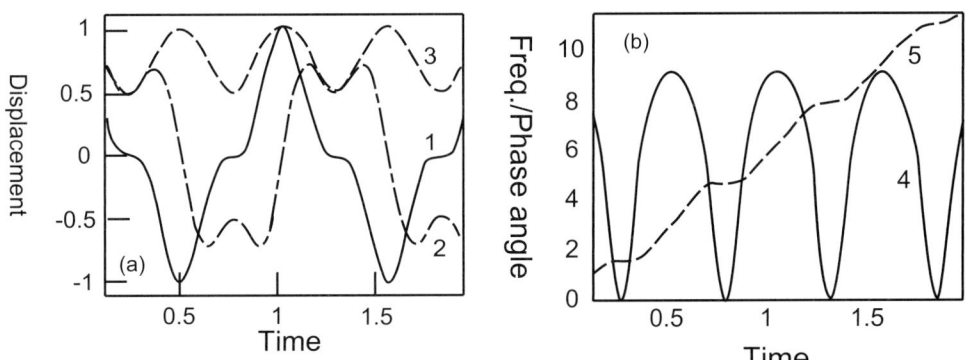

Figure 1.22 1: Input signal; 2: Hilbert transform; 3: envelope; 4: instantaneous frequency; 5: instantaneous phase.

Exercise 1.4

1. When a time series is reversed, what happens to its phase properties? Demonstrate your point using the time series (1, 2, 3, 4).

2. How can the phase of a seismic trace be advanced by 30°? How can it be advanced by any degree?

3. Describe the generalized Hilbert transform (GHT) after reading Luo *et al.*, 2003.

1.5   Data format and quality control (QC)

In order to process real seismic data, we need to know the data format, the way that seismic data are stored in computers. To assure that the data processing works, we need to conduct many quality control checks.

1.5.1   Format of digital seismic data

The **data format** is a standard for arranging digital data in the file storage units of computers. This is an essential issue for seismic data processing because a well-designed data format standard will ease the understanding of various datasets between different users, maximize efficiency in moving large amounts of data within computers and through the Internet, and be convenient for some of the quality control measures. A computer records each piece of digital information, such as a number, by its value and its address in the computer memory. One common data format is ASCII (American Standard Code for Information Interchange), which is readable because it is a character-encoding scheme based on the English alphabet. Another character encoding scheme is EBCDIC (Extended Binary Coded Decimal Interchange Code) format used in many IBM computers.

Figure 1.23 shows an example of earthquake phase data in ASCII format. The data are a portion of a dataset downloaded from the website of the Southern California Earthquake Center Data Center. In this ASCII data file, the first line contains information on the earthquake event, including its occurrence year, month, date, hour, minute, and second, its local magnitude $M_L = 3.32$, latitude = 33.580° N, latitude = 116.822° W, and depth = 7.1 km below the sea level. The quality score of this event is B, and its event number is 513 877. The last four digits in the first line are used for various statistics. Here 29 is the number of phase readings following this line, and 130 is the number of all phase readings. The rest of the lines in this file contain the information on the phase pickings. On each phase picking line, the three letters in the first column denote the station name, the three letters VHZ in the second column tell us that the recording was from the *z*-component of a high-frequency velocity meter, and the three digits in the third column are the station number. The fourth column designates the seismic phases, such as P, S, P_g, and P_n waves. The fifth column describes the sharpness and polarity of the first motion reading of the corresponding phase. The two numbers in the following columns are the source-to-receiver distance in kilometers and traveltime in seconds.

```
1981/04/01 02:05:14.60 L3.32 I 33.580 -116.822  7.1  B513877  29  130  0  0
        POB   VHZ   605   P    IU0    15.16    3.14
        KEE   VHZ   427   P    IU0    16.78    3.30
        PLM   VHZ   586   P    ID0    25.45    4.96
        VG2   VHZ   858   P    ID0    28.10    4.88
        DB2   VHZ   244   P    IU0    28.12    4.96
        SMO   VHZ   730   P    IU1    33.52    5.85
        PSP   VHZ   620   P    IU0    34.52    5.69
        MDA   VHZ   471   P    ID0    40.61    7.03
        PEC   VHZ   568   P    IU0    46.71    7.78
        WWR   VHZ   927   P    ID0    48.10    7.96
        RAY   VHZ   633   P    ID0    50.88    8.66
        COY   VLZ   224   P    IU0    52.92    9.04
        SME   VHZ   722   P    IU0    56.17    9.24
        CFT   VHZ   189   P    ID0    57.24    9.78
        MLL   VHZ   478   P    ID0    57.83    9.66
        VST   VHZ   869   P    IU0    60.44    9.98
        VST   VHZ   869   S    E 2    60.44   17.28
        SS2   VHZ   747   S    E 3    93.69   26.70
        RDM   VHZ   645   S    E 2    97.10   27.25
        BAR   VHZ   127   S    ID1   101.02   27.94
        BAR   VHZ   128   S    I     101.02   27.90
        ELR   VHZ   296   P    ID1   103.05   18.28
        PCF   VHZ   566   P    IU1   103.70   16.97
        SUN   VHZ   768   P    IU0   106.66   17.22
        HAY   VHZ   389   P    E     109.98   17.81
        CRR   VHZ   233   PG   ID1   110.10   18.86
        CRR   VHZ   233   PN   IU1   110.10   17.76
        FNK   VHZ   328   PG   E 2   111.28   18.42
        FNK   VHZ   328   PN   E2    111.28   17.76
```

Figure 1.23 Portion of the phase data of an earthquake from Southern California Earthquake Center Data center.

To achieve the goal of data processing, it is necessary to have a standard data format that is simple to understand, easy to use, efficient for transporting and storing, and versatile for different types of applications. Most industry seismic data are stored in binary format, which has the advantage of efficiency and security. Hence, we want a data format that is able to store binary data and customized for most common types of seismic applications. The move towards standard data formats started shortly after the appearance of computers and the use of digital seismic data. For many years, however, different firms developed their own 'standards' which were useful only locally. The Society of Exploration Geophysicists (SEG) has played a leading role in developing open standards for storing geophysical data. These include the SEGA, SEGB, SEGC, SEGD, and SEGY, the latest standard. The original rev 0 version of the SEGY format was developed in 1973 to store single-line seismic digital data on magnetic tapes (Barry *et al.*, 1975). Owing to significant advancements in geophysical data acquisition since rev 0, especially 3D seismic techniques and high-capacity recording, rev 1 of the SEGY format was created at the start of the twenty-first century (Norris & Faichney, 2001).

Optional SEG Y Tape Label	3200 byte Textual File Header	400 byte Binary File Header	1st 3200 byte Extended Textual File Header (optional)	≋	Nth 3200 byte Extended Textual File Header (Optional)	1st 240 byte Trace Header	1st Data Trace	≋	Mth 240 byte Trace Header	Mth Data Trace

Figure 1.24 Structure of a SEGY file, with N extended textual file header records and M trace records.

1.5.2   File header, trace header, and data traces

There are three main components of a digital seismic dataset: file headers, trace headers, and data traces. A **file header** contains the overall information about the dataset, including information on the survey, previous processing flow, and parameters of the data. A **trace header** contains the specifics of the data traces, such as the number of samples, sample rate, and number of traces in the following. Each **data trace** is simply a string of values such as the amplitude of a seismogram following the specific set of parameters in its trace header. Such a three-level structure offers many advantages. For instance, one can conduct reordering or sorting processes, such as a conversion from CSG to CMP, only using the trace headers.

As an example, Figure 1.24 shows the byte stream structure of a SEGY file. It starts with the optional SEGY tape label, followed by the 3200 byte textual EBCDIC character encoded tape header, then a 400 byte binary header. This file header can be extended with additional pieces of 3200 byte textural file headers that allow the user to store information such as an image of the survey area. The file header will specify M number of traces of the data and, as shown in the figure, there are M pairs of trace header and data trace following the file headers.

1.5.3   Data loading and quality control

In practice, the very first physical step in a seismic data processing project is data loading, the process of putting the raw data into the processing computer. The raw data are usually stored on some media device such as various types of magnetic tapes or computer storage units. Nowadays, some datasets of relatively small size may be downloadable from the Internet. Obviously we need to know the data format in order to use the data files properly. After managing to load the data into the processing computer, our first priority is to conduct QC measures to validate the data content and evaluate the quality and characteristics of the data. For example, the result of the data QC may indicate that the data loaded are not what we want. On the other hand, the characteristics of the data from the evaluation process often help us refining the processing parameters.

Data QC at the initial stage of a data processing project involves checking the survey geometry, data format, and consistency between different portions of the dataset. It is critical to assure that the quality and quantity of the data are satisfactory for our study objectives. The first task after loading the data into the computer is to conduct a series of tests to verify that all the information given about the dataset is true. The tests may simply involve making and viewing graphs of the survey geometry and shot gathers using the loaded data, or simple calculation of the number of bytes in order to check the consistency between the values in the file headers, and the values in the trace headers and data traces.

Exercise 1.5

1. Find out the meanings of the values in the earthquake phase data example shown in Figure 1.23.

2. Many data sorting procedures can be carried out by sorting the data headers. Why is this approach efficient? Can you give examples to illustrate your arguments?

3. Write a flow chart of the processing steps required to convert a seismic dataset in CSG arrangement into CMP arrangement, assuming that the data are in SEGY format. Try to write the processes and parameter values in as much detail as possible.

1.6　Summary

- Seismic data are physical observations, measurements, or estimates of seismic sources, seismic waves, and their propagating media. The processes of data acquisition, processing, and interpretation are interconnected and complement each other.

- The definitions of signal and noise depend on data quality and business objectives. A major effort of seismic data processing is to improve the signal-to-noise ratio (SNR).

- Common midpoint (CMP) stacking is used widely to generate post-stack seismic traces. Because CMP stacking assumes a layer-cake Earth model, the stacked traces have the best SNR for cases with gently dipping reflectors.

- Sampling of seismic data must meet the Nyquist condition of at least two samples per cycle of the highest signal frequency, in order to avoid aliasing artifacts due to under-sampling.

- Gain control includes a number of data processing measures to compensate for the reduction of the amplitude of seismic waves due to factors such as source radiation, attenuation of the media, and geometric spreading of the energy during wave propagation within rock strata and across interfaces.

- The phase of a single-frequency wave is the angular difference between time $t = 0$ and the nearest peak. The phase of a seismic wavelet is a function of its shape: zero-phase is symmetric, minimum-phase is front-loaded, maximum-phase is tail-loaded, and the rest are of mixed phase.

- Each seismic trace is the real component of its analytical signal, a complex trace that is constructed using the Hilbert transform. The analytical signal is useful in generating instantaneous attributes such as envelope, instantaneous phase, and instantaneous frequency.

- Data formats such as SEGY are rules about how the digital seismic data are stored in computers, and the main components of a digital seismic dataset are file headers, trace headers, and data traces.

- Different quality control (QC) measures are required throughout each application of seismic data processing. Always thinking about QC is a good habit for anyone engaged in seismic data analysis.

FURTHER READING

Caldwell, J. and W. Dragoset, 2000, A brief overview of seismic air-gun arrays, *The Leading Edge*, 19, 898–902.
Norris, M. W. and A. K. Faichney (eds.), 2001, SEG Y rev1 Data Exchange format1, SEG.
Sheriff, R. E. and L. P. Geldart, 1995, *Exploration Seismology*, 2nd edn, Cambridge University Press.

2 Preliminary analysis of seismic data

The practice of seismic data processing with digital records has been progressing for over six decades. Today all seismic processing projects are started with a set of scientific and business objectives in mind that often require specific processing flows; usually each flow involves some pre-processed data rather than the raw data. The **pre-processing** includes all preparation steps through which both major and relatively simple problems in the input data are cleaned up so that the main processing flow can function more effectively. While the pre-processing steps may be standard and even apparently routine, each step can be critical to the final result.

This chapter starts with illustrations of the most common pre-processing tasks. One important aspect of learning seismic data processing is to appreciate the physical processes that the wavelet from a seismic source has experienced, so that we may approximately undo or redo some of the processes in computers. For this reason, the filtering expression of seismic data processing is introduced. As a modern example, the processing of a multi-component dataset from vertical seismic profile is shown. This chapter examines several simple but common processing operators, including normal moveout, stacking, convolution, correlation, and Radon transform. Often the reason for using these techniques is to suppress the most common types of noise. The readers should try to envision the physical processes that each operator attempts to emulate. As an example of preliminary analysis, the effects of surface topography and near-surface velocity variations are analyzed using the concept of near-surface statics.

2.1 Pre-processing

Pre-processing refers to the preparation type of processing work that comes before the main processing task. Its purposes are to identify and fix simple problems of the dataset, and to apply some common corrections such as removing the gain factor of a seismic sensor. Although most of the tasks in pre-processing are standard and probably routine, care needs to be taken to QC the data by assessing the quality of the input dataset, checking for errors, and finding out the characteristics of the data relevant to the objectives of the processing work.

2.1.1 Traditional pre-processing steps

Typical tasks in pre-processing of seismic data include:

- Detecting errors by checking consistency between different portions of data
- Assessing uncertainties such as errors in the source and receiver positions
- Sorting data into desired form, for instance **demultiplexing** and CMP sorting
- Editing file headers and trace headers to update changes
- Merging different navigation data with seismic data
- Muting bad traces
- Amplitude correction
- Phase rotation

In the old days of exploration seismology, many pre-processing steps were necessary to deal with the special recording processes. One such process is **multiplexing**, the process of combining multiple analog ordigital data streams into a single dataset. This means that the wave field from a shot is recorded in the order of receivers. We may regard a common shot gather as a matrix of recorded amplitudes, with its column numbers denoting the orders of receivers and its row numbers denoting the order of time samples. Then a multiplexed dataset stores the data matrix into a single stream of time series row by row, or taking column number as the fast axis. So the single stream starts with the top row elements of the data matrix in their column order, followed by the next row elements in their column order, and so on. In contrast, a demultiplexed dataset stores the data matrix into a time series column by column, or taking row number as the fast axis. Hence, the demultiplexing process is equivalent to a matrix transpose operation. The process is carried out by a simple transpose sorting process.

Pre-processing may be done in order to allow workers to familiarize themselves with the data. To do so with a seismic reflection dataset, one can use a processing software package to produce a number of common offset sections or common shot gathers. Common offset sections, particularly those of small offset, give a general picture of the structural trend of the study area and variations in data quality from place to place. It is of particular importance if any key **horizons**, such as the base of the weathering zone or major unconformities, can be recognized. A horizon refers to a seismic reflection event that is a particular geologic boundary. In areas with low lateral velocity variation and little thickness variation of the rock strata, a near-offset section may be a good approximation of a **time stack**, a cross-section over the study area with the vertical axis being the two-way traveltime. Common shot gathers may allow us to decipher more detail at each location, such as the depth of water column or weathering zone, the level of surface waves, the change in velocity with depth, and the noise level and frequency content from shallow to deep depths. A scan of a series of common shot gathers or common midpoint gathers along a profile will allow the worker a quick evaluation of both the data quality and general geologic features along the profile. Often the processing strategy and parameters are formulated or refined from repeated previews of common offset sections and various types of seismic gathers.

Some problems in the data identified from the preview process may be fixable by simple **editing** of the data. For example, we may find that a value in the file header is inconsistent with the data traces because the value has not been updated during previous processing steps. If we can verify the correct value, then a simple update will fix the problem. Alternatively, when we need to create a subset of the dataset, we will design and carry out a sorting process, and then we will need to edit the file header and perhaps the trace headers of the subset data to reflect the changes. In a similar way, when we need to combine multiple datasets over the same area into a single dataset, we need to update both the headers and data traces to reflect the merged data.

2.1.2   Navigation merge

As a major pre-processing task, **navigation merge** involves checking the accuracy of different navigation measurements and combining the navigation data with the seismic data. Because the quality of seismic data is strongly correlated to the quality of navigation data, the best total quality is achieved when navigation and seismic QCs are carried out in an integrated process. A major breakthrough in seismic acquisition during the past century has been the arrival of the global positioning system (GPS), which provides the spatial and temporal coordination for all modern seismic surveys. In offshore surveys, for instance, GPS receivers are mounted on the vessels, airgun floats, and tail buoys to triangulate the positions of the vessel and streamers using signals from satellites. Acoustic or laser positioning devices are also placed locally on the vessels, airgun floats, and tail buoys to provide secondary positioning. Data may be checked in the field or in-house by merging the seismic traces with the field or processed navigation data. During the navigation merge, the geometric positions of the source and receiver of each trace are added into its trace header. In offshore surveys the navigation QC aims to detect errors such as:

- Source and receiver positioning errors
- Source timing errors
- Gun flag reversals
- Incorrect definitions
- Shift in source array centers
- Incorrect laybacks to center source
- Unacceptable radio navigation
- Multi-boat timing

2.1.3 Merging datasets

To merge multiple datasets together, one has to consider several consistency issues: (1) consistency between sources used in different datasets; (2) compatibility between receivers of different datasets; (3) consistency in the spacing and orientation of sources and receivers; and (4) consistency in sample rate, length, and gain parameters.

As shown in Figure 1.14, the amplitude of the source may vary with the azimuth and dip angles even for airguns, which are among the best seismic sources available. In contrast, onshore seismic surveys use dynamite or Vibroseis sources that are far more complicated. A frequency band of about 8 to 80 Hz, typical for most seismic surveys, has wavelengths between 250 and 25 meters if the average velocity is around 2 km/s. This means that the effective source will include the rocks and fluids in the immediate neighborhood of the 'real' dynamite or Vibroseis sources. Difference in the lithology and structure surrounding different shots is likely to be a major cause of inconsistency between the source signatures in an onshore seismic survey. A similar effect also exists for different receivers. The low repeatability of source and receiver functions for onshore seismic surveys leads to lower quality for most onshore seismic data than for offshore seismic data.

In order to merge multiple datasets, we have to make the wavelets of different datasets as similar as possible, usually through the process of **phase rotation**. If we know the **source wavelet** in the data, we can rotate the phase angle using the Hilbert transform described in the previous chapter. In practice, however, the source wavelet is usually unknown. We might think of taking a portion of the data, such as that within a small time window around the first break, as an approximation of the source wavelet. But this approximation is a source of errors because we choose the time span of the window arbitrarily and all portions of the data may contain noise. A more practical way to match the phase angle follows the concept of a matched filter.

2.1.4 Matched filter

Let us look at the process of a sending a source wavelet f_t into an Earth structure x_t, and producing a seismic trace g_t, as shown in Figure 2.1. The subscript t denotes that each function is a time series. This process can be described mathematically using the convolution operator denoted by the $*$ sign:

$$g_t = x_t * f_t + n_t \qquad (2\text{--}1)$$

where n_t is the noise term. In the frequency domain, the convolution becomes multiplication:

$$G(\omega) = X(\omega)F(\omega) + N(\omega) \qquad (2\text{--}2)$$

where the capitalized terms correspond to the Fourier transforms of the lower-case terms in the time domain.

Figure 2.1 A source wavelet f_t traverses through the Earth x_t, producing a seismic trace g_t.

Our goal is to uncover the Earth structure that is contained in the recorded trace but is distorted by the source wavelet and the noise terms. The **matched filter** of the source wavelet will allow optimal estimation of the Earth structure in the presence of stochastic noise. When a known signal is embedded in noise, its matched filter will maximize the SNR. We will examine this issue later in frequency domain deconvolution. In the field of telecommunication, a matched filter is obtained by correlating a known wavelet with a time series to detect the presence of the wavelet in the time series. This is equivalent to convolving the time series with a conjugated and time-reversed version of the wavelet, equivalent to the **cross-correlation** process. The SNR is maximized when the impulse response of the matched filter is a reversed and time-delayed version of the transmitted signal, which in our case is the source wavelet. As an example, we may use the matched filter concept to choose the best estimate of the source wavelet among all candidates, such as the first-break wavelets taken from different time windows at various offsets, by checking the SNRs after cross-correlating each candidate with all the wavelets.

2.1.5   Processing multi-component data using hodograms

As multi-component seismic surveys become more popular, let us see an example of pre-processing multi-component data. Figure 2.2a shows a sketch of an offset **vertical seismic profile** (**VSP**). A VSP survey uses surface shots and receivers placed along a wellbore to image the structure in the neighborhood of the wellbore. Its main advantages include recordings at different depths to provide time-to-depth conversion and recognition of multiple reflections, as well as having frequency content higher than surface seismic data to assist the seismic–well tie, the process of linking horizons on the surface seismic records with those from well logs. An **offset VSP** uses a fixed source-to-receiver offset, producing a common shot gather. A **walkaway VSP** uses a fixed receiver and a number of shots of different offsets, producing a common receiver gather. An **azimuthal VSP** measures seismic waves by varying the azimuthal angle between sources and receivers. We may switch sources with receivers, producing a **reverse VSP** (**RVSP**). In practice, nearly

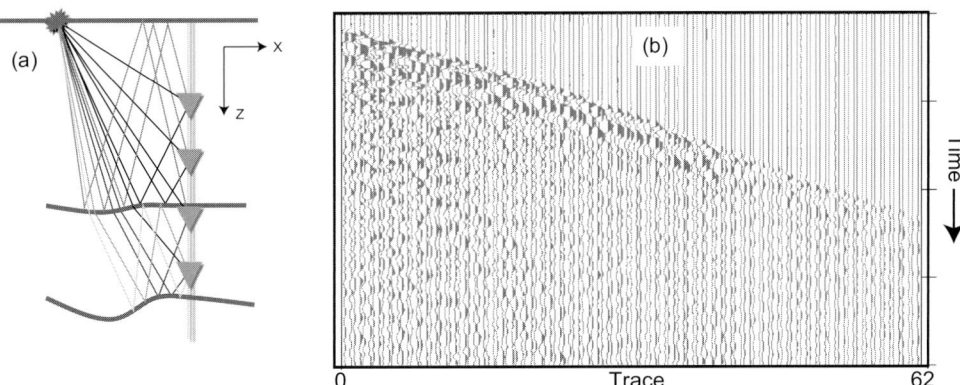

Figure 2.2 (a) A sketched cross-section of offset VSP, where raypaths show various waves from a shot (star) to the receivers (triangles) along the well bore. (b) A common shot gather of the two horizontal components from an offset VSP. For color versions see plate section.

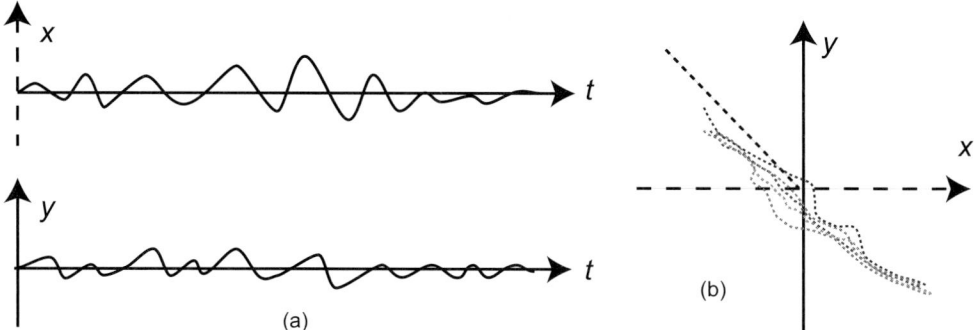

Figure 2.3 (a) A pair of x and y components of seismic traces. (b) A hodogram, a plot of the time-varying trajectory of the traces, using the amplitudes of the x and y components as the two axes. The straight dashed line denotes the average orientation of the hodogram.

all wells are deviated, so there are no truly vertical wells. We may use a combination of different source and receiver geometry to optimize the quality of VSP data and images.

Figure 2.2b shows a common shot gather of the two horizontal components of an offset VSP data. A convenient tool for data QC of such data is the **hodogram**, which is a graph of multi-component data in polar coordinates. As illustrated in Figure 2.3, for a two-component pair of seismic traces, we take each component as a coordinate axis. Then the hodogram is a graph of the time trajectory of the traces, as shown in Figure 2.3b, using the amplitudes of the traces as the coordinate axes of the graph. In this case, the x and y components represent the two horizontal directions, easting and northing, respectively. Hence the hodogram represents a map view of the ground particle motion at the receiver during the time span of the data traces shown in Figure 2.3a. The average orientation of the particle motion, denoted by a straight dashed line, represents the principal direction of the seismic waves at this particular location and time span.

The principal orientation of the particle motion as revealed by the hodogram is very useful. In this case, all three-component geophones were cemented into an old wellbore. The problem is that the orientations of the multi-component geophones are unknown. The geophones might have been rotated as they were being raised up from the bottom of the wellbore before cementation, although their vertical component is likely to be oriented vertically along the wellbore. Our aim is to determine the orientation of the two horizontal components of each geophone using hodograms.

We first select a group of shots that have almost the same source-to-receiver offset, as sketched in Figure 2.4a. If we make a hodogram using the horizontal components and a time window around the first break, the principal orientation of the hodogram will be parallel with the azimuth from the receiver to the shot, unless the velocity field is highly heterogeneous between the shot and receiver. Hence, for each receiver in the wellbore, we plot the horizontal-component hodograms of the first break traces from all shots. The angular difference between the source-to-receiver azimuths and the principal orientation of the hodograms will allow a determination of the rotation angle between the x–y axes and the easting and northing angles. We can then apply a correction for the angle rotation error for all horizontal seismic traces of each geophone. Figure 2.4b shows the resulting

(a)

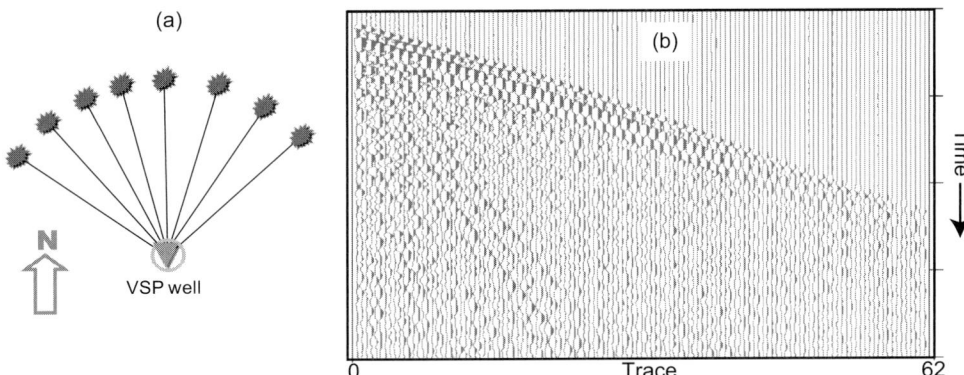

(b)

Time →

0   Trace   62

Figure 2.4 (a) A map-view sketch showing a selected group of shots (stars) of similar source-to-receiver offsets but different azimuths from the VSP well (circled triangle). (b) The common shot gather after correction for orientation errors of the geophones. For color versions see plate section.

common shot gather after such a correction of geophone rotation errors. This shot gather shows much better data quality than the version before the correction (Figure 2.2b).

Exercise 2.1

1. Search the literature to find typical surface seismic survey geometry onshore and offshore, and the typical ranges of parameters such as source depths, receiver depths, source-to-receiver offsets, source-to-receiver azimuths, sample rates, data frequencies, and recording lengths. The results may be presented using a spreadsheet with clear citation of the references.

2. Use your own words to define the concepts of: demultiplexing, time stack, matched filter, and walkaway RVSP.

3. If we have multi-component surface reflection data, can we use a hodogram to suppress ground rolls? How would you do it? What problems can you anticipate for your approach?

2.2   Normal moveout analysis

Among the first seismic processing methods that we discussed in Section 1.1 was CMP stacking, which is the most common and useful pre-processing method. CMP stacking consists of three steps: CMP binning, normal moveout (NMO) correction, and stacking of the NMO corrected traces. Given the important position of CMP stacking in seismic data processing, this section explains NMO analysis in terms of its assumptions, formulation, and usage. Two ubiquitous applications of CMP stacking are noise suppression and velocity analysis.

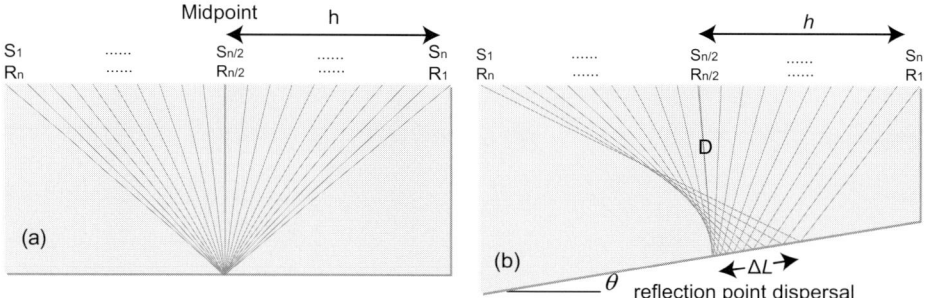

Figure 2.5 For a CMP gather in a constant velocity field the dispersal of reflection points is zero for a flat reflector (a) and minimum for a dipping reflector (b). In case (b), can you derive the amount of dispersal ΔL as a function of offset h and dip angle θ?

2.2.1   Stacking

Among common types of seismic gathers, the reflection spread, or lateral extent of reflection points across the reflector, is zero for the CMP gather in the case of a flat reflector beneath a constant velocity field or a layer-cake velocity field (Figure 2.5a). If the reflector dips, then the spread of the reflection points will be widened: this is referred to as **reflection point dispersal** (Figure 2.5b). For a dipping reflector in a constant velocity, the dispersal ΔL can be expressed in terms of offset h, midpoint norm to the reflector D, and reflector dip θ:

$$\Delta L = \frac{h^2}{D} \cos \theta \sin \theta \qquad (2\text{–}3)$$

As predicted by the above equation, the reflection point dispersal increases as the square of the offset. Even in the case of a dipping reflector, however, the CMP gather still has the lowest spread of any gather. While the NMO assumes a layer-cake velocity model, this assumption has the highest chance of being valid in a CMP gather because of its minimum reflection point dispersal. This fact makes the CMP gather the most suitable type for stacking multiple reflection traces in order to improve the SNR (Mayne, 1962).

The main reason behind many of the stacking processes in seismic processing and imaging is to improve the SNR. Stacking is probably the most common and most effective way to do this for many observational sciences, an example being the stacking of faint image signals from remote objects in astronomy. In addition to NMO stacking, other stacking processes include: (1) field stacking, or vertical stacking of traces from multiple geophones located at the same place; (2) slant stacking, which stacks dipping reflections across a seismic gather after transforms such as plane-wave decomposition; and (3) migration stacking, which stacks migrated traces belonging to the same position together.

2.2.2   Normal moveout correction

For successful stacking of reflection events, we need to align the reflection events in the gather, and this is the process of NMO. As shown in Figure 2.6a, the two-way reflection time from source S to receiver R in a constant velocity field V over a flat reflection interface can be calculated using the Pythagorean theorem for the relationship between the sides of a right triangle:

$$(vt)^2 = x^2 + (z_r + z_s)^2 \qquad (2\text{–}4)$$

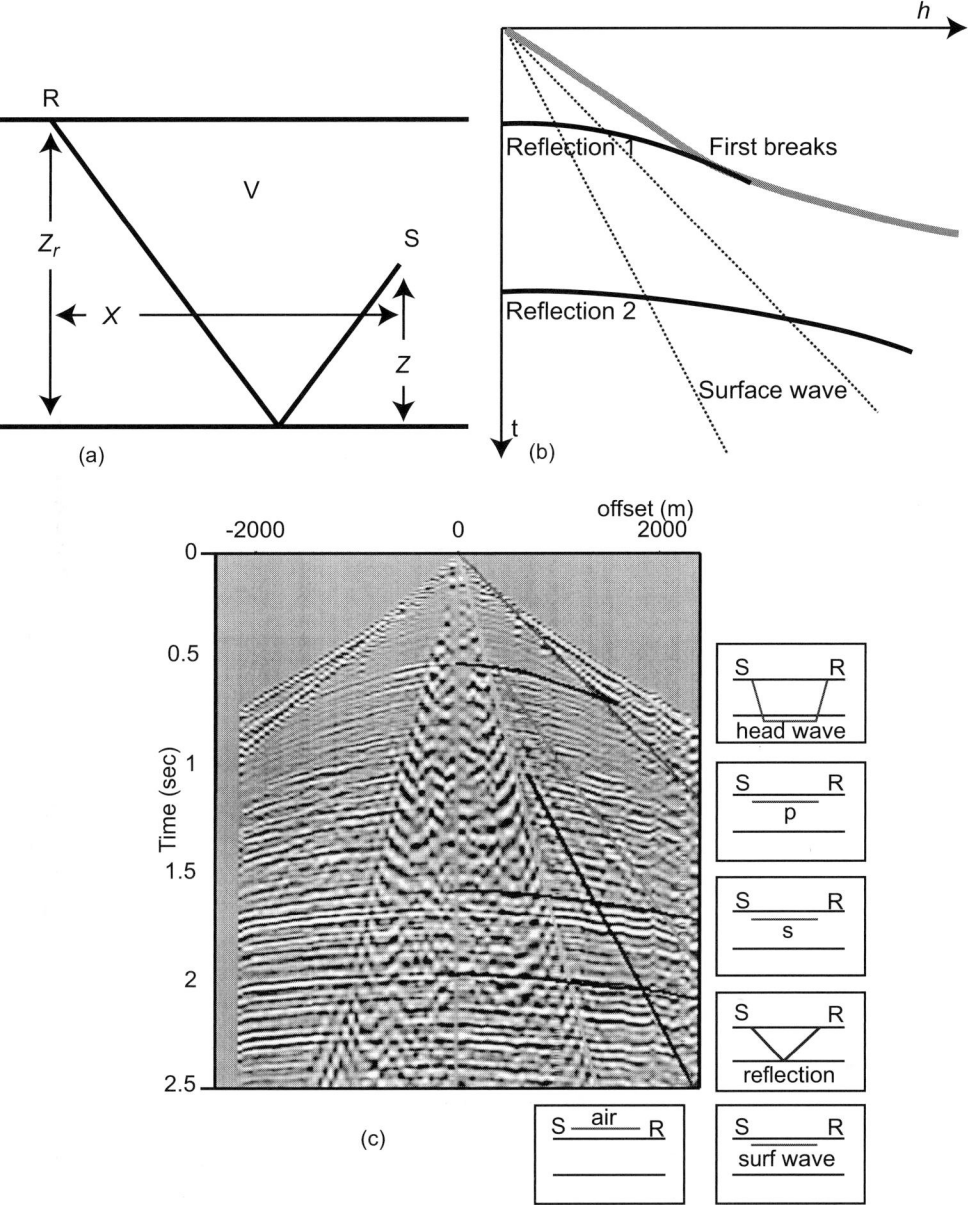

Figure 2.6 (a) A cross-section showing the model space for a NMO with receiver (R) and source (S) at different depth levels. V = velocity. (b) A sketch of common shot gather showing the data space for the NMO. Notice that the first breaks include direct wave, refraction (head wave) in the case of a layer-cake velocity model, and turning wave in the case of a gradient velocity model. What will be the trends of the events if this is a common midpoint gather? (c) A land shot record from Alaska (Liner, 1999).

where z_r and z_s are depths to the reflector from the receiver and source, respectively. When the two depths are equal to the same z, we have

$$(vt)^2 = (2h)^2 + (2z)^2 \tag{2–5}$$

where h stands for offset from the midpoint to either source or receiver. Of course, h is one-half of the source-to-receiver offset x.

Since our goal is to flatten reflections on the offset versus traveltime plane (e.g., Figure 2.7b), we need to find the relation between traveltime t and offset h. We can easily derive an expression for this from (2–5):

$$t(h) = \frac{2}{v}\left(h^2 + z^2\right)^{1/2} \tag{2–6}$$

At zero offset, we have

$$t_0 = t(h = 0) = \frac{2}{v}z \tag{2–7}$$

Thus we arrive at the NMO correction term Δt:

$$\Delta t = t(h) - t_0 = \frac{2}{v}\left[\left(h^2 + z^2\right)^{1/2} - z\right] \tag{2–8}$$

If we replace z by t_0 using (2–7) we obtain

$$\Delta t(h, t_0, v) = \left[(2h/v)^2 + t_0^2\right]^{1/2} - t_0 \tag{2–9}$$

The above equation indicates that the NMO correction term is non-linear with respect to offset h and zero-offset time t_0. This non-linear relationship is the source of NMO stretch (see Box 2.1). One can easily demonstrate using the above equation that the NMO stretch reaches its maximum extent at shallowest zero-offset time or farthest offset.

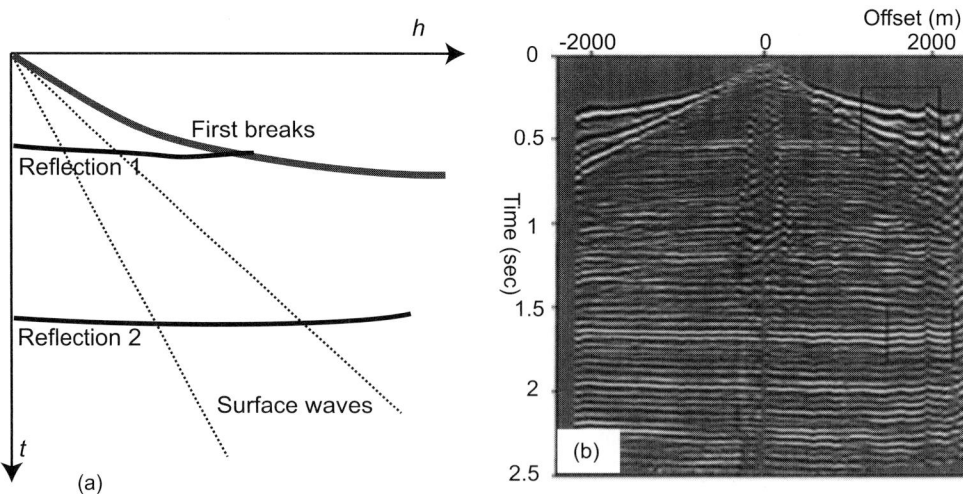

Figure 2.7 (a) A sketch of common shot gather after NMO. (b) The shot record of Figure 2.6c after NMO (Liner, 1999).

Box 2.1 Evaluation of NMO stretch

The NMO correction is non-linear with respect to the offset and zero-offset traveltime. This produces the NMO stretch, the phenomenon of stretched wavelet length after NMO. Rewrite (2–9) as

$$\tau = \Delta t\,(h, t, v) = \left[(2h/v)^2 + t^2\right]^{1/2} - t \qquad (2\text{–}10)$$

Differentiating the above equation with respect to t, the zero-offset traveltime,

$$\frac{\partial \tau}{\partial t} = \frac{t}{\left[(2h/v)^2 + t^2\right]^{1/2}} - 1 \qquad (2\text{–}11)$$

The above differentiation quantifies the NMO stretch. Evaluating (2–11) in several end-member cases, if t approaches infinity or h approaches zero, the differentiation approaches zero, or no stretch. However, when t approaches zero, or h approaches infinity, the differentiation approaches –1, its maximum value. Thus, we conclude that the NMO stretch increases at greater offset or shallower depth (smaller reflection time).

Figure 2.7 shows the consequence of applying NMO. Keep in mind that each vertical seismic trace was shifted statically (no distortion of the wiggles on the trace) by a fixed amount of time according to the NMO equation. The field data shown in the right panel display several problems, and some of them are amplified by the NMO process. Would you be able to identify them?

2.2.3 Usage of NMO stacking

The NMO stacking process is a good way to suppress coherent noise whose moveout differs from that of the primary reflection. One such example is the suppression of multiple reflections, such as the offset versus traveltime graph in Figure 2.8 (from Mayne, 1962) in which the primary reflection and second order multiple have very different moveout curves as predicted using a velocity function typical of the Gulf of Mexico.

While the main reason for NMO and the corresponding stacking process is to improve the SNR, the process requires an appropriate stacking velocity. This dependency on the stacking velocity becomes the basis for semblance velocity analysis, which is the first major step in velocity analysis of all reflection seismic data. The NMO equation (2–9) indicates that the amount of the NMO correction depends on h, the offset from the midpoint (or half the offset from shot to receiver); t_0, the zero-offset two-way reflection time; and v, the velocity of the constant velocity layer.

Despite the real-world complexity, once we identify a major primary reflection event on CMP gather, we may approximate the velocity function above this reflector with a constant velocity value, perform NMO with this velocity and then stack all traces across the offset axis into a single stacked trace. This constant velocity is the stacking velocity, which serves as a medium average of the overburden velocities for the reflector. Though we do not

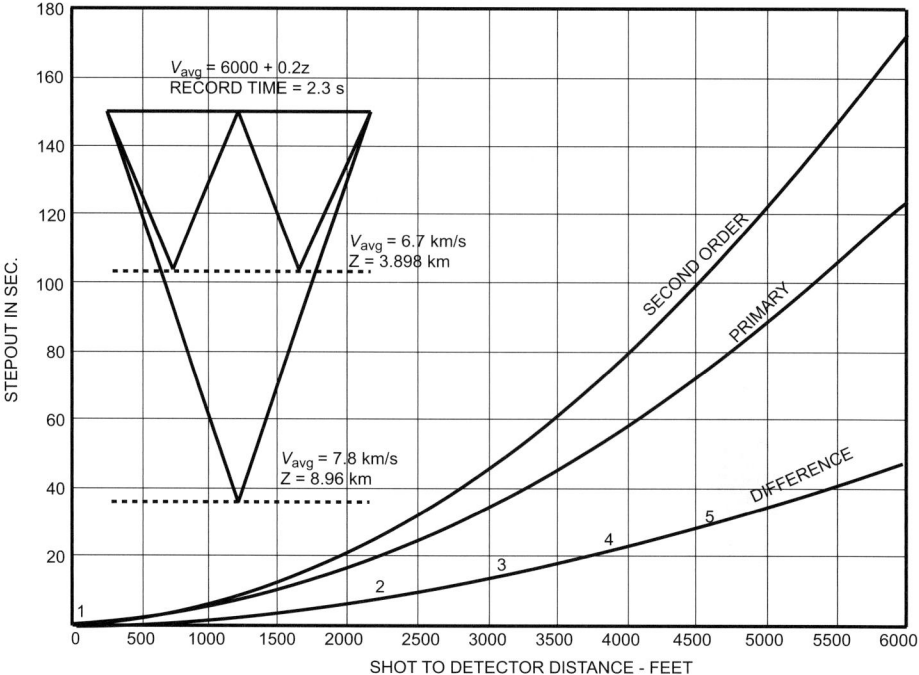

Figure 2.8 Predicted moveout trends of primary and second-order multiples, based on a velocity function typical of the Gulf of Mexico (from Mayne, 1962).

know what would be the best stacking velocity for each reflector, we can try many velocity values within a reasonable range, as shown in Figure 2.9 for a 2D reflection dataset. The semblance, which is a normalized summation of the stacked traces, is grayscale-coded on the semblance panel at each location (in terms of common depth point or CDP number) across the profile.

Notice in Figure 2.9 that the horizontal distance along the profile is expressed in terms of CDP numbers. CDP is a traditional name for the CMP (e.g., Mayne, 1962). However, as shown in Figure 2.5b, a CMP gather may not have a common depth point of reflections. Nevertheless, as one of many traditional names with incorrect meanings, CDP is a common term describing the distance along the profile. Each profile will be discretized into a number of **CDP bins**, whose width is usually half of the **station spacing**, and all source-to-receiver midpoints will then be taken into the corresponding CMP (CDP) bins to form the corresponding CMP (CDP) gathers.

2.2.4 Semblance velocity analysis

Among pioneers who attempted to find ways to quantify different levels of coherency across multi-channel traces, Neidell and Taner (1971) introduced **semblance**, a measure of cumulative amplitudes across seismic traces. The concept stems from the cross-correlation function which can only be used between two traces. Suppose we have N traces in the gather

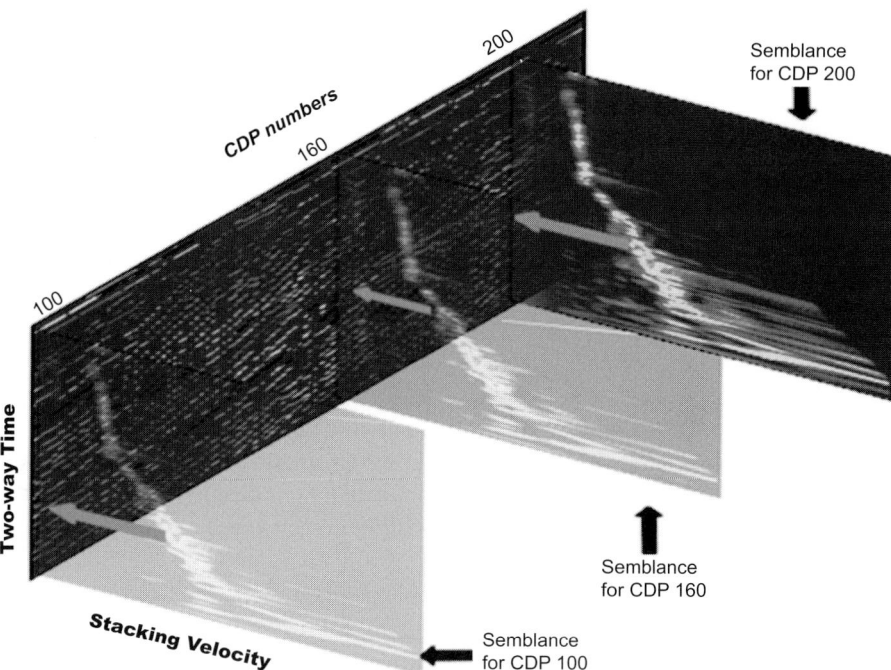

Figure 2.9 Semblance velocity analyses for a 2D dataset. The vertical slice at the back shows the stacked section as a function of CDP number and two-way reflection time. The semblance panels at three locations along the profile are functions of time and velocity.

and a_{ti} is the amplitude at time t on trace i, then the cross-correlation between trace i and trace j over a time window t to $t + m$ will be

$$c_{ij} = \sum_{t}^{t+m} a_{ti} \times a_{tj} \Big/ \left[\sum_{t}^{t+m} (a_{ti})^2 \sum_{t}^{t+m} (a_{tj})^2 \right] \qquad (2\text{--}12)$$

Similarly, the semblance over the same time window will be

$$S(t + m, N) = \sum_{t}^{t+m} \left(\sum_{i}^{N} a_{ti} \right) \Big/ \sum_{t}^{t+m} \sum_{i}^{N} (a_{ti})^2 \qquad (2\text{--}13)$$

In the above expression $\sum_{i}^{N} a_{ti}$ is the amplitude stack at time t across all N traces that measures the coherency of events across this time, and $\sum_{i}^{N} (a_{ti})^2$ is the total "energy" of all traces at time t. Thus, the numerator of the semblance is a time summation of the squared amplitude stacks, and the denominator is a time summation of total energy for normalization. The normalization by the denominator allows the semblance function to be sensitive to the relative amplitude of the traces. Notice that the cross-correlation function applies multiplication between components in its numerator and denominator. In contrast, the semblance function applies summation between components in its numerator and denominator.

Computationally, summation will have a much lower cost than multiplication. Some people have explored the application of weighting functions in the summation (e.g., Celis & Larner, 2002).

Now what about the choice of length for the time window m? Clearly the time window should be greater than the "central portion" of the wavelet in the data. The example in Figure 2.9 used about a dozen sample points as the length of the time window. In general, for data of high SNR we tend to choose a narrow window, because the narrower the time window the higher the resolution but also the less tolerance to noise and problems. The noise includes static shifts, frequency variation, and amplitude variation. The problems include NMO stretch and systematic change of noise characteristics with increasing time and offset.

Based on semblance panels like that shown in Figure 2.9, we can pick a string of semblance peaks as the stacking velocity values for the corresponding zero-offset two-way time and CDP locations. This will resulted in a stacking velocity section for a 2D profile or a stacking velocity volume for a 3D survey. In practice, rather than picking the stacking velocity profile at every CDP position, usually the profile is obtained at one out of every 10 to 50 CDPs to ensure computational efficiency without losing too much of the quality of the velocity model.

We should keep in mind that semblance as described here is not necessarily the only or the best way of velocity model building following the stacking approach. Another very good measure is **differential semblance optimization (DSO)**, proposed by Symes and Carazzone (1991). A recent paper on velocity analysis by DSO has been presented by Mulder and ten Kroode (2002).

Exercise 2.2

1. Prove or refute the statement that stacking of seismic traces has the effect of low-pass filtering.

2. Most semblance velocity analysis stacks all traces with equal weight. Should we apply variable weights as a function of offset and intersection time? Devise a way to conduct such a weighted semblance stack process. It would be great if you could write a program to realize the process.

3. The semblance velocities of multiple reflections are usually slower than the velocity of the primary reflection, as shown in Figure 2.8. Explain the reasons behind this observation.

2.3 Convolution and correlation

As two of the commonly used data processing tools, convolution and correlation are very similar in their mathematical expression. The property produced by convolution is the multiplication of the two data strings, or the result of using one data string to filter the other.

In contrast, the property resulting from correlation is the similarity function between the two input data strings.

2.3.1 Convolution

Convolution is the mathematical operation of multiplying two time series representing two input data strings. It is one of the most useful operations in seismology and digital processing because it represents the physical process of combining two or more time series. For example, filtering a signal can be expressed as the convolution of the input signal with the filter function, producing an output signal. A seismogram can be approximated as the convolution of a source wavelet with the medium function, and with the receiver function. To match the well-log measurements with a seismic reflection profile in the same area, a **synthetic seismogram** is often computed by convolving well-log traces with an appropriate wavelet.

Mathematically, multiplication of two scalars leads to another scalar, the product of the original two scalars. Multiplication of two time series leads to another time series, the convolution of the original two time series. We may express the three time series as $a_t = \{a_1, a_2, \ldots, a_N\}$, $b_t = \{b_1, b_2, \ldots, b_M\}$ and $c_t = \{c_1, c_2, \ldots, c_{N+M-1}\}$, and we may make three vectors and two matrices as follows:

$$
\mathbf{a} = \begin{pmatrix} a_1 \\ a_2 \\ \vdots \\ a_N \end{pmatrix}, \quad
\mathbf{b} = \begin{pmatrix} b_1 \\ b_2 \\ \vdots \\ b_M \end{pmatrix}, \quad
\mathbf{c} = \begin{pmatrix} c_1 \\ c_2 \\ \vdots \\ c_{N+M-1} \end{pmatrix}, \quad
\mathbf{A} = \begin{pmatrix}
a_1 & 0 & \cdots & 0 \\
a_2 & a_1 & \ddots & \vdots \\
\vdots & a_2 & \ddots & 0 \\
a_N & \vdots & \ddots & a_1 \\
0 & a_N & \vdots & a_2 \\
\vdots & \ddots & \ddots & \vdots \\
0 & \cdots & 0 & a_N
\end{pmatrix}_{(N+M-1)\times M},
$$

$$
\text{and} \quad \mathbf{B} = \begin{pmatrix}
b_1 & 0 & \cdots & 0 \\
b_2 & b_1 & \ddots & \vdots \\
\vdots & b_2 & \ddots & 0 \\
b_M & \vdots & \ddots & b_1 \\
0 & b_M & \vdots & b_2 \\
\vdots & \ddots & \ddots & \vdots \\
0 & \cdots & 0 & b_M
\end{pmatrix}_{(M+N-1)\times N}
\tag{2–14}
$$

Note that matrix $\mathbf{A}$ has M columns and each column contains the time series a_t which is shifted one element downwards sequentially from the second column on, with zeros filling the remaining portion. Similarly the matrix $\mathbf{B}$ has N columns, each column containing the time series b_t which is also shifted downwards sequentially, and with zeros filling the remaining portion. Here **A** and **B** are **Toeplitz matrices** because they are **diagonal-constant** – the diagonally oriented values are equal to each other. Then the convolution of a_t and b_t is

$$\mathbf{c} = \mathbf{Ab} = \begin{pmatrix} a_1 & 0 & \cdots & 0 \\ a_2 & a_1 & \ddots & \vdots \\ \vdots & a_2 & \ddots & 0 \\ a_N & \vdots & \ddots & a_1 \\ 0 & a_N & \vdots & a_2 \\ \vdots & \ddots & \ddots & \vdots \\ 0 & \cdots & 0 & a_N \end{pmatrix} \begin{pmatrix} b_1 \\ b_2 \\ \ddots \\ b_M \end{pmatrix} \tag{2–15a}$$

or

$$\mathbf{c} = \mathbf{Ba} = \begin{pmatrix} b_1 & 0 & \cdots & 0 \\ b_2 & b_1 & \ddots & \vdots \\ \vdots & b_2 & \ddots & 0 \\ b_M & \vdots & \ddots & b_1 \\ 0 & b_M & \vdots & b_2 \\ \vdots & \ddots & \ddots & \vdots \\ 0 & \cdots & 0 & b_M \end{pmatrix} \begin{pmatrix} a_1 \\ a_2 \\ \vdots \\ a_N \end{pmatrix} \tag{2–15b}$$

Taking the kth row of the above two equations we have

$$c_k = \sum_j a_{k-j+1} b_j \tag{2–16a}$$

$$c_k = \sum_l b_{k-l+1} a_l \tag{2–16b}$$

where k goes from 1 to $N+M-1$, and the indexes j and l scan through all the non-zero elements of the two input time series a_t and b_t. As the digital expression of convolution, the above equations show that convolution is commutable, or **a** convolved with **b** is equal to **b** convolved with **a**. Since a time series can be regarded as a digitized version of a continuous function, we can derive the continuous form of convolution:

$$c(t) = a(t) * b(t) = \int_{-\infty}^{\infty} a(t-\tau)\tilde{b}(\tau)d\tau \tag{2–17a}$$

$$c(t) = b(t) * a(t) = \int_{-\infty}^{\infty} b(t-\tau)\tilde{a}(\tau)d\tau \tag{2–17b}$$

where $*$ stands for the convolution operator, and $\tilde{a}(\tau)$ and $\tilde{b}(\tau)$ are the complex conjugates of $a(\tau)$ and $b(\tau)$, respectively. A comparison between (2–16) and (2–17) indicates that, if complex time series are used, then the first time series on the right-hand side of (2–16) will be the corresponding complex conjugates, $\{\tilde{a}_j\}$ and $\{\tilde{b}_l\}$. In other words,

$$c_k = \sum_j a_{k-j+1}\tilde{b}_j \tag{2–18a}$$

$$c_k = \sum_l b_{k-l+1}\tilde{a}_l \tag{2–18b}$$

2.3.2   Correlation

The operation of **correlation** is just a normalized **covariance**, and both of them quantify the similarity between two time series as a function of the offset between the two series. They are among the most useful operations in seismology and digital processing. As an example, the semblance process as described in the previous section is a direct application of the correlation function. We may learn the covariance relationship from a study of random variables. If we have two random variables, x and y, we define the **covariance matrix** between them as

$$C = \begin{pmatrix} \sigma_{xx} & \sigma_{xy} \\ \sigma_{yx} & \sigma_{yy} \end{pmatrix} \tag{2–19}$$

The cross-covariance is defined as

$$\sigma_{xy} = \sigma_{yx} = \int_{-\infty}^{\infty} dx \int_{-\infty}^{\infty} dy (x - \bar{x})(y - \bar{y}) f(x, y) \tag{2–20}$$

where $\bar{x}$ and $\bar{y}$ are the means of the two random variables, and $f(x, y)$ is their joint **probability density function (PDF)** which quantifies the chance of their occurrence. The cross-covariance in (2–20) is an expectation of the second moment between the two random variables. You may derive the auto-covariance σ_{xx} and σ_{yy} using the format of (2–20).

In practice the PDF of any random variable can be estimated using the **histogram** of a set of observed samples of the variable. Here, great care is needed to ensure that the set contains sufficient samples and that the samples are representative of the general characteristics of the random variable. When two random variables are independent with respect to each other, their cross-covariance becomes zero:

$$\sigma_{xy} = \sigma_{yx} = 0$$

Hence the covariance matrix becomes diagonal. This is comparable to the case when two time series are totally uncorrelated.

We define the **cross-covariance** of two time series $x(t)$ and $y(t)$ in the continuous case as

$$\gamma^{xy}(t) = \int_{-\infty}^{\infty} \tilde{x}(\tau) y(t + \tau) d\tau \tag{2–21a}$$

$$\gamma^{yx}(t) = \int_{-\infty}^{\infty} \tilde{y}(\tau) x(t + \tau) d\tau \tag{2–21b}$$

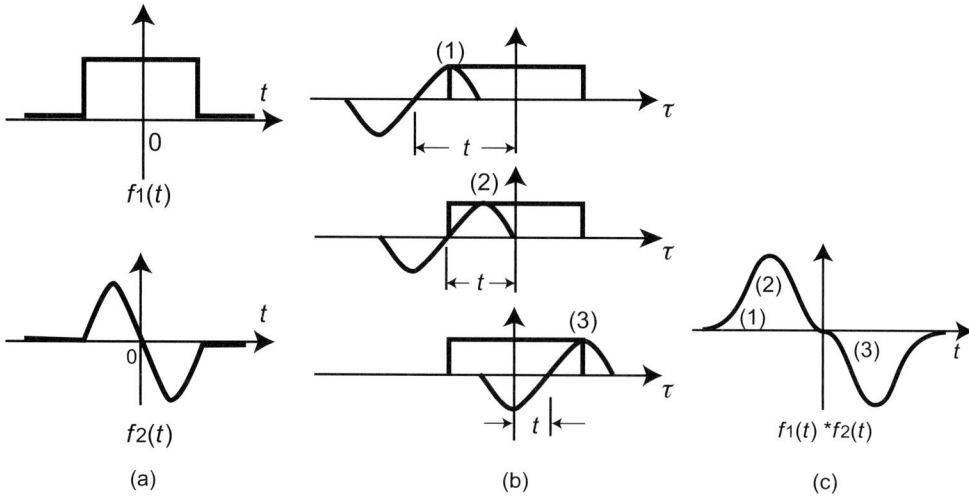

(a)   (b)   (c)

Figure 2.10 (a) A boxcar function $f_1(t)$ and a minus sine function $f_2(t) = -\sin t$. (b) Convolution between the two functions can be viewed as sliding the second function across the first function, and cumulating their overlapping area as a function of their differential time. (c) Result of convolving the two functions.

and in the discrete case as

$$\gamma_t^{xy} = \sum_j x_{t+j+1}\tilde{y}_j \qquad (2\text{–}22a)$$

$$\gamma_t^{yx} = \sum_l y_{t+l+1}\tilde{x}_l \qquad (2\text{–}22b)$$

Comparing the discrete cross-covariance in (2–22) with discrete convolution in (2–18), their mathematical difference is only in one sign on the right-hand sides. However, cross-covariance and convolution represent two very different physical processes. The former cumulates the similarity, whereas the latter quantifies the multiplication or combination of two time series. Both correlation and convolution operators have numerous applications in geophysics and other areas. For the correlation operator, a recent example is given by Shapiro *et al.* (2005) who obtained valuable data by correlating ambient noise (see Box 10.1 in Chapter 10). Much of the discussion in the chapters on filtering and deconvolution is devoted to the convolution operator.

2.3.3   Examples of convolution and correlation processes

To help us visualize the convolution process, let us see a graphic example of convolving a boxcar function $f_1(t)$ and a minus sine function $f_2(t) = -\sin t$. in Figure 2.10. Convolution between the two functions can be viewed as sliding the second function across the first function, and cumulating their overlapping area as a function of their differential time. Here the graphs in Figure 2.10b display three different stages of sliding the second function across the first. The graph in Figure 2.10c shows the convolution result, the overlapping area of the two functions at each moment.

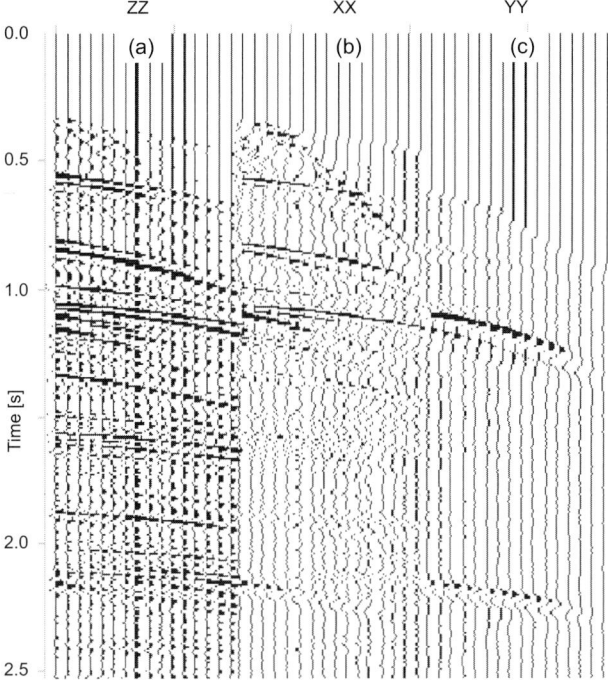

Figure 2.11 An example of multi-component data from physical modeling. (a) Record from the vertical component of receivers receiving a vertical source. (b) Record from the inline component of receivers receiving an inline source. (c) Record from the crossline component of receivers receiving a crossline source.

You may have noticed that the second function is reversed in time in Figure 2.10b with respect to that in Figure 2.10a. A time series or a discrete function is order-dependent: its first element is at its left end and its last element is at its right end. Thus, when one function interacts with another, the first element of the first function will be the first one encountering the other function; hence graphically the second function needs to be time-reversed so that its leading element first encounters the leading element of the first function. In each of the graphs in Figure 2.10b, the first function, which is a boxcar, is plotted against the integration variable τ. The second (sine) function is plotted with a reversal of the τ axis, owing to the minus sign in the convolution, and it is displaced by an amount t. For a given time shift, the product of the two functions is integrated over the interval in which they overlap, producing one point on the curve in Figure 2.10c.

Similarities between seismic waveforms are often quantified using cross-correlation. For instance, a seismometer measures particle motions in three directions: up–down (UD), north–south (NS), and east–west (EW). The transverse and radial components of seismic data from a source at an arbitrary location are generally correlated on the NS and EW coordinates. Rotation of the seismograms to align with the source–receiver azimuth could minimize the unwanted correlation between the transverse and radial components of the signals, while there may still be correlated noise. As an example, Figure 2.11 shows a multi-component dataset acquired from a physical modeling experiment. The data are common

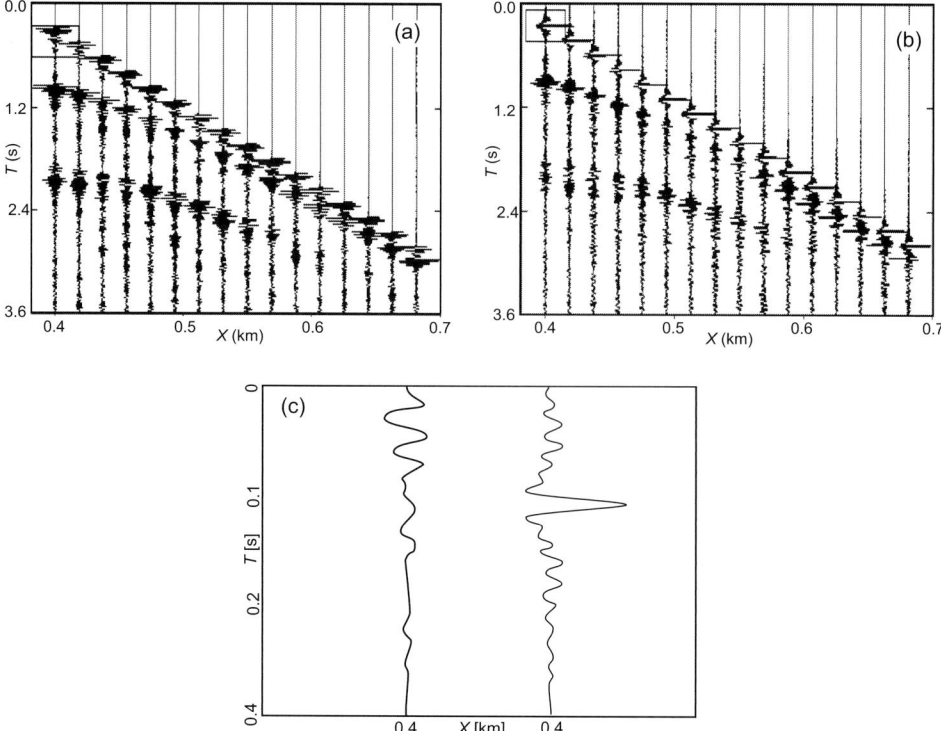

Figure 2.12 (a) A 16-trace common shot gather from a physical modeling experiment. (b) After applying a matched filter to alter the wavelet of all traces. (c) Traces within a 400-ms window shown in the dashed boxes in (a) and (b), showing the first break wavelets of the first trace.

shot gathers with 16 traces for each component. Since each component was acquired separately with matching polarity for source and receivers, the similarity between different components should be caused only by the geological structure.

An application of cross-correlation for real data is given in Figure 2.12. Figure 2.12a shows a common shot gather from a physical modeling experiment. A small window contains the time series around the first break of the first trace, which is shown in the left trace in Figure 2.12c. By auto-correlating the left trace with itself, we produce the trace shown in the right side of Figure 2.12c, and this trace is symmetric since it is a result of auto-correlation. We may take this symmetric trace as an approximated version of a zero-phase wavelet. By cross-correlating the left trace in Figure 2.12c with all the traces shown in Figure 2.12a, we produce the shot gather in Figure 2.12b. We have advanced the time in Figure 2.12b by a time interval, the difference between the onset time of the left trace and the peak time of the right trace in Figure 2.12c. This time advancement is done in order to align the time zero of the approximated zero-phase wavelet with that of the input traces. You may view the use of the cross-correlation in this case as an application of the matched filter concept to retrieve the known signal. Note that the cross-correlation of two traces is exactly the convolution of the first trace with a time-reversed version of the second trace.

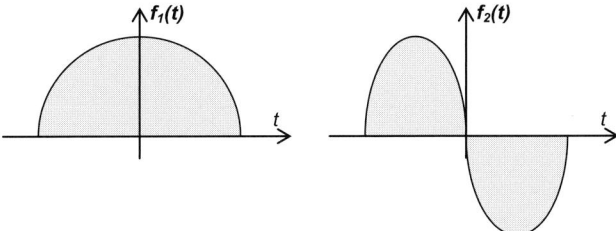

Figure 2.13 Plots of two time functions of the same length.

Exercise 2.3

1. Sketch the cross-correlation of two functions with equal time durations in Figure 2.13. (Hint: you may first discretize the functions into short time series, and then use the formula of discrete cross-correlation.)

2. Compare and contrast the convolution and cross-correlation operations. (Note: it may be best to organize your answer by using a table listing the similarities and differences.)

3. Cross-correlation is used to measure the similarity between two vectors $\mathbf{u} = (u_1, u_2, \ldots, u_N)$ and $\mathbf{v} = (v_1, v_2 \ldots, v_N)^{\mathrm{T}}$. By removing the averages $\bar{u}$ and $\bar{v}$, we obtain residual vectors $\Delta \mathbf{u} = (u_1 - \bar{u}, u_2 - \bar{u}, \ldots, u_N - \bar{u})$ and $\Delta \mathbf{v} = (v_1 - \bar{v}, v_2 - \bar{v}, \ldots, v_N - \bar{v})$. Is the cross-correlation between $\mathbf{u}$ and $\mathbf{v}$ the same as the cross-correlation between $\Delta \mathbf{u}$ and $\Delta \mathbf{v}$? Please explain your answer with evidence.

2.4   Noise suppression methods

Noise suppression or **noise attenuation** is one of the traditional goals of seismic data processing. Ideally, we first identify the main differences between the characteristics of the signal and that of the noise so that a proper data processing flow can be designed to suppress the noise with minimum impact on the signal. However, the task often becomes difficult because the signal and noise may not be precisely defined. Some processing methods assume **white noise**, with random distribution or white spectra. However, **colored noise**, having similar behavior to the signal, often exists in the real world. Onshore seismic data, for example, are usually associated with complex near-surface or subsurface geologic conditions that may mean that the seismic signal and noise share a similar range of frequencies and apparent velocities. As a result, we may divide noise suppression into two types, **data-driven** methods and **model-driven** methods.

2.4.1   Model-driven noise suppression methods

For any seismic dataset given, an experienced data processor always attempts to come up with a simple model to explain key features of the data. As an example, the common shot

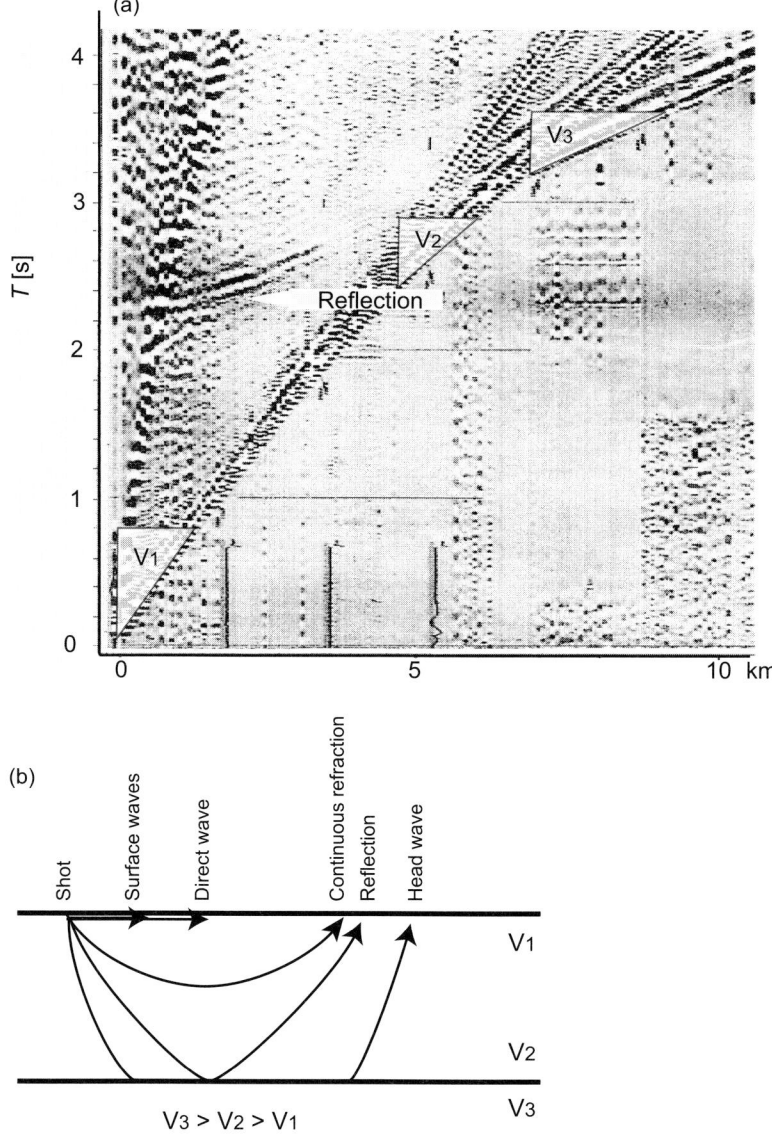

Figure 2.14 (a) Data space: A common shot gather. The slopes of the three triangles quantify the apparent velocities of the first break. (b) Model space: an interpretation of the velocity field and raypaths of various waves. (After Sheriff & Geldart, 1995.)

gather in Figure 2.14a shows a coherent first break and a strong reflection at time 2.2 s on zero-offset, which extends into a head wave beyond a distance of 6.5 km. The slopes of three triangles along the first break quantify the **apparent velocity** of the first break at these locations. An apparent velocity of a seismic event refers to its slope on a time–distance or equivalent graph. The ground rolls in this case, as an example, have very slow apparent velocity because they are within a small offset range of about 2 km. The single strong reflection leads us to come up with the simple model in Figure 2.14b, with a layer over a half space. The top and base velocities of the top model layer can be well approximated by the apparent velocities V_1 and V_2 of the two triangles at near-offset and before the **critical distance** at which the reflection joins the first break. The top velocity of the half space is approximated by the apparent velocity V_3 of the head wave, as shown by the third triangle in Figure 2.14a. An increase of velocity with depth as shown in Figure 2.14b is evident from the change of the slopes of the three triangles in Figure 2.14a. We can estimate the thickness of the top model layer using the zero-offset two-way time of 2.2 s of the reflection plus the values of the top and base velocities of the layer.

2.4.2 Data-driven noise suppression methods

Data-driven noise suppression methods are based on a difference in character between the desired signal and unwanted noise in data space. A simple example is when the signal and noise have separate frequencies, such as when we want to keep high-frequency reflections and suppress low-frequency surface waves: in this case **high-pass filtering** of the data may do the job. However, losing low frequencies will severely reduce the bandwidth and resolution of the data. Hence we have to make a balanced choice about how much low frequency we remove. Another example is that most seismic migrations use only primary reflection events, so we want to remove first arrivals and head waves. Because the events associated with first breaks fall in different locations from most reflections on CMP or common shot gathers, we may just apply a **mute**, the process by which we delete all events in a pre-defined window. In such cases, an **outer mute** is often used as a crude way to remove first breaks on pre-stack gathers. As shown in Figure 2.15, we could also use an **inner mute** to remove ground rolls that have low apparent velocities, and a **polygon mute** to "surgically" remove any unwanted portion of the data. The mute approach is certainly effective, but it is a quick-and-dirty method: it throws away both the baby and the bath water, namely the noise and the signal within the mute zones. In addition, the sharp edges of the mute zones may introduce artifacts in the subsequent seismic images.

Stacking, or summing of multiple traces along the horizontal axis into a single trace, is one of the most widely used data-driven noise suppression methods. It resembles an integration operator. Traditionally, nearly all onshore reflection seismic data are acquired using a field stack process, which deploys a group of about 10 geophones at each field location (called a station), and stacks the records of all geophones in each group into a single trace. The main purpose of such a field stack is to enhance the SNR and suppress various field noises and inconsistency between the geophones. However, the stacking may harm the recording quality of multi-component geophones. With the arrival of high durability and high fidelity of digital geophones, a trend in modern seismic surveying is to employ arrays of tens of thousands of single geophones or hydrophones without the field stack process.

2.4.3 Linear moveout and tau–p transform

An effective way to use the power of stacking is to apply a moveout, or a transform along the time axis, prior to stacking. We have already seen the example of normal moveout or NMO, which aligns reflections using an appropriate NMO velocity field. A subsequent

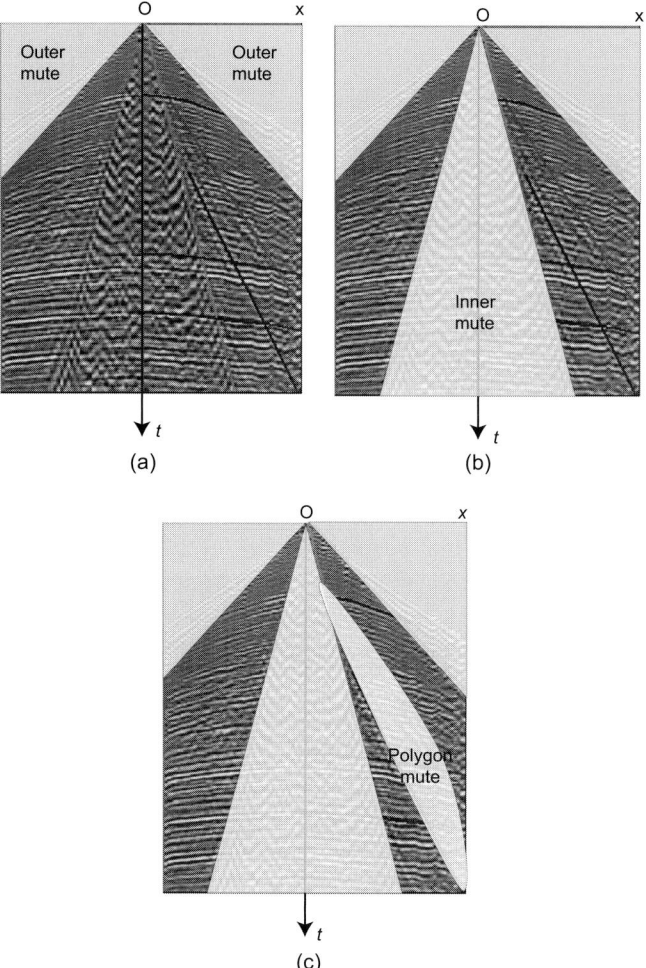

Figure 2.15 (a) Outer mute removes first breaks. (b) Inner mute removes ground rolls. (c) Polygon mute "surgically" removes various events.

stacking produces a stacked trace from each CMP gather, and thus the **pre-stack** seismic traces are converted into the **post-stack** seismic traces. Another simple transform in time is linear moveout, which converts the original time t into a **reduced time** τ using a **reduction slowness** p ($1/p$ is the reduction velocity):

$$\tau = t - px \qquad (2\text{--}23)$$

The reduction slowness is just the **ray parameter**, or **horizontal slowness**. Some people refer to the reduced time as **intercept time**.

The application of (2–23) is called a **linear moveout** (**LMO**) because the relationships between the original time, offset and reduced time are linear. In contrast, the NMO is a

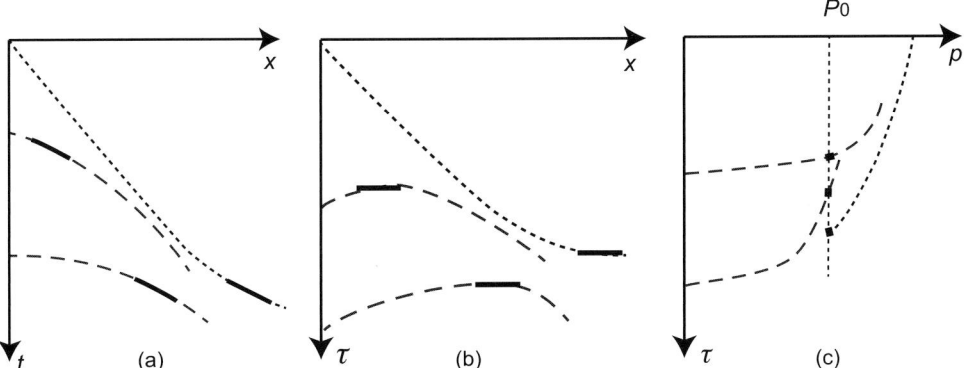

Figure 2.16 (a) A sketch of common shot gather with three bold bars denoting those portions of the first break and two reflection events that have the same slowness, p_0. (b) After a linear moveout using the reduction slowness p_0, the corresponding portions of the three events become flat on the x–τ plane. (c) Sketch of the gather after a τ–p transform.

non-linear moveout. In order to enhance an event of particular slope, we may apply a linear moveout using the corresponding slope, and then sum the entire gather along the new x-axis. Such a process is called a **slant stack**, which is used in various applications including the **Radon transform** (e.g., Sheriff & Geldart, 1995; Maeland, 2004). Radon transform is a primitive form of tomography, which assumes that the internal properties of a medium can be obtained from many measurements made along the boundary of the medium. If the slant stack process is a good approximation to the integration of the internal properties of the medium along the path of a seismic wave, then we may be able to use an inverse slant stack to uncover the internal properties of the medium. A typical implementation of the inverse slant stack formula follows the definition of 2D Fourier integration.

If we express the input gather as function of time and offset $P(t, x)$, then we can use (2–23) to transform the data as a function of the slope p and reduced time τ:

$$S(p, \tau) = \sum_x P(x, t) = \sum_x P(x, \tau + px) \qquad (2\text{–}24)$$

The above is called a **tau-p (τ-p) transform**, which is useful in many applications such as suppression of certain multiple reflections. Clearly, the transform consists of a series of slant stacks using the ray parameters across the entire input dataset. To implement the tau–p transform and its inverse in practice, it is often more stable to take the approach of a least squares inverse. Figure 2.16 shows a schematic picture of a shot gather going through the tau–p transform.

In addition to muting, filtering and stacking after linear and non-linear moveout, there are many other kinds of data-driven method to suppress various types of noise. However, the characteristics of real data are often beyond the routines of general data-driven approaches. For example, in the presence of rugged topographic variations or considerable thickness variation of the weathering zone, the regularity of all types of waves will be distorted. In

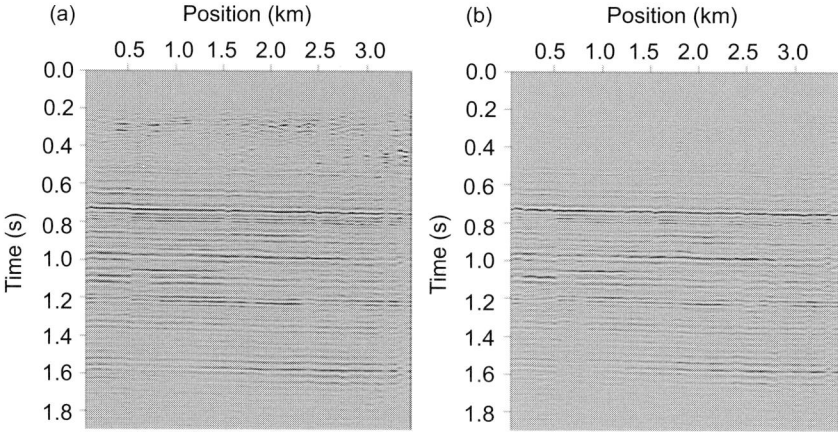

Figure 2.17 (a) Stack section after correction for geometric spreading in the *t–x* domain. (b) The same section stacked in the *τ–p* domain after applying plane-wave decomposition to remove geometric spreading. (From van der Baan, 2004.)

many cases, it is necessary to combine data-driven and model-driven methods of noise suppression. Model-driven noise suppression methods attempt to customize and adjust the parameters of the processing flow using predictions from realistic models. In the above example, if we have a good enough near-surface model that includes topography and thickness variations of the weathering zone, we should be able to make reasonable predictions about the distortions of both signals and noise. Such information serves as the base for building a customized noise suppression flow.

Figure 2.17 illustrates the effectiveness of noise suppression using the *τ–p* transform (van der Baan, 2004). The same input data were subjected to two different corrections for geometric spreading. The stack after correction in the *τ–p* domain via plane-wave decomposition yielded a stack of apparently higher SNR than that from a *t–x* domain correction. Although different approaches to data processing often yield similar results as in this case, a small but consistent improvement is often significant. We often need to analyze subtle differences between wiggle plots at known horizons and SNRs in the spectral domain.

Exercise 2.4

1. Define white noise versus colored noise, and give several examples for each type of noise. Why is it usually more difficult to suppress colored noise than white noise?

2. Discuss the meanings of signals and noise in Figures 2.14a and 2.15a.

3. Find a couple of examples in the literature on noise suppression in seismic data processing using the *τ–p* transform.

Table 2.1 P-wave velocities (V_p) of near-surface rocks (Pralica, 2005).

Rock type			V_p (km/s)
Alluvium			0.225–0.4
Loess			0.3–0.6
Weathered layer			0.3–0.9
Clay			0.3–2.5
Diluvium			0.7–1.8
Sands (0.4–2.8)	Calcareous		0.8
	Consolidated		0.61–0.82
	Loose	Above water table	1.0
		Bellow water table	1.8
	Wet		0.75–1.5
Carbonates			1.7–7.0
Reef			1.7–7.0
Tundra (permafrost)			2.3–5.9
Basalt			5.06–6.4

2.5 Correction for near-surface statics

In reflection and refraction seismology the **near-surface effect** plays a key role. It is caused by a high level of lateral variations in topography, seismic velocity, lithology, layer thickness, and seismic attenuation. Variations in topography onshore and in bathymetry offshore are strong contributors to the near-surface effect. Lateral velocity variation is another strong contributor. The presence of such a high level of lateral heterogeneities acts like a dirty glass, distorting the images of the subsurface structure produced from seismic imaging. A tradition in seismic data processing is to correct for the near-surface effect using an approximation called near-surface statics, which is introduced in this section.

2.5.1 Large lateral velocity variations near surface

A main reason for the strong lateral heterogeneity in seismic velocity is the presence of extremely slow velocities, particularly in desert land areas. The loose alluvium, for instance, may have a P-wave velocity value smaller than the speed of sound (Table 2.1).

The concern about the effect of the near surface contributes to the appreciation of surface consistency. A **surface-consistent** processing tool means it accounts for all the near-surface effects such as locations of the shot and receivers, topography, and even near-surface velocities. There are surface-consistent gain controls that associate attenuation factors with each source and geophone location. The goal of removing near-surface effects is to ensure that deeper amplitude variations are more closely associated with subsurface factors. Near-surface effects can have very adverse effects on deeper reflections for two reasons. First, velocity variation is usually at its highest near the surface. Second, the strongest near-field effects exist because both shot and geophones are located within the near-surface.

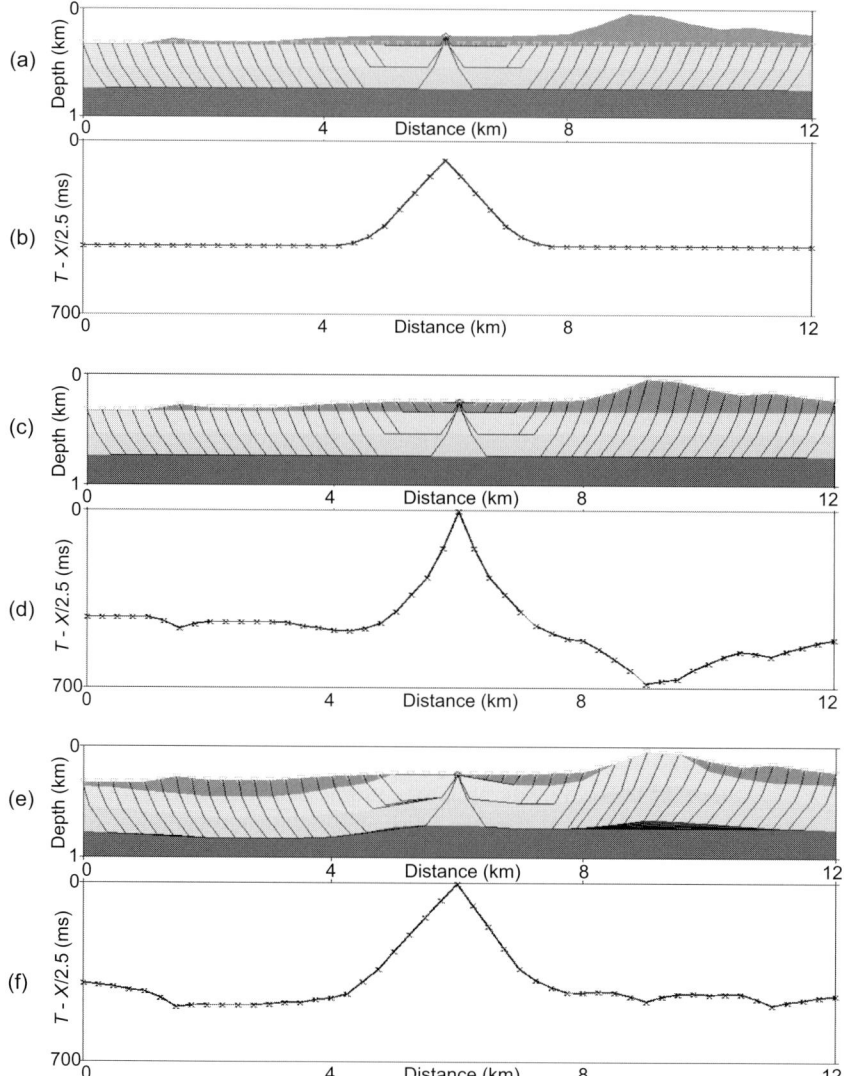

Figure 2.18 Modeling of first arrivals in three models. (a), (c), (e) Models with first-arrival rays from a single source (star) to a number of receivers (triangles). The velocities of the four model layers are 1.0, 1.5, 2.0, and 2.5 km/s from top downwards. (b), (d), (f) First arrivals plotted at horizontal positions of the receivers. The vertical axis is the reduced traveltime using a reduction velocity of 2.5 km/s.

2.5.2 The model of near-surface statics

It is customary in reflection seismology to quantify the near-surface effect on the vertical traveltime of seismic waves as **near-surface statics**. Sheriff (1991) defines static corrections as "corrections applied to seismic data to compensate for the effects of variations in elevation, weathering thickness, weathering velocity, or reference to a datum." While a

---

Box 2.2 Raypaths of a first arrival

A first arrival or first break refers to the seismic wave that arrives the earliest from a seismic source. The raypaths of the first arrival, however, depend on the velocity model. The near-offset is usually the direct wave. The far-offset could be head waves for a layer-cake model, but turning waves in a gradient model. The situation could be more complex in the presence of lateral velocity variations.

---

static correction compensates the time shifts of reflection assuming vertically traversing rays through the near-surface, the assumption is supported by the extremely slow velocities near surface. Thus, the static correction becomes an essential step in nearly all land seismic data processing (e.g., Taner *et al.*, 1998; Cox, 1999), and in many marine seismic data processing sequences to account for water velocity or tidal variations (e.g., Lacombe *et al.*, 2009). Let us see the behavior of first arrival raypaths and traveltimes in three synthetic models shown in Figure 2.18. Here each model consists of four constant-velocity layers, and the reduced traveltimes of the first arrivals are flattened by the reduction velocity of the bottom layer. The traveltime curve in Figure 2.18d shows the effect of topography. While the traveltimes for the third model may show less variation than those of the second model, the third model is actually more complicated than the second. The reason is that the topography and the undulating interfaces are compensating for each other in their impact on the traveltimes. This example warns us that, in practice, an apparently simpler dataset does not mean the corresponding model will always be simpler.

2.5.3 Corrections for near-surface statics

The goal of static correction is to transform seismic traces recorded at sources and receivers near surface into a new set of traces as if they were recorded at sources and receivers along a **datum**, a hypothetical surface that is tens to hundreds of meters below the surface (Figure 2.19). Depending on the ruggedness of topography and thickness of weathering zone, either a **flat datum** or **floating datum** can be chosen. A flat datum has a constant elevation, whereas a floating datum usually has a fairly constant depth from the surface. The depth of the datum should be lower than the base of the weathering zone to minimize the near-surface effect and to be shallower than the depth of significant reflectors. However, because it is a "static" correction (meaning a constant time shifting of each seismic trace, corresponding to moving the source and receiver vertically from their original positions), there is an error in the raypath and traveltime, as shown by the difference between the original reflection raypath and the corrected reflection raypath in Figure 2.19. Clearly, this error in static correction is smaller for shallower datum and deeper reflectors.

In practice, it is preferable to apply near-surface static correction based on real measurements. GPS and other satellite or aerial geodetic data provide precise topographic data. Uphole surveys reaching to depths of 100–200 m are often used to build up the static correction times to key markers such as the water table and base of the weathering zone. People have also constructed empirical static correction formulas at places of regularly appearing near-surface features. For instance, in desert areas covered by sand dunes, a sand dune

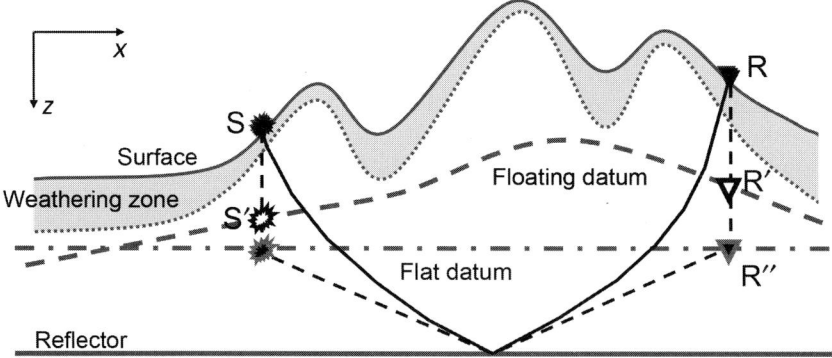

Figure 2.19 A schematic cross-section showing the weathering zone, a floating datum, a flat datum, and a reflector. With a floating datum, a static correction moves the source S and receiver R vertically down to S′ and R′. With a flat datum, a different static correction moves the source and receiver to S″ and R″. The original reflection raypath, as shown by the black curve, is approximated by the dashed black curve in the second correction.

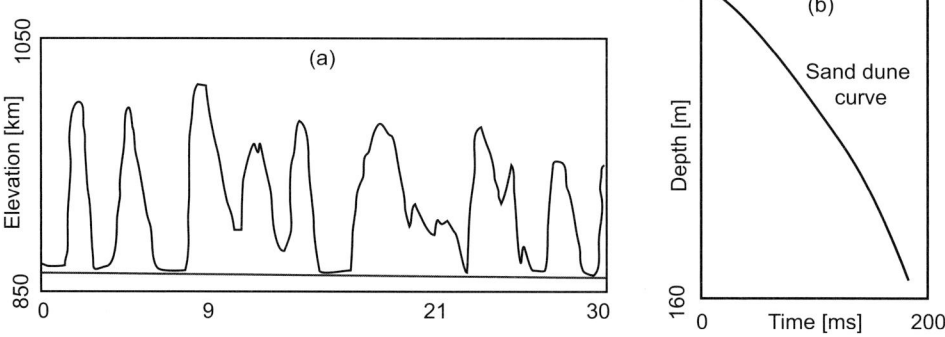

Figure 2.20 (a) Sand dune topography (upper curve) and the water table depth (lower flat curve). (b) The sand dune curve in Tarim Basin.

curve of static correction (Figure 2.20) may be established based on measurements at sample locations and applied elsewhere. Recently, remote-sensing data have been interpreted to provide lithological and static corrections.

However, it is a common practice in modern reflection seismology to determine corrections for near-surface statics, also called **residual statics**, based on traveltimes of first breaks and reflections that are readily available from seismic data (e.g., Taner *et al.*, 1974; Cox, 1999). Using refraction traveltimes, the approach of **refraction statics** includes the field statics and the **least squares inversion** and assumes a layer-cake velocity model. **Field statics** include corrections for **source statics** and **receiver statics**, as well as thickness and velocity variations of the weathering zone. Source and receiver statics refer to the time corrections due to local velocity variations at source and receiver as well as their position errors. One may take the mean value of traveltime residuals of all the traces at a source or a receiver as an estimate of the statics. A more elegant method is to formulate traveltime equations of all traces as functions of the static terms at all sources and receivers plus the structure effect, and then determine the source and receiver statics through a least squares inversion. This is a simple form of traveltime tomography (e.g., De Amorim *et al.*, 1987). Because refraction rays have a large offset in comparison with their traversing depth, they may provide a good estimation of the near-surface statics due to long-wavelength components, at a distance range comparable with the seismic cable length.

To better estimate the short-wavelength components of the near-surface statics, one may apply **reflection statics** using traveltimes of reflections. As in the determination of refraction statics, a set of traveltime equations can be formulated for traveltimes of reflection rays as functions of the static terms and the structure effect. Obviously, it is more beneficial to combine the refraction and reflection data to determine the static corrections.

Another method of correction for the near-surface statics is **trim statics**, in which small static time shifts (less than about 1/3 of the main period) are made to align the reflection wiggles of neighboring traces. Because the main period of most reflection seismic data is around 20 ms, the trim statics are usually limited to within 5–7 ms of each pair of neighboring traces. While trim statics are usually applied to pre-stack seismic gathers, they can be easily applied to post-stack seismic sections. Trim statics are typically derived from a process of cross-correlating neighboring traces of the input data. A trimmed dataset after correcting trim statics may show significantly improved alignment of reflections from the input, because the optimal alignment of reflections is the goal of the method. However, because trim statics are estimated statically with the alignment of reflection as the only objective, this method may produce wrong results, particularly in the presence of high noise level or aligned noises.

2.5.4   Tomostatics

Perhaps the most popular method of static correction is tomostatics, which first determines a near-surface velocity model using a tomographic inversion, and then corrects for the statics predicted by the velocity model. The advantage of the tomostatics approach lies in its use of a realistic near-surface velocity model to validate the static corrections and its use of global inversion to determine the velocity model using all the data. Most tomostatics methods employ first arrivals because they are widely available and the corresponding inversion is very simple. Depending on the velocity field, the raypaths of first arrivals include that of the direct wave, turning wave or refracted wave. Early studies of tomostatics were reported by Chon and Dillon (1986) and De Amorim *et al.* (1987). After the work of Zhu *et al.* (1992) and Docherty (1992), the method became widespread (e.g., Rajasekaran & McMechan, 1996; Zhang & Toksöz, 1998; Chang *et al.*, 2002; Zhou *et al.*, 2009). Two stack sections after a refraction static correction and a tomostatic correction are shown in Figure 2.21. It is often a tedious job to compare the relative quality of stack sections after static corrections because the quality may vary from place to place.

2.5.5   Wavefield datuming

As a final note on the correction for near-surface statics, its underlying model of vertical and static correction as shown in Figure 2.19 has been the subject of major debate

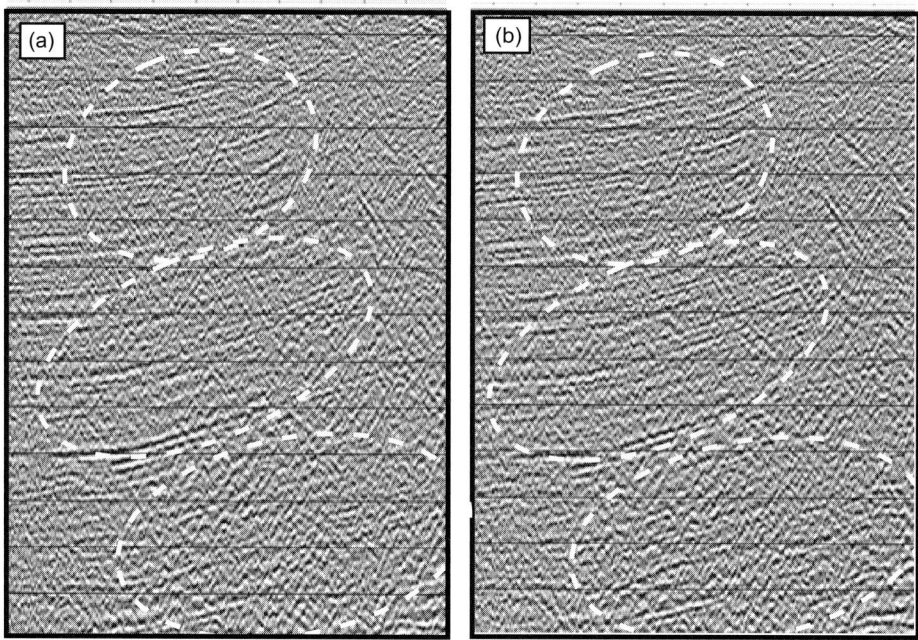

Figure 2.21 Comparison of stack sections after: (a) first-break refraction statics; and (b) tomostatics. Three dashed ellipses show areas of focused comparison.

over many years. The widespread use of static correction is due to its strong practicality. This practicality is reflected mainly in its simplicity and computational efficiency. On the other hand, if we have a sufficiently accurate near-surface velocity model, we may apply **wavefield datuming** to convert the raw data into new data as if they were recorded along a datum below the near surface (Box 2.3). Alternatively, we may run a surface-consistent depth migration by integrating the near-surface into the migration velocity model.

Box 2.3 A synthetic example of wavefield datuming

Wavefield datuming attempts to convert data recorded using surface shots and receivers into the data that would have been recorded using shots and receivers on the datum below the weathering zone. Box 2.3 Figure 1 shows a five-layer P-wave velocity model with a datum between the weathering zone and two reflectors. Box 2.3 Figure 2 shows synthetic shot records using surface shots and receivers, the static correction result, and the wavefield datuming results in the upper, middle, and lower rows, respectively. The wavefield datuming results fit the correct reflector positions much better than the results after statics correction, especially for the shallow reflector. The difference between statics correction and wavefield datuming becomes smaller for deeper reflectors.

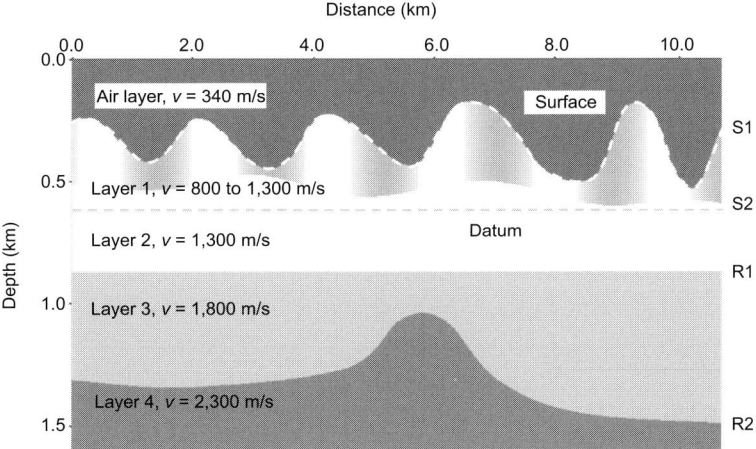

Box 2.3 Fig. 1 A model of near-surface P-wave velocities. S1 is the surface on which the shots and receivers are deployed. Layer 1 is the weathering zone whose velocity varies laterally from 800 m/s to 1300 m/s. S2 is a floating datum. R1 and R2 are two reflectors. (From Liu *et al.*, 2011.)

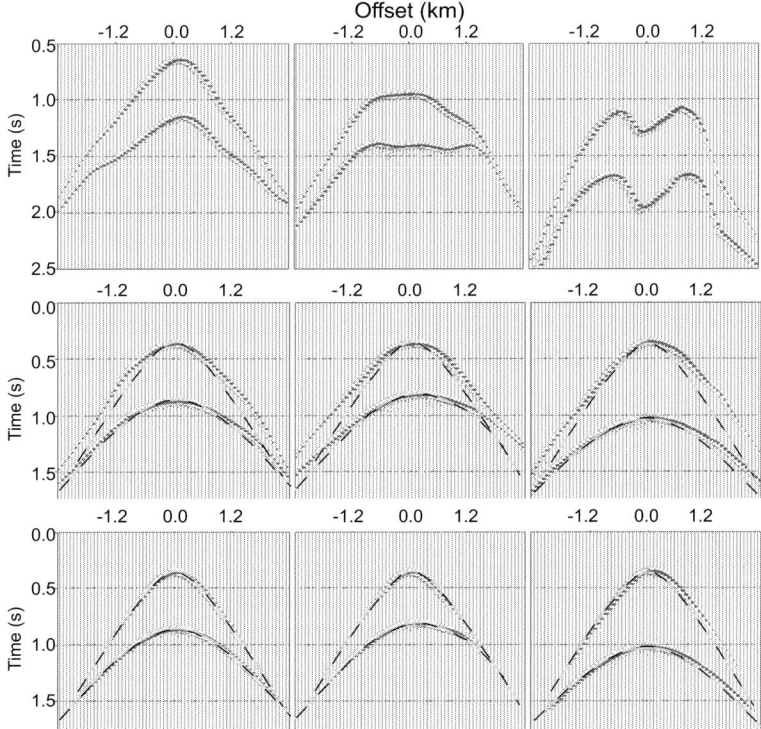

Box 2.3 Fig. 2 Shot records of two reflections R1 and R2 for three shot locations at 3 km (left column), 4.32 km (central column), and 9.24 km (right column). (Upper row) Raw data with shots and receivers on surface S1. (Middle row) After correction for statics. (Lower row) After wavefield datuming. Dashed curves are the true trajectories of reflectors. (From Liu *et al.*, 2011.)

Exercise 2.5

1. Discuss the advantages and limitations of using the concept of statics to correct for the near-surface effect on seismic data.

2. The sand dune curve shown in Figure 2.20b is a depth versus traveltime plot that can be obtained from an uphole or a check shot survey. Derive a 1D velocity function based on the depth versus traveltime plot in Figure 2.20b.

3. Search the literature to write a short description for: (1) refraction statics; (2) reflection statics; and (3) trim statics. Which method is the most effective?

2.6   Summary

- The main purposes of pre-processing are identifying problems with the dataset and applying simple corrections. We need to QC the input data by assessing its quality and finding out its characteristics relevant to our study objectives.

- Normal moveout (NMO) stacking aims at improving the SNR of primary reflections and establishing stacking velocities assuming locally layer-cake models. The stacked traces are input to post-stack data processing and imaging.

- Convolution is the mathematical operation of multiplying two time series such as two input data strings. It represents the physical process of combining two or more time series. Filtering is the convolution of the input signal with the filter function to produce an output signal.

- Correlation measures the similarity between two input time series, resulting in a new time series which is the similarity function between the two input series. The auto-correlation function is always symmetric.

- Each digital signal has a matched filter, which has the elements of the signal in reverse order. Filtering with a matched filter is equivalent to cross-correlating with the signal. A matched filter maximizes the output in response to the signal; hence it is the most powerful filter for identifying the presence of a given signal with additive noise.

- Processing of multi-component seismic data requires special care and knowledge about the orientation of the sensors and physics of wave propagation. An example is given here for identifying the components of a VSP survey using hodograms of first arrivals.

- Most noise suppression methods search for and take advantage of the difference between the characteristics of the signal and noise. Model-driven methods attempt to construct simple models to help identify the properties of signal and noise. Data-driven methods are based on obvious differences between the signal and noise.

- A linear moveout (LMO) involves time-shifting of seismic traces proportional to a linear relationship with offset, so that events of certain dips become flat after the moveout. A slant stack consists of a LMO followed by a stacking to emphasize or beam-steer events of certain dips. Slant stacking is used in various applications including the Radon transform.

- The near-surface effect due to the presence of the weathering zone and topographic variations may pose challenges to seismic processing. A traditional approximation is the static correction. Recent improvements include tomostatics and wavefield datuming.

FURTHER READING

Cox, M., 1999, *Static Corrections for Seismic Reflection Surveys*, SEG.
Sheriff, R. E., 1991, *Encyclopedic Dictionary of Exploration Geophysics*, 3rd edn, SEG.
Yilmaz, O., 1987, *Seismic Data Processing*, SEG Series on Investigations in Geophysics, Vol. 2, SEG.

3 Discrete spectral analysis

Discrete spectral analysis is a suite of classic data processing tools aiming to quantify the energy distribution of seismic data over temporal or spatial scales. This chapter starts with the law of decomposition and superposition, which is the foundation of many seismic processing methods. According to Fourier theory, a seismic trace can be expressed as a linear superposition of harmonic functions of different frequencies with appropriate amplitudes and phases, thus enabling the spectral analysis. Because seismic data are in digital form with limited time durations, classic spectral analysis is achieved using discrete Fourier transform (DFT). Readers should pay special attention to the characteristics of the DFT, as these often differ from the continuous Fourier transform. Fast Fourier transform (FFT) is described as an example to improve the computation efficiency in processing.

Discrete spectral analysis is discussed using several examples. Useful processing tricks in one type of processing are often borrowed to solve problems in another type of processing. To decompose seismic traces, for instance, we may use wavelets of fixed shape but varying amplitudes and lengths rather than harmonic functions as the basic building blocks. This enables wavelet decomposition and seismic wavelet analysis, as

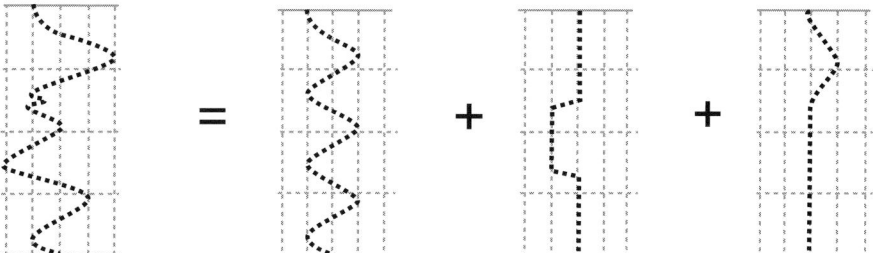

Figure 3.1 The left-most time series is the result of summing the other three time series along their amplitude direction (amplitude stacking). In other words, the left-most time series can be decomposed into the other three, and then these three can be superposed to recreate the left-most time series.

another application of the law of decomposition and superposition. Yet another usage of the law is in interpolation of digital data, which has many applications in the processing of seismic and non-seismic data.

3.1 The law of decomposition and superposition

3.1.1 Description of the law

The **law of decomposition and superposition** is a fundamental principle in physics and geophysics. That a natural system can be decomposed into components, and these components can be superposed to recreate the whole system, is the key behind many seismic data processing procedures. Figure 3.1 shows an example of decomposing a time series into three time series of the same length, and superposing these three time series into a single time series. We may, of course, superpose many time series of different lengths. The summing of the three time series in this figure is called **amplitude stacking** because the summation is done along the amplitude direction. In some refraction surveys **vertical stacking** is done, which means stacking the traces from a number of sources recorded at one location to enhance the signal-to-noise ratio.

To appreciate the physics behind the law of decomposition and superposition, let us examine an example. In 1678, the Dutch physicist Christiaan Huygens explained the propagation of light using a wave model known as the Huygens principle. This principle states that, from one propagating wavefront, we can generate the next wavefront by decomposing the first wavefront into a number of point sources emitting new waves, and then superposing the waves from these point sources. One way to prove this principle was shown in a sketch (Figure 3.2) by Isaac Newton in his famous book *Philosophiæ Naturalis Principia Mathematica*. Newton's sketch uses a wall containing a hole to decompose the first wavefront into a point source in the hole. This hole is considered as a single point source only for those waves whose first Fresnel zone, as shown by the arrow bar in the figure, is at least twice as wide as the width of the hole (the Fresnel zone is the length within which the two wavefronts constructively interfere with each other). If Huygens is correct, the wavefronts on the right side of the wall will behave as if the hole is their point source. Newton's sketch

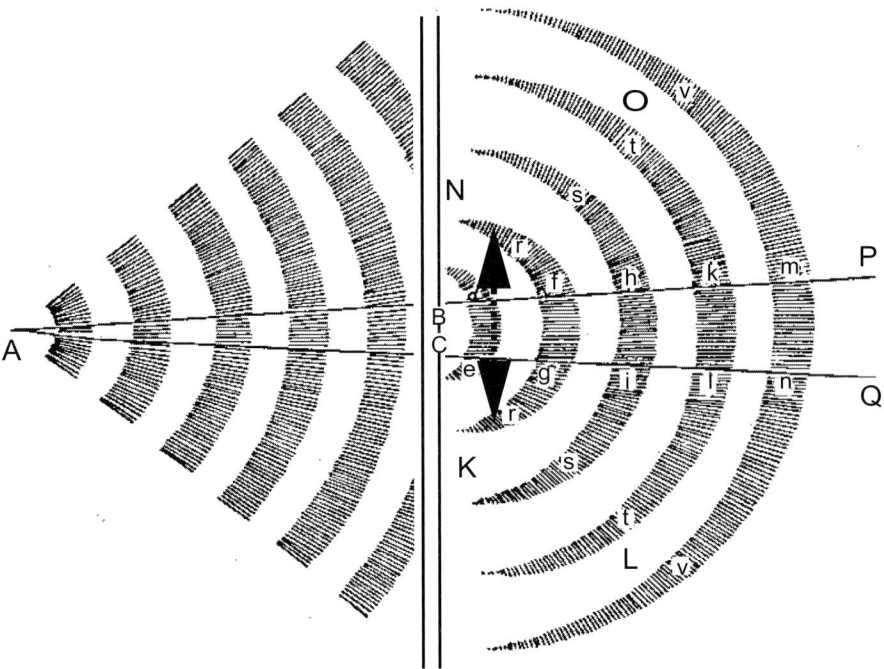

Figure 3.2 Isaac Newton used the law of decomposition and superposition to prove the Huygens principle, as shown in this sketch in his *Principia* (4th edn, 1726, reprinted 1871 by MacLehose, Glasgow, p.359). The dotted arrow bar approximates the width of the Fresnel zone of new waves near the hole.

provides a roadmap for physical experiments, such as observing the waves in a lake going through an opening along a levee.

3.1.2 Examples of the law

When pre-stack seismic data are converted into post-stack data, a NMO is done first and all the NMO-corrected traces of each CMP, such as that shown in Figure 2.7b, are stacked along the offset direction to form a single post-stack trace. The NMO stacking is an example of superposition, and going from the stacked trace to the pre-stack traces is the corresponding decomposition. Clearly, the stacking will produce a **unique** trace, whereas the decomposition does not. We can decompose a seismic trace into many possible combinations of traces. It is difficult to recover the information that was destroyed by the stacking, such as the ground rolls in a pre-stack CMP which will be largely suppressed by the NMO stacking.

A different type of superposition is provided by the convolution process. A recorded seismic trace may be regarded as the convolutional superposition of the source function, Earth function, receiver function, and noise. Hence the recorded trace can be decomposed into these functions. Since the decomposition is not unique, we must search for those decompositions that both physically make sense and are convenient for our data processing objectives. The convolution operation can be expressed as multiplying a Toeplitz matrix with a vector, as shown in equation (2–15). We may carry out the multiplication in two

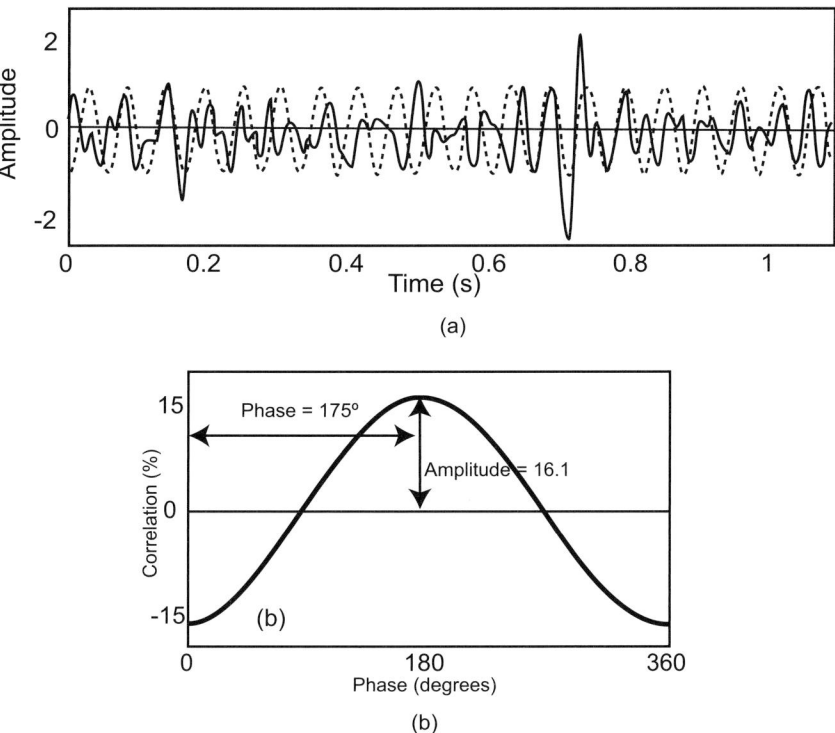

Figure 3.3 (a) An input seismic trace (solid curve) and a 20 Hz cosine wave. (b) The correlation between the input trace and 20 Hz cosine wave as a function of the phase of the cosine. The peak of the correlation (with a value of 16.1) is reached at phase 175° (Barnes, 1998).

steps. The first step is to multiply the kth column of the matrix with the kth element of the vector. We may regard this step as a weighting of the columns of the matrix using the coefficients provided by the vector. The second step is to sum or stack all weighted columns of the matrix into a single column. Hence, convolution can be regarded as a **weighted stacking**. In other words, convolving two vectors is like stacking a Toeplitz matrix made of the first vector and column-weighted by the second vector.

Fourier decomposition, as shown in Figure 1.13, is another example of the law of decomposition and superposition. As shown in the lower panel of Figure 1.13, the original time series on the left is decomposed into harmonic functions of different frequencies with different amplitudes and phase lags. A superposition or stacking of all the harmonic traces produces the original time series; this is equivalent to a discrete inverse Fourier transform. Fourier decomposition is the physical process of a Fourier transform, and the building blocks are the harmonic functions.

In practice, after we have decided to use a particular building block for the decomposition, how do we obtain the coefficients? One approach is illustrated by Barnes (1998), as shown in Figure 3.3.

Figure 3.3 illustrates, for an input seismic trace (solid curve in Figure 3.3a), the derivation of the amplitude and phase angle for a 20 Hz cosine wave as one of the building blocks. For this monochromatic cosine wave, the only two parameters that we can change are its amplitude and its phase angle, or where the time zero is defined. In this figure, the amplitude of the cosine wave is normalized at ±1. Then we can compute the cross-correlation between the input trace and the cosine wave using different phase angles. Owing to the periodicity of the harmonics, the correlation value has the same periodicity as the cosine wave. It turns out that the maximum correlation is reached at phase angle 175°, with a peak correlation value of 16.1. Figure 3.3a displays the 20 Hz cosine wave at 175° phase angle. The peak correlation of 16.1 is the value at 20 Hz on the amplitude spectra, and the phase angle of 175° is the value at 20 Hz on the phase spectra. By searching for the maximum correlation for cosine waves of each discrete frequency, we can derive all amplitude values on the amplitude spectra and all phase angles on the phase spectra for the input seismic trace. If the input trace in time domain has N sample points, we usually use N discrete frequencies in frequency domain. Because there are two variables per frequency, the number of variables in the frequency domain is $2N$.

Since the beginning of seismology, there has been an effort to decompose seismic traces into wavelets. One advantage of this is that, if the chosen wavelet represents a physical component of the data, such as the source wavelet, then the processes of decomposition and superposition give a realistic representation of the physical process. Another advantage is that the wavelet decomposition and superposition are time-domain processes, capable of showing time-domain properties and local attributes. Details on seismic wavelets and wavelet decomposition will be discussed in a later section.

Exercise 3.1

1. Find and discuss three applications of the law of decomposition and superposition from geophysics literature.

2. Conduct a modeling experiment to prove the Huygens principle by following the sketch by Isaac Newton in Figure 3.2.

3. If we alter the amplitude of the 20 Hz cosine wave in Figure 3.3a, will the pattern of the correlation function in Figure 3.3b change? Will the amplitude of the correlation change?

3.2 Discrete Fourier transform

The **Fourier transform** (FT) is an application of the law of decomposition and superposition using harmonic sine and cosine functions as the basic building blocks.

This transform enables the conversion of a function between the time domain and frequency domain. The conversion is often useful, owing to the FT's many special properties. For instance, a time-domain convolution becomes a frequency-domain multiplication, so high computational efficiency may be achievable in the frequency domain. The FT is of particular value to seismic data processing in two additional respects. First, it facilitates discrete spectral analysis, the main topic of this chapter. Second, many geophysical problems become easily solvable once the FT has been applied. For example, the wave equation as a partial differential equation may be converted using multi-dimensional FT into an

Table 3.1 Some properties of the Fourier transform

Property	Time domain		Frequency domain
Modulation	$f(t)\exp(i\omega_0 t)$	$\Leftarrow$ FT $\Rightarrow$	$F(\omega + \omega_0)$
Linear	$a_1 f_1(t) + a_2 f_2(t)$	$\Leftarrow$ FT $\Rightarrow$	$a_1 F_1(\omega) + a_2 F_2(\omega)$
Shifting	$f(t - t_0)$	$\Leftarrow$ FT $\Rightarrow$	$F(\omega)\exp(-i\omega t_0)$
Scaling	$f(at)$	$\Leftarrow$ FT $\Rightarrow$	$F(\omega/a)/|a|$
	$f(t/a)/|a|$	$\Leftarrow$ FT $\Rightarrow$	$F(a\omega)$
Convolution	$f_1(t) f_2(t)$	$\Leftarrow$ FT $\Rightarrow$	$F_1(\omega) * F_2(\omega)$
	$f_1(t) * f_2(t)$	$\Leftarrow$ FT $\Rightarrow$	$F_1(\omega) F_2(\omega)$
Parseval's theorem	$\int_{-\infty}^{\infty} |f(t)|^2\, dt$	$\Leftarrow$ FT $\Rightarrow$	$\int_{-\infty}^{\infty} |F(\omega)|^2\, d\omega$
Differentiation	$\dfrac{\partial}{\partial t} f(t)$	$\Leftarrow$ FT $\Rightarrow$	$i\omega F(\omega)$
	$\dfrac{\partial^2}{\partial t^2} f(t)$	$\Leftarrow$ FT $\Rightarrow$	$-\omega^2 F(\omega)$

ordinary differential equation that is readily solvable. In this section, the main properties of the FT will be introduced using one-dimensional continuous FT. Then the 1D discrete FT is introduced with a finite number of sample points in both time and frequency domains. Some detailed discussion on fast Fourier transform (FFT) is given to show ways to improve the computational efficiency during digital data processing. Finally, issues around multi-dimensional Fourier transform are illustrated with an example.

3.2.1 Continuous Fourier transform

A 1D continuous Fourier transform is defined as

$$\text{Forward FT:} \quad F(\omega) = \int_{-\infty}^{\infty} f(t) e^{i\omega t}\, dt \tag{3-1}$$

$$\text{Inverse FT:} \quad f(t) = \frac{1}{2\pi} \int_{-\infty}^{\infty} F(\omega) e^{-i\omega t}\, d\omega \tag{3-2}$$

Both the forward and inverse FTs are weighted integrations of the input function with the harmonic function $e^{\pm i\omega t} = \cos\omega t \pm i \sin\omega t$ as the weights. As the values of either ω or t change, the harmonic function changes its wavelength or frequency; so the output function of either forward or inverse FT is a linear superposition of the components of the input function in terms of its different frequency components. The similar forms of the forward FT and inverse FT mean that there is a duality or symmetry between the time-domain and frequency-domain expressions of a function. Note that when the input function f(t) is real, its corresponding frequency-domain term $F(\omega)$ will be complex in general. Some important properties of the FT are listed in Table 3.1.

The FT properties shown in Table 3.1 are very useful in seismic data processing. For example, the convolution property allows efficient computation of convolution through multiplication in the spectral domain. The shifting and differentiation properties are behind the frequency domain migration. As will be introduced in Chapter 7, the acoustic wave equation as a partial differential equation can be converted into an ordinary differential equation using multi-dimensional FT. In this case the FT acts as plane-wave decomposition, and the downward continuation of each plane wave is equivalent to shifting its phase angle.

3.2.2　Discrete Fourier transform

In practice, seismic data processing deals mostly with the **discrete Fourier transform** (**DFT**); see Box 3.1 for an example of this. Suppose we have N equidistant sampling points along the t-axis, with a sampling interval Δt:

$$--- \to t \quad\quad\quad\quad\quad\quad\quad\quad \to t$$
$$0\Delta t \quad 1\Delta t \quad 2\Delta t \quad \cdots\cdots \quad (N-2)\Delta t \quad (N-1)\Delta t$$

The measured function values at these N points are: $f(0\Delta t), f(1\Delta t), f(2\Delta t), \ldots$

Discretizing the continuous forward FT (3-1) using $t = n\Delta t$, we have

$$F(\omega) = \sum_{n=0}^{N-1} f(n\Delta t) e^{i\omega(n\Delta t)} \Delta t \tag{3–3a}$$

We may simplify the notation by letting $\Delta t = 1$,

$$F(\omega) = \sum_{n=0}^{N-1} f(n) e^{i\omega n} = \sum_{n=0}^{N-1} f_n e^{i\omega n} \tag{3–3b}$$

Because the range of the angular frequency ω is 2π, with N sampling points, the frequency interval will be

$$\Delta\omega = 2\pi/N \tag{3–4}$$

Thus, we obtain the value of $F(\omega)$ at a discrete frequency point $\omega = j\Delta\omega = 2\pi j/N$. Hence, the **forward DFT** is

$$F_j = F(j\Delta\omega) = \sum_{n=0}^{N-1} f_n e^{i(2\pi/N)jn} \tag{3–5a}$$

Similarly, we can discretize the continuous inverse FT (3-2) using $\omega = j\Delta\omega$ and $t = n\Delta t = n$, yielding

$$f(n) = \frac{1}{2\pi} \sum_{j=0}^{N-1} F(j\Delta\omega) e^{-i(j\Delta\omega)n} 2\pi/N$$

or

$$f_n = \frac{1}{N} \sum_{j=0}^{N-1} F_j e^{-i(2\pi/N)jn} \tag{3–5b}$$

which is the **inverse DFT** formula.

Notice in the above that we have chosen to take the same number of sampling points in the frequency domain as we have in the time domain. This is called an **even-determined** transform. Otherwise, it may become an **under-determined** transform if we are transferring

Box 3.1 An example of discrete Fourier transform

Let us use DFT to show the effect in the frequency domain of doubling the length of an N-length time series

$$x_t = \{x_0, x_1, x_2, \ldots, x_{N-1}\}$$

into a $2N$-length by padding with N zeros in the front:

$$y_t = \{0, \ 0, \ 0, \ldots, 0, \ x_0, \ x_1, x_2, \ldots, x_{N-1}\}$$
$$|\leftarrow \quad N \quad \rightarrow |\leftarrow \quad N \quad \rightarrow |$$

[Solution]
Following the forward DFT formula, we have

$$X_j = \sum_{n=0}^{N-1} x_n e^{i \frac{2\pi}{N} jn}, \ j = 0, 1, 2, \ldots, N - 1$$

Hence,

$$Y_j = \sum_{n=0}^{2N-1} y_n e^{i \frac{2\pi}{2N} jn} = \sum_{n=N}^{2N-1} x_{n-N} e^{i \frac{2\pi}{2N} jn}, \ j = 0, 1, 2, \ldots, 1N - 1.$$

Let $k = n - N$, or $n = k + N$; thus $x_{n-N} = x_k$. If $n = N$, then $k = 0$; if $n = 2N - 1$, then $k = N - 1$.

$$Y_j = \sum_{k=0}^{N-1} x_k e^{i \frac{2\pi}{2N} j(k+N)} = \sum_{k=0}^{N-1} x_k e^{i \frac{2\pi}{2N} jk} e^{i\pi j} = (-1)^j \sum_{k=0}^{N-1} x_k e^{i \frac{2\pi j}{2N} k}$$

If $j = 2m$,

$$Y_{2m} = \sum_{k=0}^{N-1} x_k e^{i \frac{2\pi}{N} mk} = X_m$$

If $j = 2m + 1$,

$$Y_{2m+1} = - \sum_{k=0}^{N-1} x_k e^{i \frac{2\pi}{N} (m+1/2)k} = - \sum_{k=0}^{N-1} x_k e^{i \frac{2\pi}{N} mk} e^{i \frac{\pi}{N} k}$$

which is a modulated version of $(-X_m)$.

Therefore, the FT of the new series is made up of the FT of the original time series interlaced by a modulated version of the FT of the original time series.

---

fewer points from the original domain to more points in the transferred domain; or an **over-determined** transform if it is the other way around.

Like the case of the continuous FT, the differences between the forward DFT (3–5a) and the inverse DFT (3–5b) lie in the sign of the exponent and scaling, which is $1/2\pi$ for FT versus $1/N$ for DFT. Consequently, the same computer program may be used for both processes with minor modification. Such a **duality** makes the meaning of "time" and "frequency" interchangeable.

One may write the forward DFT in matrix form, using a variable $W = e^{i2\pi/N}$:

$$
\begin{bmatrix} F_0 \\ F_1 \\ F_2 \\ \vdots \\ \vdots \\ F_{N-1} \end{bmatrix} =
\begin{bmatrix}
1 & 1 & 1 & \cdots & 1 \\
1 & W^{1\times1} & W^{1\times2} & \cdots & W^{1\times(N-1)} \\
1 & W^{2\times1} & W^{2\times2} & \cdots & W^{2\times(N-1)} \\
\vdots & \vdots & \vdots & \cdots & \vdots \\
\vdots & \vdots & \vdots & \cdots & \vdots \\
1 & W^{(N-1)\times1} & W^{(N-1)\times2} & \cdots & W^{(N-1)\times(N-1)}
\end{bmatrix}
\begin{bmatrix} f_0 \\ f_1 \\ f_2 \\ \vdots \\ \vdots \\ f_{N-1} \end{bmatrix}
\qquad (3\text{--}6)
$$

One can express f_n in terms of F_n using the inverse of the matrix in the above equation, which can be obtained from the inverse Fourier transform. Notice that W^{-1} is the complex conjugate of W. Since the forward DFT differs from the inverse DFT only by a change of sign in the exponent (from W to W^{-1}) and a scaling factor $1/N$ in front of the inverse DFT, it is not difficult to figure out from (3–6) that the matrix notation of the inverse DFT is:

$$
\begin{bmatrix} f_0 \\ f_1 \\ f_2 \\ \vdots \\ \vdots \\ f_{N-1} \end{bmatrix} = \frac{1}{N}
\begin{bmatrix}
1 & 1 & 1 & \cdots & 1 \\
1 & W^{-1\times1} & W^{-1\times2} & \cdots & W^{-1\times(N-1)} \\
1 & W^{-2\times1} & W^{-2\times2} & \cdots & W^{-2\times(N-1)} \\
\vdots & \vdots & \vdots & \cdots & \vdots \\
\vdots & \vdots & \vdots & \cdots & \vdots \\
1 & W^{-(N-1)\times1} & W^{-(N-1)\times2} & \cdots & W^{-(N-1)\times(N-1)}
\end{bmatrix}
\begin{bmatrix} F_0 \\ F_1 \\ F_2 \\ \vdots \\ \vdots \\ F_{N-1} \end{bmatrix}
\qquad (3\text{--}7)
$$

Since (3–6) and (3–7) are inverse to each other, the matrices on the right-hand side of these equations must form an **inverse matrix pair**. Let us prove this notion. Our proof is based on the periodic property of W, i.e. $W^N = 1$. The multiplication of these two matrices is a matrix $\mathbf{Y}$:

$$
\mathbf{Y} = \frac{1}{N}
\begin{bmatrix}
1 & 1 & 1 & \cdots & 1 \\
1 & W^{-1\times1} & W^{-1\times2} & \cdots & W^{-1\times(N-1)} \\
1 & W^{-2\times1} & W^{-2\times2} & \cdots & W^{-2\times(N-1)} \\
\vdots & \vdots & \vdots & \cdots & \vdots \\
\vdots & \vdots & \vdots & \cdots & \vdots \\
1 & W^{-(N-1)\times1} & W^{-(N-1)\times2} & \cdots & W^{-(N-1)\times(N-1)}
\end{bmatrix}
$$

$$
\times
\begin{bmatrix}
1 & 1 & 1 & \cdots & 1 \\
1 & W^{1\times1} & W^{1\times2} & \cdots & W^{1\times(N-1)} \\
1 & W^{2\times1} & W^{2\times2} & \cdots & W^{2\times(N-1)} \\
\vdots & \vdots & \vdots & \cdots & \vdots \\
\vdots & \vdots & \vdots & \cdots & \vdots \\
1 & W^{(N-1)\times1} & W^{(N-1)\times2} & \cdots & W^{(N-1)\times(N-1)}
\end{bmatrix}
$$

We just need to show that $\mathbf{Y}$ is an identity matrix $\mathbf{I}$.

Suppose y_{ij} is an element of the squared matrix $\mathbf{Y}$ on the ith row and jth column. The above matrix multiplication indicates that y_{ij} is the dot product of the ith row of the first matrix with the jth column of the second matrix.

$$y_{ij} = \frac{1}{N}(W^{-i0}, W^{-i1}, \ldots, W^{-i(N-1)}) \cdot (W^{i0}, W^{i1}, \ldots, W^{-i(N-1)})^T = \frac{1}{N}\sum_{k=0}^{N-1} W^{(j-i)k}$$

If $i = j$, $y_{ij} = \frac{1}{N}\sum_{k=0}^{N-1} W^0 = 1$;

If $i \neq j$, notice that $\sum_{k=0}^{N-1} x^k = (1 - x^N)/(1 - x)$ and let $x = W^{j-1}$ to give us

$$y_{ij} = \frac{1}{N}\sum_{k=0}^{N-1} W^{(j-i)k} = \frac{1}{N}(1 - W^{(j-i)N})/(1 - W^{(j-i)}) = 0$$

Therefore, $\mathbf{Y} \equiv \mathbf{I}$.

3.2.3 Fast Fourier transform

Since the operating matrix in (3–6) or (3–7) is an Nth-order square matrix, a straightforward implementation of the DFT demands N^2 operations, i.e. N inner products each of length N (N multiplications and N additions). The speed of this operation is too slow for general applications. In practice, **fast Fourier Transform (FFT)** is used with $N \log_k N$ operations, and $k = 2$ in most cases. If $N = 1024$, then $\log_k N = 10$; this means a tremendous saving in computation speed. The FFT algorithm was first published by Cooley and Tukey (1965), although rumor has it that Vern Herbert from Chevron had developed an FFT algorithm in 1962.

FFT takes advantage of the **periodic property** of the base functions of the FT, the trigonometric functions. When $k = 2$, this property is called **doubling**; when $k = 3$, it is called **tripling**. Let us examine the case of doubling here. The **doubling operation** is to use the FT solutions of two equal-length series

$$x_t = (x_0, x_1, x_2, \ldots, x_{N-1}) \quad \text{and} \quad y_t = (y_0, y_1, y_2, \ldots, y_{N-1})$$

In constructing the FT solution for the interlaced series of the above two,

$$z_t = (x_0, y_0, x_1, y_1, x_2, y_2, \ldots, x_{N-1}, y_{N-1})$$

Note that the DFT of the interlaced series z_t requires $(2N)^2 = 4N^2$ operations, while the DFT of the first two series requires a total of $2N^2$ operations. Therefore, if we need to do FT for a series of length $N = 2^k$, we can do the doubling by regarding the series as the interlaced series of two shorter series, and do a further doubling on the two shorter series, and thus repeat the doubling operation k times until we reach the shortest series which contains only one element. Since we go from N to two $N/2$, and to four $N/4$, and so on for a total of k divisions of the data, and each division needs N operations to update for all data points, the entire FFT process requires a total of $N \log_2 N = Nk$ operations.

Let us examine the FFT process for a series

$$f_t = (f_0, f_1, f_2, \ldots, f_{N-1})$$

where $N = 2^k$. The lengths of half-division of N for k times are all integers:

$$N/2 = K_1$$
$$K_1/2 = N/4 = K_2$$
$$K_2/2 = N/8 = K_3$$
$$\cdots$$
$$N/N = K_k = 1$$

According to (3–5a), the ordinary DFT of f_t is

$$F_j = \sum_{n=0}^{N-1} f_n e^{i\frac{2\pi}{N} jn}, \qquad \text{for } j = 0, 1, 2, \ldots, N - 1$$

Since $N = 2K_1$,

$$F_j = \sum_{n=0}^{2K_1-1} f_n V^{jn}$$

where $V = e^{i2\pi/N} = e^{i\pi/K_1}$.

Let us consider the even and odd indexed elements of f_t in two separate series:

$$x_i = f_{2i}$$
$$y_i = f_{2i+1}$$

where $i = 0, 1, \ldots, K_1$. Their FTs are, for $j = 0, 1, \ldots, K_1$,

$$X_j = \sum_{n=0}^{K_1-1} x_n V^{2jn}$$

$$Y_j = \sum_{n=0}^{K_1-1} y_n V^{2jn}$$

Going back to the original series f_t for $j = 0, 1, \ldots, K_1-1, K_1, \ldots, N-1$

$$F_j = \sum_{n=0}^{K_1-1} f_{2n} V^{2nj} + \sum_{n=0}^{K_1-1} f_{2n+1} V^{(2n+1)j}$$

$$= \sum_{n=0}^{K_1-1} x_n V^{2nj} + \sum_{n=0}^{K_1-1} y_n V^{(2n+1)j}$$

We then consider two cases as follows.

(i) When $0 \leq j < K_1$, we have

$$F_j = \sum_{n=0}^{K_1-1} x_n V^{2nj} + V^j \sum_{n=0}^{K_1-1} y_n V^{2nj}$$

$$= X_j + V^j Y_j$$

(ii) When $K_1 \leq j < N$, let $m = j - K_1$, then

$$F_j = F_{K_1+m} = \sum_{n=0}^{K_1-1} x_n V^{2n(K_1+m)} + \sum_{n=0}^{K_1-1} y_n V^{(2n+1)(K_1+m)}$$

$$= \sum_{n=0}^{K_1-1} x_n V^{2nm} V^{2nK_1} + V^m \sum_{n=0}^{K_1-1} y_n V^{2nm} V^{(2n+1)K_1}$$

$$= X_m - V^m Y_m$$

In the foregoing derivation we have used the fact that $V^{2nK1} = V^{nN} = 1$, and $V^{K1} = V^{N/2} = -1$. Combining the above two cases together

$$F_j = X_j + V^j Y_j$$

$$F_{K_1+j} = X_j - V^j Y_j$$

for $j = 0, 1, \ldots, K_1-1$. In other words, if we let

$$F_j^{(e)} = X_j$$

$$F_j^{(o)} = Y_j$$

where superscripts (e) and (o) stand for even and odd indexes of the original series, we have

$$F_j = F_j^{(e)} + V^j F_j^{(o)} \tag{3–8a}$$

$$F_{N/2+j} = F_j^{(e)} - V^j F_j^{(o)} \tag{3–8b}$$

The above equations describe a general **doubling operation**. Applying the doubling operation to the next level of division, we have

$$F_j^{(e)} = F_j^{(ee)} + V^j F_j^{(eo)}$$

$$F_{N/4+j}^{(e)} = F_j^{(ee)} - V^j F_j^{(eo)}$$

$$F_j^{(o)} = F_j^{(oe)} + V^j F^{(oo)_j}$$

$$F_{N/4+j}^{(o)} = F_j^{(oe)} - V^j F^{(oo)_j}$$

where $j = 0, 1, 2, \ldots, N/4 - 1$. The doubling operation will be applied until each subdivided series on the RHS has reduced to contain only a single element, which is f_t. The actual practice of the FFT goes from the lower level upwards.

In the general case, one iteration of the FFT relates the FT of a length-$2N$ series $\{F_k\}$ into two of its sub-series of length N, $\{X_k\}$ and $\{Y_k\}$, through the doubling algorithm,

$$F_k = X_k + V^k Y_k \tag{3–9a}$$

$$F_{k+N} = X_k - V^k Y_k \tag{3–9b}$$

where $k = 0, 1, \ldots, N-1$. This doubling operation may be represented symbolically as that in Figure 3.4. Interested readers may study Box 3.2 for further detail on the implementation of DFT.

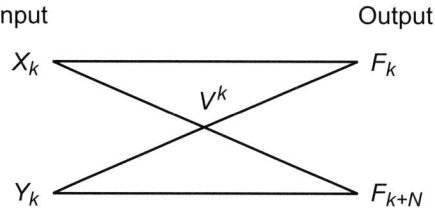

Figure 3.4 A symbolic FFT operator.

Box 3.2 Assigning coefficients of DFT via bit reverse

Let us look at an example of using the doubling operation to build up a complete DFT for a length-4 time series. Box 3.2 Figure 1 shows a sketch of such a DFT between its time-domain values $\{f_n$ and its frequency-domain values $\{F_j\}$.

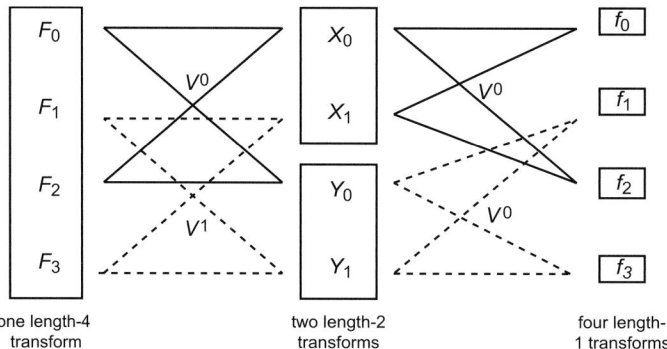

one length-4
transform

two length-2
transforms

four length-
1 transforms

Box 3.2 Figure 1 Sketch of DFT for a length-4 time series.

Note that the DFT of a length-1 series is just itself. Now we can rewrite the above system by flipping the left and the right sides of this figure to get the corresponding forward DFT in Box 3.2 Figure 2.

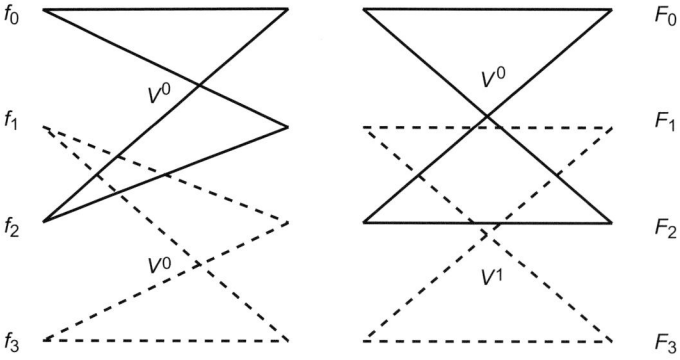

Box 3.2 Figure 2 Sketch of forward DFT for a length-4 time series.

Discrete spectral analysis

The above operation cannot be done "in place" because the doubling operations in the left side of Figure 2 are asymmetric. We may rearrange the order of the input time series to make it in place, as shown in Box 3.2 Figure 3.

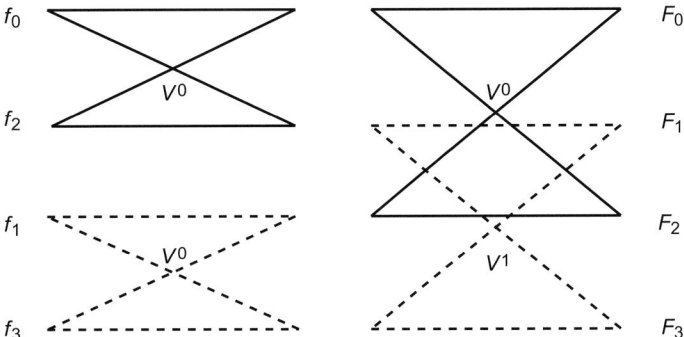

Box 3.2 Figure 3 The forward DFT for a length-4 time series with all symmetric doubling operations in place.

The reordering in the previous diagram from $\{F_0, f_1, f_2, f_3\}$ to $\{f_0, f_2, f_1, f_3\}$ is referred to as a "**bit reverse**" ordering. In other words, the difference between the orders of these two series is just to reverse their binary orders. Box 3.2 Figure 4 shows the bit-reverse ordering for the indices of a length-8 series.

Decimal	Binary		Binary	Decimal
0	000		000	0
1	001		100	4
2	010		010	2
3	011	bit	110	6
4	100	reverse	001	1
5	101		101	5
6	110		011	3
7	111		111	7
Original indexes				New indexes

Box 3.2 Figure 4 Bit-reverse ordering of a length-8 index.

The doubling diagram of FFT for this length-8 series is shown in Box 3.2 Figure 5. You may check the exponents of V in the above diagram to make sure that they follow equation (3–9). Are you able to draw such a doubling diagram for a length-16 time series?

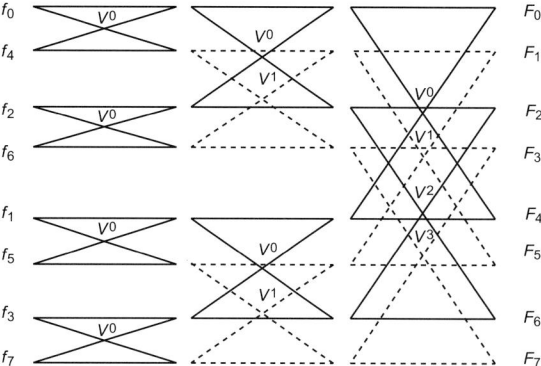

Box 3.2 Figure 5 The forward DFT for a length-8 time series.

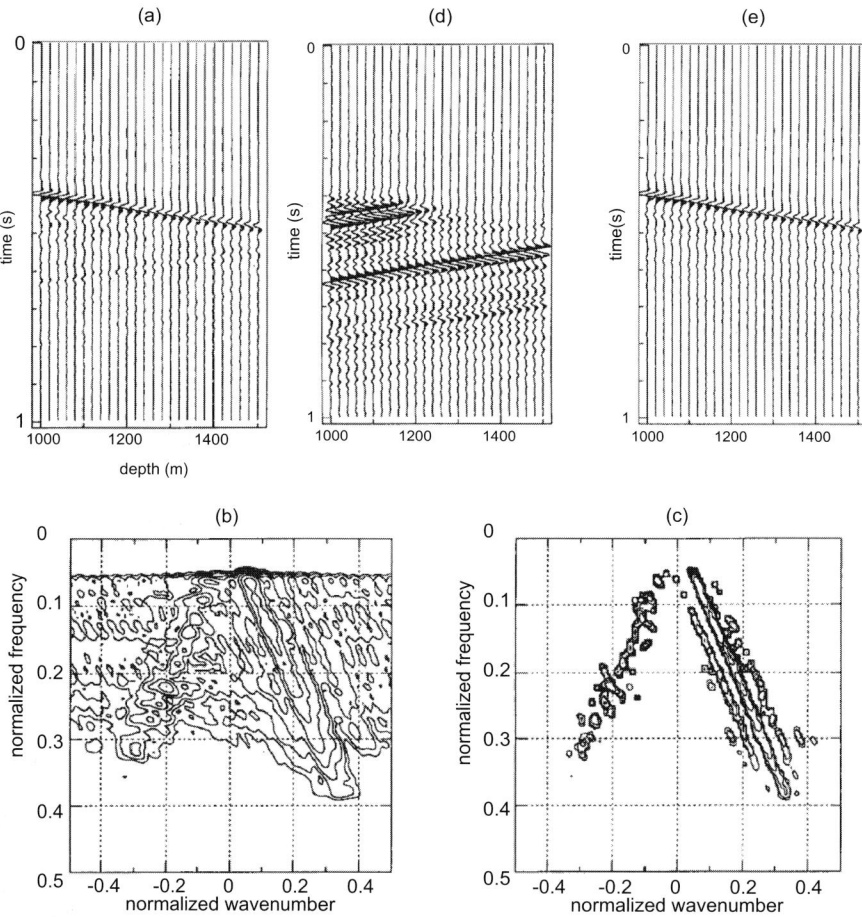

Figure 3.5 An example of 2D FT (Sacchi & Ulrych, 1996). (a) VSP data. (b) f–k spectrum from conventional DFT. (c) f–k spectrum using the Cauchy–Gauss model. (d) Upgoing waves. (e) Downgoing waves.

3.2.4 Multi-dimensional Fourier transform

Multi-dimensional Fourier transforms are among the most useful seismic data processing tools. Figure 3.5 shows an example of decomposing a VSP dataset into upgoing and down-going wavefields. In a VSP survey the geophones are located in the well bore. An **upgoing wavefield** arrives at geophones in the upwards direction, including waves reflected from horizons below the geophones. A **downgoing wavefield** arrives at geophones downwards, including direct waves from sources above the geophones and multiple reflections reflected from horizons above the geophones. Many seismic imaging methods use only the upgoing wavefield, so a separation of the upgoing and downgoing wavefields is necessary. In this example, the original data in the time-offset domain, or t–x domain, can be transformed via 2D FT into the frequency–wavenumber domain, or f–k domain. Sacchi and Ulrych (1996) used a Cauchy–Gauss model to produce the f–k spectrum in Figure 3.5c which is much less noisy than the conventionally produced f–k spectrum in Figure 3.5b.

Exercise 3.2

1. What happens to a function $F(\omega)$ if you take two consecutive inverse Fourier transforms of it? What is the implication of the result?

2. Calculate the Fourier transform of $f(t) = (4, -3, 2, -1)$ using DFT formulation. Please show the resultant real part and imaginary part separately.

3. Using DFT to show the effect in the frequency domain of interlacing zeros into a time series x_t. Here, interlacing zeros means turning an N-length $x_t = \{x_1, x_2, x_3, \ldots, x_N\}$ into a 2N-length $y_t = \{x_1, 0, x_2, 0, x_3, 0, \ldots \ldots, x_N, 0\}$.

3.3 Spectral analysis

Spectral analysis is a digital data processing method to assess the distribution of the data energy as a function of frequency and spatial scales. In practice, frequency distributions of the data energy are often used as important attributes in revealing the nature of the data. For example, spectral analysis may allow us to detect special seismic events such as explosions versus shear faulting. An example is given in Box 3.3.

3.3.1 Spectrum definitions

In general, the **spectrum** of a time series refers to a display of one of the properties under consideration as a function of frequency (a systematic change in temporal scale) or wavenumber (systematic change in spatial scale). The most commonly encountered spectra in geophysics include the **amplitude spectrum**, **phase spectrum**, and **power spectrum**

Box 3.3 Spectral analysis of the 9/11 disaster events

As a famous example, let us examine the seismic data and displacement spectra for the September 11 World Trade Center disasters, in Box 3.3 Figure 1. The data were recorded at a seismologic station in the Lamont Doherty Observatory at Columbia University, located 34 km due north from the World Trade Center. The measured seismic energy due to the collapse of the two buildings is equivalent to magnitude 2.1 and 2.3 events on the Richter scale. On the spectra the impacts and collapses stand out as anomalous spectral events in red, with respect to the background noise spectra in black.

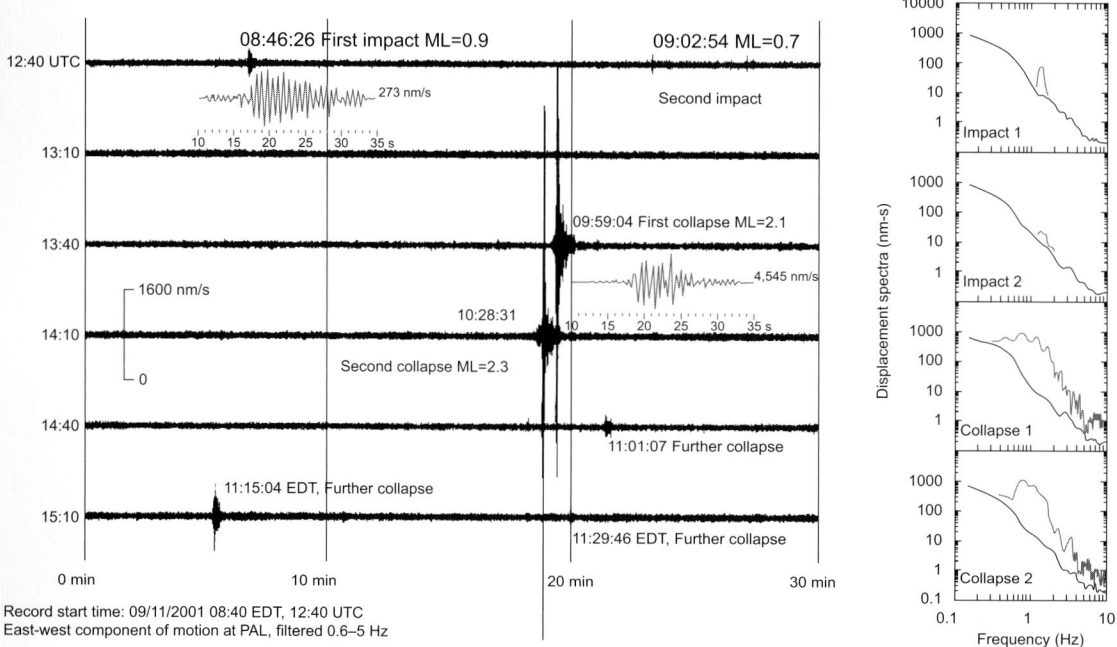

Box 3.3 Figure 1 Seismic record at Palisades, NY, 34 km north of the World Trade Center during the 9/11 disaster. (Left) East–west component of time record started at 8:40 EDT, or 13:40 WTC, on 9/11/2001. Two inserted seismograms are zoom-in plots of the first impact and the first collapse. (Right) Displacement spectra [nm s]. In each panel the upper curve is the signal spectrum, and the lower curve is the noise spectrum (from Kim *et al.*, 2001). For color version see plate section.

(squared amplitude spectrum). Let us illustrate the issues of spectral analysis using some simple examples. Suppose we have a time series:

$$a_t = (35, -12, 1)$$

In the z-transform notation,

$$a(z) = 35 - 12z + z^2$$

Discrete spectral analysis

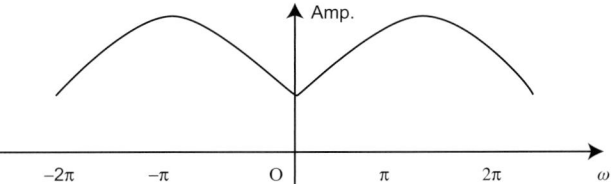

Figure 3.6 An example amplitude spectrum.

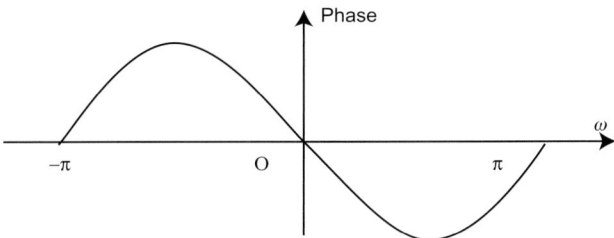

Figure 3.7 An example phase spectrum.

To denote it as a function of frequency, we insert $z = e^{i\omega\Delta t}$,

$$a(\omega) = 35 - 12e^{i\omega\Delta t} + e^{i2\omega\Delta t}$$
$$= \underbrace{(35 - 12\cos\omega\Delta t + \cos 2\omega\Delta t)}_{\text{Re}\,[a(\omega)]} + \underbrace{i(-12\sin\omega\Delta t + \sin 2\omega\Delta t)}_{\text{Im}\,[a(\omega)]}$$

Figure 3.6 shows the amplitude spectrum, which is a plot of the function's amplitude with respect to angular frequency ω. Here we have

$$\text{Amplitude}[a(\omega)] = \{\text{Re}[a(\omega)]^2 + \text{Im}[a(\omega)]^2\}^{1/2}$$
$$= [(35 - 12\cos\omega\Delta t + \cos 2\omega\Delta t)^2 + (-12\sin\omega\Delta t + \sin 2\omega\Delta t)^2]^{1/2}$$
$$= [1300 - 864\cos\omega\Delta t + 140\cos^2\omega\Delta t]^{1/2}$$
$$= [(1300^{1/2} - 140^{1/2}\cos\omega\Delta t)^2]^{1/2}$$
$$= 36.1 - 11.8\cos\omega\Delta t \tag{3-10}$$

The phase spectrum is the phase of the time series as a function of the frequency. In this example, we have

$$\text{Phase}[a(\omega)] = \tan^{-1}\{\text{Im}[a(\omega)]/\text{Re}[a(\omega)]\}$$
$$= \tan^{-1}\{(-12\sin\omega\Delta t + \sin 2\omega\Delta t)/(35 - 12\cos\omega\Delta t + \cos 2\omega t)\} \tag{3-11}$$

which is shown in Figure 3.7.

Sometimes it is convenient to use the square of the amplitude to represent the power distribution, since the squared amplitude has a dimension of energy. This is called the **power spectrum**, or simply **spectrum**.

$$\text{Spectrum}[a(\omega)] = |\text{Amplitude}[a(\omega)]|^2 \qquad (3\text{--}12a)$$

Since the spectrum is the distribution of power over the frequency range, it can also be expressed as the magnitude square of the Fourier transform:

$$\text{Spectrum}[a(\omega)] = A^*(\omega)A(\omega) \qquad (3\text{--}12b)$$

where $A(\omega)$ is the FT of a_t.

3.3.2   Amplitude spectrum and Wiener–Khinchin theorem

The amplitude spectrum displays the energy distribution over the frequency range of the series of interest, and thus provides important physical insight into the process under investigation. However, the spectrum obtained is based on a finite sampling of the physical process; this may cause error in the spectrum estimation. Let us see some ways to estimate the spectrum of some discrete geophysical data. Figure 3.8 shows sketches of some typical time series and their amplitude spectra. The horizontal scale of the spectral plots goes from zero to Nyquist frequencies. Of course, each sampled time series and its spectrum are just different expressions of the same physical phenomena.

Prior to the invention of FFT, the traditional spectrum estimations for discrete time series were summarized by Blackman and Tukey (1959). The base of these estimations is the **Wiener–Khinchin theorem** (see Box 3.4) which states that the **(power) spectrum equals the Fourier transform of the auto-covariance function of the signal**:

$$\text{Spectrum}(x_t) = F\left[\int_{-\infty}^{\infty} x^*(\tau)x(t+\tau)d\tau\right]$$

$$= \int_{-\infty}^{\infty}\left[\int_{-\infty}^{\infty} x^*(\tau)x(t+\tau)d\tau\right]e^{i\omega t}dt \qquad (3\text{--}13)$$

The term inside the bracket of the above expression is the auto-covariance function, denoted by $\gamma(t)$. For a continuous time series,

$$\gamma_x(t) = \int_{-\infty}^{\infty} x^*(\tau)x(t+\tau)d\tau \qquad (3\text{--}14)$$

The auto-covariance function $\gamma_x(t)$ is a measure of coherent components within a signal $x(t)$. For example, if the signal $x(t)$ is a real function composed of a source wavelet $b(t)$ and its echo,

$$x(t) = b(t) + 0.5\,b(t - t_0)$$

which is shown in Figure 3.9.

Discrete spectral analysis

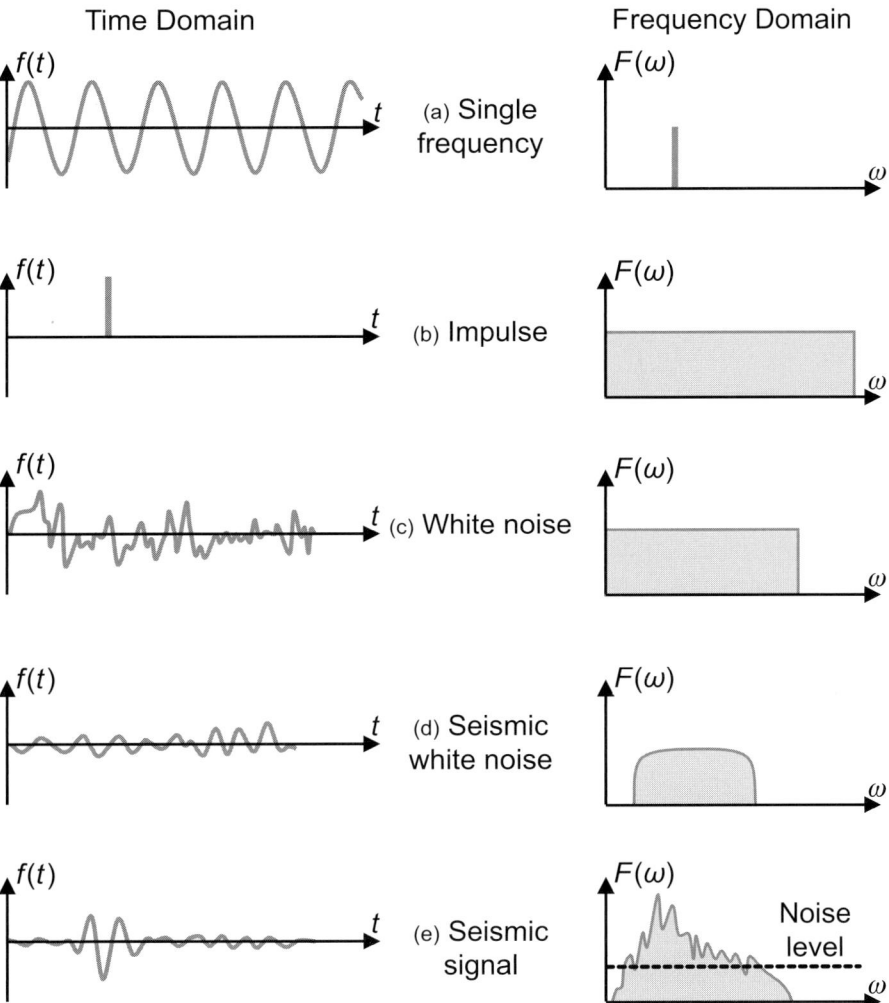

Figure 3.8 Some seismic time series and their amplitude spectra.

Following the definition of auto-covariance (3–14), we have

$$\gamma_x(t) = \int_{-\infty}^{\infty} x^*(\tau)x(t+\tau)d\tau$$

$$= \int_{-\infty}^{\infty} [b(t) + 0.5b(t-t_0)][b(t+\tau) + 0.5b(t+\tau-t_0)]d\tau$$

$$= \int_{-\infty}^{\infty} [b(t)b(t+\tau) + 0.5b(t)b(t+\tau-t_0) + 0.5b(t-t_0)b(t+\tau)$$

$$+ 0.25b(t-t_0)b(t+\tau-t_0)]d\tau$$

Box 3.4 Proof of the Wiener–Khinchin theorem

The Fourier transform of the auto-covariance function is

$$\int_{-\infty}^{\infty} \gamma_x(\tau)e^{i\omega\tau}d\tau = \int_{-\infty}^{\infty}\left[\int_{-\infty}^{\infty} x^*(\tau)x(t+\tau)d\tau\right]e^{i\omega t}dt$$

$$= \int_{-\infty}^{\infty} x^*(\tau)\left[\int_{-\infty}^{\infty} x(t+\tau)e^{i\omega t}dt\right]d\tau \qquad (3\text{–}13')$$

Substitute $q = t + \tau$ into the inner integral

$$\int_{-\infty}^{\infty} \gamma_x(\tau)e^{i\omega\tau}d\tau = \int_{-\infty}^{\infty} x^*(\tau)\left[\int_{-\infty}^{\infty} x(q)e^{i\omega q}dq\right]e^{-i\omega\tau}d\tau$$

$$= \int_{-\infty}^{\infty} x^*(\tau)X(\omega)e^{-i\omega\tau}d\tau$$

$$= \int_{-\infty}^{\infty} x^*(\tau)e^{-i\omega\tau}d\tau\, X(\omega)$$

$$= X^*(\omega)X(\omega)$$

This proves that the FT of auto-covariance equals the (power) spectrum of the signal.

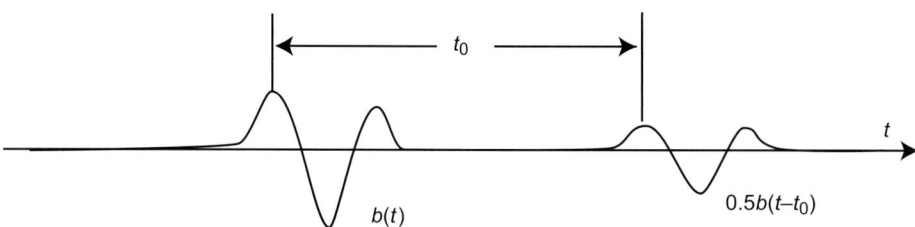

Figure 3.9 An example time series of a wavelet and its echo.

Notice that

$$\gamma_b(t) = \int_{-\infty}^{\infty} b(\tau)b(t+\tau)d\tau = \int_{-\infty}^{\infty} b(\tau - t_0)b(t + \tau - t_0)\, d\tau$$

Thus

$$\gamma_x(t) = (1 + 0.25)\gamma_b(t) + 0.5\,\gamma_b(t - t_0) + 0.5\,\gamma_b(t - t_0)$$

Graphically in Figure 3.10, we have

We therefore can find the spectrum by examining the auto-covariance function. The commonly used **auto-correlation** function is actually the normalized auto-covariance function. Figure 3.11 shows an example of **multiples**, or multiple reflections, and use of auto-correlation as a means to recognize multiple reflections.

The traditional approach to estimate the spectrum of a sampled time series is to estimate the auto-covariance function first, then multiply the estimate by a taper and finally apply

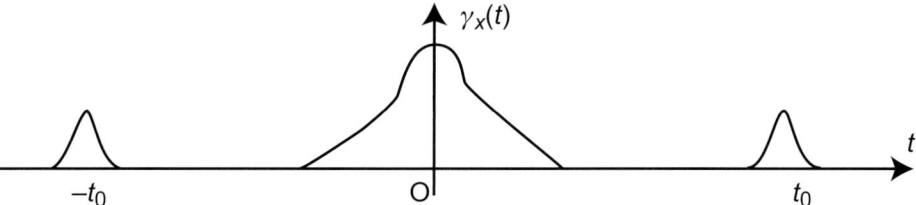

Figure 3.10 Auto-covariance of the time series shown in the previous figure.

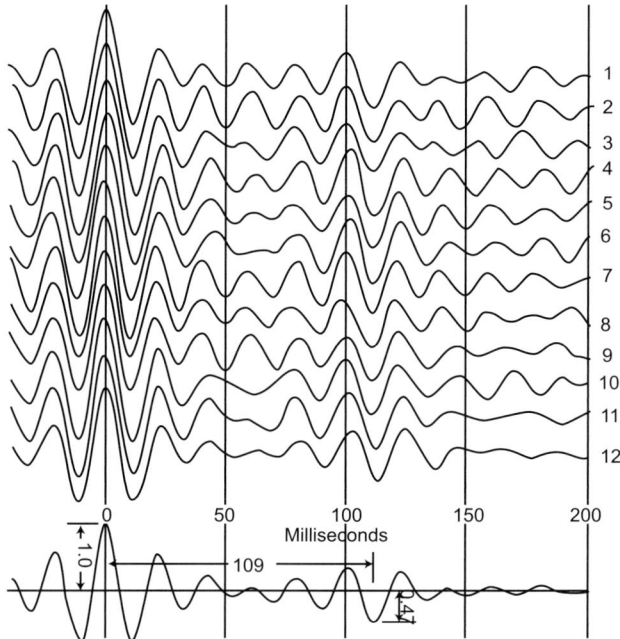

Figure 3.11 Twelve traces of a primary reflection and a multiple. The sum of the traces is shown as the bottom trace.

the Fourier transform to get the spectrum. With FFT, however, the spectrum can be directly estimated using the squared magnitude of the transform function.

3.3.3   The uncertainty principle

Figure 3.12 from Claerbout (1992) shows auto-correlations and amplitude spectra of some common signals. Only one-half of the graphs for auto-correlations and amplitude spectra are shown because these functions are symmetric. The auto-correlation function focuses the energy towards the origin, making it a noise-resistant operator. The only signal which preserves its form in the time domain, auto-correlation domain, and spectral domain is the Gaussian function. Notice in this figure a time series of longer duration in the time domain usually has shorter duration in frequency domain, and vice versa. In other words,

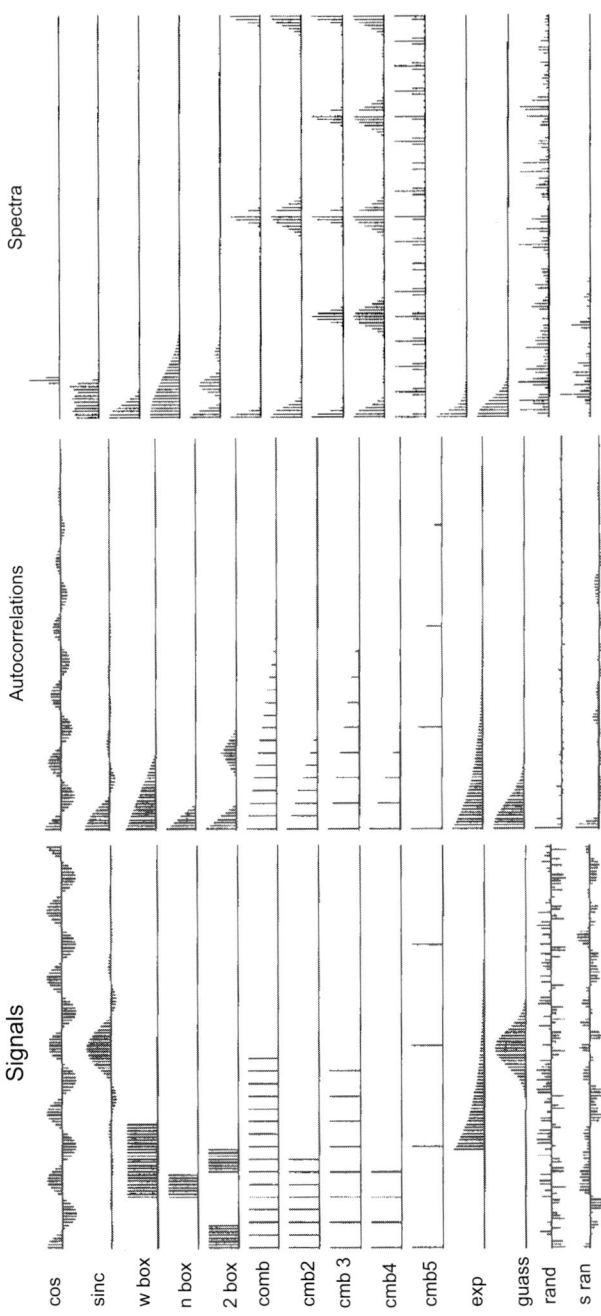

Figure 3.12 Some common signals, their autocorrelations and amplitude spectra (from Claerbout, 1992).

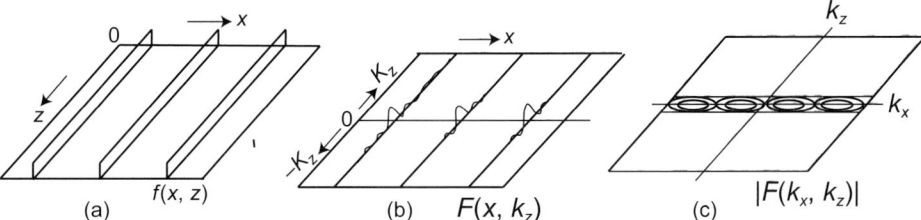

Figure 3.13 A sketch of 2D FT of three bounded vertical reflectors (Chun & Jacewitz, 1981). (a) Line spike model in the depth domain. (b) Fourier transform in the depth direction. (c) Typical amplitude contours in the 2D frequency domain.

the product between the temporal duration and spectral duration is more or less a constant; some call this property **the uncertainty principle**.

We can see the uncertainty principle in a sketch of 2D Fourier transform (Chun & Jacewitz, 1981) shown in Figure 3.13. Note that the vertical elongation of the three events in the space domain turns into the horizontal elongation of a single event in the frequency domain. The uncertainty principle implies that elongated events after multi-dimensional Fourier transform will become elongated events in a direction perpendicular to the elongated direction in the original domain. On the other hand, the width of the events in the frequency domain is dictated by the **bandwidth**, or the spectral duration.

We usually prefer data with a broad bandwidth because the corresponding temporal duration will be short, or of higher temporal resolution. The bandwidth is usually quantified by **octave**. One octave of bandwidth covers an interval between any two frequencies having a ratio of 2 to 1. Hence a power spectrum with energy from 10 to 20 Hz has a bandwidth of one octave, and another power spectrum with energy from 10 to 40 Hz has a band width of two octaves, and yet another power spectrum with energy from 1 to 4 Hz also has a band width of two octaves. Two or more octaves of bandwidth are often required for many processing tasks.

3.3.4 Practical issues of spectral analysis

When we compare spectra of different time series at various scales, we need to normalize the spectrum. The normalized spectrum is called the power spectrum, which is often denoted by $P(\omega)$:

$$P(\omega) = X^*(\omega)X(\omega) \tag{3–15}$$

The normalization of the auto-covariance function is the auto-correlation function. In continuous form, it is

$$r(t) = \gamma(t)/\gamma(0) \tag{3–16}$$

In discrete form, it is

$$r(t) = \frac{1}{N} \sum_{k=0}^{N-1-t} x_k^* \, x_{k+t} \tag{3–17}$$

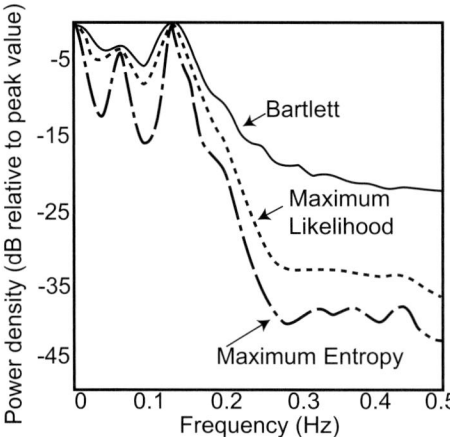

Figure 3.14 Spectra of horizontal component of 17 minutes of long-period seismic noise using three different estimates (Lacoss, 1971).

By the Wiener–Khinchin theorem, the power spectrum of a time series equals the Fourier transform of its auto-correlation function.

Estimation of the power spectrum for real geophysical data requires further modification of the sample squared magnitude of the Fourier transform. To suppress the effect of noise, one may average the squared magnitude for different data samples, or smooth the raw spectrum in the frequency domain.

Perhaps the most critical issue is the choice of the window over which the spectrum is estimated. For many applications, the data series is assumed to be **stationary**, which means statistical equilibrium. In layperson's terms, a stationary time series consist of no trends or no statistical difference from window to window. Hence, a stationary process may be adequately described by the lower moments of its probability distribution, such as mean, variance, covariance, and power spectrum. Of course, the validity of the stationary assumption depends not only on the real physical processes, but also on the size of the window and the amount of data available. A synthetic illustration shown below is taken from Jenkins and Watts (1968).

If the interpretation of the spectrum is indeed critical, an adaptive non-linear analysis may be necessary. In the presence of high noise levels, there are techniques such as the **maximum likelihood method** (MLM) and the **maximum entropy method** (MEM) that will help maintain good spectral resolution. These methods are described in a review by Lacoss (1971), and an example is shown in Figure 3.14. The unit of amplitude spectrum as shown in this figure is in **dB** (decibel). According to Webster's dictionary, dB is a unit for expressing the ratio of the magnitudes of two electric voltages or currents or analogous acoustic quantities equal to 20 times the common logarithm of the voltage or current ratio. For analysis of amplitude spectrum,

$$dB = 20\log[A(\omega)/A_{max}] \tag{3–18}$$

where $A(\omega)$ is the amplitude as a function of the angular frequency ω, and A_{max} is the maximum amplitude. Because $A(\omega) \leq A_{max}$, the dB value of amplitude spectrum is always negative. Using a logarithmic scale in a spectral plot with dB helps to elaborate the detail of the spectra.

Exercise 3.3

1. Find three examples showing the benefits of the Fourier transform. How are the benefits realized? What are the assumptions? Are there any limitations or drawbacks?

2. When applying AGC to a seismic trace, will the phase spectrum be changed? Is it possible to alter amplitude without changing phase spectrum?

3. Figure 3.15 shows the power spectrum of a 30 Hz Ricker wavelet denoted with the center frequency, the bandwidth, and the root-mean-square frequency. The two dashed lines are equidistance from the centralfrequency, and the distance between them is twice the spectral bandwidth.

   (a) How many octaves of this spectrum are there in your estimate?

   (b) Why is the center frequency not at the position of the peak amplitude?

   (c) How would you define the spectral bandwidth for this case?

   (d) How is the rms frequency defined?

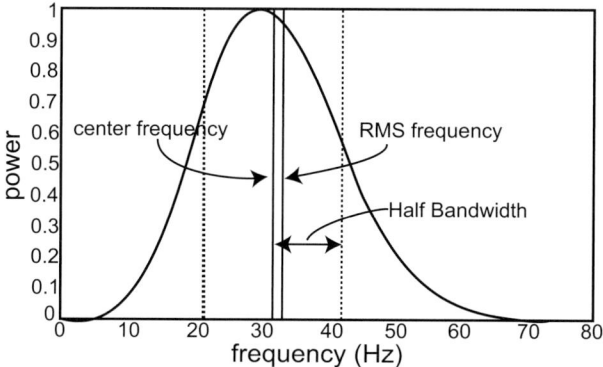

Figure 3.15 An example power spectrum.

3.4   Seismic wavelets

3.4.1   Wavelet definition

A **wavelet** is generally defined as a transient signal with two properties. First, it is a one-sided function. In other words, the first condition that a discrete time function w_t is a wavelet is that all of its coefficients are zero before the time origin:

$$w_t = 0 \quad \text{for } t < 0 \tag{3–19}$$

Second, a wavelet must be a stable function, meaning it has a finite energy:

$$\sum_t |w_t|^2 < \infty \qquad (3\text{–}20)$$

In seismic data processing, a stable time series also indicates its time duration is relatively short.

The idea of a **seismic wavelet** comes from a long-standing model in seismology that a recorded seismic trace is a convolution of the source function, Earth structure function, and receiver function. The wavelet is usually a seismic pulse consisting of a few cycles only. Hence we may define a seismic wavelet as a short time series with finite amplitude range and finite time duration. An unusually long seismic wavelet is the **chirp** or **sweeping** signal, which has many cycles and in which the frequency increases (**upsweeping**) or decreases (**downsweeping**) with time, such as the Vibroseis source wavelet. In exploration geophysics, the receiver function and general trend of geometric spreading are measurable and correctable in most cases. Hence we often approximate a seismic trace as the source wavelet convolved with Earth reflectivity and plus noise, as shown in Equation (2–1) and Figure 2.1.

A physically meaningful approach is to take a seismic wavelet as the far-field response of an impulsive source. Following this notion, Norman Ricker conducted a series of investigations on the form and nature of seismic waves from the 1930s to 1950s (Ricker, 1940, 1953a, 1953b). He studied the expressions of seismograms in terms of displacement, velocity, and acceleration. Clearly, velocity and acceleration are the second and third temporal derivatives of displacement. Typically, structural geologists are mostly interested in displacement data measured by strainometers or extensometers; engineers are mostly interested in strong motion data, in other words accelerations measured by accelerometers; and exploration geophysicists are mostly interested in velocity data measured by geophones or hydrophones.

3.4.2   Common seismic wavelets

As shown in Figure 3.16, a **Ricker wavelet** is a zero-phase wavelet resembling the far-field **impulse response** on the velocity record. Ricker published the mathematical expression of the Ricker wavelet in 1940 (Ricker, 1940), and a VSP field study that confirmed the wavelet in 1953 (Ricker, 1953a). In the time domain, the mathematical formula for a Ricker wavelet is

$$\text{Ricker}(t) = (1 - 2\pi^2 f^2 t^2)\exp(-\pi^2 f^2 t^2) \qquad (3\text{–}21)$$

where f is its peak frequency, which uniquely defines a Ricker wavelet.

The **impulse response** of a system is the output of the system after an impulsive input signal. The significance of an impulse response is that we may use it as a natural building block to decompose or superpose the system. For instance, an impulse response of an elastic system is called a **Green's function**. Using an impulsive source and an impulsive receiver, we can measure the Green's functions of an elastic system in terms of changing source positions, receiver positions, and frequency. In another example, the impulse responses between the data space and model space are used to gauge the resolution level in seismic imaging.

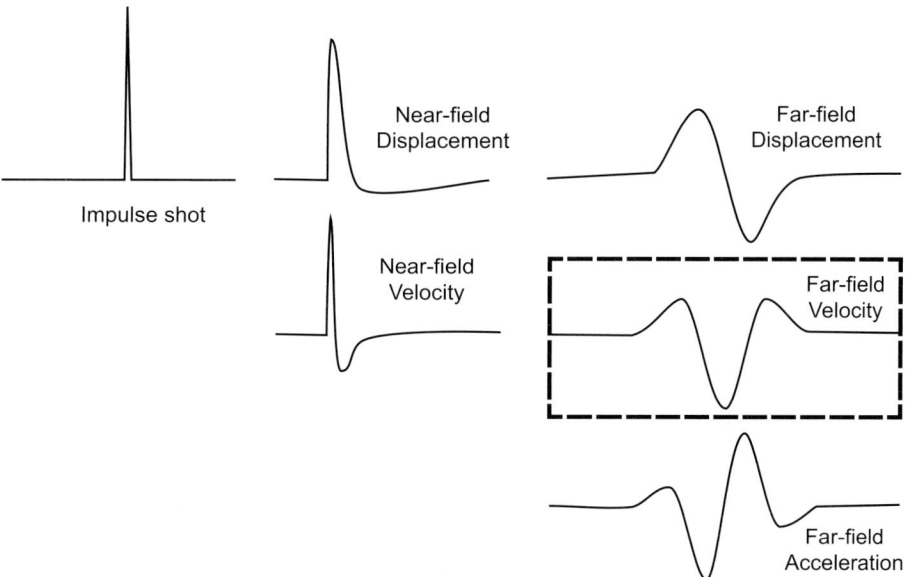

Figure 3.16 A Ricker wavelet resembles the far-field velocity record of an impulse source (after Ricker, 1953a).

Considering the non-uniqueness in decomposing seismic traces, the most natural way to decompose seismic traces is using the true impulse responses, or true seismic wavelets. Therefore, its resemblance to the far-field seismic impulse response makes the Ricker wavelet the most popular seismic wavelet. In order to tie reflection seismic imageries with well logs, a general practice is to generate a reflectivity function based on well logs, and convolve it with a Ricker or other suitable wavelet to product a **synthetic seismogram**, which will be compared closely with seismic sections around the well bore.

Figure 3.17 shows the time expressions and amplitude spectra of a Ricker wavelet, Ormsby wavelet, and Klauder wavelet. **Ormsby wavelets** are zero-phase wavelets defined as the result of applying a trapezoidal-shaped filter (**Ormsby filter**) to a unit impulse function. An Ormsby filter is specified by its four corner frequencies in the form of f_1–f_2–f_3–f_4. These four corner frequencies are called the low-cut, low-pass, high-pass, and high-cut frequencies, respectively. The filter is 1 from f_2 to f_3, linear from f_1 to f_2 and from f_3 to f_4, and zero below f_1 and beyond f_4. Unlike the Ricker wavelet with just two side lobes, an Ormsby wavelet has many side lobes. The number of side lobes increases as the slope of the sides of the trapezoidal filter gets steeper, or as its bandwidth narrows.

A **Klauder wavelet** is another zero-phase wavelet representing the autocorrelation of a chirp or linear sweep signal such as that used in Vibroseis. The real part of the following complex trace will generate a Klauder wavelet:

$$\text{Complex Klauder}(t) = \frac{\sin(\pi kt(T - t))}{\pi kt} \exp(i2\pi f_0 t) \qquad (3\text{--}22)$$

where k is rate of change of frequency with time, T is the time duration of the input signal, and f_0 is the middle frequency. Note in Figure 3.17 that the Klauder wavelet and Ormsby wavelet are similar in the time domain and in their amplitude spectra.

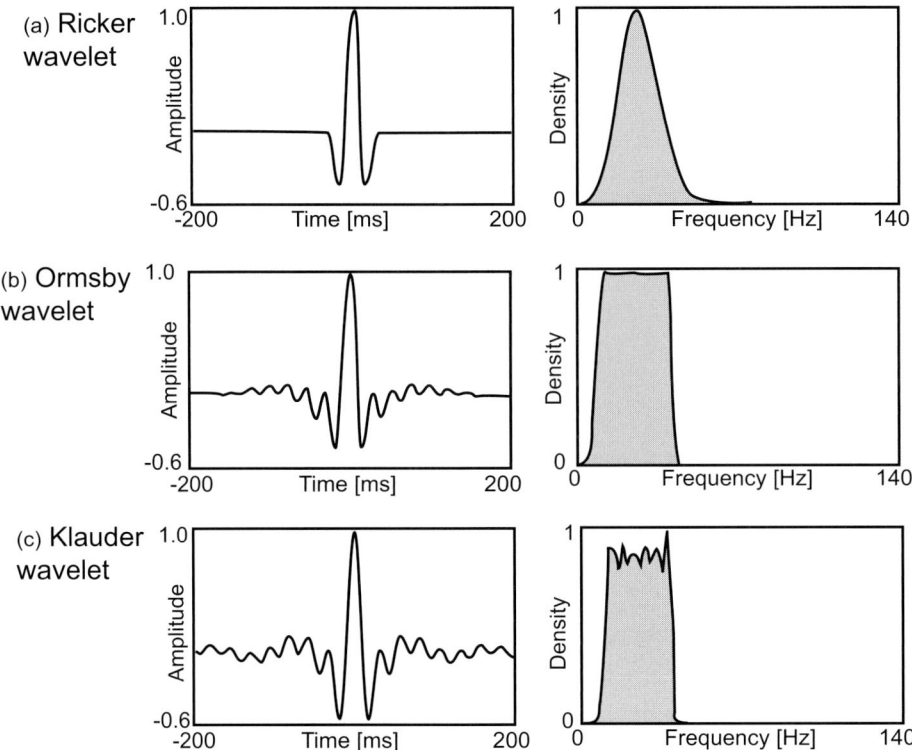

Figure 3.17 Time expressions and amplitude spectra of: (a) Ricker wavelet; (b) Ormsby wavelet; and (c) Klauder wavelet (Ryan, 1994).

3.4.3   Wavelet transform

The use of the convolution model in making the synthetic seismogram leads to the idea of **local decomposition**, in contrast to **global decomposition**. A global decomposition, such as Fourier decomposition, aims to measure global properties such as amplitude and phase values across the entire length of the input trace for each wavelength-fixed building block, such as a cosine function of a fixed frequency. Such global decompositions provide spectral properties that are global, or relevant to the entire input trace. In contrast, a local decomposition intends to quantify the properties at local points of the input trace. For instance, the reflectivity function at a location may be characterized by several peaks and troughs corresponding to major interfaces of elastic impedance in the subsurface. Seismic imaging aims at achieving the best resolution of these interfaces. Local decomposition, through the use of a **wavelet transform**, has an advantage over global decomposition in capturing both the local and spectral information. Basically, a wavelet transform expresses an input trace as a superposition of different expressions of a chosen wavelet whose length can be stretched continuously or discretely. Figure 3.18 shows the Daubechies 12-coefficient wavelet at two different scales of stretching, with their amplitude spectra.

In seismic data processing, a **discrete wavelet transform** (DWT) is suitable and the wavelets are discretely sampled. The first DWT was proposed by Alfréd Haar (Mallat,

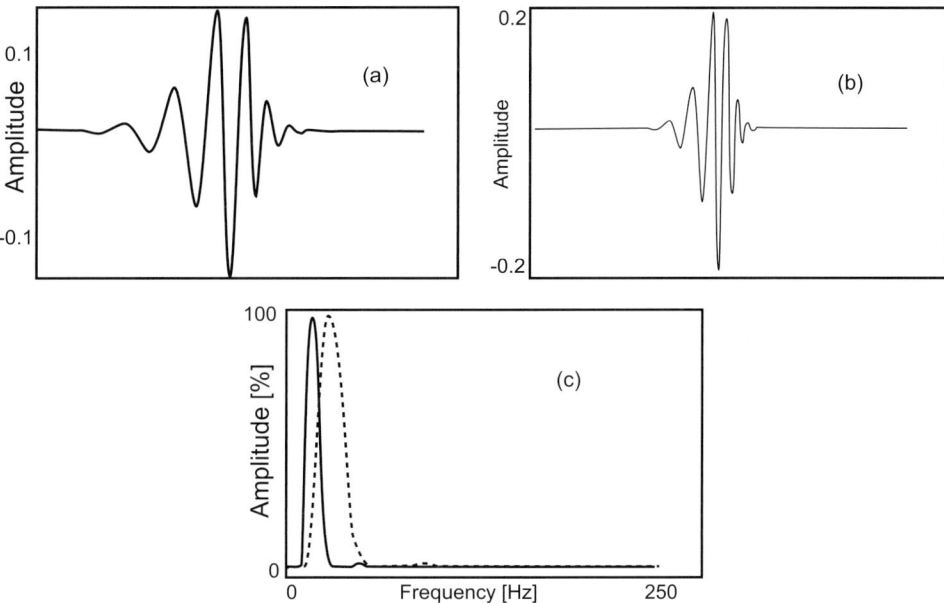

Figure 3.18 Daubechies 12-coefficient wavelet. (a) Scale 5. (b) Scale 4. (c) Amplitude spectra of scale 5 (solid curve) and scale 4 (dotted curve). (After Daubechies, 1988.)

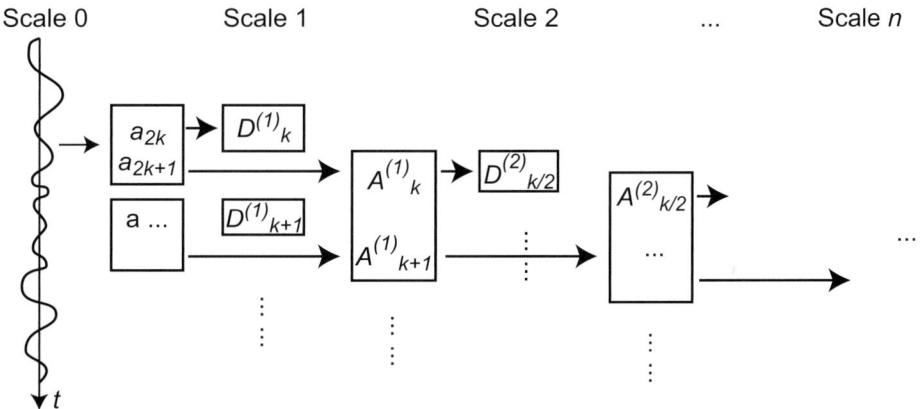

Figure 3.19 Illustration of Haar wavelet transform of a seismic trace. For each pair of neighboring values of a scale, their difference (D) is the value of the current scale and their sum or approximation (A) is the input for the next scale.

1999). Similar to the fast Fourier transform, the **Haar wavelet transform** requires the input sample number to be a power of two, such as 2^n. The first iteration of the Haar transform pairs up input values, storing the difference and passing the sum to the next iteration. For each pair of values, their difference (D) becomes the value for the current scale and their sum becomes the approximation (A) that is passed on as the value in the next iteration. In the next iteration on a new scale, all the neighboring sums from the previous iteration

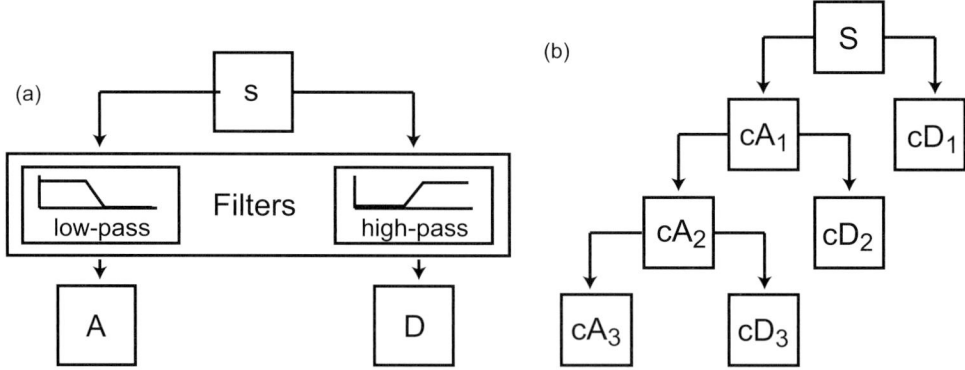

Figure 3.20 (a) Decomposing signal (S) into its approximation (A) and detail (D) components. (b) Decomposing a signal into subset components of approximations (cA) and details (cD).

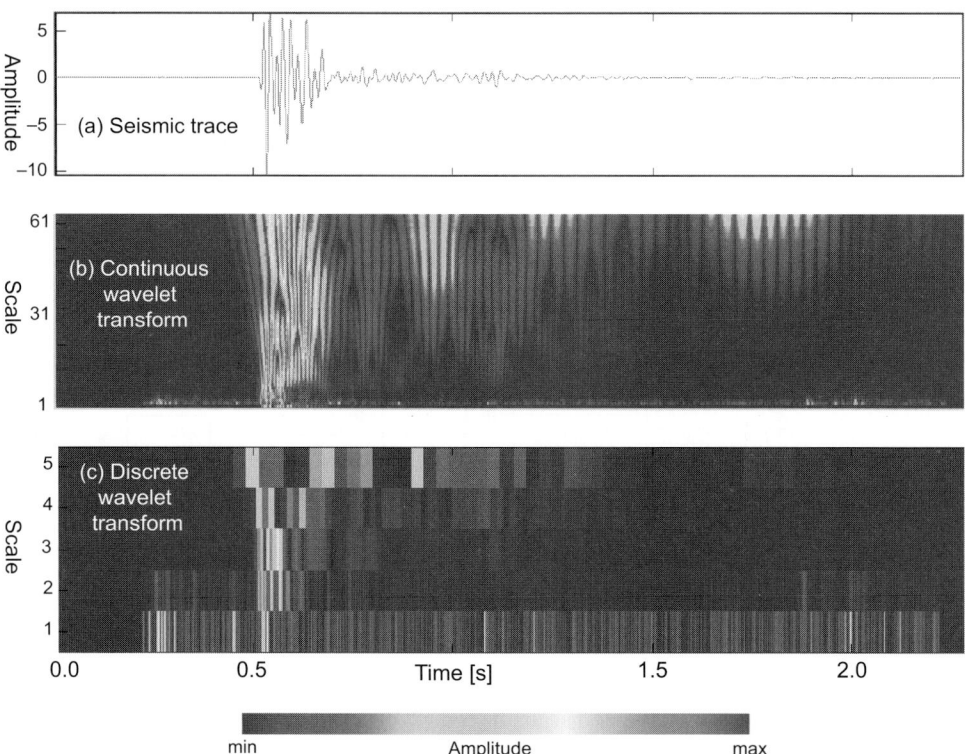

Figure 3.21 (a) A seismic trace. (b) Continuous wavelet transform. (c) Discrete wavelet transform. For color versions see plate section.

form a new set of pairs to provide the differences as the values of the scale and sums for the next iteration. After n iterations, the transformation results in $2^n - 1$ differences and a final sum. The process is illustrated in Figure 3.19. Clearly, the sums obtained from each iteration form a smoothed version of the original trace, and the final sum equals the sum of the original input values.

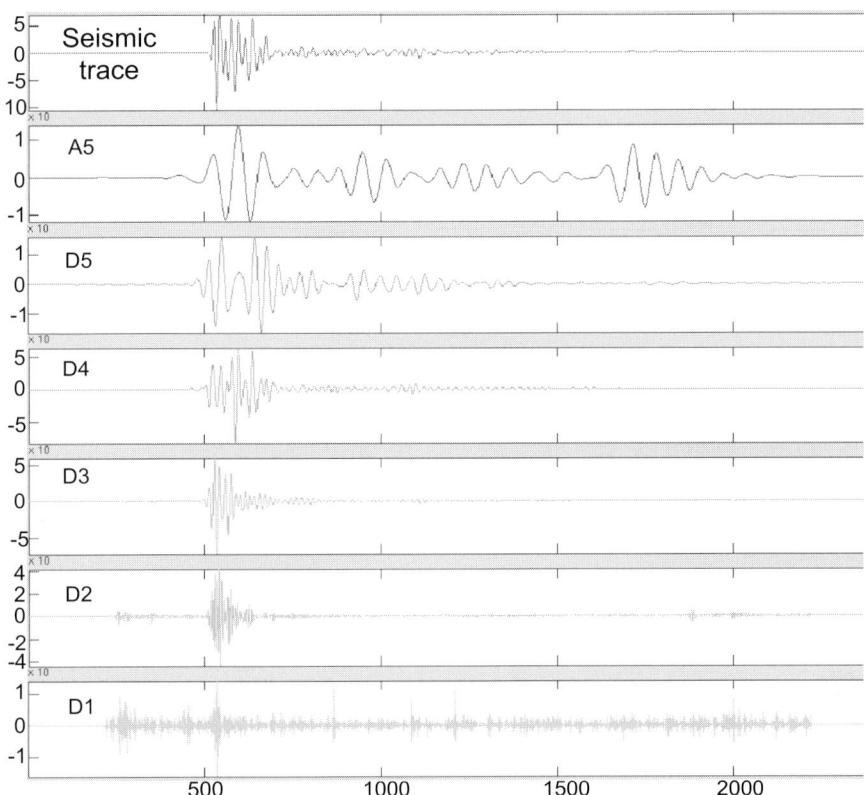

Figure 3.22 Discrete wavelet transform of seismic signal. Upper panel shows input seismic trace. A5 – approximation coefficients; D5, D4, D3, D2, and D1 – detail coefficients at scales 5, 4, 3, 2, and 1, respectively.

The ordering of the scales in the Haar wavelet transform starts from the finest detail of the input trace, and the higher the scale the longer the stretching of the wavelet. Alternatively, we may start from the long-wavelength features before moving to the finer details. This approach is more adequate for seismic applications where the details are contaminated much more by noise than long-wavelength components. Two schematic plots of this kind of decomposition are shown in Figure 3.20. In Figure 3.20a, the signal S is low-pass filtered into its approximation A and high-pass filtered into the detail D. Similarly in Figure 3.20b, the signal can be decomposed into a cascading series of approximations and details of different scales. Following this approach, a practical way of wavelet decomposition for seismic applications is the **matching pursuit** algorithm (Davis *et al.*, 1994). Starting from the long-wavelength scales, the algorithm matches the corresponding wavelet with the input trace, and takes the misfit residuals as the input in the processing of the next scale.

3.4.4 Examples of wavelet transform

In practice, most discrete wavelet transforms follow the formulation of Ingrid Daubechies (1988) using recurrence relations to generate progressively finer discrete samplings of an

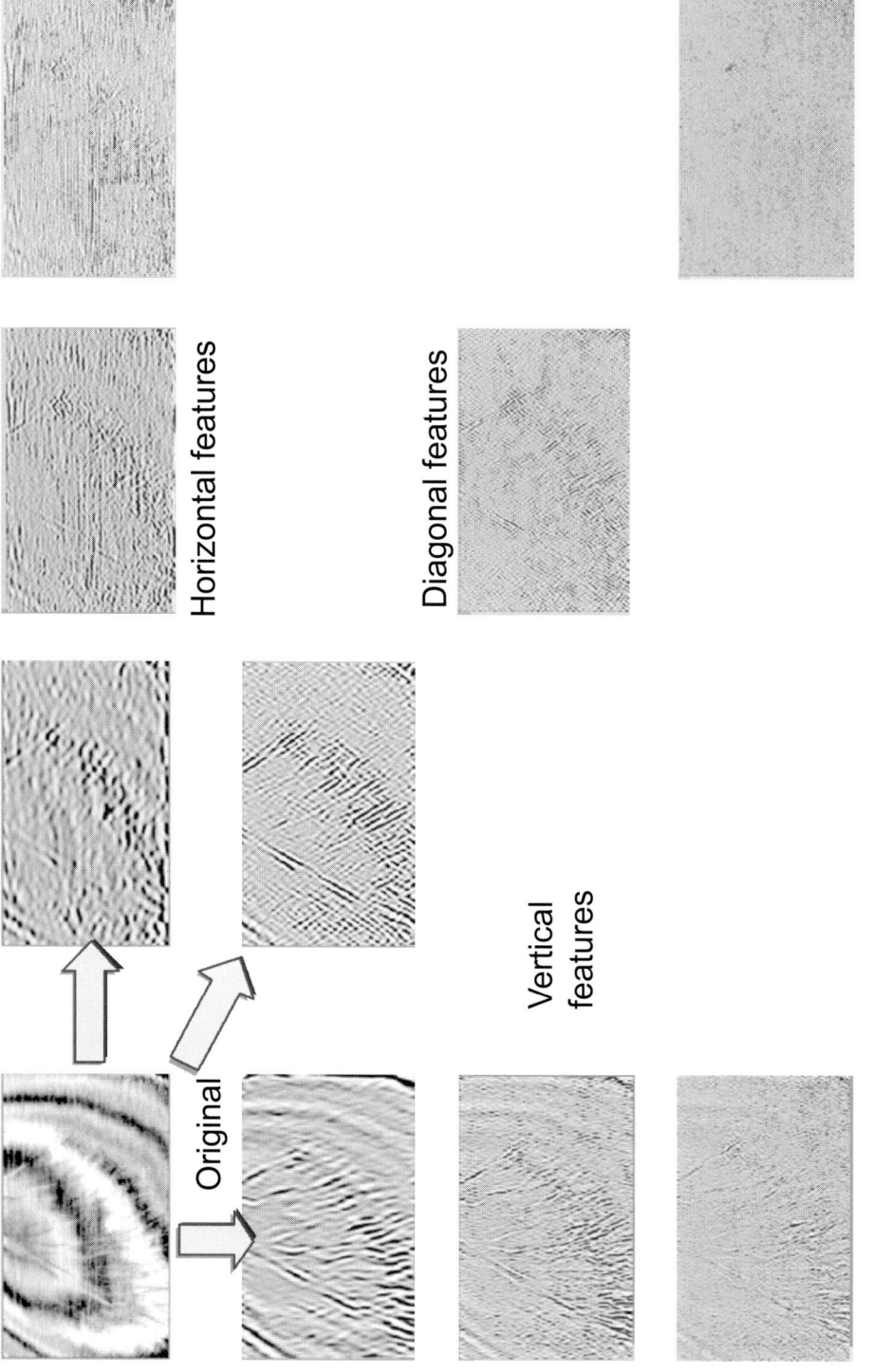

Figure 3.23 Two-dimensional wavelet decomposition of a seismic time slice in the upper-left panel into three levels or scales. Higher scales are of lower spatial frequencies. Panel of each scale can be included in the superposition, or inverse wavelet transform.

Horizontal features

Diagonal features

Vertical features

Original

implicit mother wavelet function. This approach allows Daubechies to derive a family of **orthogonal wavelets**: the first one is the Haar wavelet, and each resolution is twice that of the previous scale. Examples of the **Daubechies wavelet** are shown in Figure 3.18 in the previous subsection. Figure 3.21 shows a seismic trace with its continuous wavelet transform and discrete wavelet transform. The detail of the discrete wavelet transform is shown in Figure 3.22, showing the approximation coefficients at scale 5 and detail coefficients at scales 5 to 1. Note that the higher the scale, the lower the frequency.

The practice of wavelet transforms is still a new subject in seismic data processing. This means the existence of many obstacles as well as opportunities. Some possibilities for using wavelet transforms include:

- Data compression
- Estimation of seismic wavelet
- Enhancement of seismic resolution
- Noise suppression by separating particular signals from noise
- Enhancement of seismic imaging quality and efficiency
- Extraction of seismic attributes, such as localized spectra and AVO due to thin beds

Figure 3.23 shows an example of decomposing a time slice of a seismic volume using a 2D discrete wavelet transform. Spatial components of the imagery of horizontal, vertical, and diagonal orientations are decomposed into three levels. The usefulness of such a decomposition relies on geologic constraints determining the characteristics of the signal, such as beddings, interfaces, faults, and fractures, as well as noise such as footprints, spatial aliasing, and other artifacts.

Another example of noise suppression is shown in Figure 3.24, from the work of Deighan and Watts (1997). A common shot gather in panel (a) shows the presence of ground

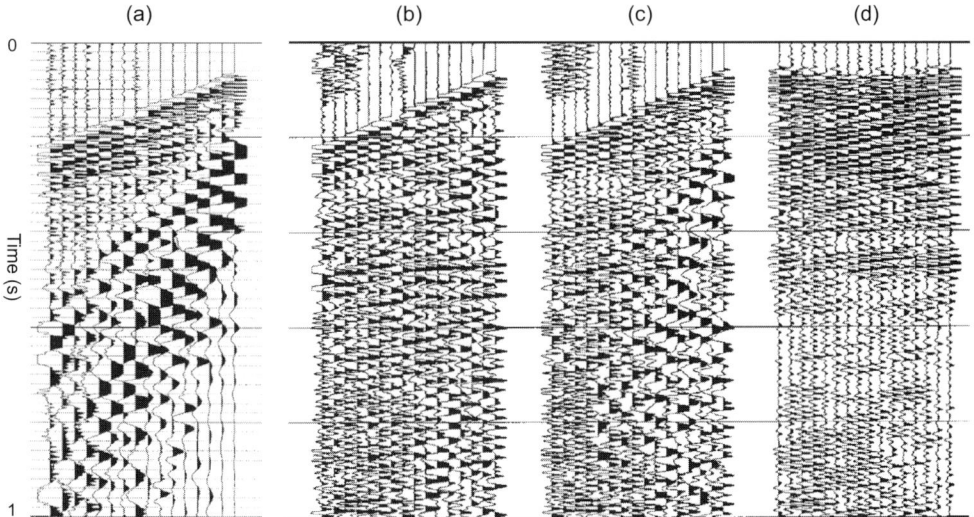

Figure 3.24 (a) A vertical-component shot gather. Removal of the ground rolls using: (b) a 1D wavelet transform filter; (c) a 40–70 Hz band-pass filter with a 100 dB/octave slope; and (d) an *f–k* filter. (Deighan & Watts, 1997.)

rolls that severely mask the reflections. The result of applying a 1D wavelet transform, which is a trace-by-trace operation, is compared with the results of a band-pass plus f–k filter. The wavelet transform result appears to be slightly better than the other two filtered results.

Exercise 3.4

1. Discuss wavelet transform in terms of what, why, and how. What are the main benefits and limitations of the method?

2. Make a table to compare and contrast Fourier decomposition and wavelet decomposition.

3. Comment on the quality of the three filtering results shown in Figure 3.24. What are the signals and noises? What are the elements that high-quality signals should possess in such cases?

3.5   Interpolation

3.5.1   Orthogonal function series

A function series defined on $[a, b]$

$$\phi_0(x), \phi_1(x), \phi_2(x), \ldots, \phi_n(x), \ldots \tag{3–23}$$

is called a **orthogonal function series** if all elements of the series satisfy

$$\delta_{mm} = \frac{1}{b-a} \int_a^b \phi_m(x)\phi_n^*(x)dx \tag{3–24}$$

where $\phi_n^*(x)$ is the conjugate of $\phi_n(x)$ in the case that those are complex functions, and δ_{mn} is the **Kronecker delta** function which equals 1 if $m = n$ and 0 if $m \neq n$. We can always normalize the coefficient so that the integral (3–24) equals 1 if $m = n$. An orthogonal function series can be used as the base for an expansion, if we are interested in the global interpolation of a function based on a finite number of measurements.

A good example of an orthogonal function series is the trigonometric functions

$$1, \cos \omega x, \cos 2\omega x, \ldots, \cos n\omega x, \ldots, \sin \omega x, \sin 2\omega x, \ldots, \sin n\omega x, \ldots \tag{3–25}$$

where $\omega = 2\pi / T$. These functions are orthogonal in the interval $[-T/2, T/2]$.

If a function $f(x)$ can be integrated over the interval $[-\pi, \pi]$, the trigonometric series (3–25) forms the base of the Fourier series of the function

$$a_0/2 + \sum_{n=1}^{\infty} (a_n \cos nx + b_n \sin nx) = \sum_{n=-\infty}^{\infty} c_n e^{inx} \tag{3–26}$$

where the coefficients are defined as

$$a_n = \frac{1}{\pi} \int_{-\pi}^{\pi} f(t) \cos nt \, dt \quad (n = 0, 1, 2, \ldots) \tag{3-27a}$$

$$b_n = \frac{1}{\pi} \int_{-\pi}^{\pi} f(t) \sin t \, dt \quad (n = 1, 2, \ldots) \tag{3-27b}$$

$$c_n = \frac{1}{2\pi} \int_{-\pi}^{\pi} f(t) e^{-\text{int}} dt \quad (n = \ldots, -2, -1, 0, 1, 2, \ldots) \tag{3-27c}$$

When the series (3–26) is convergent at a value of x, we can use it to approximate $f(x)$ and call it the **Fourier expansion** of the function. This expansion leads to the Fourier transform that we have discussed previously. Notice that the expansion is a periodic function.

The Fourier expansion was historically the first example of an expansion of a function into orthogonal functions and has retained its supreme importance as the most universal tool of applied mathematics. The expansion has the nice property that the error oscillation spreads uniformly throughout the interval of the expansion. If $f(x)$ is an even function, then all the coefficients in (3–27b) are zero so we get a Fourier cosine expansion. Similarly, if $f(x)$ is an odd function we get a Fourier sine expansion. When the function is defined over an interval different from $[-\pi, \pi]$, one can always convert the scale of the variable to make the expansion work. Around the point of a first-order discontinuity of the original function, or when the expansion is terminated with too few terms, the Fourier expansion has an unusually high fitting error known as **Gibbs oscillations**.

3.5.2   Interpolation

We now turn to the issue of interpolation. Most geophysical data are in tabular form, i.e. they are time series of equal sampling interval. For instance, conventional 2D or 3D seismic reflection surveys use regular spacing between geophones and surveying lines, and the sampling rate is also a constant. Obviously, sampling data at equal distance not only simplifies the sampling process but also eases data storage as well as the subsequent processing and analysis. Occasionally, we face irregularly sampled data, such as the geographic locations of many gravity survey occupations. In reflection seismology, data regularization is itself an important research area. Irregular seismic data are commonly seen in cases of **dead traces**, **crooked lines**, **suture zones**, and other inaccessible locations within the survey area. Dead traces are erroneous records due to causes such as a bad sensor or poor coupling between the sensor and the medium. Crooked lines are irregular 2D seismic survey lines due to limits in accessibility, such as survey lines following a zigzag road or river. With a crooked 2D line, the midpoints of reflection will not fall on a straight line, causing a number of irregularities for seismic data processing. Suture zones refer to overlapping areas between two or more separate seismic surveys.

According to Lanczos (1961) the art of interpolation goes back to the early Hindu algebraists. The idea of **linear interpolation** was known by the early Egyptians and Babylonians and belongs to the earliest arithmetic experiences of mankind. But the science of interpolation in its more intricate forms starts with the time of Newton and Wallis. The estimation of error of interpolation came only after the establishment of the exact "limit concept" at the beginning of the nineteenth century, through the efforts of Cauchy and Gauss. The true nature of equidistant interpolation was established even later, around 1900, through the investigation of Runge and Borel.

The essence of the theory of interpolation may be illustrated through the well-known Taylor expansion, an extrapolation of a function's value at the neighborhood of a known value. If a function $f(z)$ is analytic, meaning that the function's derivatives of all orders exist, in the neighborhood of a certain point $z = a$, we then can represent the function's value at an arbitrary point z on the complex plane in the neighborhood by the infinite Taylor series (Laurent series, if a central circular region has to be excluded)

$$F(z) = f(a) + \frac{f'(a)}{1!}(z - a) + \frac{f''(a)}{2!}(z - a)^2 + \cdots + \frac{f^{(k)}(a)}{k!}(z - a)^k + \cdots \quad (3\text{--}28)$$

Although by formal differentiation on both sides we can show that $F(z)$ coincides with $f(z)$ in all its derivatives at the point $z = a$, this does not prove that the infinite series (3–28) represents $f(z)$ at all values of z. In fact, the theory of complex variables says that $F(z)$ represents $f(z)$ only in the domain of convergence, which is a circle centered at point $z = a$ on the complex plane with a radius extending to the nearest "singular" point of the function.

The Taylor expansion is seen on the one hand as an infinite series and on the other hand as a finite series with a remainder term for error, i.e.,

$$F(x) = f_n(x) + \frac{f^{(n)}(\bar{x})}{n!}(z - a)^n \quad (3\text{--}29)$$

where

$$f_n(x) = f(a) + \frac{f'(a)}{1!}(x - a) + \frac{f''(a)}{2!}(x - a)^2 + \cdots + \frac{f^{(n-1)}(a)}{(n - 1)!}(x - a)^{n-1} \quad (3\text{--}30)$$

and $\bar{x}$ is some unknown point in the convergence interval. Notice that we denote the variable here by x, though it could also be z as a complex variable. The finite expansion (3–29) with the remainder term is much more useful than the infinite expansion (3–28). Since derivatives of higher than nth order do not appear in either $f_n(x)$ or the remainder term, the finite expansion does not demand the existence of derivatives of all orders, hence there is no longer a restriction that the function should be analytic.

3.5.3   Linearization via perturbation theory

The usefulness of the finite Taylor expansion shown in equation (3–30) can be viewed in two ways. First, it facilitates an extrapolation of the function with error estimation. Second, the polynomial approximation thus derived, especially the lower-order terms, has wide application in **perturbation theory**. To the first point, we should add that the remainder error estimation term does not represent a guarantee for the convergence of the expansion with increasing n; it merely provides error estimation for the given expansion. It could happen that the reminder term decreases up to a certain n and then increases again, or it may increase to infinite with increasing n. Yet we may obtain a very accurate value of $f(x)$ if we choose an appropriate value of n. In the convergent case, we denote the remainder term by $O[(x - a)^n]$.

Let us see an example using the first-order Taylor expansion to linearize a modeling process. In most geophysical modeling, we attempt to choose a particular model estimation $\mathbf{m}^{est}$ that will optimize the prediction $\mathbf{d}^{est}$ for the observation $\mathbf{d}$ through the known physical relationship $f(.)$ between data and model

$$\mathbf{d} = f(\mathbf{m}) \tag{3–31}$$

To analyse the trend, we may want to examine the linear component of the above relationship, which is generally continuous and non-linear. If there are sufficient data to enable an inversion for the model parameters, it is definitely desirable to discretize the function to formulate the discrete inverse problem, which is manageable by digital computers. If the function is already known, the simplest way of linearization is to expand (3–31) into a first-order finite Taylor series with respect to a known reference model $\mathbf{m}_0$ and $\mathbf{d}_0 = f(\mathbf{m}_0)$

$$\mathbf{d} = \mathbf{d}_0 + \frac{\partial f}{\partial \mathbf{m}} f(\mathbf{m} - \mathbf{m}_0) + O(\mathbf{m} - \mathbf{m}_0)^2 \tag{3–32}$$

where we denote the remainder term by $O(\mathbf{m} - \mathbf{m}_0)^2$ to assure the convergence, and $\frac{\partial f}{\partial \mathbf{m}}$ are the **Frechet differential kernels**.

The reference model, such as the constant a in (3–28) and (3–29), also contributes to the convergence of the expansion. If we have a good reference model, the relationship between the perturbations of data and model will be more linear, i.e.

$$\Delta \mathbf{d} \approx \frac{\partial f}{\partial \mathbf{m}} \Delta \mathbf{m} \tag{3–33}$$

where $\Delta \mathbf{d} = \mathbf{d} - \mathbf{d}_0$ and $\Delta \mathbf{m} = \mathbf{m} - \mathbf{m}_0$. One can predict the trend of the data from that of the model, and vice versa. Many other techniques of interpolation exist. For example, one can conduct a forward Fourier transform using irregularly sampled data as input, followed by an inverse Fourier transform with regularly sampled output. The recently developed wavelet transform, as discussed in the previous section, opens another door for many new ways of interpolation.

Exercise 3.5

1. Regularization of irregularly sampled data is a common need in geophysics. One may regularize data by first taking a forward Fourier transform of irregularly sampled data and then taking an inverse Fourier transform using uniformly sampled output. Write a special DFT formula using irregular sampling intervals for the input, and then comment on the choice of sampling interval in the frequency and output time domains.

2. Explain why a time series may lose its causality after a forward DFT plus an inverse DFT.

3. We may need to quantify the similarity between the original dataset and interpolated dataset. Explain how to quantify the similarity between digital data of two or higher dimensions. Please give your procedure.

3.6   Summary

- That a natural system such as a section of a seismic record can be decomposed into components, and these components can be superposed to form the original or even the entire system, is the basis of the law of decomposition and superposition that is useful in physics and geophysics. The Fourier transform and wavelet transform are examples of this.

- The Fourier transform (FT) allows digital data to be converted back and forth between the time and frequency spaces. It provides the framework for spectral analysis. Many problems become easily or efficiently solvable using FT.

- Owing to the finite sample numbers of digital seismic data, their discrete Fourier transform (DFT) may have different properties from that of continuous FT.

- Fast Fourier transform (FFT) is a classic way to improve computation efficiency by taking advantage of the cyclic nature of harmonic functions. The idea and tricks involved in FFT can be useful in other data processing areas.

- Spectral analysis assesses the energy distribution of seismic data over frequency or wavelength scales. It is useful in many data processing applications.

- For any time series, its duration in temporal space is inversely proportional to its duration in spectral space; this is known as the uncertainty principle. This principle may help us in developing intuition about the relationship between temporal and spectral spaces using the multi-dimensional Fourier transform.

- A seismic wavelet represents the fundamental signal originating from the seismic source. By the law of decomposition and superposition, we may be able to express a seismic record as a superposition of seismic wavelets of different wavelengths and amplitudes. We may also express the seismic record as a convolution of the source wavelet with the media, the wave propagation, and receiver functions.

- Wavelet transform enables a local decomposition of the input data trace to capture its local as well as spectral information. It has high potential for many future applications.

- Interpolation takes information from known locations to estimate it at unknown locations. There are a number of different approaches to interpolation, with a broad range of current and potential applications in seismic data processing and imaging.

FURTHER READING

Claerbout, J.F., 1992, *Earth Sounding Analysis: Processing versus Inversion*, Blackwell.

Hatton, L., Worthington, M.H. and Makin, J., 1986, *Seismic Data Processing: Theory and Practice*, Chapter 2, Blackwell.

Press, W.H, Teukolsky, S.A., Vetterling, W.T. and Flannery, B.P., 1992, *Numerical Recipes in C*, 2nd edn, Chapters 12 and 13, Cambridge University Press.

Ricker, N., 1940, The form and nature of seismic waves and the structure of seismograms, *Geophysics*, 5, 348–366.

4 Seismic resolution and fidelity

Seismic resolution and fidelity are two important measures of the quality of seismic records and seismic images. **Seismic resolution** quantifies the level of precision, such as the finest size of subsurface objects detectable by the seismic data. Several definitions of seismic resolution are introduced in this chapter. **Seismic fidelity** quantifies the truthfulness, such as the genuineness of the data or the level to which the imaged target position matches its true subsurface position. Since seismic data are band-limited, seismic resolution is proportional to the frequency bandwidth of the data or the resulting images. If the bandwidth is too narrow, the resolution will be poor because a single subsurface reflector may produce a number of indistinguishable wiggles on the seismic traces. For multiple datasets with the same bandwidth, it is easier in practice to recognize or resolve events with the zero phase wavelet rather than the minimum phase or mixed phase wavelets. Seismic fidelity is about global resolution, the resolution in the big picture.

In principle, the highest-quality seismic imagery requires the highest level of seismic resolution and fidelity. However, in real cases the seismic resolution and fidelity are always limited because of limited seismic illumination, producing various types of

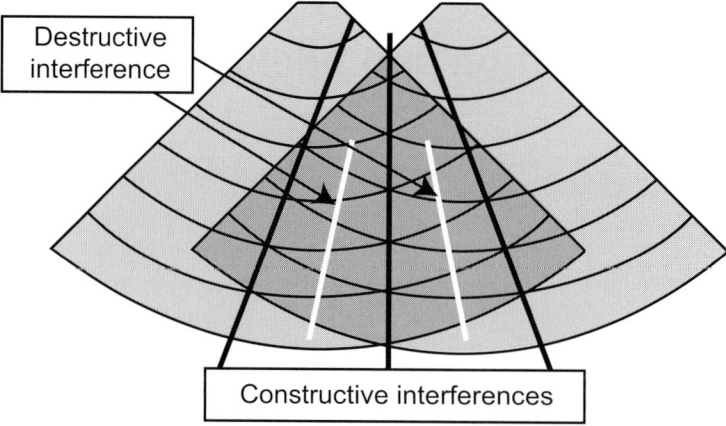

Figure 4.1 The superposition of two monochromatic wavefields from two point sources results in areas of destructive interferences and constructive interferences.

seismic artifacts. It is therefore important to assess the level of seismic resolution and fidelity in each seismic processing project, as illustrated by examples in this chapter. One of the physical reduction factors for seismic resolution is seismic attenuation, the loss of energy of the seismic signal with the increase of traversing distance. Hence the final topic discussed here is on the estimation and correction for seismic attenuation in processing seismic data.

4.1 Vertical and horizontal resolution

As one of the most essential parameters in assessing the quality of seismic data, **seismic resolution** is a measurement of how small a feature can be detected or resolved from seismic data. The use of the concept in practice is complicated by the fact that seismic data are made of remotely sensed wiggles combining signal and noise. As an example of the challenges, the superposition of multiple wavefronts, owing to the presence of either velocity inhomogeneity or multiple sources, produces **constructive interference** at places of phase alignment and **destructive interference** at places of phase differences (Figure 4.1). Such interferences take place for both signal and noise. It is difficult to infer the geology hiding behind the seismic wiggles if you do not know what you are looking for.

In the following, different definitions of seismic resolution are discussed, with illustrations. These concepts are of important practical value because most geologic layers containing oil and gas are of sub-wavelength resolution, meaning the thickness of these layers is smaller than the wavelength of the main frequency of seismic data. To help the readers gain more intuition on resolution, the classic wedge model is introduced to assess the behavior of seismic reflections when the thickness of a thin bed is reduced from more than a wavelength of the data down to zero. "**Tuning**" behavior due to the interference of the reflections from the top and bottom of the thin bed is seen.

Vertical resolution

Applying the concept of resolution to seismic wiggles, **vertical resolution** quantifies the resolvability of seismic waves along the direction of wave propagation, which is usually the vertical direction. Hence it is also called temporal resolution, depth resolution or 1D resolution, depending on the application. Sheriff (1991) gives the following three definitions of vertical resolution:

- **The smallest change in input that will produce a detectable change in output.** This is the most generic definition geared towards quantifying a system from its input and output. It defines the resolution of the system, which can be a piece of hardware or software. This definition of resolution is applicable to all cases, including all types of seismic resolution.
- **The ability to distinguish two points that are near each other.** This definition is the most commonly used in exploration geophysics because it helps to define the thickness of a thin bed (Widess, 1982). This definition was derived for vertical or temporal resolution. However, the concept can be generalized for horizontal resolution as well.
- **The ability to localize an event seen through a window, usually taken as the half width of the major lobe.** This definition specifies the vertical resolution (or temporal resolution) of a 1D seismic trace. Such a specification was pioneered by Ricker (1953b), using the concept of Ricker wavelet that was discussed in the previous chapter. Hence, the major lobe referred to that of the underlying seismic wavelet. This definition implicitly assumes the convolution model, that the seismic trace is a convolution of the seismic wavelet with the Earth reflectivity function plus noise.

Figure 4.2 shows examples of vertical resolution. Following the above second definition of seismic resolution, the upper panel shows four cases of two events close together, with

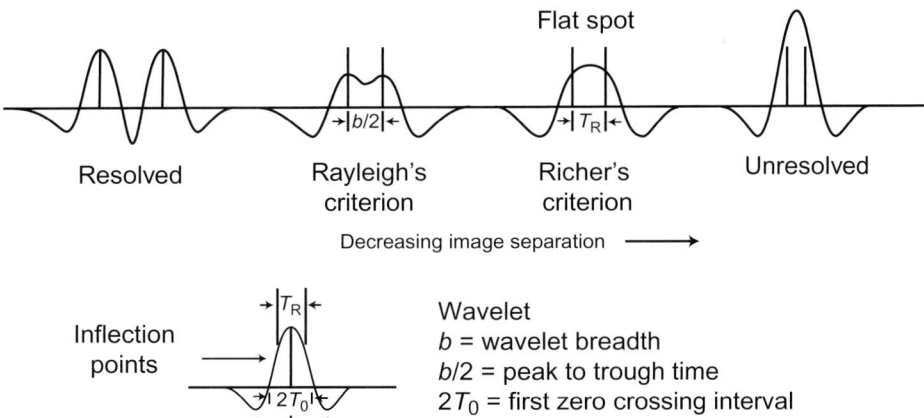

Figure 4.2 (Upper) Four cases of vertical resolution defined by the detectability of two events close together. (Lower) Ricker's resolution limit is the separation interval between inflection points of the seismic wavelet. (After Kallweit & Wood, 1982.)

Box 4.1 Conventional-resolution versus high-resolution seismic data

As an example, Box 4.1 Figure 1 shows two time sections taken from 3D migrated seismic volumes in the same area. The left panel is a section from the dataset produced from a conventional seismic survey, and the right panel is from a separated high-resolution seismic survey using much denser shot and receiver spacing and sample rate. The geology of the area consists of a series of dipping beds of sands and shales. We can see in the time sections that the pulse width is sharper for the high-resolution data than the conventional-resolution data. The spectra show that the section from the high-resolution data has a wider bandwidth than that of the conventional-resolution data. The high-resolution data reveal more detail about some gas-bearing sands with high reflectivity. In terms of money and time, however, the high-resolution data cost is more than 10 times that of the conventional-resolution data.

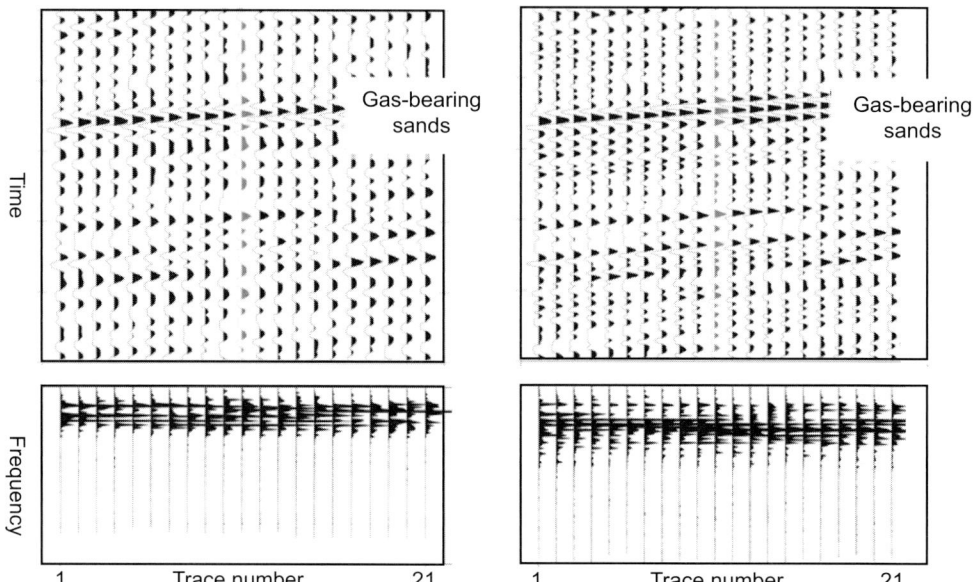

Box 4.1 Figure 1 (Left) A conventional-resolution seismic section and power spectra of all traces. (Right) A high-resolution seismic section and its power spectra.

their separation decreasing from left to right. The second case is the **Rayleigh criterion** which defines the resolution limit at 1/4 of the main seismic wavelength, which is half of the dominant seismic period for two-way reflection time. This criterion defines the **tuning limit**, the maximum thickness for a thin bed with constructive interference between its top and bottom reflections. The lower panel in Figure 4.2 shows the third definition of seismic resolution, or **Ricker's criterion**. So if we know the seismic wavelet, the resolution limit is finer than the Rayleigh criterion.

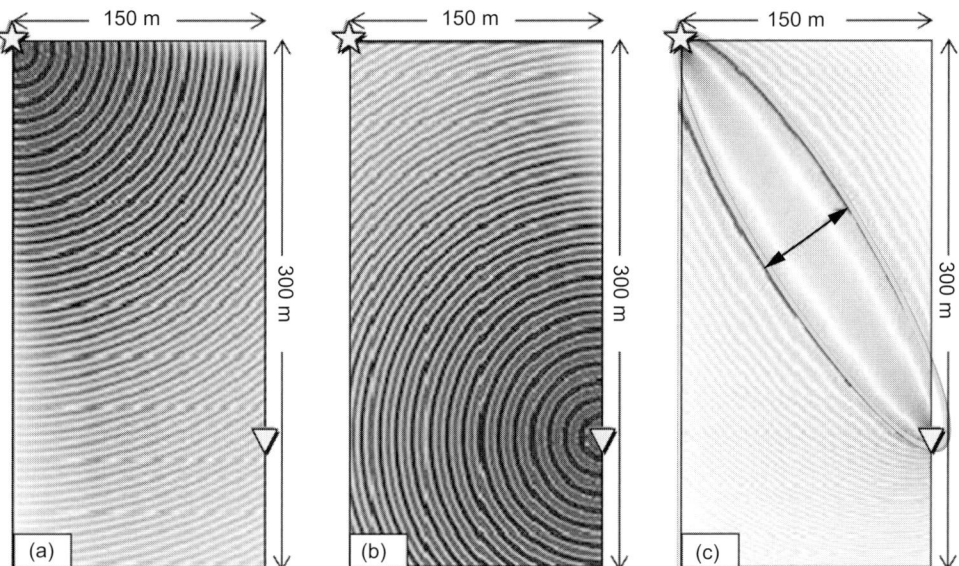

Figure 4.3 (a) Transmitted wave of 300 Hz from a shot (star). (b) Transmitted wave of 300 Hz from a receiver (triangle). (c) Scatter wavefield from the shot to the receiver. Here the first Fresnel zone as depicted by the innermost gray ellipse has a width of 73.8 m (Pratt, 2005).

4.1.2 Horizontal resolution

The **horizontal resolution** quantifies the resolvability of seismic wave perpendicular to the direction of wave propagation. The horizontal resolution is practically the **Fresnel resolution**, defined as the width of the first Fresnel zone due to the interference of spherical waves from the source and from the receiver. Figure 4.3 illustrates the situation with monochromatic waves in a constant velocity field of 3 km/s. Panel (a) shows a 300 Hz wave transmitted from a source denoted by a star, and panel (b) shows another 300 Hz wave from a receiver denoted by a triangle; here we treat the receiver as another source. The scatter wavefield from the source to the receiver is a superposition of the two transmitted wavefields, as shown in panel (c). This scattered wavefield consists of a number of elliptical wavefronts, which are Fresnel zones due to interference between the wavefields from the source and receiver. The width of the first Fresnel zone as shown in panel (c) is

$$W = \sqrt{2d\lambda + \lambda^2/4} \tag{4–1}$$

where λ is the wavelength, and d is the distance between the source and receiver. In this case, the wavelength is 10 m, and the source-to-receiver distance is 271.2 m. This leads to 73.8 m for the width of the Fresnel zone.

We can generalize the definition of the horizontal resolution for reflection case, as shown in Figure 4.4, showing spherical waves from a point source reflected from a plane interface back to a receiver at the source position. The first Fresnel zone is the portion of the interface

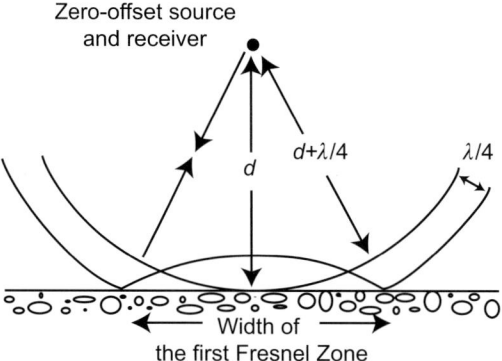

Figure 4.4 Spherical waves from a point source reflected from a plane interface back to a receiver at the source position. The first Fresnel zone is a portion of the interface which reflects energy back to the receiver within half a cycle of the primary reflection. Owing to the two-way reflection time, half a cycle behind the primary wavefront is another wavefront one-quarter of a wavelength ($\lambda/4$) behind the primary wavefront.

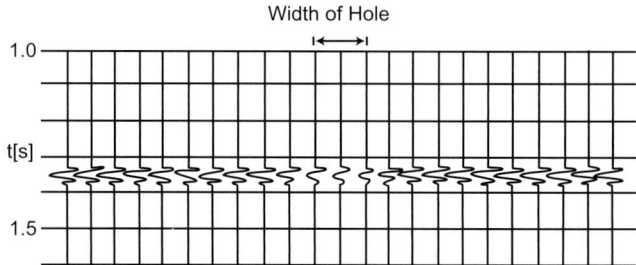

Figure 4.5 Reflection from a flat reflector containing a hole. Reflection is also observed at the hole because the hole width is smaller than the Fresnel zone.

that reflects energy back to the receiver within one-quarter of a wavelength behind the primary wavefront. One can easily derive equation (4–1) using the geometry shown in Figure 4.4. In the presence of velocity variation, the geometry of the Fresnel zone will be distorted accordingly. In practice, it is difficult to detect horizontal features that are smaller than the Fresnel zone, as shown by a synthetic example in Figure 4.5.

4.1.3   The wedge model and tuning

Following the definition of resolution as the ability to separate two features that are close together, a systematic study on the resolution of thin bed was pioneered by Widess (1973, 1982) using **wedge models**. Such studies are of critical importance because most petroleum-bearing reservoir beds are thinner than the wavelength of most reflection data. As shown in

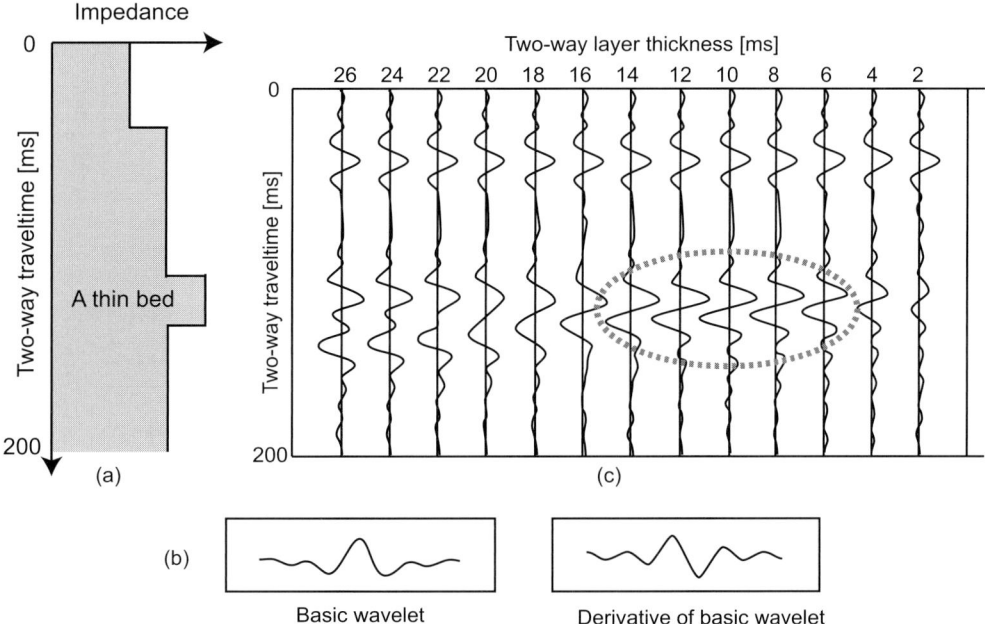

Figure 4.6 (a) An elastic impedance profile with a thin bed. (b) A basic wavelet and its time derivative. (c) A 2D wedge model of reflectivity spikes based on profile (a) with the thin bed thickness decreasing to zero at the right end, and synthetic seismograms of the wedge model using the basic wavelet in (b). Dashed ellipse shows a zone of tuning due to constructive interference between the top and bottom reflections of the thin bed. (After Kallweit & Wood, 1982.)

Figure 4.6, when the thickness of a thin bed is around one-quarter wavelength ($\lambda/4$) of the basic wavelet, the reflections from the top and bottom edges of the thin bed will interfere constructively, producing an increase in amplitude called **tuning**. Consequently, a **thin bed** is defined as having a thickness less than one-quarter of the main wavelength of the data, or one-quarter of the basic wavelet if it is known. Widess (1973) observed that as the bed thickness decreases beyond the **tuning thickness**, the composite wavelet approaches the derivative of the basic wavelet (Figure 4.6b). The characteristic response of thin beds is one of the "evergreen" topics in exploration seismology (e.g., Knapp, 1990; Okaya, 1995).

Figure 4.7 shows a modeling study of the wedge model by Partyka *et al.* (1999). The reflectivity of a low-velocity thin bed model is in reverse polarity from that in Figure 4.6. The wedge here thickens from 0 ms on the left to 50 ms on the right. After an 8–10–40–50 Hz **Ormsby filtering** (a trapezoidal-shaped filter specified by four corner frequencies), the band-limited reflectivity is shown in Figure 4.7b. On the spectral amplitudes plot in Figure 4.7c, the black/white stripes of spectral peaks and troughs shift towards lower frequencies from left to right, corresponding to the lowering of frequencies with the thickening of the wedge.

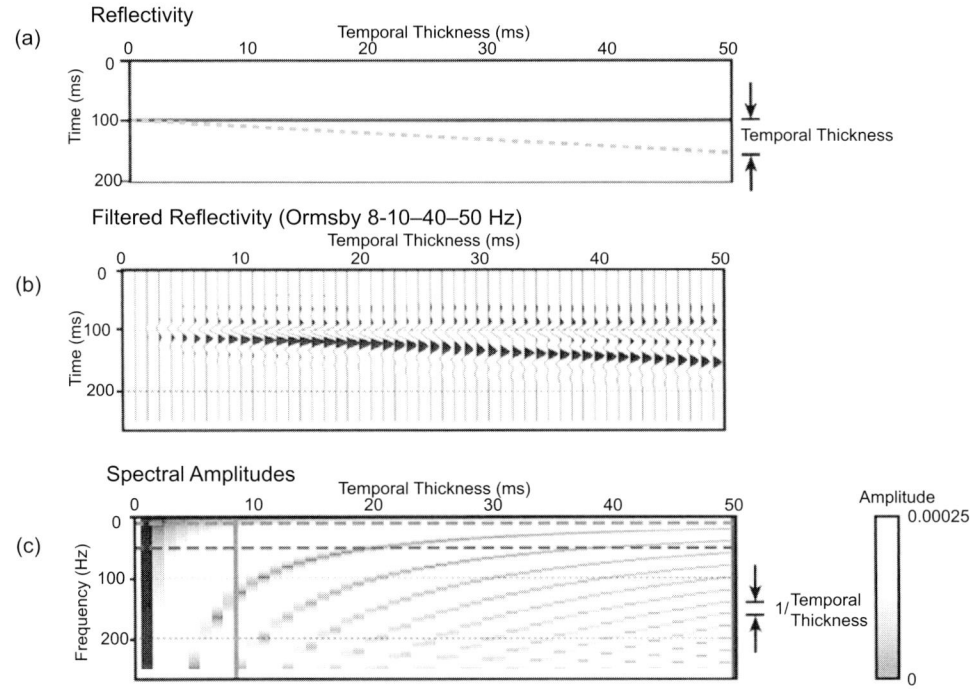

Figure 4.7 For a blocky wedge model: (a) reflectivity; (b) filtered reflectivity; (c) spectral amplitudes. (After Partyka *et al.*, 1999.)

Exercise 4.1

1. Following Figure 4.6, make synthetic seismograms for a wedge model containing a low-velocity thin bed of decreasing thickness. Please demonstrate: (1) tuning; and (2) that the combined wavefield approaches the derivative of the input wavelet below the tuning thickness.

2. Based on Figure 4.7, derive a relationship between the thickness of the thin layer in time domain and the spacing of the notches in the amplitude spectrum.

3. Make a table to compile the ranges of vertical resolution of various seismic imaging applications such as near-surface geophysics, petroleum exploration in onshore and offshore cases, crustal seismology, and mantle tomography.

4.2   Resolution versus bandwidth

4.2.1   Frequency bandwidth

A direct consequence of the uncertainty principle that we discussed in Section 3.3 is the proportionality between the temporal resolution and the **frequency bandwidth**. In

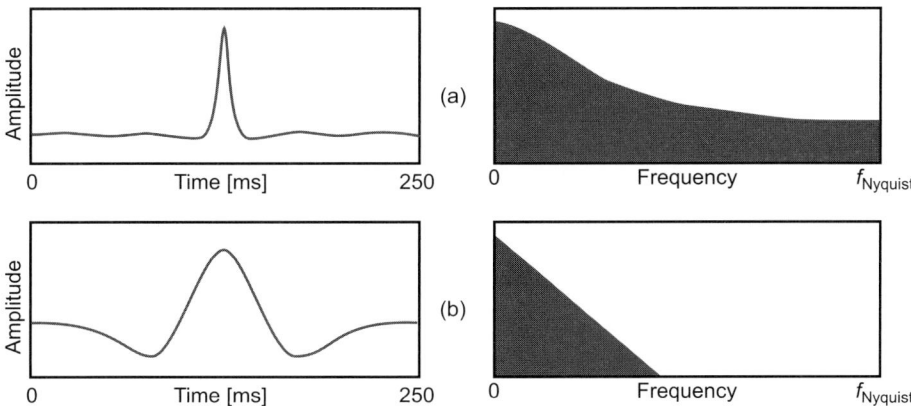

Figure 4.8 Time domain and amplitude spectrum of: (a) a Sinc wavelet; (b) a truncated Ricker wavelet. The vertical range of the spectra is from 0 to –100 dB.

other words, a higher temporal resolution requires a broader bandwidth. We have learned previously that the bandwidth in frequency is quantified by **octave**, and the number of octaves for an amplitude spectral band with a low-corner frequency f_{LC} and a high-corner frequency f_{HC} is quantified by

$$\text{Number of octaves} = \log_2 \frac{f_{HC}}{f_{LC}} \tag{4–2}$$

In the following we will see several examples of the relationship between temporal resolution and spectral bandwidth, examine some practical limitations on the bandwidth, and check the relationship between resolution and phase.

You may wonder which wavelet offers the broadest bandwidth. A theoretically correct answer is a spike in the time domain. In practice, seismic data are digital and their entire time duration is usually less than several thousands of sample points. A practical answer to the above question is the Sinc wavelet defined by

$$\text{Sinc}(t) = \frac{\sin(ct)}{ct} \tag{4–3}$$

where c is a constant, which equals π for a normalized Sinc function. Figure 4.8 shows a Sinc function and its amplitude spectrum, in comparison with that of a truncated Ricker wavelet. The amplitude spectrum of the Sinc function has high amplitude from zero frequency all the way to the Nyquist frequency $f_{Nyquist}$. The truncation of the side lobes of the Ricker wavelet has boosted its low frequencies, but its amplitude decreases to zero at a frequency around 43% of the Nyquist frequency.

4.2.2   Resolution versus bandwidth

In practice, we can apply trapezoidal filtering of the Sinc wavelet to produce an Ormsby wavelet with our desired bandwidth. Figure 4.9 shows the convolution of a reflectivity sequence with four Ormsby wavelets. The high corner frequencies of these wavelets are the same, 62 Hz as the high-pass and 70 Hz as the high-cut frequencies. Their lower pass

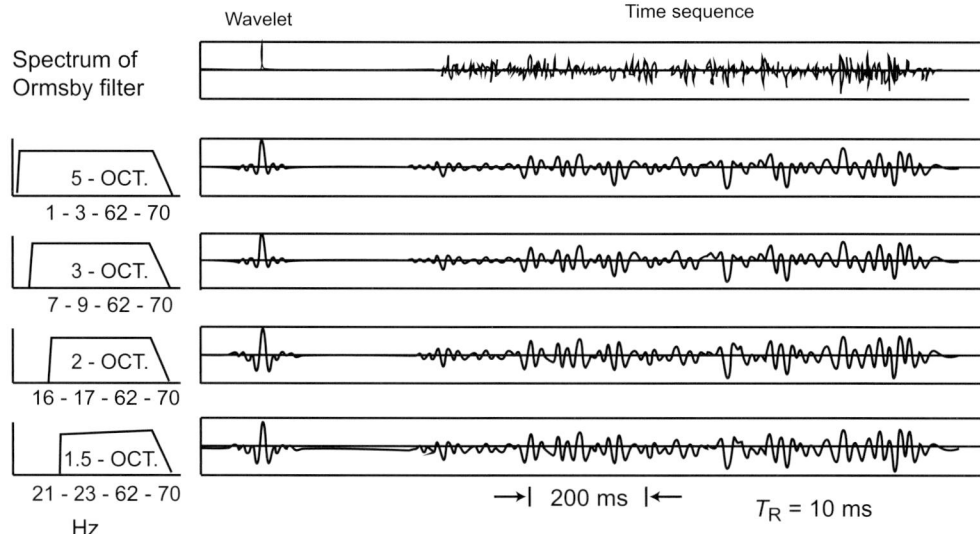

Figure 4.9 A suite of synthetic seismograms produced by convolving a reflectivity sequence derived from a well log with a Ormsby wavelet. The upper terminal frequency is kept constant while the lower terminal frequency varies. The interval between inflection points of the main lobe, T_R, is 10 ms (Kallweit & Wood, 1982).

frequencies are 3, 9, 17, and 23 Hz, respectively, giving bandwidths of approximately 5, 3, 2, and 1.5 octaves, respectively. As the bandwidth decreases, the wavelet becomes "leggier" or gains more side lobes of higher amplitudes, and the resolution decreases in the time domain.

One of the key seismic data processing tasks in exploration seismology today is inversion of migrated seismic data into **acoustic impedance**, which is the product of P-wave velocity with density. The resultant impedance sections or volumes are valuable assets for interpreters because such data carry information about the velocity and density with much higher accuracy than the seismic data. However, owing to the limited bandwidth of seismic data, the inverted acoustic impedance is also band-limited. As the bandwidth decreases (Figure 4.10), the inverted acoustic impedance departs from the measured well acoustic impedance. Because the real world has multiple rock beds, the combination of the side lobes due to narrow bandwidth produces severe artifacts. This example underscores the importance of maintaining low frequencies in seismic data processing.

The bandwidth of seismic data in the real world is constrained by the bandwidth of the source signal, the effect of Earth filtering, and the bandwidth of the receiver. Among these three factors, the Earth filtering effect is the most challenging because it is a physical fact that the loss may not be recoverable. Most young sedimentary rocks, such as fluvial and deltaic sands and shale sequences, have very low quality (Q) factor; hence they will attenuate much of the high frequencies of seismic waves. A historic rule of thumb in seismic data processing is that the SNR of seismic data propagated through such soft rocks will be too low after 100 wavelengths of propagation. By this rule, if the average velocity is

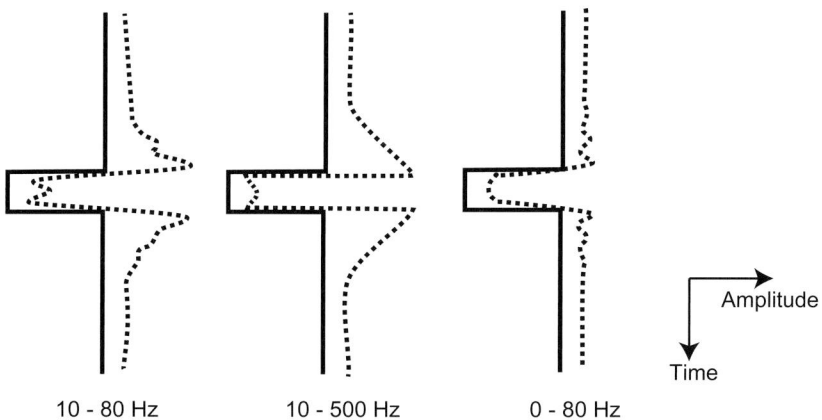

Figure 4.10 Impact of low and high frequencies on vertical resolution, as shown by comparing well acoustic impedance (solid curves) with inverted acoustic impedance (dotted curves) with different bandwidths (Latimer *et al.*, 2000).

4 km/s, the highest useful frequency of reflections from a reflector at 4 km below the sources and receivers is 50 Hz, and the corresponding wavelength is as large as 4 km/50 = 80 m.

In practice, it is more feasible to increase our ability to record low-frequency components of seismic data. Most seismic sources such as dynamite, Vibroseis and earthquakes contain very low frequencies that will not be attenuated much by the Earth filtering effect. Modern broadband seismometers used in solid Earth geophysics can easily record seismic frequencies as low as 0.01 Hz (a period of 100 s). A practical challenge is the high cost for low-frequency seismometers. However, unlike the Earth filtering of high frequencies which is mostly unrecoverable, the recording of low frequencies is achievable and hence the only feasible way to improve the bandwidth of seismic data. Many examples, such as Figure 4.10, have shown the value of retaining the low frequencies.

4.2.3 Resolution versus wavelet phase

Another factor affecting seismic resolution is the phase of the seismic wavelet. Most seismic data processing flows require the use of the minimum-phase wavelet because it has minimum time duration in the presence of noise and it is **causal**, or a one-sided function. In contrast, a zero-phase wavelet that is symmetric with a vertical axis at time zero cannot be created from a real source. Historically, people have thought that a minimum-phase wavelet would give the best resolution based on the argument of its minimum time duration. This view was challenged by Schoenberger (1974) who conducted a series of synthetic modeling experiments comparing a minimum-phase wavelet and a zero-phase wavelet that have nearly identical amplitude spectra (Figure 4.11).

As shown in Figure 4.12, Schoenberger demonstrated that the zero-phase wavelet resolved several pairs of spiky reflectors much better than the minimum-phase wavelet. The zero-phase wavelet also allows interpreters to easily tie the peaks of seismic wiggles with the corresponding reflecting interfaces. Although a zero-phase wavelet is not

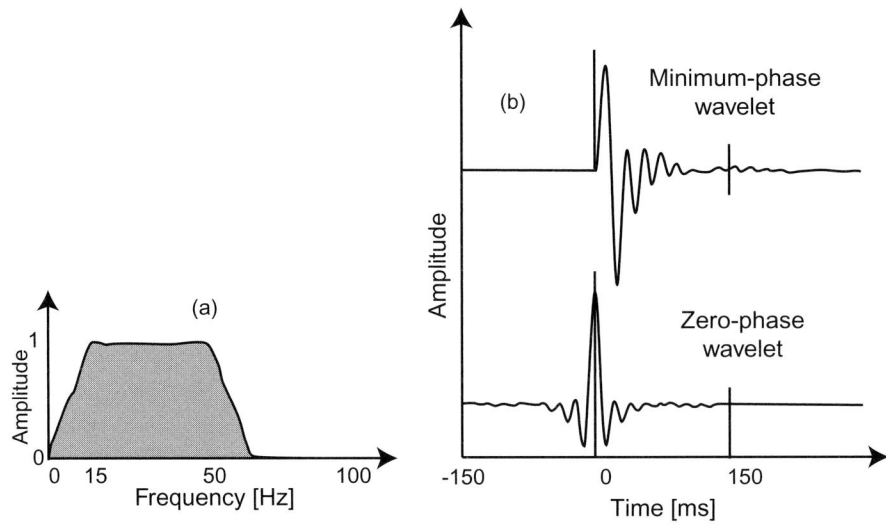

Figure 4.11 Comparison between minimum-phase and zero-phase wavelets in: (a) amplitude spectra; and (b) time domain (Schoenberger, 1974).

Box 4.2 Hypothetical bandwidth changes in processing

Box 4.2 Figure 1 shows three hypothetical cases of losing bandwidth due to processing. We may regard the light gray areas as the amplitude spectra of an input signal, and the dark gray areas as the spectra of outputs from three different processing works. In case (a), the processing reduces the magnitude of the amplitude; hence there is no loss in bandwidth, although the SNR may decrease. In case (b) there is a loss in the high-frequency components. In case (c) the entire band is shifted toward high frequency, meaning a loss in the low frequencies and gain in high frequencies. However, in terms of reduction in bandwidth, case (c) is much worse than case (b) if the loss in the low frequencies and gain in the high frequencies are of the same number of hertz.

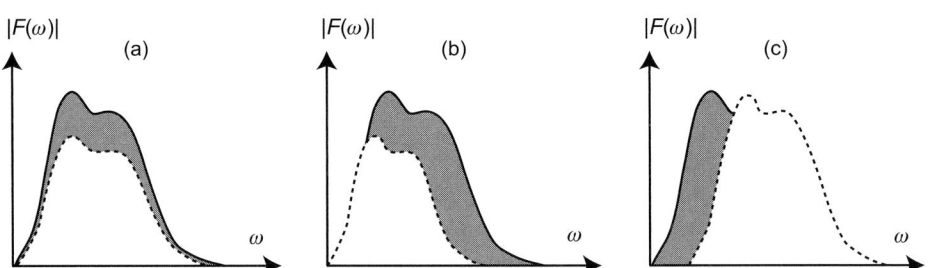

Box 4.2 Figure 1 Three hypothetical cases of amplitude spectra for the same input (gray areas) and outputs (white areas outlined by dashed curve) of three different processing procedures.

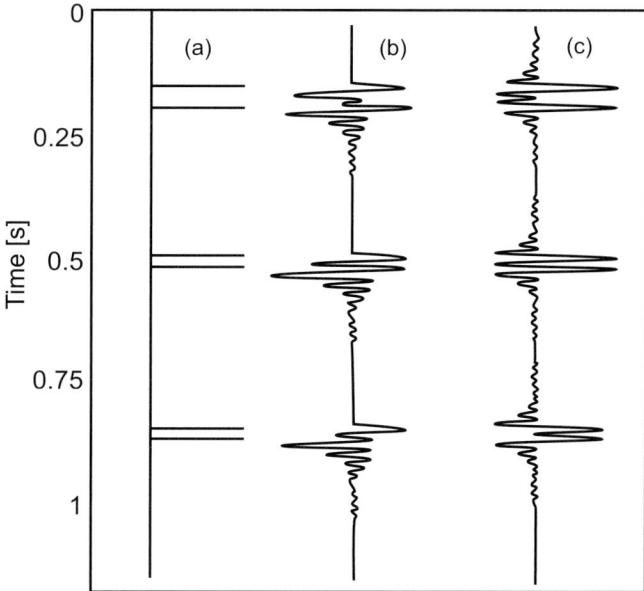

Figure 4.12 Comparing vertical resolution of minimum-phase and zero-phase wavelets of nearly the same amplitude spectra, like that shown in Figure 4.11. (a) A reflectivity function with three pairs of double spikes. Convolving the reflectivity function with: (b) a minimum-phase wavelet and (c) a zero-phase wavelet (Schoenberger, 1974).

physically possible near a source, the far-field seismic response due to an impulse source does resemble a zero-phase wavelet, as shown by Ricker (1953a).

Exercise 4.2

1. Elaborate the reasons for preserving low-frequency components in seismic data acquisition and data processing.
2. Read the following paper and write a short summary: Schoenberger, 1974.
3. Each geophone has its marked frequency. What is the position of the marked frequency on the geophone's amplitude spectrum? Is it the central frequency of the geophone? (Hint: search literature to find the answer.)

4.3   Resolution versus fidelity

4.3.1   Fidelity

Fidelity means truthfulness. Thus, seismic fidelity means the truthfulness of the processed result, the targeted signal. For a targeted signal as the objective of seismic data processing, fidelity measures how well this signal has been extracted from the processing. In seismic imaging, on the other hand, fidelity specifies how accurately an event has been resolved at the correct location on the imaging result. Seismic fidelity consists of two measures: **seismic resolution** specifying how much detail there is in the processing results, and **seismic positioning** specifying the temporal or spatial position errors. These two measures often overlap in seismic data processing.

Most tasks in seismic data processing are categorized into two classes, **time processing** and **seismic imaging** (or seismic mapping). Historically, much of the effort in time processing has aimed at improving seismic resolution, and much of the effort in seismic mapping has aimed at improving both seismic positioning and seismic resolution. Examples of time processing include:

- Demultiplexing and editing
- Various data sorting and binning
- Gain corrections
- Phase rotations
- Temporal filtering
- Deconvolution
- Temporal inversions

Examples of seismic mapping include:

- QC and correction for position errors of sources and receivers
- Stacking and velocity estimates
- Time migration
- Depth migration
- Tomography
- Seismic modeling
- Seismic waveform inversion

4.3.2   Resolution versus fidelity

What happens if we have poor seismic resolution but good positioning? In this case, we have an unclear but well-positioned picture about our target. An example is shown in Box 4.1 Figure 1 near the beginning of this chapter which compares two seismic sections, one with conventional resolution and the other with high resolution. In that example, even if we only have the conventional resolution result and if the position error is acceptable, we can still use it to achieve many scientific and/or business objectives.

On the other hand, what happens if we have good seismic resolution but poor positioning? In this case, a well-resolved seismic imagery with expected targets at wrong positions is one of the worst artifacts from seismic data processing, because it is likely to mislead interpreters. An example is given in Figure 4.13, comparing results from two different pre-stack depth migrations, one using an isotropic velocity model and the other an anisotropic model. The horizon marked by the dotted curve moves by almost 450 ms between the two results. If the actual velocity field has anisotropy but this is ignored in pre-stack depth migration, the position error will be comparable to what is shown here. Clearly, an important

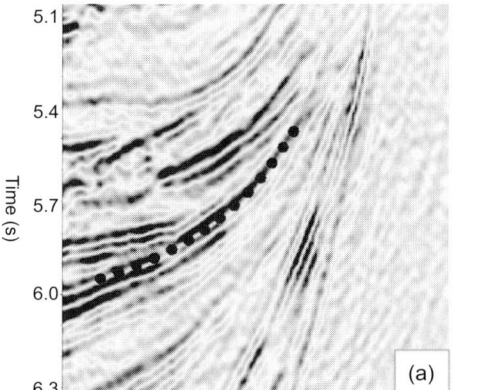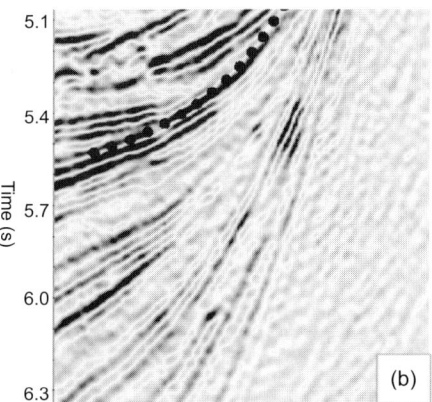

Figure 4.13 Comparison between results of two pre-stack depth migrations using (a) isotropic and (b) anisotropic velocity models (Zhou *et al.*, 2004).

issue of seismic fidelity is to better understand various errors and artifacts from different types of seismic data processing methods.

As will be discussed later (Section 4.4.1), in most applications of seismic imaging the resolution is proportional to the spatial coverage of the given data. Many imaging artifacts are caused directly by poor data coverage. In contrast, fidelity requires sufficient quality in the spatial coverage as well as the accuracy of the data. We will not be able to achieve good fidelity if the given dataset only has good coverage over the targeted area but poor SNR.

4.3.3   Assessing resolution and fidelity

Several techniques exist to quantify seismic resolution for most seismic data processing methods, as will be discussed in the next section. However, it is not so straightforward to assess the level of mis-positioning because in the real world the truth is usually unknown. Several checks may help remediate the situation:

- Comparing the processed results with known geology to evaluate their geologic plausibility: do the results make sense geologically?
- Comparing independently derived results based on independent datasets;
- Conducting re-sampling tests, by dividing the original data into subsets, running the subsets through the processing flow, and then checking the consistency between the solutions;
- Quantifying the potential impact of error-bearing factors using synthetic modeling tests with realistic value ranges of the factors.

As an example for the first two checks, Figure 4.14 shows the result of a crustal tomography study using first arrivals from earthquake data and a regional seismic survey using active shots and receivers. The velocities in color are produced from a deformable layer tomography method which determines the best data-fit geometry of some constant-velocity layers (Zhou, 2004b; 2006). Following the first check, we can verify whether the basins are underlain by basin-shaped slow velocities and whether the mountains correspond to fast

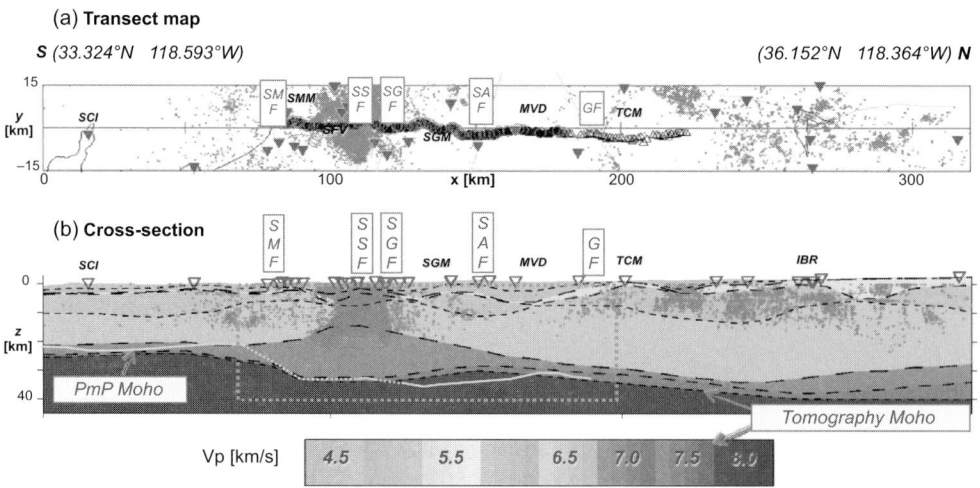

Figure 4.14 (a) Map of the profile in southern California showing earthquake foci (small crosses), seismic stations (triangles), faults (lines), shots (circles), and receivers (triangles). (b) Cross-section of tomographic velocity model. The tomography Moho at the top of the layer with 8 km/s velocity is compared with the PmP Moho in light color. Locations: Santa Catalina Island (SCI), Santa Monica Mountains (SMM), San Fernando Valley (SFV), San Gabriel Mountains (SGM), Mojave Desert (MVD), Techchapi Mountains (TCM) and Isabella Reservoir (IBR). Faults in boxes: Santa Monica (SMF), Santa Susana (SSF), San Gabriel (SGF), San Andreas (SAF), and Garlock (GF). For color versions see plate section.

velocity anomalies in the upper crust. Following the second check, the Moho discontinuity, which marks the base boundary of the crust, is interpreted here as the interface between the two layers with velocities of 7.5 and 8.0 km/s. This tomographic Moho is compared with a PmP Moho that is interpreted based on the **PmP waves**, or reflections from the Moho (Fuis *et al.*, 2007). The two Moho interpretations agree well in their general trend, showing a thinning of the crust toward the ocean on the south side. The depths of the PmP Moho vary from one side to the other of the tomographic Moho. The difference between the depths of the two Moho interpretations is about several kilometers. Considering that the main frequencies of both the first-arrival waves and PmP reflections are less than 5 Hz around the Moho depth with a velocity of nearly 8 km/s, their wavelengths must be more than 1.5 kilometers.

Figure 4.15 compares three seismic sections for the same area using different methods and datasets. Panel (a) shows a **reflection stack** together with a tomographic velocity model shown in color. The stacking is among the simplest methods to produce an image using seismic reflection data. However, it suffers from various sources of noise such as multiple reflections and off-line reflections and scatterings which are difficult to remove. Panel (b) shows the result of a **pre-stack depth migration** using the same seismic reflection data as used to create panel (a). Although this method suffers from the same noise sources as the reflection stack, the **imaging condition** necessary for all pre-stack depth migration methods has a noise removal effect. In this case, the imaging condition requires that the amplitudes of events in the image are proportional to the level of cross-correlation between the reflection data and modeled wavefield. Hence the image quality of panel (b) is higher than that in (a),

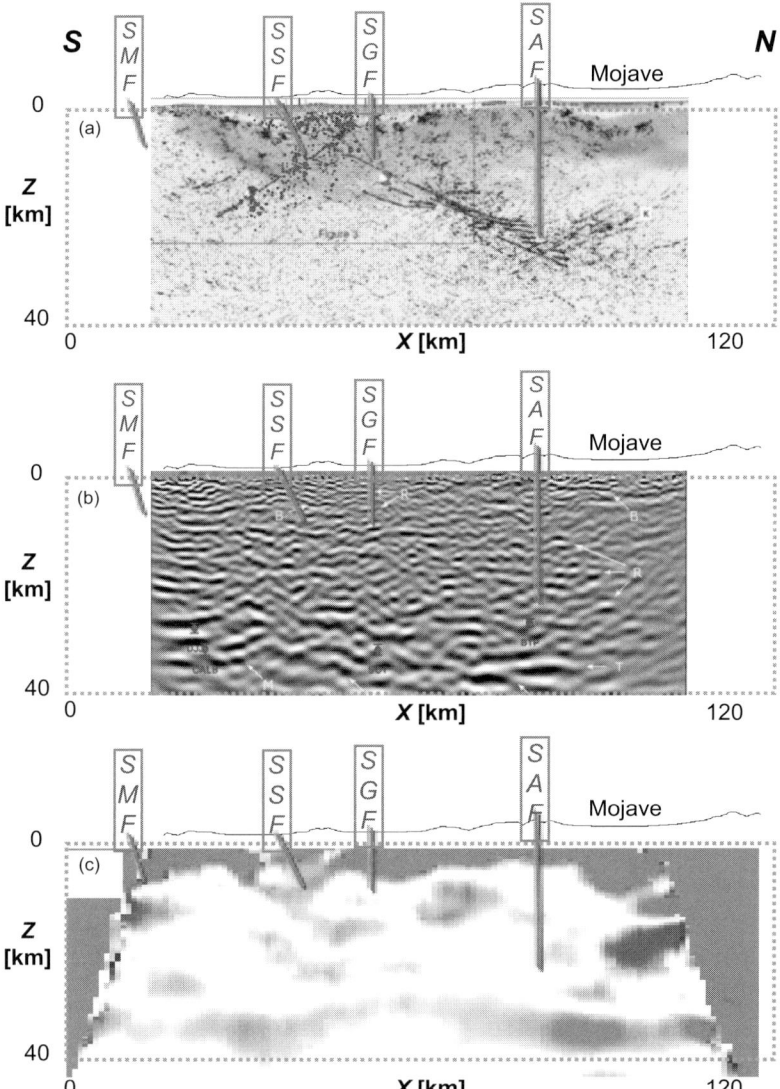

Figure 4.15 Comparison of three seismic cross-sections enclosed by the dashed box that is also shown in Figure 4.14. (a) Reflection stack (Fuis *et al.*, 2003). (b) Pre-stack depth migration (Thornton & Zhou, 2008). (c) Receiver functions (Zhu, 2002). Faults denoted: Santa Monica (SMF), Santa Susana (SSF), San Gabriel (SGF), and San Andreas (SAF). For color versions see plate section.

although both used the same input data. Panel (c) shows an image of **receiver functions** using teleseismic earthquake data recorded by mobile broadband seismometers. This is a stack of many receiver functions, and each was produced by deconvolving converted waves from the vertical component of records of teleseismic earthquakes. Although the wavelength and hence the resolution of receiver functions are lower than the surface reflection images, the trends of events are comparable amongst them.

A commonality in the three imaging methods shown in Figure 4.15 is their dependency on velocity model, which is a requirement of all reflection imaging methods. Error in the velocity model is a major source of mis-positioning for these methods. As a result, an important topic in exploration geophysics is how to improve the velocity model building and how to cope with its associated uncertainties.

Exercise 4.3

1. Explore the relationship between resolution and fidelity. Which one is easier to assess in the practice of seismic data processing? In practice, do the objectives of improving resolution and improving fidelity conflict with or complement each other?

2. A major objective of the work shown in Figures 4.14 and 4.15 is to probe the nature of the Moho discontinuity. Draw your own interpretation of the Moho based on these figures. What can you say about resolution and fidelity of your Moho interpretation? How sharp is the Moho discontinuity in this area?

3. In Figure 4.15, what kinds of signal can you identify from each imaging result? In your interpretation of each result, what may be missing and what may be misinterpreted? Is there anything extra to be gained from looking at all results together?

4.4 Practical assessments of resolution and fidelity

4.4.1 Factors affecting resolution and fidelity

We have shown in the previous section that seismic fidelity includes aspects of resolution and positioning. In real cases, errors exist in both seismic resolution and positioning. It is therefore necessary to assess the level of error in both aspects in order to quantify the fidelity of the product of a seismic data processing project. Factors to be analyzed include:

- The highest resolution level of the data, based on analysis of frequency bandwidth and SNR of datasets from the original input all the way to the final result;
- Reduction of the resolution by all processing methods, including that during time processing and seismic imaging;
- Possible causes of mis-positioning due to errors in the positions of sources and receivers, insufficient data quality, and insufficient data coverage;
- Possible causes of mis-positioning due to factors ignored by the processing, such as the presence of anisotropy and error in the velocity model;
- Artifacts from the processing.

Assessment of the resolution limit for time processing relies on analysing many of the issues discussed in the first two sections of this chapter, namely vertical resolution, bandwidth, phase, and the influence of noise. For most seismic imaging processes, the

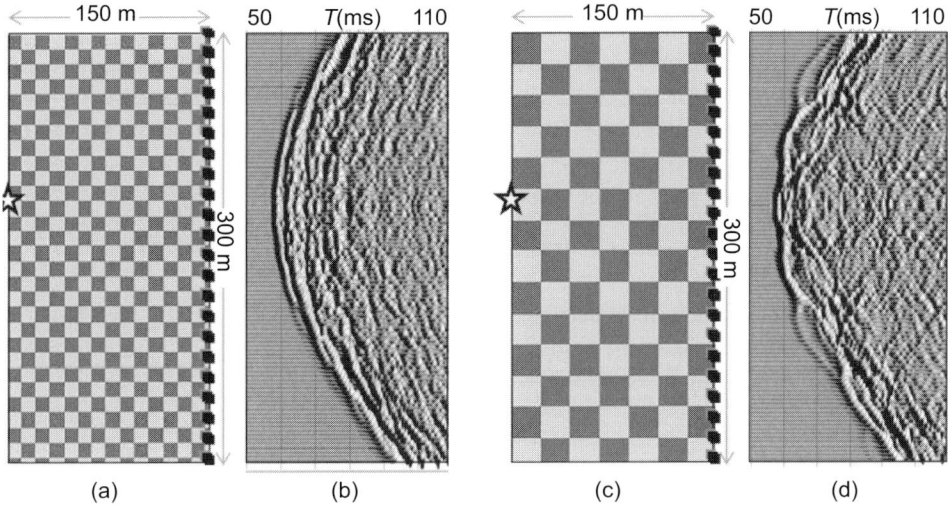

Figure 4.16 Waveform modeling in a cross-well setup (from R. G. Pratt, unpublished data). A shot (star) is placed along the left edge, and the geophones (boxes) are placed along the right edge. (a) A checkerboard model with cell size 10.5 × 10.5 m and velocities at 2.3 and 3.0 km/s. (b) Shot gather from the setup of (a). (c) Another checkerboard model with cell size 21 × 21 m and velocities at 2.3 and 3.0 km/s. (d) Shot gather from the setup of (c).

resolution depends on the extent of **seismic illumination**, or **data coverage**, over the targeted area, while the positioning of the result depends on correctness of the velocity model, processing methodology, data coverage, and SNR. For a given location, seismic illumination or data coverage refers to the number and angular range of seismic waves traversing through the location. In high-frequency seismic studies using traveltime data rather than waveform data, seismic illumination is simplified into **raypath coverage**, which is the number and angular variation of seismic raypaths traversing through the given location.

The extent of seismic illumination depends on the distribution of sources and receivers, the frequency bandwidth of data, and velocity variations. The dependency on velocity variations is due to the fact that seismic waves tend to bend away from areas of lower velocities. To appreciate this dependency, let us see a pair of seismic modeling results in Figure 4.16, from Pratt (2005). There are two shot gathers in two models, with a single shot on the left side and 101 geophones on the right side. The two velocity models have fast and slow anomalies arranged in a checkerboard pattern with two different cell sizes, 10.5 × 10.5 m and 21 × 21 m, respectively. The bandwidth of the wavelet is very broad, at 50–500 Hz. Most parts of the waveforms consist of scatters from the edges of the model cells. The input to waveform tomography is the first portion of records like that shown in panels (c) and (d). Notice that the effect of "**wavefront healing**" (e.g., Wielandt, 1987), which is a gradual disappearance of disturbances as the wave propagates, is more obvious in the case of smaller cell size in panels (a) and (b). Owing to the effect of wavefront healing, a model with more detail, as in panel (a), may give smoother seismic data (panel (b)) than the seismic data (panel (d)) from a model with less detail.

We realize that the difference between the two wavefields in response to the same source signal in Figure 4.16 is due to the relative difference between the frequencies of the signal and sizes of the checkerboard velocity anomalies. If we change the bandwidth of the signal, we expect a change in the pattern of the shot gathers even if the velocity models remain the same. This reflects the dependency of seismic illumination on the frequency bandwidth of the input signal.

In summary, seismic resolution and fidelity are limited by:

- Signal bandwidth in the available data
- Data SNR
- Distribution of sources and receivers with respect to the target location
- Correctness of the velocity model and variation level of the velocity field
- Processing methodology

4.4.2 Restoration tests on resolution

Among methods to assess the fidelity of seismic imaging, a practically useful way is a **restoration test**. In the test a synthetic true model is used to generate data for the given source and receiver locations. These data with added noise are used by the seismic imaging method under test. Finally the resolution is assessed by how well the true model is replicated by the seismic imaging method. Advantages include the ease of conducting the test and the fact that it assesses both the level and the resolution and positioning of the intended targets. A common version of the method for seismic tomography is the **checkerboard resolution test**, in which the synthetic true model consists of velocity anomalies in checkerboard pattern like those shown in the previous figure. In such a test, the velocity anomalies are expressed in terms of **lateral velocity variation**, which is typically quantified by the magnitude of deviation from the layer average velocity (V_{1D}). If the velocity value is V_i at a model location, then the lateral velocity variation is $V_i / V_{1D} - 1$.

Figure 4.17 shows a checkerboard resolution test for a cell tomography method using the source–receiver positions shown in Figure 4.14. The synthetic true velocity model consists of some isolated fast and slow velocity blocks with lateral velocity variations of up to 5%. In addition to examining how well the method replicates the checkerboard anomalies, we also examine how well the method replicates the layer velocity averages in this case. The results of the test indicate that the cell tomography method has successfully replicated the layer average velocities, as demonstrated by the curves and numbers in the two panels shown on the right side of the figure. Most of the velocity anomalies shallower than 20 km in depth are recovered well. The recovery is poor for anomalies deeper than 20 km and many along-raypath smear artifacts are present. Poor ray angle coverage and parallel raypaths cause along-raypath smear. Raypath smear artifacts are especially insidious because of the tendency for researchers to interpret linear anomalies in seismic images as real features in the crust and mantle (e.g., Humphreys *et al.*, 1984; Zhou, 1988). A simple rule of thumb is that all linear tomographic anomalies are questionable if they show geometric pattern similar to that of the raypaths.

Another restoration test is shown in Figure 4.18 for a deformable layer tomography method. The synthetic true model contains variations in velocity interfaces including

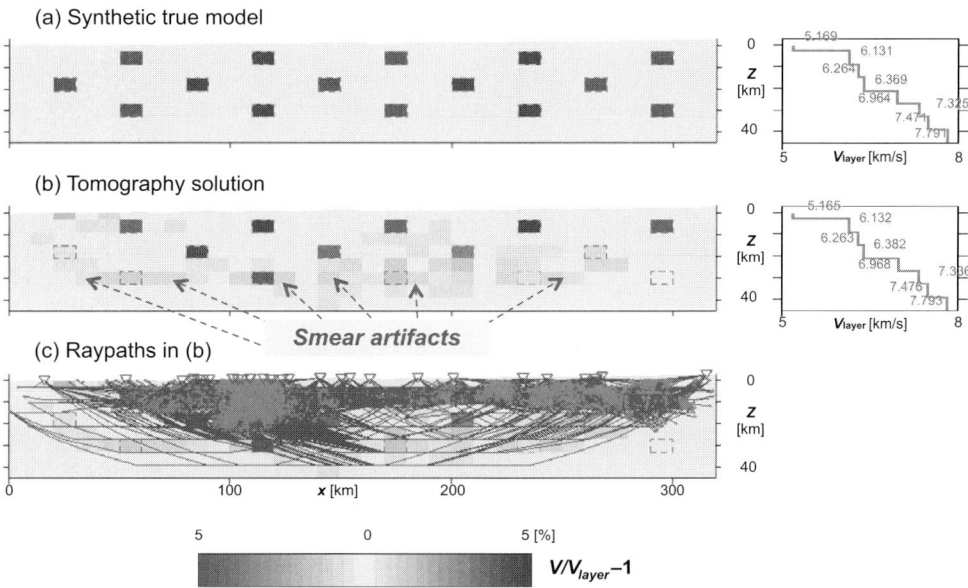

Figure 4.17 Checkerboard resolution test for a traveltime tomography. The layer velocity averages are shown in the right panels and the numbers are in km/s. The left panels are the lateral velocity variations after removal of layer velocity averages. (a) The synthetic true model. (b) Tomography solution. The small dashed boxes outline the correct positions of the checkerboard anomalies. (c) Raypaths for model (b). For color versions see plate section.

pinchouts. Nineteen stations, 906 sources, and 2034 first-arrival rays are taken from real earthquake data to create a synthetic dataset by computing traveltimes in the true model. Zero-mean Gaussian noise with a standard deviation of 0.11 s was added to the synthetic data. The initial reference model in panel (b) has eight layers with layer velocities matching those of the true model, but with flat velocity interfaces whose geometry differs greatly from that of the true model. The final solution as shown in panel (c) has a strong similarity to the true model. From the initial model to the final deformable layer tomography (DLT) solution, the average of the traveltime residuals was reduced from −2.219 s to 0.003 s, and the standard deviation of the traveltime residuals was reduced from 1.417 s to 0.157 s. Note that the standard deviation of the added Gaussian noise is 0.11 s. The correlation with the true model was increased from 72.8% for the initial model in Figure 4.18b to 98.8% for the final solution in Figure 4.18c. The correlation values were calculated using lateral velocity variations, which were obtained by mapping the velocities of each deformable layer model into a fine mesh of regular spatial grids and removing the average velocities of all flat fine layers.

4.4.3   Re-sampling tests on fidelity

Another assessment of seismic fidelity is to verify the level of consistency between multitudes of measurements on the same subject. If only a single but large dataset is available,

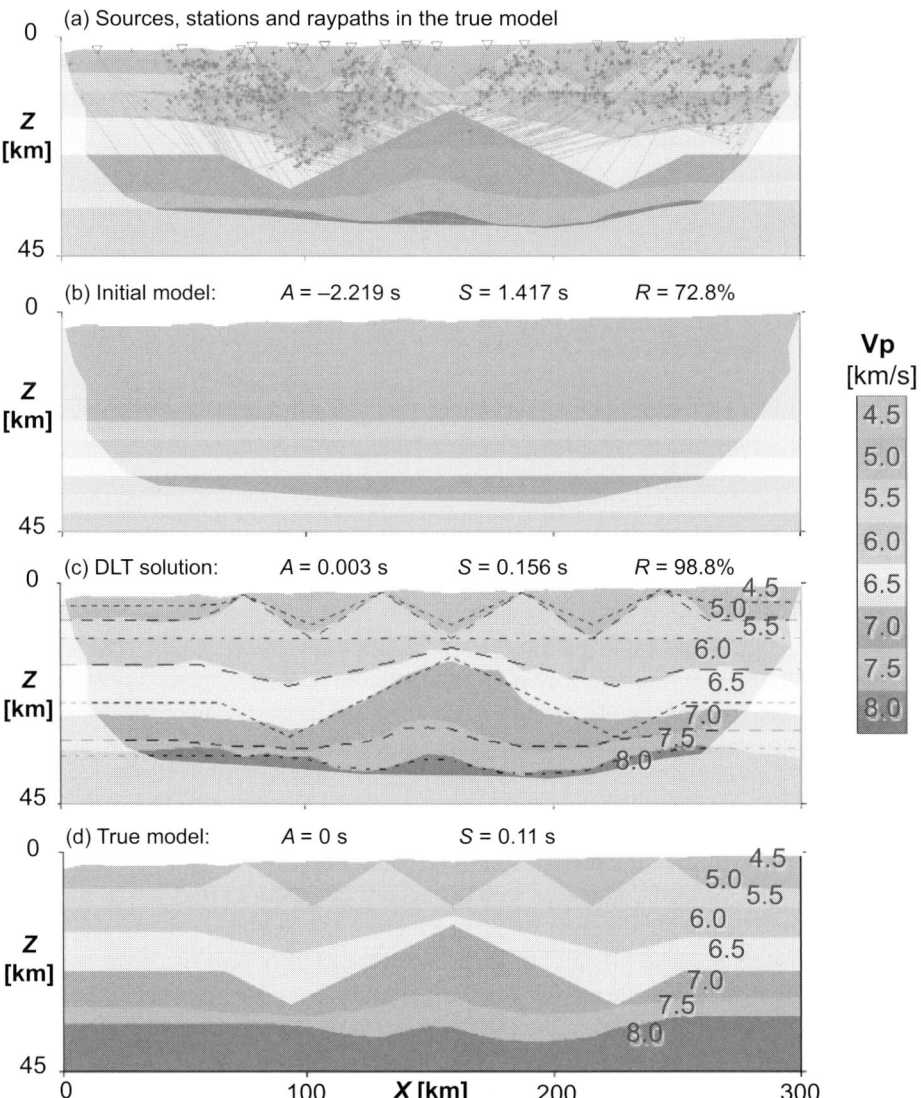

Figure 4.18 Restoration test of a deformable layer tomography. (a) Stations (triangles), sources (crosses), and rays (curves) in the synthetic true model. (b) The initial reference model. (c) Solution velocities in comparison with true model interfaces denoted by dashed curves. (d) True model. In (c) and (d) the numbers denote layer velocities in km/s. A and S are the average and standard deviation of traveltime residuals, and R is the correlation with the true model. Areas without ray coverage are shadowed. For color versions see plate section.

the assessment can be carried through **re-sampling**. Re-sampling is a set of statistical techniques based on the notion that we can repeat the experiment by constructing multiple datasets from a single large dataset. These techniques are easy to use and offer great promise for estimating the best model and model variance in linear and non-linear problems. The main reason for considering such re-sampling techniques is that many geophysical data are not reproducible. For example, it is not possible to directly verify any assumptions or estimations made with respect to the probability distribution of the data. The difficulty is further increased when the physical relation between data and model is non-linear in many applications. Re-sampling techniques are insensitive to assumptions made with respect to the statistical properties of the data and do not need an analytical expression for the connection between model variance and data variance.

Two re-sampling techniques are introduced here: **jackknifing** and **bootstrapping** (Tichelaar & Ruff, 1989). The re-sampling is applied to digital data and model, say $\mathbf{d} = (d_1, \ldots, d_n)^{\mathrm{T}}$ and $\mathbf{m} = (m_1, \ldots, m_{\mathrm{p}})^{\mathrm{T}}$, and the ith datum is expressed as

$$d_i = f_i(\mathbf{m}) + \varepsilon_i \tag{4-4}$$

where f_i is the kernel function and ε_i represents the noise. Using the least squares, the model estimate $\mathbf{m}^{\mathrm{est}}$ is the one that minimizes $(\mathbf{d} - \mathbf{f}(\mathbf{m}))^{\mathrm{T}} (\mathbf{d} - \mathbf{f}(\mathbf{m}))$.

Because any statistic estimate is robust only when it is drawn from a large number of data, the key requirement is that the original data can be re-sampled to form a number of subsets of data, and the subsequent multiple estimates of the model give information on model variance. In other words, each subset of data is a "copy" of the original data in terms of the statistics that we are trying to assess. The re-sampled dataset $\mathbf{d}^*$, with a new length k, can be expressed using a matrix $\mathbf{D}$ called the re-sampling operator:

$$\mathbf{d}^* = \mathbf{D}\mathbf{d} \tag{4-5}$$

where $\mathbf{D}$ has n columns and k rows. Each re-sample now defines a new model estimate $\mathbf{m}^{*\mathrm{est}}$ that minimizes $(\mathbf{d}^* - \mathbf{f}^*(\mathbf{m}))^{\mathrm{T}} (\mathbf{d}^* - \mathbf{f}^*(\mathbf{m}))$, where $\mathbf{f}^* = \mathbf{D}\mathbf{f}$. The difference between jackknifing and bootstrapping is in the choice and dimension of the re-sampling operator.

A **jackknifing re-sampling** is extracted from the original data by deleting a fixed number, say j, of the n original data points ($k = n - j$). It is therefore called "**delete-j**" **jackknifing**. Each row of $\mathbf{D}$ here contains only one element of value 1 and the rest are of value 0; thus an original datum is never copied into a re-sample more than once. The total number of possible jackknifing re-sample data (and hence the number of model estimates) is C_n^K. For example, the total number of re-sample data for "delete-1" jackknife is $C_n^1 = n$. See Box 4.3 for an example of delete-half jackknifing.

A statistic of interest for the original dataset can then be obtained from the re-sampled data. The statistic could be the mean, medium, variance, etc. The estimating procedure is illustrated in the following with the standard deviation as an example, as this is of wide interest in geophysics. For a mean $\hat{x}$ defined as

$$\hat{x} = \frac{1}{n} \sum_{i=1}^{n} x_i \tag{4-6}$$

the corresponding standard deviation is usually given by

$$\hat{\sigma} = \left(\frac{1}{n-1} \sum_{i=1}^{n} |x_i - \hat{x}|^2 \right)^{1/2} \tag{4-7}$$

Since there are many re-sampled datasets, one may use some type of averaging scheme to represent the new dataset. Suppose we consider a new dataset as the sample average of the dataset deleting the ith datum ("delete-1" jackknifing):

$$\hat{x}_i^* = \frac{n\hat{x} - x_i}{n - 1} = \frac{1}{n - 1} \sum_{\substack{j \neq i}}^{n} x_j \qquad (4\text{--}8)$$

The average of all n jackknife averages, each defined by (4–8), is

$$\bar{x} = \frac{1}{n} \sum_{i=1}^{n} \hat{x}_i^* \qquad (4\text{--}9)$$

which equals the average of the original full dataset. Furthermore, the "delete-1" jackknife estimator of the corresponding standard deviation is given by

$$\hat{\sigma}_J = \left(\frac{n - 1}{n} \sum_{i=1}^{n} \left| \hat{x}_i^* - \bar{x} \right|^2 \right)^{1/2} \qquad (4\text{--}10)$$

which can be shown to be equivalent to the usual expression (4–7). The advantage of using the above expression is that it can be generalized to an estimator of the standard deviation for any statistical parameter θ that can be estimated from the data by replacing $\hat{x}_i^*$ with $\hat{\theta}_i^*$ and $\bar{x}$ with $\bar{\theta}$. $\hat{\theta}_i^*$ is an estimator of θ, calculated for the dataset with the ith datum deleted.

A **bootstrap re-sampling** is a random selection of a set of n data out of n original data. In contrast with the jackknife, the re-sampling operator is a square matrix and each column may contain more than a single "1", which means that a re-sample may contain a certain original datum more than once. Just like the jackknife, the bootstrap estimator of standard deviation $\hat{\sigma}_B$ can be calculated without knowing an analytical expression that relates the statistic of interest to the data. Suppose that $\hat{\theta}_i^*$ is an estimator of the statistic θ of interest (say the average, as used above, for example), calculated for the bootstrap re-sample i. To do a Monte Carlo approximation of $\hat{\sigma}_B$, a large number L of bootstrap estimators $\hat{\theta}_i^*$ need to be calculated. The bootstrap estimate of the standard deviation of θ is

$$\hat{\sigma}_B = \left(\frac{1}{L - 1} \sum_{i=1}^{L} \left| \hat{\theta}_i^* - \bar{\theta} \right|^2 \right)^{1/2} \qquad (4\text{--}11)$$

where

$$\bar{\theta} = \frac{1}{L} \sum_{i=1}^{L} \hat{\theta}_i^* \qquad (4\text{--}12)$$

As was the case for jackknifing, there are different bootstrap estimators of standard deviation.

Box 4.3 An example of delete-half jackknifing

Box 4.3 Figure 1 shows horizontal slices from two P-wave velocity models in southern California. Panel (a) shows the 300 km by 480 km study area with major faults and coastlines. Panels (b) and (c) are horizontal slices of two 3D P-wave velocity models that were derived from two different halves of the first arrival data covering the periods from 1981 to 1987, and from 1988 to 1994, respectively. The two models are generated in order to examine the consistency of the data. The patterns of the two models are mostly similar. The overall correlation between the slowness perturbations reaches 75%, which is significantly high because each of the correlating vectors has 18 720 unknowns and layer averages have been removed. Such a high correlation between solutions using non-overlapping data suggests that the prominent features in these models come from consistent signals in the data.

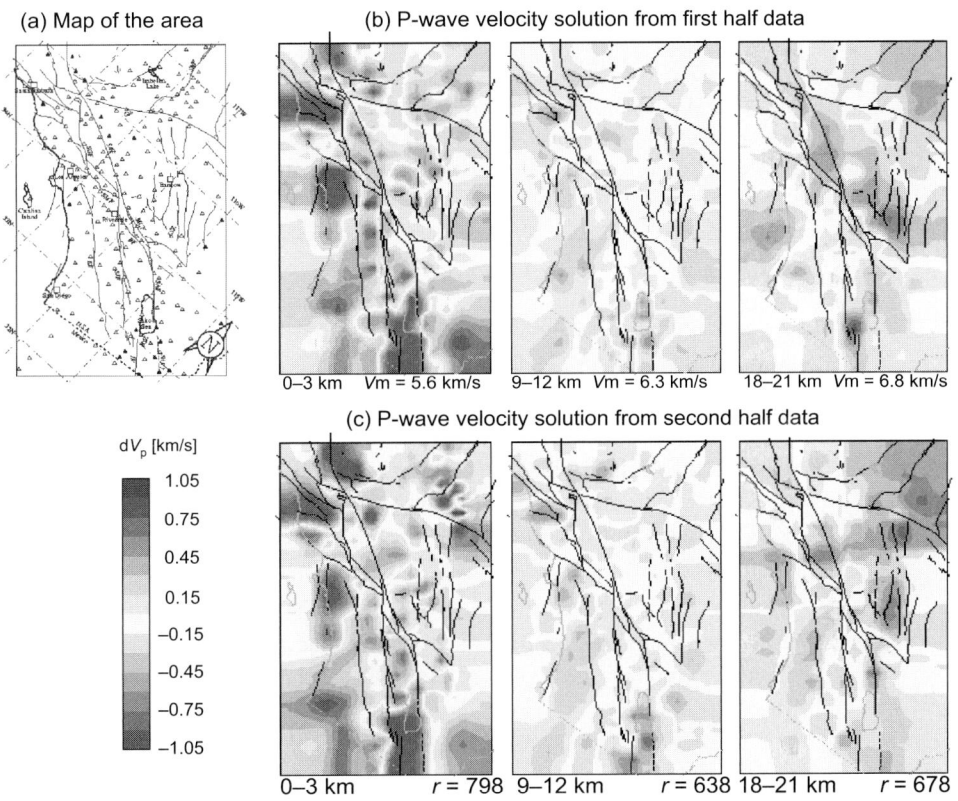

Box 4.3 Figure 1 A delete-half jackknife test for crustal P-wave tomography in southern California (Zhou, 2004a). (a) Map view of major faults and seismologic stations (triangles). (b) Horizontal slices of the velocity model from the first half of the data. (c) Horizontal slices of the velocity model from the second half of the data. The dataset consists of more than 1 million P-wave arrivals from local earthquakes to local stations. V_m is the average velocity of the layer, and r is the correlation coefficient between two model slices in the same depth range. For color versions see plate section.

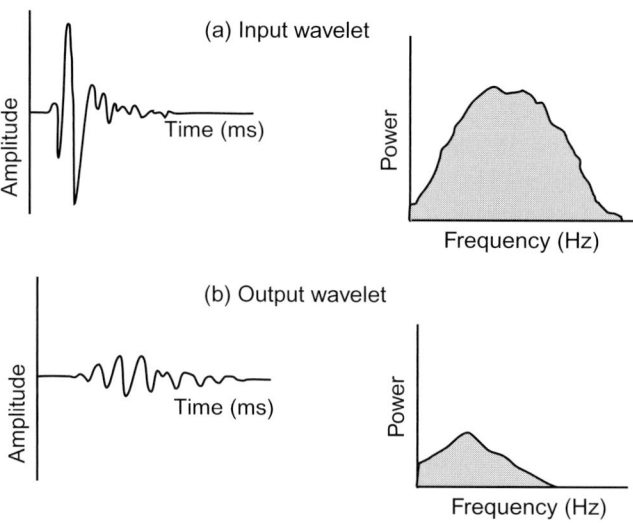

Figure 4.19 A sketch of attenuation of a seismic wavelet in fractured rock, in time and frequency domains (Young & Hill, 1986).

Exercise 4.4

1. For a seismic wave from a point source, its vertical resolution will be finer than its horizontal resolution. Does this mean that the resolution of reflection is finer than that of the direct wave of the same wavelet? How should we define resolution for refraction? What about a turning wave?

2. Discuss the limitations of the checkerboard resolution test shown in Figure 4.17b. Will the test yield good results if the given dataset has good ray coverage but poor SNR? For such a dataset will the layer velocity averages in the right panels be resolved well or not?

3. The two models shown in Figure 4.19 used different datasets to study the same area. If the two model results differ at a certain place, are these results wrong at that place? If the two models show similar anomalies at the same place, are these anomalies real?

4.5   Correction for seismic attenuation

The seismic attenuation property of real materials sets a physical limit for seismic wave propagation and therefore seismic resolution and fidelity. In this section, the quality factor Q is introduced along with one of the common ways to estimate the Q value of media using spectral ratio between two seismic measurements. Two examples of Q estimation and correction are discussed.

4.5.1   The quality factor

One important factor reducing temporal resolution of seismic data is **seismic attenuation**, which causes a decay of seismic energy as a function of traveling distance. There are propagation effects, such as geometric spreading for a point source and energy partitioning across an interface of acoustic impedance, that attenuate the amplitude of seismic waves. There is also an **intrinsic attenuation** of seismic waves due to **anelasticity** of the medium. A plane wave traveling in a medium still experiences decay from its amplitude at time zero, $A_0(f)$, to its decayed amplitude at time t:

$$A_t(f) = A_0(f) \exp\left(\frac{-\pi f t}{Q}\right) \qquad (4-13)$$

where f is frequency and Q is the **quality factor**. In analogy to releasing a basketball to bounce on the ground and measuring the height the ball reaches during subsequent bounces, seismic attenuation is measured in terms of the ratio of amplitude decay through each period or each wavelength.

The Q value of media is usually assumed to be independent of frequency. The high-frequency components of a seismic wave decay faster than its low-frequency components, because the former have more cycles than the latter over a fixed propagating distance. Consequently, a seismic wavelet undergoes **pulse broadening** as it travels, as shown in Figure 4.20. The pulse broadening in the time domain is associated with a bandwidth narrowing in the spectral domain. For a causal time series, such as a real signal, we can prove that seismic attenuation will co-exist with **seismic dispersion**. Seismic dispersion is simply the variation of seismic velocity of different frequency components. The normal case is that the low frequencies or long wavelengths will travel at faster speed in a realistic medium, forming a **normal dispersion** trend. The reverse case is called **inverse dispersion**, or reverse dispersion. The common occurrence of the normal dispersion is due to the fact that seismic velocity typically increases with depth, because of the increase of pressure and greater compaction and solidification of the rocks with depth. The long-wavelength components will sense a greater depth than the short-wavelength components of a wave.

Correction for the attenuation using estimates of **effective Q** is an important topic in seismic data analysis. Attenuation has been suggested as an attribute for quantifying pore fluid and saturation (e.g., Winkler & Nur, 1982; Best et al., 1994; Gurevich et al., 1997). It also provides information on lithology and structure (Young & Hill, 1986; Peacock et al., 1994) and improve the quality of migrated images (Deal et al., 2002). Quantification and compensation for attenuation is a necessary step in AVO analysis (Estill & Wrolstad, 1993). Although estimates of the Q value can be drawn from surface seismic data, wellbore and VSP seismic data offer higher quality Q estimates. In the following, the estimation of effective Q by the spectral ratio method and an analysis of its accuracy (White, 1992) are reviewed.

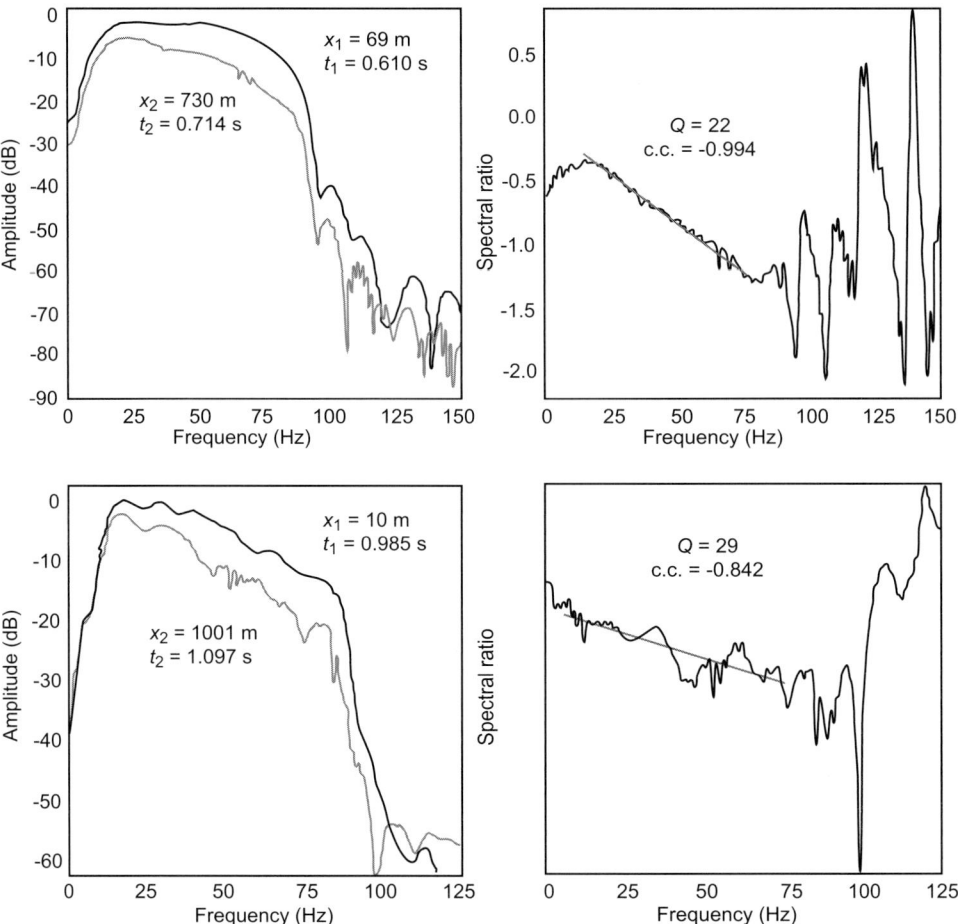

Figure 4.20 Estimation of effective Q using walkaway VSP data (Guerra & Leaney, 2006). (Upper) An effective $Q = 22$ is estimated for the offset range 60–730 m. The amplitude spectra of records at 60 m and 730 m offsets are shown on the left, and their spectral ratio is shown on the right. (Lower) An effective $Q = 29$ is estimated for the offset range 10–1001 m. The amplitude spectra of records at 10 m and 101 m offsets are shown on the left, and their spectral ratio is shown on the right.

4.5.2   Estimating Q using spectral ratio

Using power spectra of surface reflection data in two time intervals, t_1 and t_2, Q can be estimated from

$$\ln\left[P_2(f)/P_1(f)\right] = 2\ln\left[A_2(f)/A_1(f)\right] = 2\pi f(t_2 - t_1)/Q \qquad (4\text{--}14)$$

The power spectra can be estimated in various ways, such as by using multiple coherence analysis (White, 1973). The method separates the signal and noise spectra on the basis of the short-range, trace-to-trace coherence of their spectral components. The estimated spectra $\hat{A}_1(f)$ and $\hat{A}_2(f)$ are independent assuring they come from separate time gates.

White (1992) showed that the accuracy of the estimated spectral ratio can be quantified by its variance

$$\text{var}\{\ln[\hat{A}_2(f)/\hat{A}_1(f)]\} = \frac{1}{2(f_H - f_L)T} \tag{4-15}$$

where f_H and f_L are the high- and low-corner frequencies, and T is the duration of the data segment.

With two pieces of wellbore seismic data, Q can be estimated from the amplitude spectra $\hat{A}_{12}(f)$ and $\hat{A}_{21}(f)$ of two matching filters, matching the deeper to the shallower and shallower to deeper recording. Both filters are biased estimates of the absorption response because they contain a noise suppression filter. Random error and noise are reduced in the ratio

$$|\hat{A}_{12}(f)|/|\hat{A}_{21}(f)| = P_2(f)/P_1(f) \tag{4-16}$$

which is precisely the spectral ratio of the trace segments, leading to the Q estimate using (4–14). This ratio is unbiased if the SNR of the two recording segments is the same. In this case, White (1992) showed that the accuracy of the estimated spectral ratio is quantified by

$$\text{var}\{\ln[\hat{A}_2(f)/\hat{A}_1(f)]\} = \frac{1 - \gamma^2}{2(f_H - f_L)T} \tag{4-17}$$

where $\gamma^2(f)$ is the spectral coherence between the two recording segments.

4.5.3   Examples of Q estimation and correction

Figure 4.20 shows two effective Q estimates from an application of the spectral ratio method to walkaway VSP data (e.g., Guerra & Leaney, 2006). The correlation coefficients (c.c.) in the right panels quantify the fit of the data with the straight lines, which are predictions from the modeled Q values within their respective frequency ranges.

Figure 4.21 compares two inline seismic sections without and with a correction for effective Q estimated from the surface reflection seismic data (Deal et al., 2002). The dimming or low-amplitude zone in the central part of the section on the left panel is due to shallow gas in the first 400 ms. After the Q correction as shown in the right panel the amplitudes of reflections are restored beneath the gas zone. Although only a 1D Q model was used in this study, the benefit of the correction is clear. At this moment, however, accurate estimation of Q model is still a research topic in most cases.

Exercise 4.5

1. Search the literature to compile the Q values of common rocks into a table, including the physical conditions of the Q measurements or estimates.

2. Discuss factors that may cause error in the estimations of Q using the spectral ratio method.

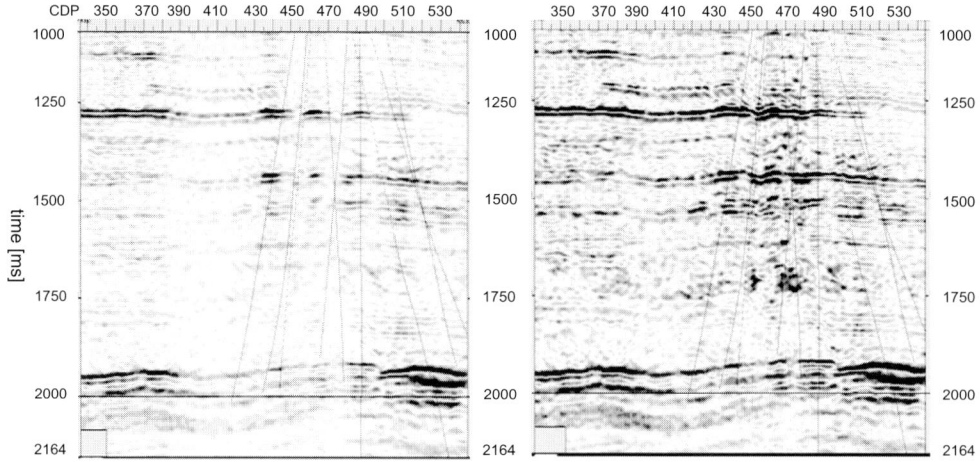

Figure 4.21 Inline sections through a migrated seismic volume without (left) and with (right) a correction for estimated effective Q based on surface reflection data. Vertical axis is in milliseconds. The main reservoir is immediately above 2 seconds. The dimming in the central part of the section is due to shallow gas in the first 400 ms. The nearly vertical lines indicate existing production wells in the region. (From Deal *et al.*, 2002.)

3. A common notion states that Q is independent of frequency. In this case the spectral ratio will be linear, like the straight lines shown in the right panels of Figure 4.21. Discuss the validity of the above notion in light of the observed spectral ratio in Figure 4.21.

4.6 Summary

- Seismic resolution quantifies the finest scale of subsurface objects detectable by the seismic data. Vertical resolution measures the resolvability of seismic waves along the direction of wave propagation, which is typically in near-vertical directions. Horizontal resolution measures the resolvability of the seismic wave perpendicular to the direction of wave propagation, which is typically in sub-horizontal directions.

- Seismic fidelity is another quality measure which quantifies the truthfulness, such as the accuracy of seismic data or the correctness of the imaged target position.

- Because seismic data are band-limited, seismic resolution is proportional to the frequency bandwidth of seismic data. If the bandwidth is too narrow, the resolution will be poor because a single subsurface reflector may correspond to a number of indistinguishable wiggles on the seismic traces.

- For multiple datasets with the same bandwidth, it is easier in practice to recognize events with zero-phase wavelets than wavelets with non-zero phase, such as the minimum-phase wavelet preferred by many data processing operations.

- It is important to assess the level of seismic resolution and fidelity in every seismic processing project. Seismic resolution and fidelity are limited in real cases by the seismic illumination. Poor resolution and fidelity may produce various types of seismic artifacts.

- Seismic resolution is proportional to seismic illumination provided by the given data coverage. In practice we may conduct several types of restoration test to assess the resolution level of given data coverage. However, the real resolution may be hampered by many factors such as inaccuracy in data coverage due to unknowns in the model.

- Seismic fidelity depends on both data coverage and signal to noise ratio. It is usually much more difficult to assess fidelity than resolution. One way to assess fidelity is to check seismic data against direct measurements that may be available. Another way is to check the level of consistency between multitudes of measurements on the same subject through re-sampling. We may also infer the fidelity level by comparing results from different studies using different types of data and methods.

- Seismic attenuation, the phenomenon of losing energy as seismic wave travels in real media, sets physical limits on seismic resolution and fidelity. Correction for the quality factor Q is important for many seismic studies, although much of this topic is still in the research stages today.

FURTHER READING

Schoenberger, M., 1974, Resolution comparison of minimum-phase and zero-phase signals, *Geophysics*, 39, 826–833.

Sheriff, R.E. and L.P. Geldart, 1995, *Exploration Seismology*, 2nd edn, Section 6.4, Cambridge University Press.

Tichelaar, B.W. and L.J. Ruff, 1989, How good are our best models? Jackknifing, bootstrapping, and earthquake depth, *Eos*, May 16, 593–606.

Widess, M.B., 1982, Quantifying resolving power of seismic systems, *Geophysics*, 47, 1160–1173.

5 | Digital filters

Chapter contents

Digital filtering is a very commonly used seismic data processing technique, and it has many forms for different applications. This chapter begins by describing three ways to express digital filtering: the rational form, recursive formula, and block diagram. The names of the filters usually come from their effects on the frequency spectrum. In the rational form of a filter, the zeros are the roots of the numerator, and the poles are the roots of the denominator. Using the zeros and poles we can make the pole–zero representation on the complex *z*-plane as a convenient way to quantify the effect of a digital filter as a function of frequency. The rule of thumb is: poles add, zeros remove, and the magnitude of the effect of the pole or zero depends on their distance from the unit circle. Different types of filtering in seismic data processing are discussed in the chapter using several examples. In particular, *f–k* filtering is discussed in detail with its typical processing flow. Owing to the widespread application of inverse problem in geophysics, much of the attention is given to inverse filtering, which requires that the corresponding filter be invertible. It can be proven that a minimum-phase filter is always invertible because all of its zeros and poles are outside the unit circle on the complex *z*-plane. This notion means that the minimum-phase filters occupy an important position in seismic data processing. In general, a minimum-phase wavelet is preferred in seismic data processing because of stability concerns, while a zero-phase wavelet is preferred in seismic interpretation to maximize the seismic resolution. The final section of the chapter prepares the reader with the physical and mathematical background materials for inverse filtering. These materials are fundamental to the understanding of deconvolution, an application of inverse filtering, in the next chapter.

5.1   Filtering of digital data

5.1.1   Functionality of a digital filter

A digital filter is represented by a sequence of numbers called weighting coefficients, which can be expressed as a time series or denoted by the z-transform. When the filter acts on an input digital signal which can be expressed as another time series, the filter functions as a convolution with the input signal (Figure 5.1).

For example, we have a filter $h(z)$ with the weighting coefficients

$$h(z) = \left(\frac{1}{16}, \frac{1}{4}, 1, \frac{1}{4}, \frac{1}{16} \right) \tag{5–1}$$

This filter can be graphed using a **block diagram** as shown in Figure 5.2.

Convolving any filter $f(z)$ with a unit impulse time series, which is zero everywhere except for having value 1 at one location, produces the same filter $f(z)$. Thus, we can input a unit impulse to any unknown digital filter to produce the **impulse response** as the output of the filter. Clearly, such an impulse response is just the sequence of weighting coefficients of the filter. In general, the impulse response will serve as a good characterization of any filter or any digital system that can be characterized as a combination of digital filters.

Figure 5.1 A digital filter acts like a "black box" convolving with the input time series.

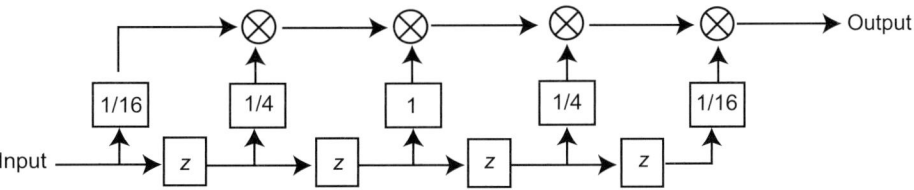

Figure 5.2 A block diagram for the filter (1/16, 1/4, 1, 1/4, 1/16). The ⊗ symbols in the top row are summing operators, the numbers on the middle row are for scaling, and the z values in the bottom row are unit-delay operators.

Unless specified, we usually consider causal digital filters in the context of this book. A **causal digital filter** could have the following three functionalities:

- Storage (delay)
- Multiplication by a scale constant
- Addition

Rational form of a digital filter

Mathematically, a digital filter $H(z)$ can be expressed in a rational form as

$$H(z) = \frac{a_0 + a_1 z + a_2 z^2 + \cdots}{b_0 + b_1 z + b_2 z^2 + \cdots} \qquad (5\text{--}2)$$

If the denominator is a constant, i.e. $b_i = 0$ for all $i > 0$, then the filter is called **non-recursive**. A non-recursive filter has two properties: it has a finite weighting coefficient sequence, and is generally stable. For example, $h(z)$ in (5–1) is a non-recursive filter.

On the other hand, recursive filters have a non-constant denominator that causes **feedback loops**. Such filters may become unstable although they generally involve few storage units. For example, we have a recursive filter:

$$h(z) = (1 + az)/(1 + bz) \qquad (5\text{--}3)$$

If the input is $x(z)$, then the output is

$$y(z) = x(z)(1 + az)/(1 + bz)$$

Let

$$w(z) = x(z)/(1 + bz)$$

i.e.

$$x(z) = w(z)(1 + bz)$$

or

$$w(z) = x(z) - w(z)bz$$

This last expression is the feedback loop. The filter output is

$$y(z) = w(z)(1 + az)$$

Recursive formula for filters

The general case of using a recursive filter can be denoted as

$$x(z) \Rightarrow \boxed{N(z)/D(z)} \Rightarrow y(z)$$

For stability, $D(z)$ must be **minimum phase**. There are two ways to perform the filtering; the hard way is to divide $D(z)$ into $N(z)$ and come up with an infinitely long filter, and the easy way is to do recursion. For instance, if

$$y(z) = \frac{n_0 + n_1 z + n_2 z^2}{1 + d_1 z + d_2 z^2} x(z) \tag{5-4}$$

we can express $x(z) = \sum_t x_t z^t$, $y(z) = \sum_t y_t z^t$, then multiply by $D(z)$ on both sides. Taking all the coefficients of the term z^t, on both sides of the equation, or at time t, we have

$$y_t + d_1 \, y_{t-1} + d_2 \, y_{t-2} = n_0 x_t + n_1 \, x_{t-1} + n_2 \, x_{t-2}$$

Note that in the above equation the sum of the subscripts for each combined term is always t. Hence

$$y_t = n_0 x_t + n_1 \, x_{t-1} + n_2 \, x_{t-2} - d_1 \, y_{t-1} - d_2 \, y_{t-2} \tag{5-5}$$

The above equation (5–5) is a **recursive formula** of a digital filter. The last two terms on the right-hand side of the equation represent **recursive feedback**.

5.1.4   Examples of filter block diagrams

Example 1

What is the block diagram for $y(z) = x(z)\frac{1+az}{1+bz}$?

(i) First, for $y(z) = w(z)(1 + az)$, the block diagram is shown in Figure 5.3.

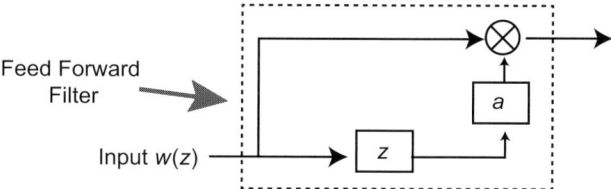

Figure 5.3 Block diagram of $y(z) = w(z)(1 + az)$.

(ii) Next, for $w(z) = x(z) - w(z) bz$, the block diagram is shown in Figure 5.4.

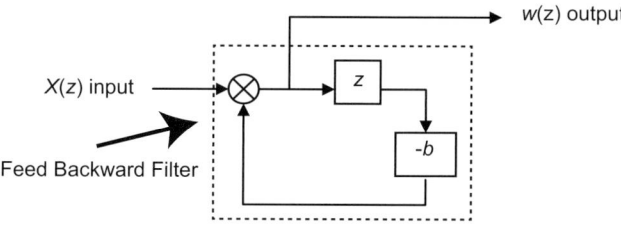

Figure 5.4 Block diagram of $w(z) = x(z) - w(z) bz$.

Note the negative sign of coefficient b in the feedback loop. Now combine (i) and (ii) together as in Figure 5.5.

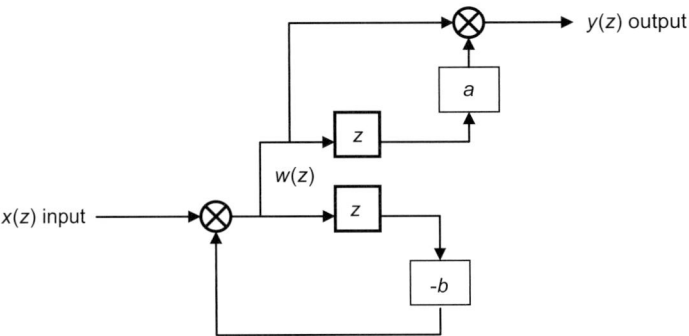

Figure 5.5 Block diagram of $y(z)$.

This can also be shown as in Figure 5.6.

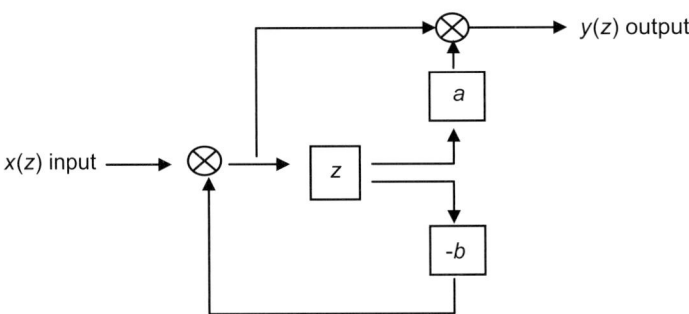

Figure 5.6 Block diagram of Example 1.

Example 2

For $H(z) = \frac{a_0 + a_1 z + a_2 z^2 + a_3 z^3}{1 + b_1 z + b_2 z^2}$, the block diagram is shown in Figure 5.7.

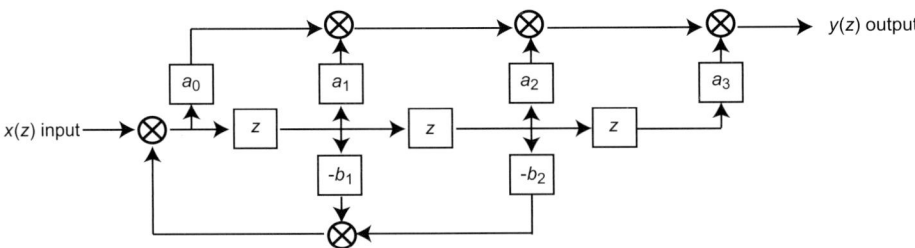

Figure 5.7 Block diagram of Example 2.

Example 3

For $H(z) = \frac{5 + z - 3z^2}{2 + z - 6z^2 + z^3} = \frac{(5/2) + (1/2)z - (3/2)z^2}{1 + (1/2)z - 3z^2 + (1/2)z^3}$, the block diagram is shown in Figure 5.8.

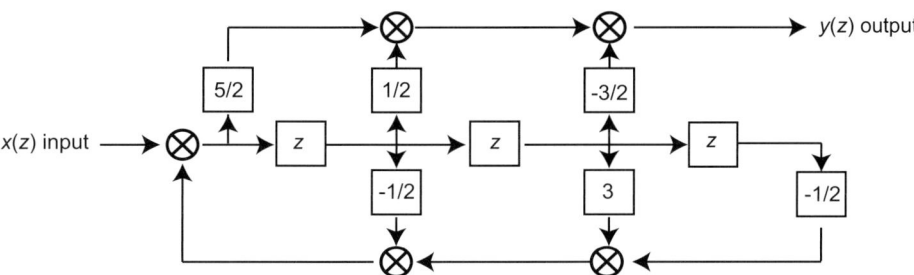

Figure 5.8 Block diagram of Example 3.

Example 4

For $H(z) = \frac{1+z^4}{1-0.656z^3}$, the block diagram is shown in Figure 5.9.

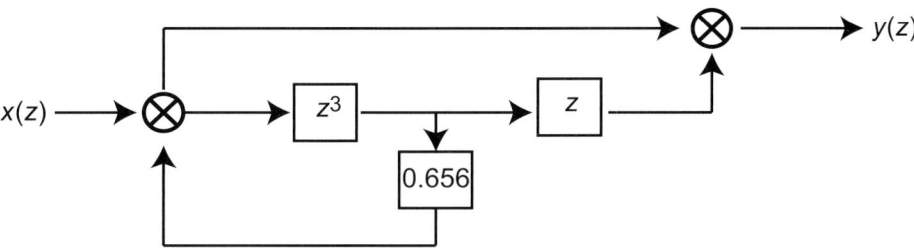

Figure 5.9 Block diagram of Example 4.

Box 5.1 Amplitude and phase responses of a Hanning function

A **Hanning function** is used to smooth the cutoff of a window in data processing to avoid the undesirable effects of sharp truncation. Let us compute the amplitude and phase responses of a discrete **Hanning filter** ($\frac{1}{4}$, $\frac{1}{2}$, $\frac{1}{4}$), or $H(t) = (\frac{1}{4}, \frac{1}{2}, \frac{1}{4})$. Using the z-transform we have

$$H(z) = \frac{1}{4}(1 + 2z + z^2)$$

Take $z = e^{i\theta}$, where $\theta = \omega n \Delta t$,

$$H(e^{i\theta}) = \frac{1}{4}(1 + 2e^{i\theta} + e^{i2\theta})$$

$$= \frac{1}{4}e^{i\theta}(e^{-i\theta} + 2 + e^{i\theta})$$

$$= \frac{1}{2}e^{i\theta}(1 + \cos\theta)$$

Hence, the amplitude response is $|H| = \frac{1}{2}(1 + \cos\theta)$ and the phase response is $\text{Arg}\{H\} = \theta$.

Suppose the sampling interval is 4 ms. Then in time domain for this filter, $\theta = 2\pi f \Delta t = 2\pi f(4 \times 10^{-3})$. We can compute the values of the spectra as shown in the following table.

f (Hz)	0	25	50	75	100	125
$\lvert H \rvert$	1	0.90451	0.65451	0.34549	0.09549	0
θ (degree)	0	36	72	108	144	180

The responses are graphed in Box 5.1 Figure 1.

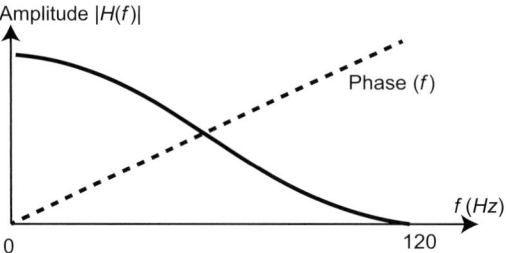

Box 5.1 Figure 1 Sketches of the amplitude and phase spectra of a Hanning filter (¼, ½, ¼).

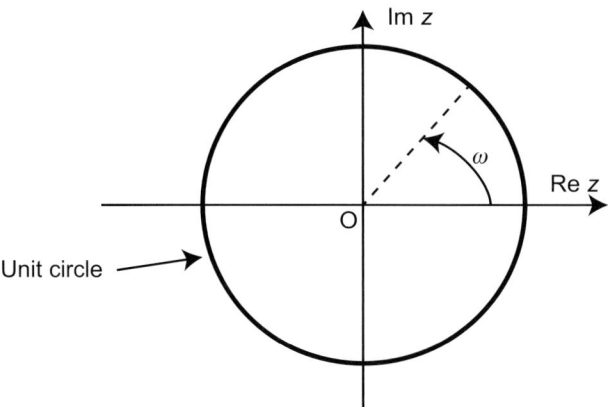

Figure 5.10 An unit circle on the complex z-plane.

Exercise 5.1

1. Along the unit circle in the complex z-plane shown in Figure 5.10, indicate the angle of:

   (a) Zero frequency;

   (b) Nyquist frequency;

   (c) 100 Hz if the time sampling rate is 4 ms.

2. For each of the four example block diagrams shown in Section 5.1.4:
    (a) What is the recursive formula?
    (b) Compute the amplitude spectrum.
    (c) What is the effect of the filter?
3. Find the z-transform formula and recursive formula of the filter whose block diagram is shown in Figure 5.11.

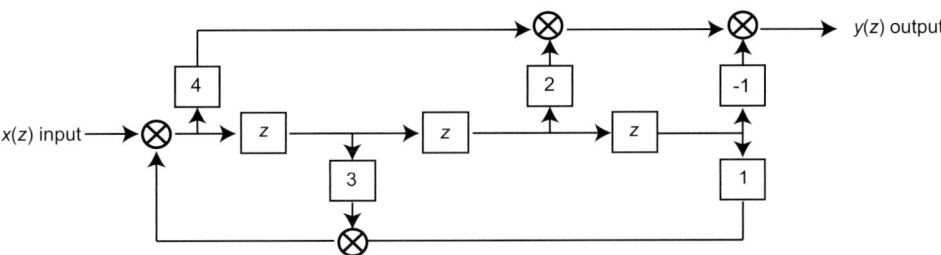

Figure 5.11 Block diagram of a filter.

5.2 Types of filters: pole–zero representation

Perhaps the best way to classify types of filters is to see their spectral configuration, as shown in Figure 5.12. Another important way to quantify the characteristics of filters is to use the distribution of their **poles** (denoted by X) and **zeros** (denoted by O) on the complex z-plane. A zero is the zero root of the numerator of the filter, and a pole is the zero root of the denominator of the filter. A plot of the digital filters' zeros and poles on the complex z-plane is called a pole–zero representation of the filters.

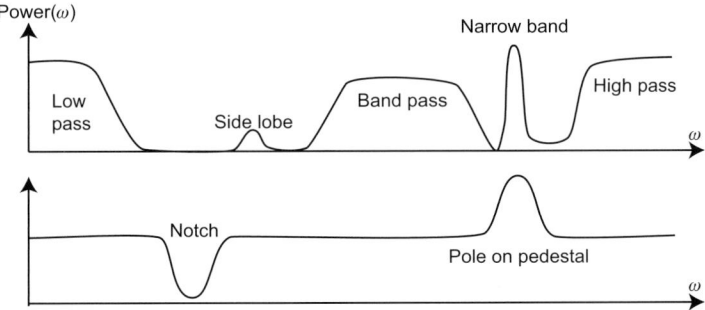

Figure 5.12 Sketches of power spectra of some common filters.

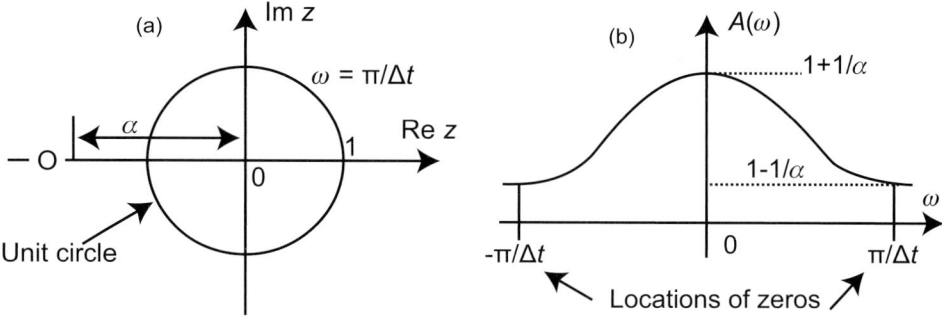

Figure 5.13 Filter $F(z) = 1 + z/\alpha$ on: (a) complex z-plane; (b) amplitude spectrum.

5.2.1   Zeros of a filter

Let us look at a simple **low-pass filter**

$$F(z) = 1 + \frac{z}{\alpha} \tag{5-6}$$

which becomes zero at $z = -\alpha$. We say that filter $F(\;)$ has a zero at $-\alpha$ because $F(-\alpha) = 0$. As shown in Figure 5.13, the location of the zero is denoted by an O on the **complex z-plane**. In the frequency domain $F(\omega) = 1 + e^{i\omega\Delta t}/\alpha$. Thus, the amplitude spectrum $A(\omega) = |F(\omega)|$ is

$$A(\omega) = \left[\left(1 + \frac{1}{\alpha}\cos\omega\Delta t\right)^2 + \left(\frac{1}{\alpha}\sin\omega\Delta t\right)^2\right]^{1/2}$$

$$= \left(1 + \frac{1}{\alpha^2} + \frac{2}{\alpha}\cos\omega\Delta t\right)^{1/2}$$

which is shown in Figure 5.13b. By dividing the amplitude at the highest frequency by that at the zero frequency, we find that $|F(z)|$ decreases with frequency by a factor of

$$|F(\omega = \pi)|/|F(\omega = 0)| = \left(1 - \frac{1}{\alpha}\right)\bigg/\left(1 + \frac{1}{\alpha}\right) \tag{5-7}$$

Another example of a filter has two zeros at the same place:

$$F(z) = (1 + z/\alpha)(1 + z/\alpha) \tag{5-8}$$

This will happen when two filters are in series:

$$\text{Input} \Rightarrow \boxed{1 \text{ zero}} \Rightarrow \boxed{1 \text{ zero}} \Rightarrow \text{output}$$

Based on the fact that the spectrum of a filter $F(z) = A(z)B(z)$ is equal to the spectrum of $A(z)$ times the spectrum of $B(z)$, the spectrum of the filter in this example is

$$F(\omega) = 1 + \frac{1}{\alpha^2} + \frac{2}{\alpha}\cos\omega\Delta t$$

Its graph is shown in Figure 5.14. From the lowest to the highest frequencies, the amplitude spectrum decreases by a factor of

$$|F(\omega = \pi)|/|F(\omega = 0)| = \left[\left(1 - \frac{1}{\alpha}\right)\bigg/\left(1 + \frac{1}{\alpha}\right)\right]^2 \tag{5-9}$$

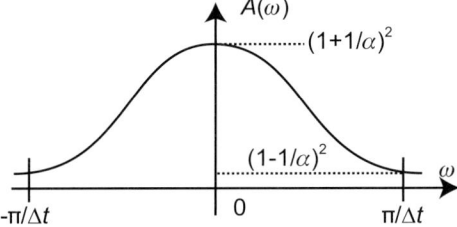

Figure 5.14 Amplitude spectrum of filter $F(z) = (1 + z/a)^2$.

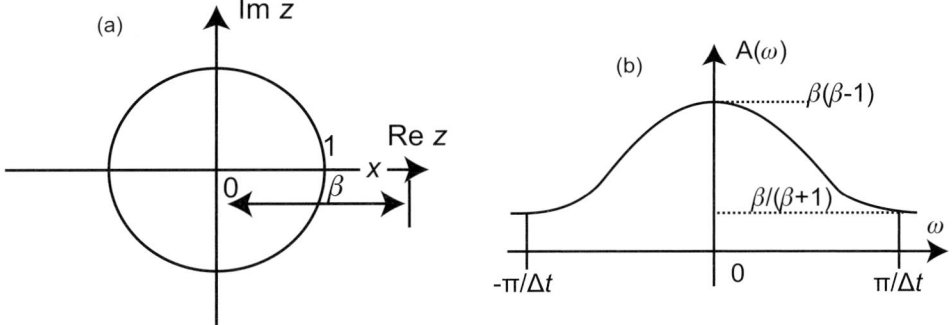

Figure 5.15 Filter $F(z) = 1/(1 - z/\beta)$ on: (a) complex z-plane; (b) amplitude spectrum.

We may conclude from the above two examples that **the closer the zero is to the unit circle on the complex z-plane, the greater the filter's effect** (in removing the high-frequency energy in this case).

5.2.2 Poles of a filter

We now turn to a filter with poles

$$F(z) = 1/(1 - z/\beta) \qquad (5-10)$$

The zero root of the denominator of the filter is referred to as the pole of the filter, as said above. In this case, a pole is located at $z = \beta$. The amplitude spectrum is

$$F(\omega) = \left(1 + \frac{1}{\beta^2} - \frac{2}{\beta}\cos\omega\Delta t\right)^{-1/2}$$

From the lowest to the highest frequencies, the amplitude spectrum decreases by a factor of

$$|F(\omega = \pi)|/|F(\omega = 0)| = \frac{1 - \beta}{1 + \beta} \qquad (5-11)$$

Figure 5.15 shows this filter on the complex z-plane and its amplitude spectrum. The cross on the complex z-plane to the right indicates the location of the pole.

Now we combine pole and zero into a filter

$$F(z) = \frac{1 + z/\alpha}{1 - z/\beta} \qquad (5-12)$$

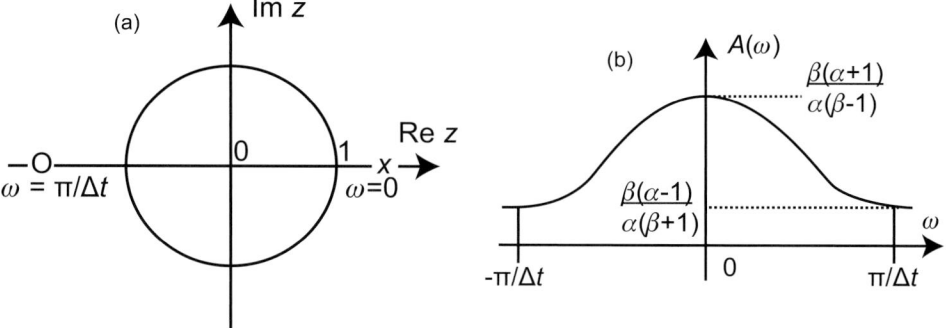

Figure 5.16 Filter $F(z) = (1 + z/\alpha)/(1 - z/\beta)$ on: (a) complex z-plane; (b) amplitude spectrum.

The zero (O) and pole (X) are shown on the complex z-plane in Figure 5.16. The amplitude spectrum is

$$F(\omega) = \left(\frac{1 + \frac{1}{\alpha^2} + \frac{2}{\alpha} \cos \omega \Delta t}{1 + \frac{1}{\beta^2} - \frac{2}{\beta} \cos \omega \Delta t} \right)^{1/2}$$

which is displayed on the right side of the figure.

Let us summarize the observation based on the previous examples: **Poles add, zeros remove. The magnitude of the effect of the pole or zero depends on their distance from the unit circle**.

The distance of a zero or a pole to the unit circle is measured by comparing its modulus in polar coordinates with 1, the radius of the unit circle. Suppose there are two zeros at locations $z_1 = r_1 \exp(i\omega_1)$ and $z_2 = r_2 \exp(i\omega_2)$ on the complex z-plane. When both zeros are outside the unit circle, their moduli r_1 and r_2 are both greater than 1; hence the smaller one of the two choices $r_1 - 1$ and $r_2 - 1$ is closer to the unit circle. When the two zeros are inside the unit circle, their modules r_1 and r_2 are both smaller than 1; then the smaller one between $1/r_1 - 1$ and $1/r_2 - 1$ is closer to the unit circle. Finally, if z_1 is inside the unit circle and z_2 is outside, we can convert the distance of the inside zero to its outside distance using the reciprocal of its modulus, $1/r_1$. Hence we can find the smaller one between $1/r_1 - 1$ and $r_2 - 1$ that is closer to the unit circle. The distance measure for poles is exactly the same as that for zeros.

By using one zero and one pole of the same angular frequency ω together, we can create a notch filter by placing the zero slightly closer to the unit circle than the pole; we can also create a "pole on pedestal" filter by placing the pole slightly closer to the unit circle than the zero. For example, suppose that we want to make a pole on pedestal filter at 50 Hz for sampling rate at 4 ms. Then the sampling frequency is 1 s / 4 ms = 250 Hz, so the Nyquist frequency is 125 Hz. Since the Nyquist frequency 125 Hz corresponds to the Nyquist angular frequency of π or 180°, then our desired frequency of 50 Hz corresponds to 180° × 50 Hz/125 Hz = 72°. We may choose to use 1.1 as the modulus of the pole, and a slightly large value of 1.2 as the modulus of the zero; then on the complex z-plane the pole is at $z_p = 1.1 \exp(i72°)$ and $z_o = 1.2 \exp(i72°)$. This pole on pedestal filter is

$$F(z) = \frac{z - z_o}{z - z_p} = \frac{z - 1.2e^{i72°}}{z - 1.1e^{i72°}}$$

Box 5.2 Minimum delay interpretation of minimum-phase wavelet

We want to examine the minimum phase concept using the complex z-plane. As has been discussed in Sections 1.4 and 4.2, minimum phase is a short form for minimum phase delay, meaning a minimum change in phase angle for a wavelet over a full cycle on the phase spectrum. Consider four zeros:

$$A(z - 2),\ B(z - 3),\ C(2z - 1),\ \text{and}\ D(3z - 1).$$

As shown in Box 5.2 Figure 5.1, A and C, B and D are polar reciprocals of each other. We make four wavelets by cascading pairs of zeros:

AB: $(z - 2)(z - 3) = 6 - 5z + z^2$, minimum phase
AD: $(z - 2)(3z - 1) = 3 - 7z + 2z^2$, mixed phase
CB: $(2z - 1)(z - 3) = 2 - 7z + 3z^2$, mixed phase
CD: $(2z - 1)(3z - 1) = 1 - 5z + 6z^2$, maximum phase

Notice that we have labeled the phase property of these four wavelets, and the only minimum-phase wavelet AB has both of its zeros located outside the unit circle.

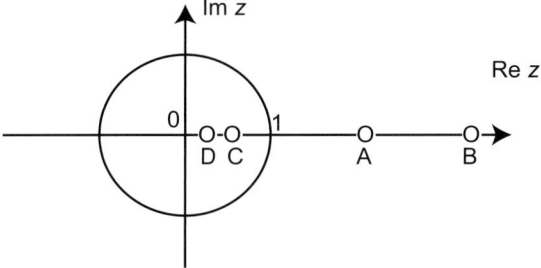

Box 5.2 Figure 1 Four zeros, A, B, C and D, on complex z-plane.

We can quantify the energy distribution with time by defining an accumulated energy at time step j as

$$E_j = \sum_{k=0}^{} |f_k|^2$$

The accumulated energies for the four wavelets are:

Wavelet	E_0	E_1	E_2	Phase	Line type in Box 5.2 Fig. 2
AB	36	61	62	minimum	thick line
AD	9	58	62	mixed	dashed line
CB	4	53	62	mixed	thin line
CD	1	26	62	maximum	dotted line

The accumulated energies at four time steps for these wavelets are shown in Box 5.2 Figure 2.

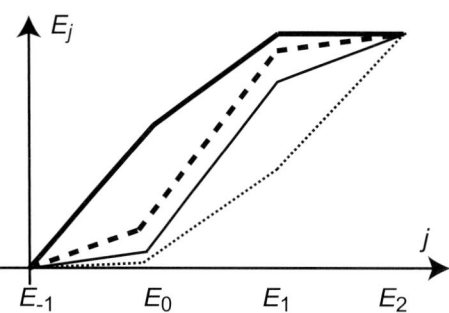

Box 5.2 Figure 2 Accumulated energies of four wavelets.

We see from this example that, of all the possible phases for a given spectrum, the minimum phase has its energy arrive the earliest, or front-loaded in the time domain.

Q: Why isn't $\phi = 0$ the minimum phase?
A: With $\phi = 0$, the signal will not be causal.

5.2.3   Stability of a filter

A digital filter is **stable** if its time duration covers just a short time span and **unstable** if it has a long time span. The difference between short and long is relative. In practical seismic data processing, because the total sample number is usually less than 10 000 points, the number of points for a stable filter is typically less than 50–100.

We call a filter **invertible** if its inverse (1/filter) is a stable filter. It can be proven that a minimum-phase filter is always invertible because all of its zeros and poles are outside the unit circle on the complex z-plane. Owing to the widespread application of the inverse problem in geophysics, minimum-phase filters have high significance in seismic data processing. Figure 5.17 shows the complex z-planes and amplitude spectra of several **minimum-phase** filters. Naturally all of them are invertible.

If we want to design a filter that is not complex but real, we can put poles and zeros in the conjugate positions. For example, the following filter

$$F(z) = (1 - z/z_0)(1 - z/z_0^*)$$

has two zeros in conjugate positions, as shown in Figure 5.18. We can see that

$$F(z) = 1 - \left(\frac{1}{z_0} + \frac{1}{z_0^*} \right) z + \frac{1}{|z_0|^2} z^2$$

where $\left(\frac{1}{z_0} + \frac{1}{z_0^*} \right)$ is certainly real.

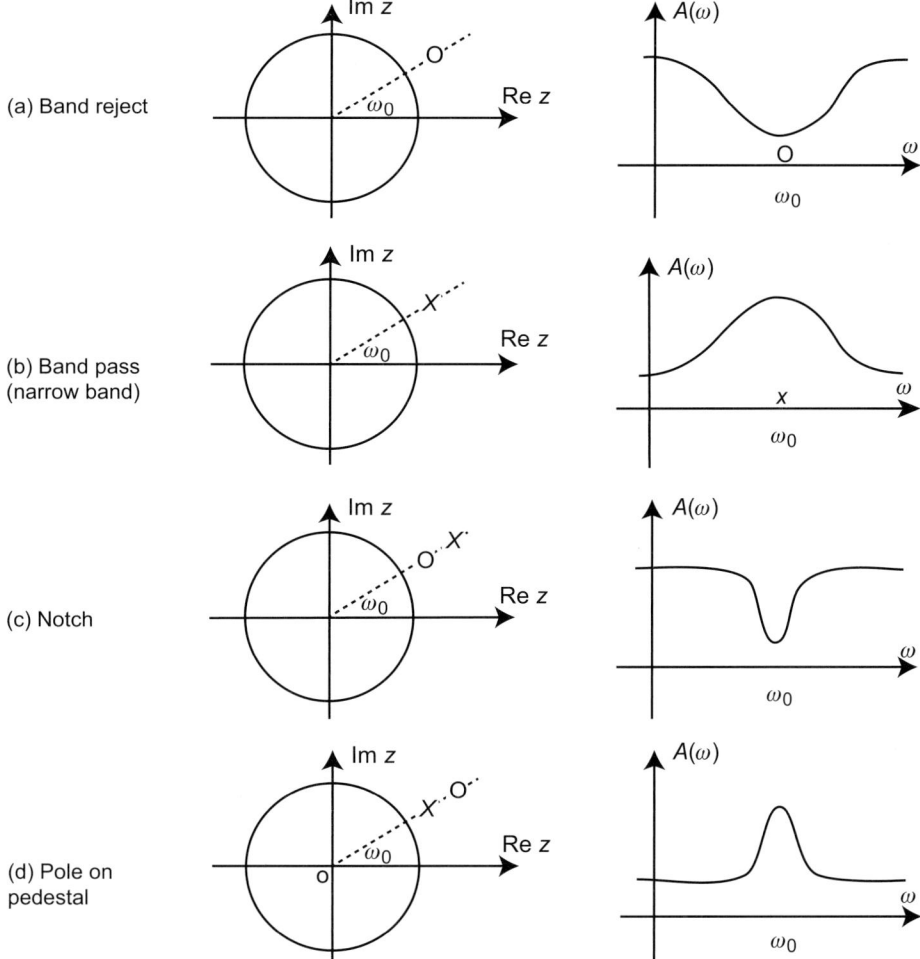

Figure 5.17 Plots on z-plane and amplitude spectra of some minimum-phase filters.

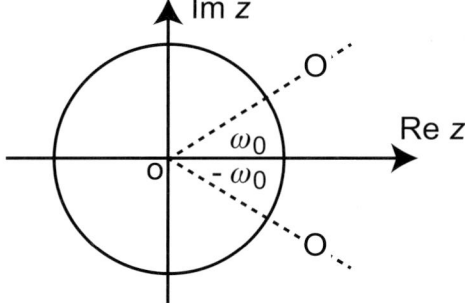

Figure 5.18 A real-number filter on complex z-plane.

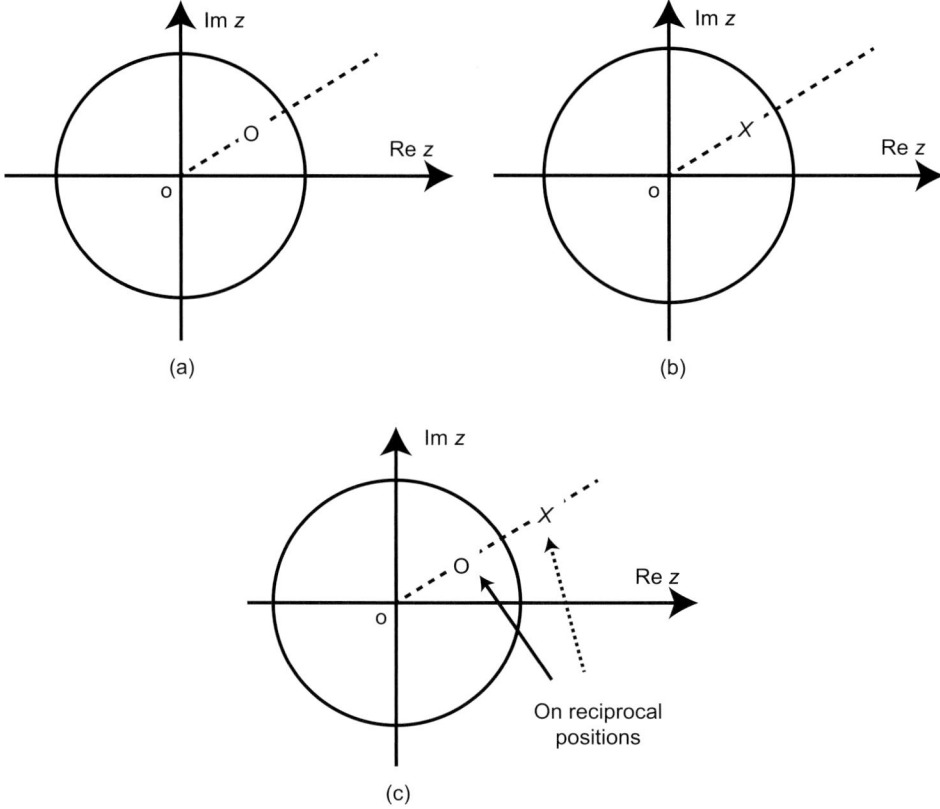

Figure 5.19 Properties of three filters as shown on complex z-plane. (a) Band reject: stable but not invertible. (b) Pole inside unit circle: this type of filter is useless because it is always unstable. (c) All-pass filter: the poles and zeros are polar reciprocal pairs. This filter does not alter the amplitude spectrum, but may alter the phase spectrum. It is stable but not minimum phase.

Three other types of filters are shown in Figure 5.19. When a filter has only zeros and all of them are inside the unit circle, it is also a stable band reject filter, but not invertible. When a filter has only poles and all of them are inside the unit circle, it is always unstable.

5.2.4　All-pass filters

Let us see an example of an all-pass filter

$$F(z) = \frac{1/\alpha - z}{1 - z/\alpha}, \ (\alpha > 1)$$

Note that

$$|F(z)|^2 = \left(\frac{1/\alpha - z}{1 - z/\alpha} \right) \left(\frac{1/\alpha - z^*}{1 - z^*/\alpha} \right) = \frac{1/\alpha^2 - z/\alpha - z^*/\alpha + 1}{1 - z/\alpha - z^*/\alpha + 1/\alpha^2} = 1$$

To check the phase spectrum of this all-pass filter,

$$\text{Phase } \{F(z)\} \equiv \Phi(\omega)$$

$$= \Phi_{\text{num}}(\omega) - \Phi_{\text{den}}(\omega)$$

$$= \text{phase } \{1/\alpha - e^{i\omega\Delta t}\} - \text{phase}\{1 - e^{i\omega\Delta t}/\alpha\}$$

$$= \tan^{-1}\left\{\frac{-\sin\omega\Delta t}{1/\alpha - \cos\omega\Delta t}\right\} - \tan^{-1}\left\{\frac{\frac{1}{\alpha}\sin\omega\Delta t}{1 - \frac{1}{\alpha}\cos\omega\Delta t}\right\}$$

Now checking two limiting cases:

(i) $\alpha \to \infty$ (zero and pole getting further apart)

$$\Phi(\omega) \to \tan^{-1}\left\{\frac{-\sin\omega\Delta t}{-\cos\omega\Delta t}\right\} - 0 = \omega\Delta t \text{ (phase delay)}$$

(ii) $\alpha \to 0$ (zero and pole becoming closer together)

$$\Phi(\omega) \to 0 - \omega\Delta t = -\omega\Delta t$$

Hence the phase spectrum is not flat, although the amplitude spectrum is.

Exercise 5.2

1. What type of filter is

$$y_t = \frac{1}{18}x_t - \frac{1}{6}x_{t-1} + \frac{1}{2}x_{t-2} + \frac{1}{2}y_{t-1} - \frac{1}{4}y_{t-2}?$$

   For a signal sampled at 125 sample/second, at what frequency does this filter have its maximum effect?

2. For a seismic dataset sampled at 2 ms, design a minimum-phase and real-number filter that will suppress noise associated with AC power at 60 Hz. What is the recursive formula of this filter?

3. For the following filter

$$f(z) = \frac{1.21 + 1.1\sqrt{2}z + z^2}{1.44 + 1.2\sqrt{2}z + z^2}$$

   (a) Sketch the pole–zero locations and identify the filter type;
   (b) Sketch the amplitude response and label the frequency axis;
   (c) State the recursive formula;
   (d) Draw the block diagram.

5.3 Geophysical models for filtering

Table 5.1 lists many types of filtering. All of them are based on the convolution model. One of the interesting ones is **Wiener filtering**, which will be discussed with two examples.

Table 5.1 Types of filtering (after Sheriff, 1991).

Name	Function
High-pass (low cut) filter Low-pass (high cut) filter Band-pass filter	Attenuate high and/or low frequencies
Notch filter	Attenuates narrow band of frequencies
Hi-line balancing	Adjusts resistive/reactive impedance
Spike deconvolution Whitening	Builds up all frequency components within specified band-pass to same amplitude
Predictive deconvolution	Removes repetitive aspects after time lag
Optimum filtering Wiener filtering	Produces results as close as possible to some desired output subject to constraints
Wavelet processing	Determines or changes embedded wavelet
Maximum entropy filter	Produces result as unpredictable as possible
Minimum entropy filter	Maximizes spiky character of output
Median filter	Suppresses noise by taking the median value from a predefined neighborhood
Homomorphic deconvolution	Lifters in the cepstral domain
Stacking	Attenuates out-of-register components
Velocity filter f–k filter	Multichannel filter to attenuate events of certain apparent velocities or dips
Tau–p (τ–p) filter	Multichannel filter to attenuate certain events
Time-variant filter	Changing filter parameters with time, usually a linear mix of processing with different parameters
Coherency filter	Multichannel filter to attenuate where certain coherence tests are not satisfied
Automatic picking	Multichannel filter to eliminate data that fail certain coherency and amplitude tests
Spatial filter	Performs discrete sampling in space

5.3.1 Wiener filtering

In reflection seismology, we often assume the reflection sequence g_t to be an Earth-filtered version of the wavelet x_t that was sent into the Earth:

$$x_t \Rightarrow \boxed{f_t} \Rightarrow g_t \tag{5–13}$$

This can be expressed as a convolution of the source wavelet x_t with the Earth **reflectivity function** f_t:

$$x_t * f_t = g_t \tag{5–14}$$

Our objective is to construct the Earth reflectivity function from the reflection data.

One particular assumption we may invoke is that the reflection function is totally unpredictable. In other words, our knowledge of the amplitudes and traveltimes of the first k reflections does not permit us to make any deterministic statement about the amplitude and traveltime of the $(k + 1)$th reflection. This also means that the locations of reflectors are uncorrelated. With this assumption, we can express the misfit e_t between the output g_t and the prediction by an optimum filter f_t^0

$$e_t = \left| g_t - x_t * f_t^0 \right| \tag{5–15}$$

It turns out that f_t^0 is actually a good approximation of the reflection function.

The filter that will most nearly produce a desired output is called an **optimum filter**, which is usually determined by least squares fitting. In the least squares sense, we simply minimize the error function (a single scalar)

$$\left(e_t \right)^2 = \left(g_t - x_t * f_t^0 \right)^2 \tag{5–16}$$

The resultant filter f_t^0 is called a **least squares filter**, or a **Wiener filter** after Norbert Wiener (1947) who devised the method. Some more detail on this subject can be found in Sheriff and Geldart (1995, p. 295).

5.3.2   Modeling far-field seismic body waves

The observed waveforms of seismic body waves are thought to be the result of the source function (dislocation function and finiteness function) going through Earth filters (Green's functions in terms of geometric spreading, velocity heterogeneities, and attenuation), and the recording instrumentation. Mathematically, these filtering processes are expressed in terms of convolutions.

Figure 5.20 illustrates the convolution model for far-field seismic body waves emitted from an earthquake. Specifically for body waves, the far-field geometric spreading $G(t)$ is modeled as the convolution of source dislocation function (a linear ramp function having a finite rise time) with a finiteness function (a boxcar); the result is a trapezoidal function. This trapezoidal function will then be convolved with the Q structure (thought of as a low-pass filter) and geometrical radiation pattern (thought of as a spatial filter). It will be further convolved with Earth structure (often as a series of spikes), and finally with the instrumental filter.

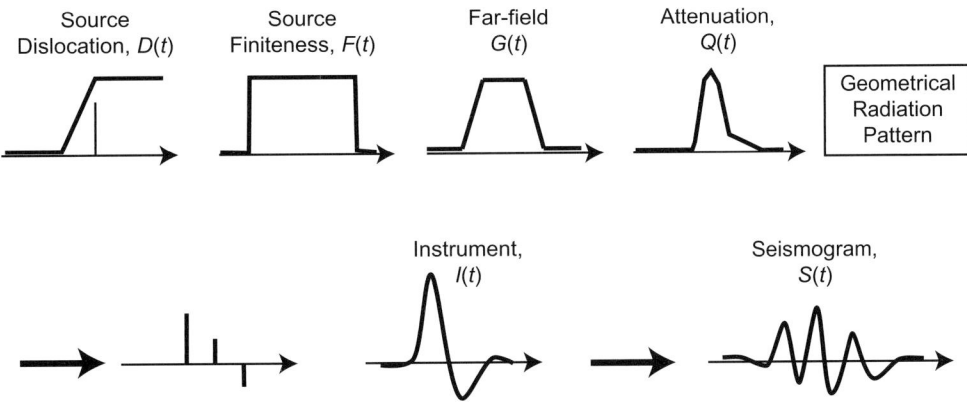

Figure 5.20 Seismograms of an earthquake is the result of convolving many processes.

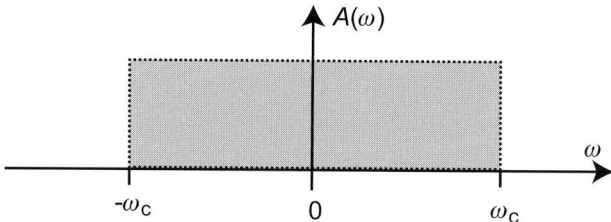

Figure 5.21 A box function in frequency domain.

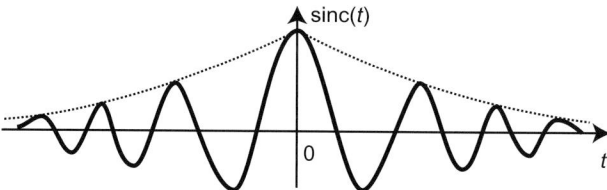

Figure 5.22 A sinc function in time domain.

5.3.3 Filtering with DFT

In general, the objective of filtering is to modify the frequency content of a signal. Perhaps the simplest and most straightforward way to achieve this is to apply the filter in the frequency domain. The procedure is:

1. Forward FFT the data to frequency domain;
2. Multiply each frequency component by the corresponding filter amplitude;
3. Inverse FFT the modulated data back to time domain.

There are two advantages for this DFT approach:

- Apparent ability to specify the filter response exactly;
- Low cost. The cost of applying the filter is $(N + 2N \times \log N)$ operations, where N is the length of the data.

The main disadvantage of the DFT approach is that in practice not all filter shapes can be used. For example, an ideal low-pass filter should have a spectrum like that in Figure 5.21.

However, the Fourier transfer of the above filter, a **sinc function**, is usually unacceptably long in the time domain (see Figure 5.22).

This tends to spread the energy out in time and may lead to wrap-around. If we truncate the impulse response in time and transfer it back to the frequency domain, we may end up with a filter whose spectrum looks like that in Figure 5.23.

The oscillations around the sharp corners in the above figure are the **Gibbs ears**. Even in the case of an infinitely long signal the overshoot will converge to about 8% of the jump in the original signal. Note that the frequencies near the cutoff points of the filter are enhanced, which is usually undesirable.

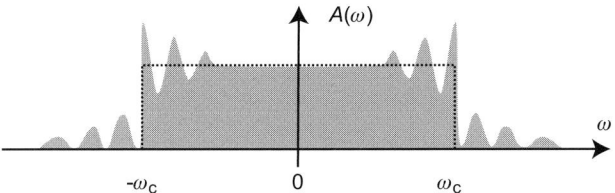

Figure 5.23 A frequency-domain box function after an inverse Fourier transform and a forward Fourier transform.

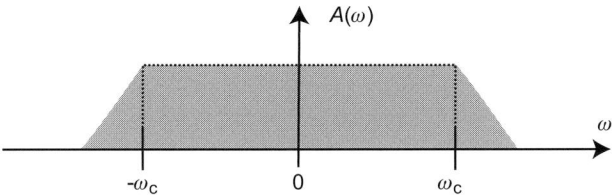

Figure 5.24 A box function in frequency domain with tapers applied to its corners.

A remedy for the above problem is to remove the discontinuity in the filter by using **tapers**, or making dipping slopes beyond the sharp corners (Figure 5.24).

Below are several comments on filtering with DFT:

- The Fourier transform of this filter is the multiplication of two sinc functions in the time domain, which means the decay rate in time is $1/t^2$.
- Causality and minimum phase properties are usually not preserved with frequency-domain filtering (owing to altering of the amplitude spectrum).
- In terms of computation cost, for short time-domain filters it is cheaper to apply a filter directly in the time domain rather than the frequency domain. The cost of the convolution in the time domain is $N \times N_f$, where N_f is the length of the filter.
- DFT methods generally cannot be used in real time.

Exercise 5.3

1. The acquisition of seismic data for petroleum exploration usually applies a low-cut filter in the field. Find the reasons for using this low-cut filter by searching the literature.

2. A geophone is specified by a frequency, such as 14 Hz, 4.5 Hz, etc. What is the meaning of the specified frequency? Is it possible for a geophone to record signals of frequencies lower than its specified frequency?

3. Explain the benefit of using tapers in applying filtering in frequency domain. Is it possible to remove the entire Gibbs artifact associated with a DFT?

5.4   Frequency–wavenumber (*f–k*) filtering

5.4.1   The idea of *f–k* filtering

An *f–k* **filter** is designed to suppress unwanted events in the frequency-wavenumber (*f–k*) domain. When applying multidimensional Fourier transforms, such as from (t, x) to (f, k_x), linear events in the original domain will also be linear events in the transformed domain, except that the orientations of each event in the two domains are perpendicular to each other (see, e.g., Chun & Jacewitz, 1981). If there are linear noises, or if there are noises with dip (offset/time) less than a certain angle, such as ground rolls, we can mute such noise in the *f–k* domain, and then transfer the remaining data back to the *t–x* domain. Hence *f–k* filtering is also called **dip-filtering** when it is used to remove linear events of certain dip angle.

Let us look at the behavior of a linear event after Fourier transform. As shown in Figure 5.25a, suppose that in the $(t–x)$ domain we have a linear event $t = x \tan \alpha + b$. If the wavelet of this event is $w(t)$, the event will be the following convolution in the (t, x) domain,

$$f(t, x) = w(t) * \delta(t - x \tan \alpha - b) \tag{5–17}$$

In Figure 5.25b, the Fourier transform of the above function is

$$F(\omega, k_x) = W(\omega) \exp(-i\omega b)\, \delta(\omega - k_x \cot \alpha) \tag{5–18}$$

Notice on the right-hand side that convolution becomes multiplication, the middle term is due to the shifting property of the Fourier transform, and the last term reflects the fact that linear events run perpendicular to the original orientation.

Let us see a classic example of *f–k* filtering in Figure 5.26 given by Embree *et al.* (1963). In the *t–x* space (panel (a)), the signal of reflection events has higher apparent velocities than that of the ground roll and most high-velocity noise such as refraction and scattering. In the *f–k* space (panel (c)), events of different slopes are separated except near the low-frequency center where everything collapses on top of each other. A horizontal dashed line in panel (b)

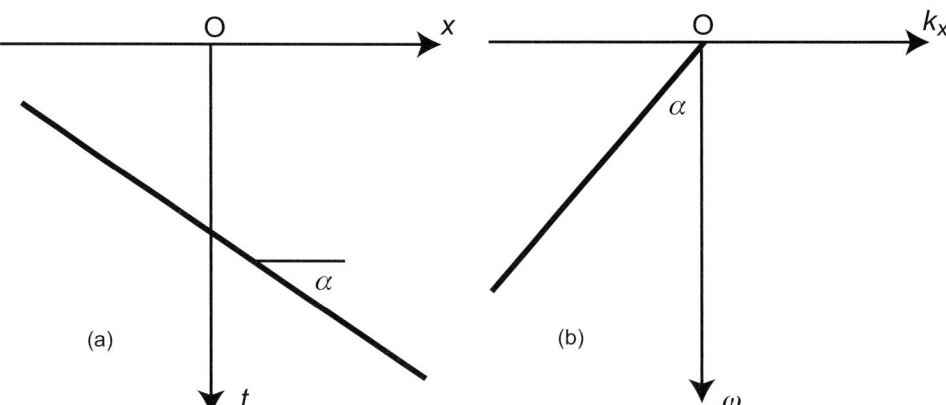

Figure 5.25 Linear events before and after a Fourier transform.

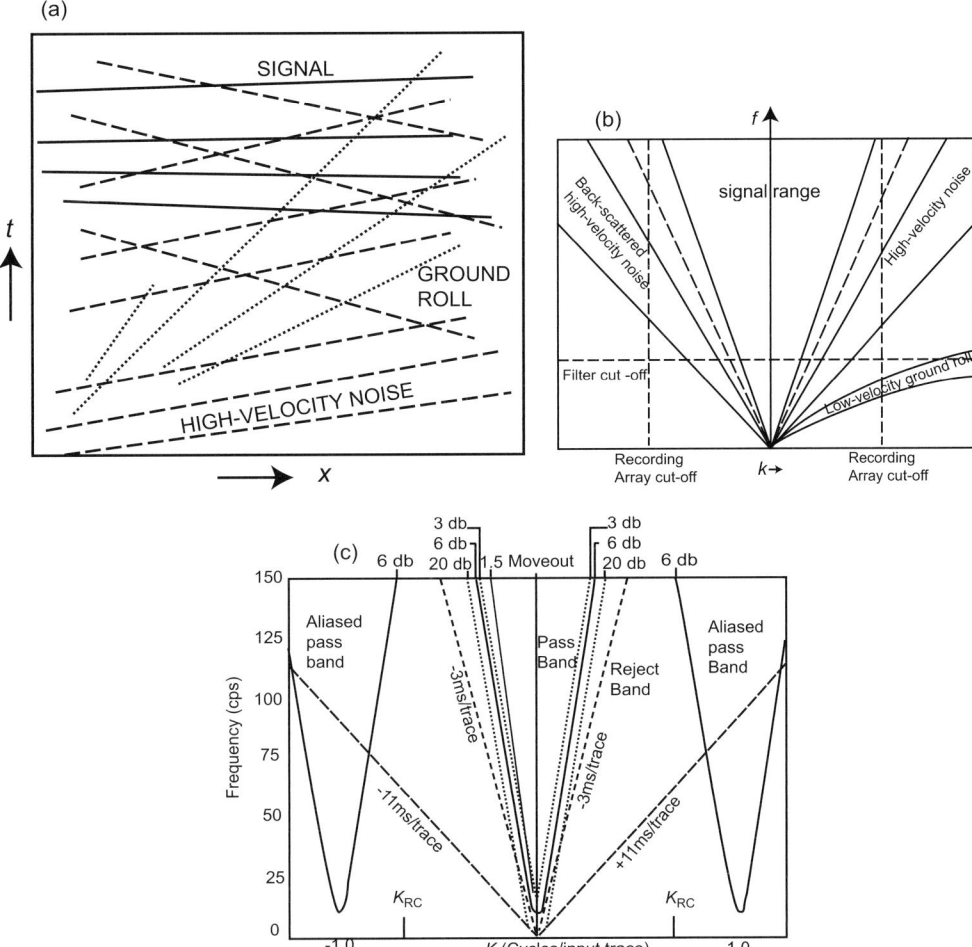

Figure 5.26 A schematic diagram of f–k filtering. (a) Time versus offset plot showing reflection signal in solid lines, ground rolls in dotted lines, and high-velocity noise in dashed lines. (b) Plot of f versus k showing spaces occupied by signal, low-velocity ground roll, and high-frequency noise and its back scatters. (c) A central pass band and two adjacent aliased pass bands for a 12-trace, ± 2 ms/trace f–k filter. (After Embree *et al.*, 1963.)

denotes the limit of a low-cut filter that is typical of seismic data acquisition for petroleum exploration. The length of the recording cable, which is indicated by two vertical dashed lines in panel (b) and by K_{RC} in panel (c), produces aliasing for longer events. Consequently a central pass band and two adjacent aliased pass bands are used by the f–k filter shown in panel (c).

5.4.2 Processing flow of f–k filtering

A common usage of f–k filtering in exploration seismology is to remove ground rolls, which are linear surface waves of very low velocity, or steep dip angle on an x–t plot like that in

Figure 5.26a. A good discussion of the processing flow of *f–k* filtering is given in Yilmaz (1987). Other references on the subject include Embree *et al.* (1963), Wiggins (1966), and Treitel *et al.* (1967). Ground rolls are usually of lower frequency than shallow reflections. However, since surface wave decay is much slower than that of body waves, sometimes surface waves are of higher amplitude and frequencies than deep reflections. Here is an *f–k* processing flow for filtering of ground rolls:

1. Input CSG, CMP, or CMP stack;
2. 2D Fourier transform;
3. Define a fan rejection zone for the amplitude spectrum;
4. Mute the transform within the rejection zone;
5. 2D inverse Fourier transform.

Another application of *f–k* filtering runs into the design of **median filtering** (e.g., Stewart, 1985; Duncan & Beresford, 1995). A more recent application of *f–k* filtering is in removal of multiples (e.g., Zhou & Greenhalgh, 1994). Below is an example *f–k* processing flow from these studies:

1. Input CSG or CMP gathers;
2. Wave-extrapolation of the input to generate multiple model traces;
3. Apply NMO correction to the input and multiple model traces using an intermediate velocity function;
4. 2D Fourier transform of the NMO-corrected input and NMO-corrected multiple model traces;
5. Compute the non-linear filter in the *f–k* domain according to the spectra obtained in Step 4;
6. Apply the filter in Step 5 to the spectrum of the input in Step 4;
7. 2D inverse Fourier transform of the filtered data obtained in Step 5;
8. Inverse NMO correction of the result in Step 6 using the same velocity function in Step 3;
9. Proceed to the next processing procedure.

5.4.3   An example of *f–k* filtering

Figure 5.27 shows stacked seismic sections with a linear *f–k* filtering and a median *f–k* filtering to suppress noises with low apparent velocities (Duncan & Beresford, 1995). The section in panel (a) without post-stack filtering shows many steeply dipping linear noises that are likely due to ground rolls and off-line scatterings. Band-pass filtering will not be effective because the frequency of the noises is comparable with the frequency of the gently dipping reflectors. The steeply dipping linear noises are well suppressed by the two *f–k* filters as shown in panels (b) and (c). At shallow depths, however, some of the dipping events may still be noise whose apparent velocity is higher than that used in the *f–k* filters.

Reflection events after the median *f–k* filtering have a rougher or more random appearance than that after the linear *f–k* filtering, although the differences between the two filtered results are not significant. Some of this roughness may be due to near-surface statics. A time migration of the data may improve the image more significantly.

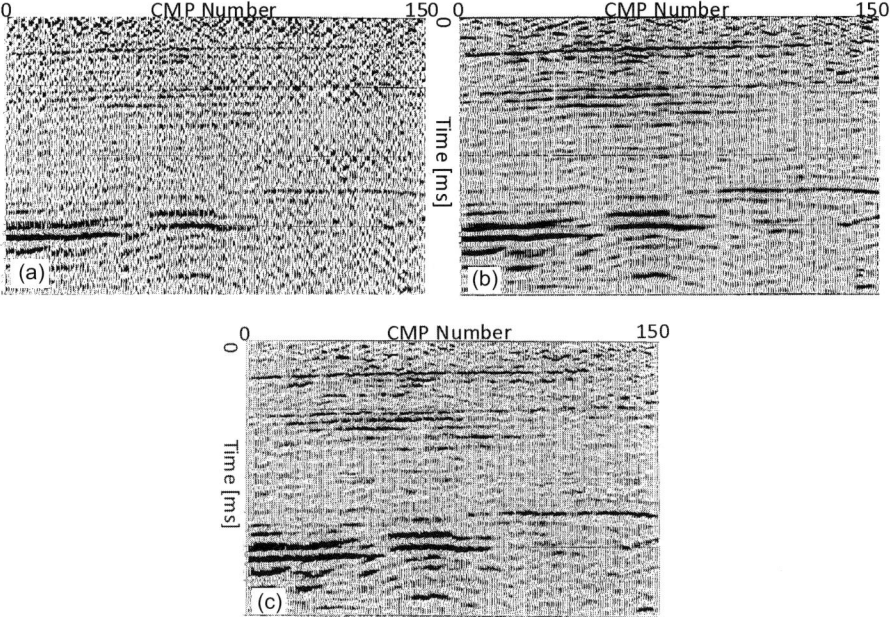

Figure 5.27 Stacked seismic field data from the Surat Basin, Australia. (a) Stacked data without post-stack filtering. (b) Stacked data after linear f–k filtering. (c) Stacked data after median f–k filtering. (From Duncan & Beresford, 1995.)

Exercise 5.4

1. Frequency–wavenumber or f–k filtering is an effective way to remove linear noise. Please research this topic and then write a short report describing f–k filtering in terms of: (1) the idea; (2) application procedure; (3) pitfalls; and (4) an example.

2. Explain why a smooth transition between the rejection zone and pass zone is necessary in f–k filtering.

3. Explain or refute the following statement: For broad band input, the cross-correlation function of the input and output of a system is the impulse response of the system.

5.5 Inverse filtering

5.5.1 Convolution and deconvolution

In reflection seismology, seismic waveform data are regarded as the convolution of the source wavelet, Earth's noise function, attenuation function, reflectivity function, and recording instrument response. We are interested in "deconvolving" the data to obtain the Earth reflectivity function (impulse response of the Earth), which will be the topic of the next chapter. As a consequence of learning the filtering theory, however, we would like to check out this issue from the inverse filtering point of view.

Following the first example in the previous section, we assume a simple case where, after a removal of the source and receiver effects, the reflection data y_t are a result of the Earth reflectivity function x_t filtered by f_t, which is a combination of the Earth's attenuation and scattering functions:

$$x_t \Rightarrow \boxed{f_t} \Rightarrow y_t \qquad (5\text{–}19)$$

Since the reflection data are given, we wish to find an inverse filter f_t^{-1} so that we can pass the reflection sequence through it to recover x_t:

$$y_t \Rightarrow \boxed{f_t^{-1}} \Rightarrow x_t \qquad (5\text{–}20)$$

The first difficulty in finding the inverse filter is that we usually do not know the filter f_t, or at least we do not know it with high accuracy. In other words, a common situation is that we only know y_t among the three factors x_t, f_t and y_t in the convolution model. We therefore have to make some assumptions about the characteristics of x_t and f_t in order to separate them. One way is to assume that the Earth's reflectivity function x_t is random and uncorrelated with itself. This technically means that x_t is white noise, although it is what we are after. Thus,

$$E(x_t, x_{t+\tau}) = \sigma_x^2 \delta(\tau) \qquad (5\text{–}21)$$

where $E(\ )$ is the expectation operator, σ_x^2 is the variance of the noise sequence x_t, and $\delta(\tau)$ is the delta function.

The above is actually the autocorrelation function in discrete form:

$$r(\tau) = \frac{1}{N} \sum_{t=0}^{N-1-\tau} x_t * x_{t+\tau}$$
$$= E(x_t, x_{t+\tau}) = \sigma_x^2 \delta(\tau) \qquad (5\text{–}22)$$

Following the Wiener–Khinchin theorem, the spectrum is equal to the Fourier transform of the auto-covariance function, hence

$$R_X(z) = X^*(1/z)X(z) = \int N\sigma_x^2 \delta(\tau)e^{i\omega\tau} d\tau = N\sigma_x^2 \qquad (5\text{–}23)$$

Now rewrite (5–19) in the frequency domain using the z-transform:

$$Y(z) = F(z)X(z)$$

The spectrum is

$$\begin{aligned}
R_Y(z) &= Y^*(1/z)Y(z) \\
&= F^*(1/z)X^*(1/z)F(z)X(z) \\
&= R_F(z)R_X(z) \\
&= N\sigma_x^2 R_F(z)
\end{aligned} \qquad (5\text{–}24)$$

In other words,

$$R_Y(z) = \text{const } R_F(z) \qquad (5\text{–}25)$$

Box 5.3 Spectral dominance of the wavelet

Since the Fourier transform of auto-covariance is the power spectrum, the dominance of the wavelet with respect to the input reflectivity function can be seen in the spectral domain. Box 5.3 Figure 1 shows an example of convolving a reflectivity with a wavelet to produce the output. Note that the spectrum of the output is nearly the same as that of the wavelet.

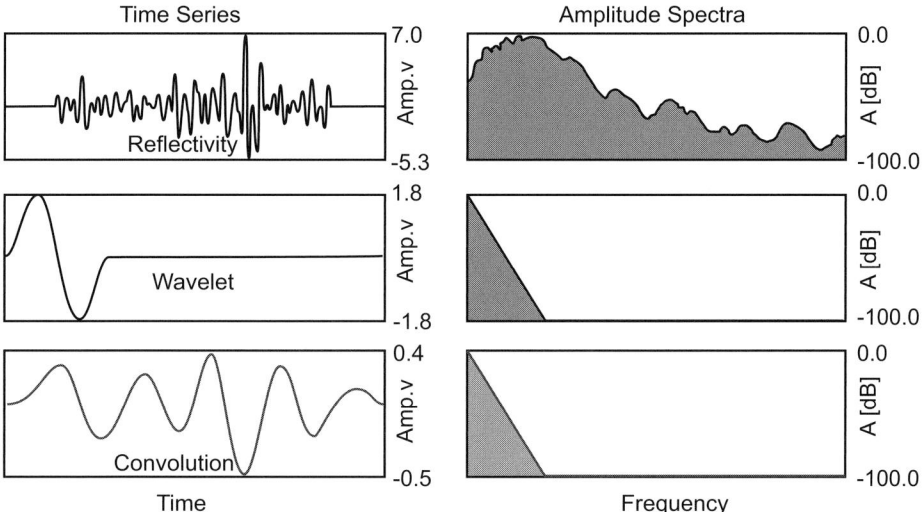

Box 5.3 Figure 1 A reflectivity function, a wavelet, and their convolution in time and spectral domains.

as a result of assuming that x_t is random in comparison with f_t in the model of (5–19). The above equation states:

The auto-covariance of the output is a scaled version of the auto-covariance of the filter itself when the input can be regarded as a random sequence.

This result suggests a way to construct the inverse filter by decomposing the auto-covariance of the data.

5.5.2 Spectral factorization

This is the typical way of decomposing the auto-covariance of the data. Because the phase information is lost in the auto-covariance, the factorization is non-unique.

The quest is to find $F(z)$ from the auto-covariance

$$R_F(z) = F^*(1/z)F(z)$$

We will require that the resultant $F(z)$ be minimum phase, because we plan to use it to construct the inverse filter. There are two methods of spectral factorization that are of interest to us, as described below.

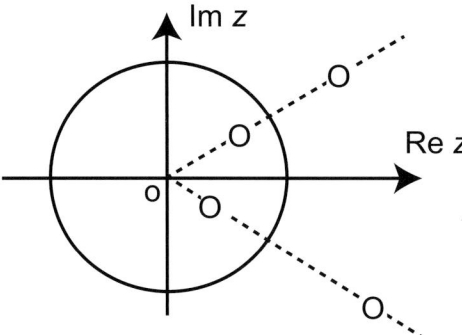

Figure 5.28 Zeros of $R_F(z)$ on the complex z-plane.

5.5.2.1 Root method

The root method was described by Wold (1938). The zeros of $R_F(z)$ are the zeros of $F(z)$ and $F^*(1/z)$. For instance, if

$$F(z) = z - z_0$$

then

$$F*(z) = 1/z - z_0^*$$

Hence if the zero of $F(z)$ is at z_0, then the zero of $F^*(1/z)$ is at $1/z_0^*$.

If we let

$$z_0 = \rho e^{i\omega_0}$$

then

$$z_0^* = (1/\rho)e^{i\omega_0}$$

Thus the zeros of $F^*(1/z)$ occur at the polar reciprocal position of the zeros of $F(z)$. This means that the zeros of $R_F(z)$ are in polar reciprocal pairs, as shown in Figure 5.28.

The root method is simply factorizing $R_F(z)$ into its zeros and then constructing $F(z)$ using all zeros outside the unit circle. This will produce a minimum phase $F(z)$ that has the correct spectrum $R_F(z)$.

However, the root method is difficult to use in practice because:

- It is complicated, expensive, and inaccurate to factorize a long $R_F(z)$ into its zeros;
- It is not clear what the effect of noise will be when using this method.

5.5.2.2 Toeplitz method

The Toeplitz method (Grenander & Szego, 1958; Atkinson, 1964) is the only practical way to solve the spectral factorization problem. We want the factorized $F(z)$ to be minimum phase, so that the inverse of $F(z)$

$$A(z) = 1/F(z)$$

is stable and well-defined.

$$R_F(z) = F^*(1/z)F(z) = F^*(1/z)/A(z)$$

or

$$R_F(z)A(z) = F^*(1/z) \qquad \cdot \qquad (5\text{--}26)$$

Suppose that we have the following expansion of each factor in the above equation:

$$F^*(1/z) = f_0^* + f_1^* z^{-1} + f_2^* z^{-2} + \cdots \qquad (5\text{--}27\text{a})$$

$$R_F(z) = \cdots + r_{-2} z^{-2} + r_{-1} z^{-1} + r_0 + r_1 z^1 + r_2 z^2 + \cdots \qquad (5\text{--}27\text{b})$$

$$A(z) = a_0 + a_1 z + a_2 z + \cdots \qquad (5\text{--}27\text{c})$$

Then the left-hand side of (5–26) becomes

$$\begin{aligned}
R_F(z)A(z) = \cdots \quad \cdots + (\cdots \quad \cdots)z^{-2} \\
+ (\cdots \quad \cdots)z^{-1} \\
+ (r_0 a_0 + r_{-1} a_1 + r_{-2} a_2 + \cdots)z^0 \\
+ (r_1 a_0 + r_0 a_1 + r_{-1} a_2 + \cdots)z^1 \\
+ (r_2 a_0 + r_1 a_1 + r_0 a_2 + \cdots)z^2 \\
+ \cdots \quad \cdots
\end{aligned} \qquad (5\text{--}28)$$

If we match the coefficients of each power of z for the LHS and RHS, we have

$$\begin{bmatrix}
r_0 & r_{-1} & r_{-2} & & \\
r_1 & r_0 & r_{-1} & \ddots & \\
r_2 & r_1 & r_0 & \ddots & r_{-2} \\
\ddots & \ddots & \ddots & & r_{-1} \\
& & r_2 & r_1 & r_0
\end{bmatrix}
\begin{bmatrix}
a_0 \\ a_1 \\ a_2 \\ \vdots \\ \vdots
\end{bmatrix}
=
\begin{bmatrix}
f_0^* \\ 0 \\ 0 \\ \vdots \\ 0
\end{bmatrix} \qquad (5\text{--}29)$$

This is known as the Toeplitz system. The matrix is called the **Toeplitz matrix** and all of its diagonal terms are equal.

To take a three-equation case, we have

$$\begin{bmatrix}
r_0 & r_{-1} & r_{-2} \\
r_1 & r_0 & r_{-1} \\
r_2 & r_1 & r_0
\end{bmatrix}
\begin{bmatrix}
a_0 \\ a_1 \\ a_2
\end{bmatrix}
=
\begin{bmatrix}
f_0^* \\ 0 \\ 0
\end{bmatrix} \qquad (5\text{--}30)$$

Those three equations contain four unknowns:

$$a_0, a_1, a_2, \text{ and } f_0^*$$

Hence, we have to give or assume one of the unknowns. Using the argument that we are not much interested in the amplitude-scaling factor of the filter, we can define a constant

$$v = f_0^*/a_0 > 0 \qquad (5\text{--}31)$$

Thus, by dividing (5–30) by a_0, we have $a'_k = a_k / a_0$, and

$$
\begin{bmatrix}
r_0 & r_{-1} & r_{-2} \\
r_1 & r_0 & r_{-1} \\
r_2 & r_1 & r_0
\end{bmatrix}
\begin{bmatrix}
1 \\
a'_1 \\
a'_2
\end{bmatrix}
=
\begin{bmatrix}
v \\
0 \\
0
\end{bmatrix}
\tag{5–32}
$$

For real filters, the Toeplitz matrix is symmetric ($r_{-k} = r_k$) because that the auto-covariance matrix $R(z)$ is symmetric, i.e., $F^*(1/z) = F(z)$.

The Toeplitz method of spectral factorization then is to solve the Toeplitz system (5–32). If N is the number of unknowns, a direct solution to this linear system by least squares requires N^3 operations. Such a solution does not guarantee the minimum-phase property of the inverse filter $A(z)$.

5.5.3   The Levinson recursion

Levinson in 1947 published an algorithm which produces a minimum phase $A(z)$ and requires only N^2 operations in solving the Toeplitz system. The Levinson approach solves the Toeplitz system by recursion (not iteration!), taking advantage of the highly structured form of the system. To show the recursion, we assume that we know the solution to the nth order system:

$$
\begin{bmatrix}
r_0 & r_1 & \ddots & r_n \\
r_1 & r_0 & \ddots & \ddots \\
\ddots & \ddots & \ddots & r_1 \\
r_n & \ddots & r_1 & r_0
\end{bmatrix}
\begin{bmatrix}
1 \\
a_1^{(n)} \\
\vdots \\
a_n^{(n)}
\end{bmatrix}
=
\begin{bmatrix}
v^{(n)} \\
0 \\
\vdots \\
0
\end{bmatrix}
\tag{5–33}
$$

where $A^{(n)}(z) = 1 + a_1^{(n)}z + a_2^{(n)}z^2 + \ldots$ is the nth order solution. From this solution, we will try to find the solution to the $(n + 1)$th order system

$$
\begin{bmatrix}
r_0 & r_1 & \ddots & r_n & r_{n+1} \\
r_1 & r_0 & r_1 & \ddots & r_n \\
\ddots & r_1 & r_0 & \ddots & \ddots \\
r_n & \ddots & \ddots & \ddots & r_1 \\
r_{n+1} & r_n & \ddots & r_1 & r_0
\end{bmatrix}
\begin{bmatrix}
1 \\
a_1^{(n+1)} \\
\vdots \\
a_n^{(n+1)} \\
a_{n+1}^{(n+1)}
\end{bmatrix}
=
\begin{bmatrix}
v^{(n+1)} \\
0 \\
\vdots \\
0 \\
0
\end{bmatrix}
\tag{5–34}
$$

where $A^{(n+1)}(z) = 1 + a_1^{(n+1)}z + a_2^{(n+1)}z^2 + \ldots$ is the $(n + 1)$th order solution.

We note that the $(n + 1)$th order Toeplitz system is simply the nth order system with an additional row and column added. In fact, the new system has the old system imbedded in either the upper-left corner or the lower-right corner, because the matrix is diagonally symmetric. To show the recursion, we rewrite the $(n + 1)$th order system as

$$
\begin{bmatrix}
r_0 & r_1 & \ddots & r_n & r_{n+1} \\
r_1 & r_0 & r_1 & \ddots & r_n \\
\ddots & r_1 & r_0 & \ddots & \ddots \\
r_n & \ddots & \ddots & \ddots & r_1 \\
r_{n+1} & r_n & \ddots & r_1 & r_0
\end{bmatrix}
\left\{
\begin{bmatrix} 1 \\ a_1^{(n)} \\ \vdots \\ a_n^{(n)} \\ 0 \end{bmatrix}
+ c
\begin{bmatrix} 0 \\ a_n^{(n)} \\ \vdots \\ a_1^{(n)} \\ 0 \end{bmatrix}
\right\}
=
\begin{bmatrix} v^{(n)} \\ 0 \\ \vdots \\ 0 \\ e \end{bmatrix}
+ c
\begin{bmatrix} e \\ 0 \\ \vdots \\ 0 \\ v^{(n)} \end{bmatrix}
\tag{5–35}
$$

Notes: <A>   <B>   <C>   <D>   <E>

<A> Notice that the up-left $(n + 1) \times (n + 1)$ system is the same as the lower-right $(n + 1) \times (n + 1)$ system;

<B> The nth order solution with a zero added to the end;

<C> A upside-down version of <B> with a scaling constant c;

<D> This vector is the result of applying the matrix <A> to vector <B>. Since the first $n + 1$ equations are the same as the nth order system (5–35), the first $n + 1$ elements of vector <D> are the same as in the nth order system. Only the last element e is different:

$$
e = r_{n+1}\,1 + r_n\,a_1^{(n)} + r_{n-1}\,a_2^{(n)} + \cdots + r_1\,a_n^{(n)} = \sum_{j=0}^{n} r_{n+1-j}\,a_j^{(n)}
$$

This shows that e depends only on the nth order solution;

<E> A upside-down version of <D> scaled by a constant c. The value of e is the same as that in <D>, because the top row of <A> is just its bottom row backwards.

To solve the above system, we require that the right-hand side of (5–34) and (5–35) be equal, hence

$$
\begin{bmatrix} v^{(n)} \\ 0 \\ \vdots \\ 0 \\ e \end{bmatrix}
+ c
\begin{bmatrix} e \\ 0 \\ \vdots \\ 0 \\ v^{(n)} \end{bmatrix}
=
\begin{bmatrix} v^{(n+1)} \\ 0 \\ \vdots \\ 0 \\ 0 \end{bmatrix}
\tag{5–36}
$$

Thus, $e + c\,v^{(n)} = 0$, or

$$
c = -e/v^{(n)}
\tag{5–37}
$$

and

$$v^{(n+1)} = v^{(n)} + ce = v^{(n)} - e^2/v^{(n)} = v^{(n)} \left[1 - (e/v^{(n)})^2\right]$$

i.e.   (5–38)

$$v^{(n+1)} = v^{(n)}(1 - c^2)$$

Note that $v^{(n+1)} > 0$ if and only if $v^{(n)} > 0$ and $|c| < 1$. This is equivalent to saying that if $v^{(n+1)}$ is positive then $|c| < 1$.

We now can write the updated inverse filter as

$$
\begin{bmatrix}
1 \\
a_1^{(n+1)} \\
\vdots \\
a_n^{(n+1)} \\
a_{n+1}^{(n+1)}
\end{bmatrix}
=
\begin{bmatrix}
1 \\
a_1^{(n)} + ca_n^{(n)} \\
\vdots \\
a_n^{(n)} + ca_1^{(n)} \\
c
\end{bmatrix}
\tag{5–39}
$$

The coefficient c is known in the literature as the **partial correlation coefficient**, or the 'reflection coefficient'.

To initiate the recursion, we start with the 0th order system

$$r_0 1 = v^{(0)}$$

hence

$$v^{(0)} = r_0$$

Let us address two questions on the Levinson recursion in the following.

Question 1: Is $A(z)$ thus obtained really minimum phase?

We first show that $v^{(n)}$ is positive. Consider the system in vector form

$$Ra = v$$

where $v = (v, 0, \ldots, 0)^T$. Multiply both sides of the above equation by a^T,

$$a^T Ra = a^T v = v \text{ (because } a_1 = 1)$$

Because R is positive definite (its eigenvalues are the power spectrum), the quadratic form

$$a^T Ra \geq 0, \quad \text{for any } a$$

This proves that $v > 0$. Therefore, because of (5–38),

$$1 - c > 0, \quad \text{or } |c| < 1$$

Using the z-transform, the operation of updating the filter, (5–39), can be written as

$$A^{(n+1)}(z) = A^{(n)}(z) + cz^n A^{(n)}(1/z) \tag{5–40}$$

where $z^n A^{(n)}(1/z)$ is just writing $A^{(n)}(z)$ backwards.

For instance, if $A(z) = a_0 + a_1 z^1 + a_2 z^2 + \cdots + a_n z^n$, then $z^n A^{(n)}(1/z) = a_0 z^n + a_1 z^{n-1} + a_2 z^{n-2} + \cdots + a_n$.

Assume the previous recursion gives us a minimum phase $A^{(n)}(z)$, and we may rewrite the updated filter as

$$A^{(n+1)}(z) = A^{(n)}(z)\left[1 + cz^n A^{(n)}(1/z)/A^{(n)}(z)\right] \tag{5-41}$$

Note that the amplitude spectra of $A^{(n)}(1/z)$ and $A^{(n)}(z)$ are identical and the spectrum of z^n is 1. Based on the additive property of minimum phase signals, the fact that $|c| < 1$ and that the spectrum of $z^n A^{(n)}(1/z)/A^{(n)}(z)$ is 1 indicates that $A^{(n+1)}(z)$ is minimum phase. Because at the starting step of the recursion the inverse filter is $A^{(0)}(z) = 1$, which is minimum phase, all the later $A^{(n)}(z)$ will be minimum phase.

Question 2: What happens if $F(z)$ is not minimum phase? (This can be the case for real data.)

Let us decompose the filter in the original forward problem as

$$F(z) = F_{in}(z)F_{out}(z) \tag{5-42}$$

where $F_{out}(z)$ has all its zeros outside the unit circle and hence is minimum phase; and $F_{in}(z)$ has all its zeros inside the unit circle and hence is not minimum phase. The auto-covariance of $F(z)$ will then be

$$R_F(z) = F_{in}^*(z)F_{out}^*(z)F_{in}(z)F_{out}(z) \tag{5-43}$$

with positions of the four zeros as follows:

$$\text{out, in, in, out.}$$

With either the root method or the Toeplitz method, we can find an inverse that uses only the zeros outside the unit circle, i.e.,

$$A(z) = 1/\left[F_{in}^*(1/z)F_{out}(z)\right] \tag{5-44}$$

If we apply this inverse filter to the original $F(z)$, we have

$$F(z)A(z) = F_{in}(z)F_{out}(z)/\left[F_{in}^*(1/z)F_{out}(z)\right] = F_{in}(z)/F_{in}^*(1/z) \tag{5-45}$$

which is a filter with its poles and zeros at polar reciprocal positions, hence an all-pass filter. This filter is stable because the poles are outside the unit circle. The amplitude spectrum of this filter is flat with frequency variation. Therefore, if we apply $A(z)$ to $Y(z)$, the output that we recorded, the effect is that the input sequence $X(z)$ has passed through an all-pass filter.

5.5.4   An example of the inverse filter

The most common reason to find an inverse filter is to "undo" the effect of an unwanted filter. In this example, we have two transient time series (with definite beginning and end), $a(t)$ and $b(t)$, of length N:

$$a(t) = a_0, a_1, a_2, \ldots, a_{N-1}$$

and

$$b(t) = b_0, b_1, b_2, \ldots, b_{N-1}$$

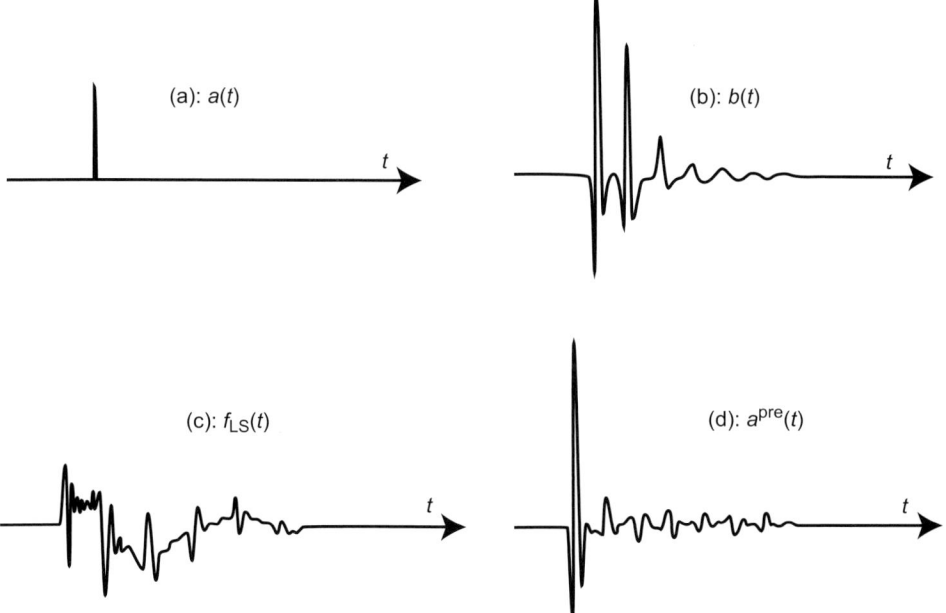

Figure 5.29 (a) A delta function $a(t)$; (b) The airgun signal $b(t)$, which is related to other functions by the convolution $a(t) = f(t) * b(t)$. (c) Estimated filter $f_{LS}(t)$ from a least squares inversion using $a(t)$ and $b(t)$. (d) The predicted signal, $a^{pre}(t) = f_{LS}(t) * b(t)$, is not exactly a delta function. (After Menke, 1989.)

Our objective is to find a filter $f(t)$ of length M:

$$f(t) = f_0, f_1, f_2, \ldots, f_{M-1} \text{ (usually } M \ll N)$$

so that

$$a(t) = f(t) * b(t) \tag{5–46}$$

An example of this model is shown in Figure 5.29 from Menke (1989). In this case $b(t)$ is the observed airgun signal with reverberations and $a(t)$ is a delta function. According to the above equation, $f(t)$ is the inverse filter of $b(t)$. The motivation here is to make the source wavelet as simple as a spike, hence the output of applying such a source wavelet to the observed signal will have a high resolution and therefore be more informative. In other words, after the processing the output will contain more components of the true Earth reflectivity rather than noises such as airgun echoes and ocean bottom reverberations or multiples. In terms of theory, we can find an estimated $f_{LS}(t)$ using least squares inversion of the following equation

$$f(t) * b(t) = \delta(t) \quad \text{(the spike delta function)}$$

Suppose that the Earth's reflectivity is $e(t)$, so the actual recorded data are

$$y(t) = b(t) * e(t)$$

Then we can apply $f_{LS}(t)$ to the data to get an estimated $e_{LS}(t)$:

$$f_{LS}(t) * y(t) = [f_{LS}(t) * b(t)] * e(t) = e_{LS}(t)$$

In terms of computation, putting (5–46) into the discrete convolution formula

$$a_i = \sum_{j=0}^{M-1} f_j b_{i-j} \tag{5–47}$$

Combining all equations ($i = 0, 1, 2, \ldots, N$) into a matrix form:

$$
\begin{bmatrix} a_0 \\ a_1 \\ \vdots \\ \vdots \\ a_{N-1} \end{bmatrix}
=
\begin{bmatrix}
b_0 & 0 & \cdots & 0 \\
b_1 & b_0 & \ddots & \vdots \\
b_2 & b_1 & \ddots & 0 \\
\vdots & b_2 & \ddots & b_0 \\
& & \ddots & b_1 \\
\vdots & & & \vdots \\
b_{N-1} & b_{N-2} & \cdots & b_{N-M}
\end{bmatrix}
\begin{bmatrix} f_0 \\ f_1 \\ \vdots \\ f_{M-1} \end{bmatrix}
\tag{5–48}
$$

That is

$$\mathbf{a} = \mathbf{Bf} \tag{5–49}$$

The system in (5–49) is linear for the unknown vector $\mathbf{f}$, and it is over-determined ($N > M$). Therefore it can be inverted by a least squares inversion that converts the system into

$$\mathbf{F}_{LS} = (\mathbf{B}^T\mathbf{B})_g^{-1}\mathbf{B}^T\mathbf{a} \tag{5–50}$$

where the subscript g stands for generalized inverse, the best possible result even if $(\mathbf{B}^T\mathbf{B})$ is not invertible in the exact sense.

We can find that $(\mathbf{B}^T\mathbf{B})$ is a $M \times M$ matrix containing the coefficients of auto-correlation of the observed signal $b(t)$:

$$
(\mathbf{B}^T\mathbf{B}) =
\begin{bmatrix}
\sum_{i=0}^{N-1} b_i b_i & \sum_{i=0}^{N-2} b_{i+1}b_i & \sum_{i=0}^{N-3} b_{i+2}b_i & \cdots & \sum_{i=0}^{N-M} b_{i+M-1}b_i \\
\sum_{i=0}^{N-2} b_i b_{i+1} & \sum_{i=0}^{N-2} b_i b_i & \sum_{i=0}^{N-3} b_{i+1}b_i & \cdots & \sum_{i=0}^{N-M} b_{i+M-2}b_i \\
\sum_{i=0}^{N-3} b_i b_{i+2} & \sum_{i=0}^{N-3} b_i b_{i+1} & \sum_{i=0}^{N-3} b_i b_i & \ddots & \vdots \\
\vdots & \vdots & \ddots & \ddots & \vdots \\
\sum_{i=0}^{N-M} b_i b_{i+M-1} & \sum_{i=0}^{N-M} b_i b_{i+M-2} & \cdots & \cdots & \sum_{i=0}^{N-M} b_i b_i
\end{bmatrix}
\tag{5–51}
$$

and $(\mathbf{B}^T\mathbf{a})$ is an $M \times 1$ vector containing the coefficients of cross-correlation between $a(t)$ and $b(t)$:

$$(\mathbf{B}^T\mathbf{a}) = \begin{bmatrix} \sum_{i=0}^{N-1} b_i a_i \\ \sum_{i=0}^{N-2} b_i a_{i+1} \\ \sum_{i=0}^{N-3} b_i a_{i+2} \\ \vdots \\ b_0 a_{M-1} \end{bmatrix} \tag{5-52}$$

As we now know, the system (5–49) can be solved either by least squares (Wiener filtering) or by the Levinson recursion.

According to Menke (1989), the result shown in Figure 5.29 was obtained using the least squares inversion, hence $f_{LS}(t)$ is called the least squares filter here. In this case $N = 240$ and $M = 100$. Notice in the figure that the predicted signal $a^{pre}(t)$ is not exactly a delta function. The reverberations of the airgun signal are certainly reduced, but not removed completely.

Exercise 5.5

1. For a single seismic trace $y(t)$ as the input, write a computer program (or a flowchart of descriptions) of the first five iterations of the inverse filtering method using the Levinson recursion.

2. The auto-covariance of a length-3 trace data trace $y(z)$ is (6, 35, 62, 35, 6).

   (a) Find the filter $F(z)$ using the root method;

   (b) Form the Toeplitz normal equations;

   (c) Find the inverse filter $A(z)$.

   (d) Is $A(z)$ minimum phase? Compare it with $F(z)$ and explain.

3. In the deconvolution example shown in Figure 5.29, the predicted signal shown in panel (C) is much longer than the airgun signal in panel (B). Explain the origin of the long tails of the predicted signal.

5.6   Summary

- Digital filters either pass or reject the frequency components of the input data according to the desired frequency content of the output. They provide an effective means to enhance the signal and/or suppress noise. The names of the filters are usually associated with their effects on the frequency spectrum.

- Each digital filter has at least three expressions: (1) the rational form with z-transform; (2) the recursive formula; and (3) the block diagram. In the rational form of a filter, the zeros are the roots of the numerator, and the poles are the roots of the denominator.

- Using the zeros and poles on the complex z-plane, we can quantify or design the effect of a digital filter following a rule of thumb that *poles add, zeros remove, and the magnitude of the effect of the pole or zero depends on their distance from the unit circle.*

- A minimum-phase filter is invertible because all of its zeros and poles are outside the unit circle on the complex z-plane. In general, a minimum-phase wavelet is preferred in seismic data processing owing to stability concerns, while a zero-phase wavelet is preferred in seismic interpretation to maximize the seismic resolution.

- f–k filtering is an effective way to separate events of different apparent velocities in the traveltime versus distance plot. After a 2D or 3D FFT, we need to design the pass bands for the signal and reject bands for the noise; tapers between the pass bands and reject bands need to be applied before the inverse 2D or 3D FFT.

- Inverse filtering is at the core of predictive deconvolution, which attempts to extract the input trace from the output trace of a filter without knowing the filter. It assumes that the input is random in comparison with the filter, so that the auto-covariance of the output is a scaled version of the auto-covariance of the filter itself. The filter is the solution of a normal equation system created by auto-covariance of the output trace.

- The Levinson recursion or Levinson–Durbin recursion is a procedure to recursively solve the normal equation system of a Toeplitz matrix. A Toeplitz matrix is a diagonal-constant matrix, in which each descending diagonal from left to right has a constant value. The inverse filter resolved by the Levinson recursion will be minimum phase.

FURTHER READING

Claerbout, J. F., 1985, *Fundamentals of Geophysical Data Processing*, Blackwell.
Hatton, L., Worthington, M. H. and Makin, J., 1986, *Seismic Data Processing: Theory and Practice*, Section 2.5, Blackwell.

6 Deconvolution

Deconvolution means to "undo" a convolution process. We may view each seismic trace as the result of convolving the subsurface seismic reflectivity with a seismic wavelet. Deconvolution can then be used to remove the seismic wavelet from the input seismic trace in order to yield the seismic reflectivity as the output. As a common time processing method, the main benefits of deconvolution include increasing data bandwidth and therefore resolution, suppressing periodicity such as multiples, and removing known wavelets. In practice we often only have the input seismic trace and want to find both the wavelet and the reflectivity. This non-uniqueness problem leads to the approach of predictive deconvolution, which assumes that the predictable components of the input trace belong to the seismic wavelet and the unpredictable components of the input trace belong to the reflectivity. To remove the effect of a known filter, we may use a frequency domain deconvolution which employs a "water level" to prevent division by zero.

As the amplitude and phase of real data vary with time, the deconvolution operator may be applied within a time window of the data. An adaptive deconvolution is a practical way to divide the data trace into time windows that overlap with each

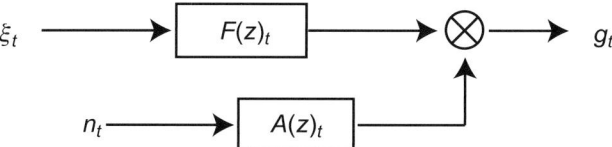

Figure 6.1 A general convolution model. ξ_t is called the driving noise, white noise, or convolutional noise. $F(z)_t$ is the process model, and $A(z)_t$ is the filter applied to "color" the additive noise n_t. The subscript t indicates that the coefficients of $F(z)_t$ and $A(z)_t$ may change with time. g_t is the output.

other, to apply deconvolution for each window, and then to integrate the deconvolved results together. By quantifying the distribution of seismic wiggles using the concept of entropy, minimum entropy deconvolution seeks to minimize the number of spikes on a seismic trace; this method works well in cases of few major reflectors. Finally, a method called extrapolation by deterministic deconvolution (EDD) is shown as a way to take predictions from sites of joint observations and to extrapolate into nearby sites that have only a single observation. This method provides the possibility of using seismic data to anticipate filtered versions of wellbore measurements.

6.1   Predictive deconvolution

6.1.1   Models of the convolution theory

From the viewpoint of convolution theory, we regard each recorded seismic trace as a superposition of the variants of a wavelet that vary randomly in amplitude with time. This means that the timing and amplitude of the wavelet are sequences of random variables. In such a view, deconvolution is a process of extracting the shape of the wavelet from the seismic trace, leaving the amplitudes of the wavelet at their respective arrival times.

Deconvolution is an act of inverse filtering, introduced at the end of the previous chapter. There are many kinds of deconvolution methods, and each is designed to eliminate a particular type of noise following a specific model. For instance, the spectral factorization method discussed in the previous chapter belongs to a typical deconvolution process to separate a random spiky sequence from repetitive events.

Let us first discuss some models for a continuous process, or time series. Probably the most general model is shown in Figure 6.1.

Recall from (5–22) that if x_t is white noise, then its statistical expectation is

$$E(x_t, x_{t+\tau}) = \frac{1}{N} \sum_{t=0}^{N-1-\tau} x_t * x_{t+\tau} = \sigma_{x^2}\delta(\tau) \tag{5–22'}$$

Thus for the model shown in Figure 6.1, we have

$$E(\xi_t\ \xi_{t+\tau}) = \sigma_\xi^2\delta(\tau) \tag{6–1}$$

$$E(n_t\ n_{t+\tau}) = \sigma_n^2\delta(\tau) \tag{6–2}$$

and the above two types of noise are uncorrelated:

$$E(\xi_t \, n_{t+\tau}) = 0 \tag{6–3}$$

The model in Figure 6.1 is too general, or over-parameterized. Although the model can fit any time series by adjusting the coefficients of $F(z)$ and $A(z)$ for each output point, it is not practical to determine these coefficients from the data. In the following we simplify this model to yield some practical models.

6.1.1.1 Time-invariant model

This is a simplified case of the general model in Figure 6.1 when both $F(z)$ and $A(z)$ are time invariant, as shown in Figure 6.2. This model assumes that the statistics of output g_t are independent of the time at which we choose to measure it; so it is a stationary process. In data processing, properties that are invariant with time are called **stationary**, and those variant with time are called **non-stationary**. Unfortunately, this model is still too general for most seismic applications, although some types of signal, such as sinusoids, use this model.

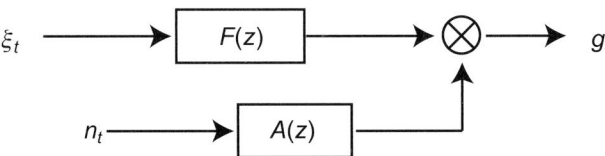

Figure 6.2 A time-invariant convolution model. ξ_t is the driving noise, $F(z)$ is the process model, and $A(z)$ is the filter applied to "color" the additive noise n_t. Both $F(z)$ and $A(z)$ are invariant with time. g_t is the output.

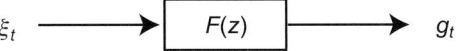

Figure 6.3 An ARMA convolution model. ξ_t is the driving noise, $F(z)$ is the process model, and g_t is the output.

6.1.1.2 Auto-regressive and moving average (ARMA) model

The auto-regressive and moving average (ARMA) model as shown in Figure 6.3 is obtained by removing the additive noise from the time-invariant general model. A more rigorous definition of this model can be found in Wold (1938) and Wei (1990).

A general form of $F(z)$ is

$$F(z) = N(z)/D(z) \tag{6–4}$$

If we let

$$N(z) = \beta_0 + \beta_1 z^1 + \beta_2 z^2 + \cdots + \beta_q z^q \qquad (6\text{--}5)$$

$$D(z) = \alpha_0 - \alpha_1 z^1 - \alpha_2 z^2 - \cdots - \alpha_p z^p \qquad (6\text{--}6)$$

then the recursion for output g_t can be written explicitly as

$$g_t = \sum_{k=1}^{p} \alpha_k g_{t-k} + \sum_{l=0}^{q} \beta_l \xi_{t-l} \qquad (6\text{--}7)$$

In the right-hand side of the above equation, the first summation is the auto-regressive (AR) part containing p parameters, and the second summation is the moving average (MA) part containing $(q + 1)$ parameters. Fitting data to solve for those parameters of the model is still a challenging task because of the large number of parameters.

6.1.1.3   Auto-regressive (AR) model

A further simplification is to drop the MA part of the ARMA model to leave

$$g_t = \sum_{k=1}^{p} \alpha_k g_{t-k} + \xi_t \qquad (6\text{--}8)$$

Then the parameters of this model are $\{\alpha_1, \ldots, \alpha_p, \text{ and } \sigma_\xi^2\}$.

6.1.2   Determination of the prediction error operator

In **predictive deconvolution**, we assume the observed data can be modeled as an AR process with the source wavelet as the deterministic or predictable portion of the data, and the Earth's response as the innovation, or the unpredictable portion of the data. Hence α_k become the coefficients of the prediction operator, and the **prediction error operator** is

$$D(z) = 1/F(z) = 1 - \alpha_1 z^1 - \alpha_2 z^2 - \cdots - \alpha_p z^p \qquad (6\text{--}9)$$

The AR coefficients can be retrieved by one of the algorithms in the following sections.

See Box 6.1 for an example of predictive deconvolution.

6.1.2.1   Yule–Walker method

To determine the auto-covariance of g_t, we multiply both sides of (6–8) by $g_{t-\tau}$ and then take the expectation operation, yielding

$$r_\tau = r_{-\tau} = E(g_t g_{t-\tau}) = \sum_{k=1}^{p} \alpha_k E(g_{t-k} g_{t-\tau}) + E(\xi_t g_{t-\tau}) \qquad (6\text{--}10)$$

Box 6.1 Predictive deconvolution for image enhancement

Because predictive deconvolution is a single-trace operator, it can be applied to either pre-stack gathers or stacked sections. For extremely noisy datasets, it can be used as an effective image enhancement tool. The left panel in Box 6.1 Figure 1 shows a common-receiver gather from an ocean bottom seismometer (OBS) recording an airgun source during an offshore seismic survey, off the coast of southern California. The combination of a high-impedance ocean floor plus rugged bathymetry produced strong and irregular multiple reflections and scattering noises that render the seismic gather nearly useless. After a predictive deconvolution, as shown in the right panel, considerable improvement is seen in reducing the multiple reflections. The presence of multiples with extremely long duration required an extremely long time gate of about 1600 ms for the predictive deconvolution in this case.

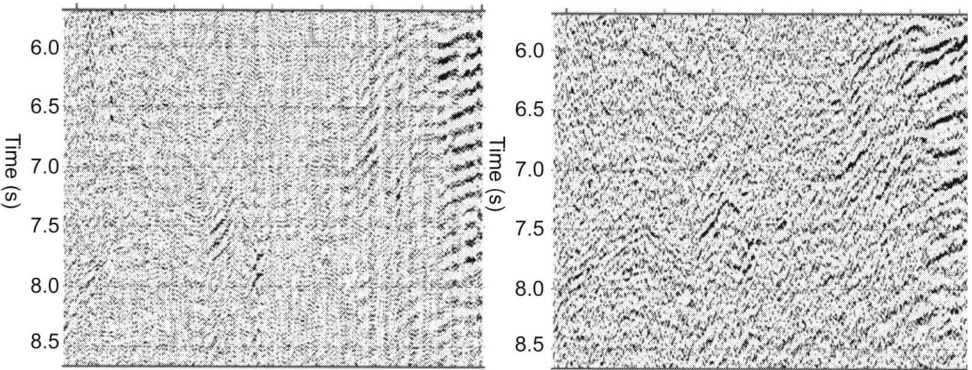

Box 6.1 Figure 1 An ocean bottom seismometer (OBS) gather recorded in offshore California before (left) and after (right) a predictive deconvolution using a nearly 1600-ms-long time gate.

For $\tau > 0$ we know that $E(\xi_t\, g_{t-\tau}) = 0$, because current values of the innovation ξ_t are uncorrelated with the past values of g_t. For $\tau = 0$ we have $E(\xi_t\, g_{t-\tau}) = \sigma_\xi^2$. Thus, for $\tau \geq 0$ we have

$$r_\tau = \sum_{k=1}^{p} \alpha_k r_{\tau-k} + \sigma_\xi^2 \delta(\tau) \qquad (6\text{--}11)$$

Note that $E(\xi_t\, g_{t-\tau}) \neq 0$ for $\tau < 0$. Now if we define

$$D(z) = \sum_{k=0}^{p} \gamma_k z^k \qquad (6\text{--}12)$$

and compare with (6–9), we have $\gamma_k = -\alpha_k$ and $\gamma_0 = 1$.

Hence, from (6–11)

$$r_\tau + \sum_{k=1}^{p}(-\alpha_k)r_{\tau-k} = \sum_{k=0}^{p}\gamma_k r_{\tau-k} = \sigma_\xi^2 \delta(\tau) \tag{6–13a}$$

This equation system forms a set of Toeplitz equations

$$\begin{bmatrix} r_0 & & & r_p \\ & \ddots & \ddots & \\ & \ddots & \ddots & \\ r_p & & & r_0 \end{bmatrix} \begin{bmatrix} 1 \\ \gamma_1 \\ \vdots \\ \gamma_p \end{bmatrix} = \begin{bmatrix} \sigma_\xi^2 \\ 0 \\ \vdots \\ 0 \end{bmatrix} \tag{6–13b}$$

where the diagonal-constant matrix is the Toeplitz matrix. Then $\{\gamma_k\}$ can be solved by the Levinson algorithm in Section 5.5.3 which guarantees that the solution $\{\gamma_k\}$ is of minimum phase.

However, in order to use the Yule–Walker method, we need to estimate the auto-covariance $\{r_t\}$ from a short sample $\{x_t\}$ taken from g_t. We may use

$$r_t = \frac{1}{N}\sum_k x_t x_{t+k} \tag{6–14a}$$

or

$$r_t = \frac{1}{|N-K|}\sum x_t x_{t+k} \tag{6–14b}$$

The above equations (6–14a, b) give poor estimates when the data segment is short. This problem motivates the following two methods that directly estimate the AR coefficients from the data rather than its auto-covariance.

6.1.2.2 Unconstrained least-squares method

We can determine the AR parameters $\{\alpha_k\}$ by least-squares fitting of the model to the sampled data $x(t)$ with the criterion of minimizing the variation of the innovation noise σ_ξ^2. This criterion means that we are putting as much energy as possible into the deterministic part of the model, the α_k, and as little as possible into the non-deterministic part, ξ_t.

In addition, we can simultaneously minimize the variances of both forward and reverse models. This will not be necessary for a perfectly stationary time sequence. On the other hand, it will help in stabilizing the processing for non-stationary time sequences.

In a least-square sense, the objective function of the combined forward and reverse prediction errors is

$$E = \sum_{t=p}^{N-1}\left(x_t - \sum_{k=1}^{p}\alpha_k x_{t-k}\right)^2 + \sum_{t=0}^{N-p-1}\left(x_t - \sum_{k=1}^{p}\alpha_k x_{t+k}\right)^2 \tag{6–15}$$

Differentiating with respect to α_j and setting the equation to zero, we obtain for $j = 1, 2, \ldots, p$,

$$0 = \frac{\partial E}{\partial \alpha_j} = 2\sum_{t=p}^{N-1}\left(x_t - \sum_{k=1}^{p}\alpha_k x_{t-k}\right)(-x_{t-j}) + 2\sum_{t=0}^{N-p-1}\left(x_t - \sum_{k=1}^{p}\alpha_k x_{t+k}\right)(-x_{t+k})$$

or

$$\sum_{k=1}^{p} \alpha_k \left(\sum_{t=p}^{N-1} x_{t-k}x_{t-j} + \sum_{t=0}^{n-p-1} x_{t+k}x_{t+j} \right) = \sum_{t=p}^{N-1} x_t x_{t-j} + \sum_{t=0}^{N-p-1} x_t x_{t+j}$$

Now let

$$\sigma_{kj} = \sum_{t=p}^{N-1} x_{t-k}x_{t-j} + \sum_{t=0}^{n-p-1} x_{t+k}x_{t+j} \qquad (6\text{--}16\text{a})$$

and

$$s_j = \sum_{t=p}^{N-1} x_t x_{t-j} + \sum_{t=0}^{N-p-1} x_t x_{t+j} \qquad (6\text{--}16\text{b})$$

Then we have

$$\sum_{k=1}^{p} \alpha_k \sigma_{kj} = s_j, \quad \text{for } j = 1, 2, \ldots, p \qquad (6\text{--}17)$$

Putting the above equation system into a matrix notation, we have

$$\begin{bmatrix} \sigma_{11} & \cdots & \sigma_{1p} \\ \vdots & \ddots & \vdots \\ \sigma_{p1} & \cdots & \sigma_{pp} \end{bmatrix} \begin{bmatrix} \alpha_1 \\ \vdots \\ \alpha_p \end{bmatrix} = \begin{bmatrix} s_1 \\ \vdots \\ s_p \end{bmatrix} \qquad (6\text{--}18)$$

The above system is symmetric, but not Toeplitz. Note that the expectation of σ_{ij} is

$$E(\sigma_{ij}) = r_{i-j} \qquad (6\text{--}19)$$

The unconstrained least-squares method determines the $\{\alpha_k\}$ by solving the above linear system. Unfortunately, this method cannot guarantee that the resulting **prediction error operator** $(1, -\alpha_1, -\alpha_2, \ldots, -\alpha_p)$ will be of minimum phase. This means the method is not very useful for constructing the prediction operator, though it can be used in estimating power spectra.

6.1.2.3 Constrained least-squares method: Burg's algorithm

In 1974 John Burg developed an algorithm to determine the AR coefficients with the property that the prediction error filter is minimum phase. The method is presented in his PhD thesis from Stanford University (Burg, 1975). Burg's method is also based on simultaneously minimizing the objective function of the combined forward and reverse prediction errors

$$E = \sum_{t=p}^{N-1} \left(x_t - \sum_{k=1}^{p} \alpha_k^{(p)} x_{t-k} \right)^2 + \sum_{t=0}^{N-p-1} \left(x_t - \sum_{k=1}^{p} \alpha_k^{(p)} x_{t+k} \right)^2 \qquad (6\text{--}15')$$

We have replaced α_k by $\alpha_k^{(p)}$, because we will develop it as a recursion from $\alpha_k^{(p-1)}$.

Let us switch back to the prediction-error notation $\gamma_k = -\alpha_k$ and $\gamma_0 = 1$,

$$E = \sum_{t=p}^{N-1} \left(\sum_{k=0}^{p} \gamma_k^{(p)} x_{t-k} \right)^2 + \sum_{t=0}^{N-p-1} \left(\sum_{k=0}^{p} \gamma_k^{(p)} x_{t+k} \right)^2 \qquad (6\text{--}20)$$

Defining the forward prediction error as

$$f_t^{(p)} = \sum_{k=0}^{p} \gamma_k^{(p)} x_{t-k} \tag{6-21a}$$

and the reverse prediction error as

$$r_t^{(p)} = \sum_{k=0}^{p} \gamma_k^{(p)} x_{t+k} \tag{6-21b}$$

we have

$$E = \sum_{t=p}^{N-1} \left(\sum_{k=0}^{p} \gamma_k^{(p)} x_{t-k} \right)^2 + \sum_{t=0}^{N-p-1} \left(r_t^{(p)} \right)^2 \tag{6-22}$$

We intend to find a recursion for f_t and r_t based on the Levinson algorithm.

We can decompose the prediction error operator into

$$\gamma_k^{(p)} = \gamma_k^{(p-1)} + c\gamma_{p-k}^{(p-1)} \tag{6-23}$$

This means

$$\begin{pmatrix} 1 \\ \gamma_1^{(p)} \\ \vdots \\ \gamma_{p-1}^{(p)} \\ \gamma_p^{(p)} \end{pmatrix} = \begin{pmatrix} 1 \\ \gamma_1^{(p-1)} \\ \vdots \\ \gamma_{p-1}^{(p-1)} \\ 0 \end{pmatrix} + c \begin{pmatrix} 0 \\ \gamma_{p-1}^{(p-1)} \\ \vdots \\ \gamma_1^{(p-1)} \\ 1 \end{pmatrix} \tag{6-24}$$

That is

$$\gamma_p^{(p)} = c \tag{6-25}$$

Thus,

$$f_t^{(p)} = \sum_{k=0}^{p} \gamma_k^{(p)} x_{t-k} = \sum_{k=0}^{p-1} \gamma_k^{(p-1)} x_{t-k} + c \sum_{k=1}^{p} \gamma_{p-k}^{(p-1)} x_{t-k} \tag{6-26}$$

The first term on the right-hand side is just $f_t^{(p-1)}$. For the second term, let $s = p - k$, so that $k = p - s$, hence

$$f_t^{(p)} = f_t^{(p-1)} + c \sum_{s=p-1}^{0} \gamma_s^{(p-1)} x_{t-p+s}$$

Reversing the order of summation for the second term on the right-hand side,

$$f_t^{(p)} = f_t^{(p-1)} + c \sum_{s=0}^{p-1} \gamma_s^{(p-1)} x_{(t-p)+s}$$

Checking the definition of $r_t^{(p)}$ in (6-21b), we have

$$f_t^{(p)} = f_t^{(p-1)} + c r_{t-p}^{(p-1)} \tag{6-27}$$

Similarly, the reverse prediction recursion can be established to be

$$r_t^{(p)} = r_t^{(p-1)} + cf_{t+p}^{(p-1)} \tag{6–28}$$

The total error can be written as

$$E = \sum_{t=p}^{N-1} \left(f_t^{(p-1)} + cr_{t-p}^{(p-1)} \right)^2 + \sum_{t=0}^{N-p-1} \left(r_t^{(p-1)} + cf_{t+p}^{(p-1)} \right)^2 \tag{6–29}$$

If we assume $f_t^{(p-1)}$ and $r_t^{(p-1)}$ are known from the previous recursion, then E is a function of a single parameter $c = \gamma_p^{(p)}$. Minimizing the error with respect to c, we have

$$\frac{\partial E}{\partial c} = 0 = \sum_{t=p}^{N-1} f_t^{(p-1)} r_{t-p}^{(p-1)} + c \sum_{t=p}^{N-1} \left(r_{t-p}^{(p-1)} \right)^2 + \sum_{t=0}^{N-p-1} r_t^{(p-1)} f_{t+p}^{(p-1)} + c \sum_{t=0}^{N-p-1} \left(r_{t+p}^{(p-1)} \right)^2$$

After some algebra, we find

$$c = 2 \sum_{t=p}^{N-1} f_t^{(p-1)} r_{t-p}^{(p-1)} \left/ \sum_{t=p}^{N-1} \left(f_t^{(p-1)} + r_{t-p}^{(p-1)} \right)^2 \right. \tag{6–30}$$

Using the value of c, we can update f_t and r_t, and hence the filter $\{\gamma_k\}$. One can prove that the final prediction error filter $\{\gamma_k\}$ is indeed of minimum phase.

6.1.3   A synthetic example of Vibroseis processing

Figure 6.4 shows a synthetic example of processing with **Vibroseis** data (Yilmaz, 1987). The first three traces show the convolution of a reflectivity function with a Vibroseis **sweep signal**. Another convolution with a minimum-phase source wavelet (d) produces the synthetic data in (e). In practice, we will first make a cross correlation between each data trace with the known Vibroseis signal (b), resulting in (f). This cross correlation will eliminate much of the impact of the Vibroseis signal because its amplitude spectrum is white. In fact, as discussed in Section 3.4.2, the auto-correlation of the Vibroseis signal will resemble the Klauder wavelet in (g). Hence the synthetic data (f) contains the zero-phase Klauder wavelet.

Deconvolution is a division between two time series. In Figure 6.4, trace (i) is the result of deconvolution because it is produced by dividing trace (h) by trace (g). We can also regard trace (i) as a minimum-phase filter because convolving it with Klauder wavelet will result in a minimum-phase version of the wavelet. Hence, convolving this minimum-phase filter (i) with (f) produces trace (j), the minimum-phase version of (f). The next two traces in (k) and (l) are results of **spiking deconvolution**, which is a specific predictive deconvolution to maximize the predicted components, so that the output will be of the spikiest form. Because traces (f) and (j) are the zero-phase and minimum-phase versions of the same time series, respectively, the resulting spiky traces (k) and (l) are also the zero-phase and minimum-phase versions of the same time series.

Trace (m) in Figure 6.4 is a band-pass filtered version of the true reflectivity function. This is the best possible solution of the reflectivity function from the data, because the passing band of the filter is the same as that of the Vibroseis signal. In comparison with

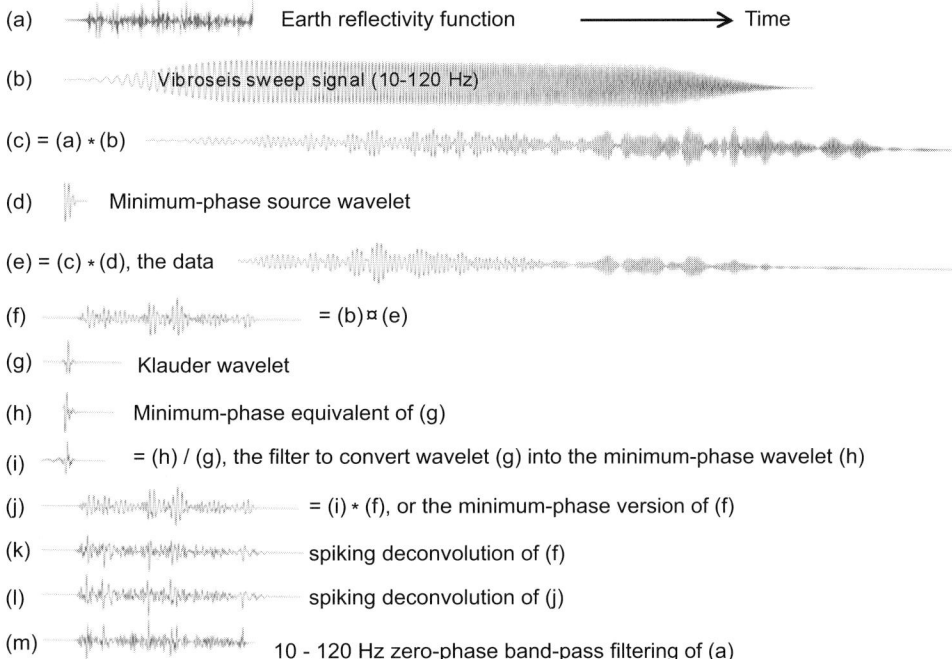

Figure 6.4 Processing with Vibroseis data, where ∗ denotes convolution and ¤ denotes cross-correlation (after Yilmaz, 1987).

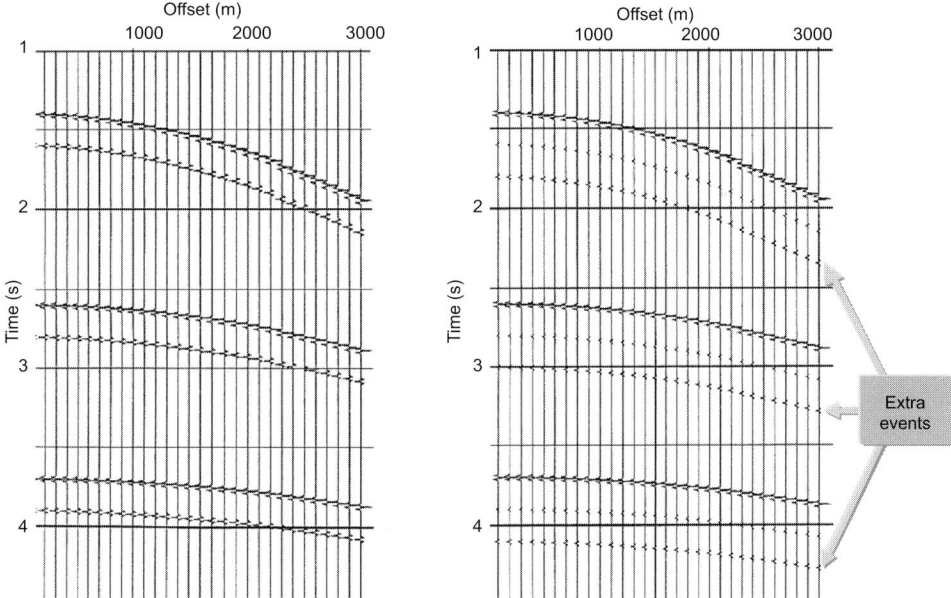

Figure 6.5 A synthetic model data after application of a stationarity transform (left), and then a predictive deconvolution (from Schoenberger & Houston, 1998).

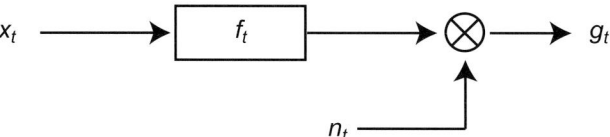

Figure 6.6 A filtering model. x_t is the Earth's reflectivity function, f_t is the source function, n_t is the noise, and g_t is the output seismic data.

trace (m), the two results from the processing in traces (k) and (l) show more similarity than traces (f) and (j). The main improvement is an increase in the frequency content; this is usually the main benefit from a deconvolution operator. The minimum-phase version in (l) appears to be slightly spikier than the zero-phase version in (k). However, both results from the processing missed the phase of the true reflectivity function in (m). A major drawback of most statistical deconvolution methods is the lack of constraints on the phase.

Exercise 6.1

1. Use your own words to define predictive deconvolution. What are the assumptions, general procedure, and solution? What is the prediction error operator?

2. Given a data series $\mathbf{d}(t) = \{d_0, d_1, d_2, \ldots, d_N\}$ and its estimated wavelet $\mathbf{w}(t) = \{w_0, w_1, w_2, \ldots, w_M\}$, how would you deconvolve $\mathbf{w}(t)$ from $\mathbf{d}(t)$? Please provide as many approaches as possible.

3. Figure 6.5 shows the input and output data for a predictive deconvolution. Explain the origin of the extra events in the output panel. What will happen if random noise is added in the input?

6.2   Frequency domain deconvolution

6.2.1   Division in frequency domain

The simplest way to deconvolve a known or an estimated filter from a seismic trace is to divide it in the frequency domain. The model shown in Figure 6.6 fits such a situation.
We may express the above model in the time domain as

$$g_t = x_t * f_t + n_t \tag{6–31}$$

and in the frequency domain as

$$G(\omega) = X(\omega)F(\omega) + N(\omega) \tag{6–32}$$

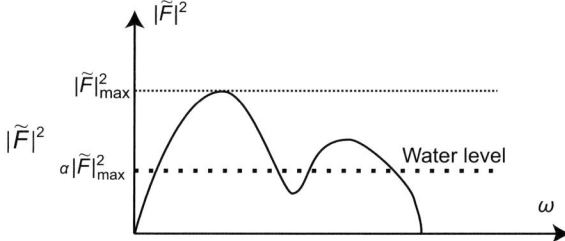

Figure 6.7 Maximum and "water level" of a spectrum.

Assuming that we have an estimate of F, say $\tilde{F}$, we can estimate $X(\omega)$ with $\tilde{X}(\omega)$

$$
\begin{aligned}
\tilde{X}(\omega) &= G(\omega)/\tilde{F}(\omega) \\
&= \frac{\tilde{F}*(\omega)}{\left|\tilde{F}(\omega)\right|^2} G(\omega) \\
&= \frac{\tilde{F}*(\omega)}{\left|\tilde{F}(\omega)\right|^2} [X(\omega)F(\omega) + N(\omega)] \\
&= \frac{\tilde{F}*(\omega)}{\left|\tilde{F}(\omega)\right|^2} F(\omega)X(\omega) + \frac{\tilde{F}*(\omega)}{\left|\tilde{F}(\omega)\right|^2} N(\omega) \\
&= X(\omega) + \frac{\tilde{F}*(\omega)}{\left|\tilde{F}(\omega)\right|^2} N(\omega)
\end{aligned}
\tag{6-33}
$$

The coefficient in front of the noise term $N(\omega)$ is $O(|\tilde{F}(\omega)|^{-1})$. Hence the noise will dominate the estimate when $|\tilde{F}(\omega)| \approx 0$. So we have to clamp the division by $|\tilde{F}(\omega)|^2$. The clamping may be done by letting $|\tilde{F}|_{max}$ be the maximum amplitude of $\tilde{F}(\omega)$, and

$$
\tilde{X}(\omega) = \frac{\tilde{F}*(\omega)}{\max\left[\left|\tilde{F}\right|^2, \left(\alpha \left|\tilde{F}\right|_{max}\right)^2\right]} G(\omega)
\tag{6-34}
$$

where $\alpha \in [0, 1]$ is a "**water level**" parameter, as sketched in Figure 6.7. Note that as $\alpha \to 0$, we get to the true deconvolution, while as $\alpha \to 1$, we get to cross-correlation, i.e. $\tilde{X} = \text{const} \times G\tilde{F}*$ (matched filter of F).

6.2.2   Spectral extension

In general, the frequency domain deconvolution is successful only in a limited bandwidth. For example, if the amplitude spectrum of output in the model $g_t = x_t * f_t$ is like that shown in Figure 6.8, then we will only be able to reconstruct $\tilde{X}(\omega)$ in the band $\omega \in [\omega_L, \omega_H]$. We would probably set $\tilde{X}(\omega)$ to zero outside this band.

However, if we know *a priori* that x_t is a sum of impulses (e.g., layered reflections) such as

$$
x_t = \sum_n A_n * \delta(t - t_n)
\tag{6-35}
$$

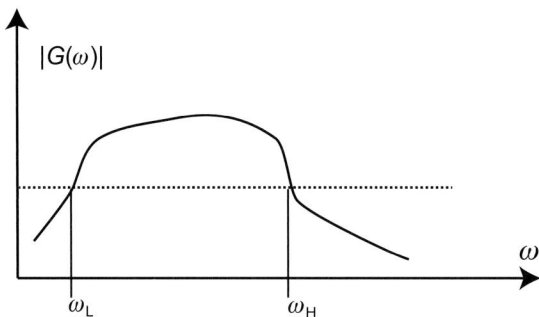

Figure 6.8 An example of amplitude spectrum with low and high corner frequencies, ω_L and ω_H.

we can improve the resolution by "predicting" the regions of $\tilde{X}(\omega)$ outside $[\omega_L, \omega_H]$. We call this a "**spectral extension**" procedure, as described below.

The Fourier transform of the model for x_t is

$$X(\omega) = \sum_n A_n \exp(i\omega t_n) \tag{6–36}$$

This means that the part of $\tilde{X}(\omega)$ that we estimate with deconvolution is actually a finite sample of a continuous process. We can therefore extend the spectrum by modeling it with an AR model.

We have at least two options:

1. Fit the AR model to $\tilde{X}(\omega)$ in the region $\omega \in [\omega_L, \omega_H]$, and then use the prediction from the AR model to extend the spectrum for $\omega_L \to 0$ and $\omega_H \to \omega_{Nyquist}$. The last step is to inversely transform the extended spectrum back to the time domain.
2. Fit the AR model to $\tilde{X}(\omega)$ in the region $\omega \in [\omega_L, \omega_H]$, and then directly use the spectrum of the AR model itself for the entire frequency range. This will give positive peaks at the impulse locations. However, the sign of the impulse is lost.

The spectral extension procedure described above depends heavily on the assumption that x_t is impulsive. Since we are fitting an AR model, some variations in the assumption can be tolerated. In addition, if x_t contains too many impulses, the ability of the AR modeling to pick up many sinusoids from a short part of $\tilde{X}(\omega)$ will be limited.

Figure 6.9 shows an example of spectral extension. The model is

$$\text{Data} = \text{source wavelet} * (\text{reflectivity} + \text{noise}) \tag{6–37a}$$

Or

$$g_t = f_t * (x_t + n_t) \tag{6–37b}$$

In the frequency domain:

$$G(\omega) = F(\omega)[X(\omega) + N(\omega)] \tag{6–38}$$

An estimated wavelet $\tilde{F}(\omega)$ will result in an estimated reflectivity $\tilde{X}(\omega)$

$$\tilde{X}(\omega) = G(\omega)/\tilde{F}(\omega) \tag{6–39}$$

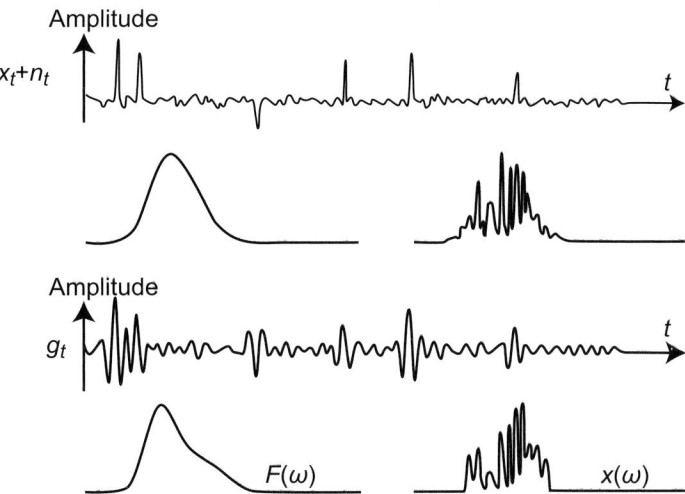

Figure 6.9 Ideal reflectivity function with noise, amplitude spectrum of source wavelet, and the spectrum and time plots of the resultant synthetic trace. The lower two plots are spectra of estimated wavelet and estimated reflectivity. All spectra extend from zero to Nyquist frequencies.

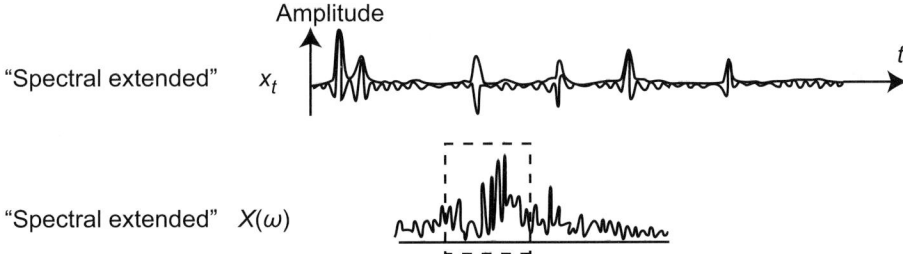

Figure 6.10 A procedure of spectral extension was applied to the synthetic trace in the previous figure. The upper plot shows the enhanced reflectivity function and its envelope. The lower plot shows its amplitude spectrum after spectral extension.

Figure 6.10 shows the estimated reflectivity and its amplitude spectrum after the spectral extension. Note that the polarity of the fourth reflection peak has the wrong polarity as compared with the first trace in Figure 6.9.

Exercise 6.2

1. Why do we prefer broadband data for deconvolution? How broad does the frequency band need to be?

2. When applying deconvolution in the frequency domain to remove the effect of a known broadening filter, any noise that came with the filter will be greatly amplified by the division. The typically small high-frequency components of the broadening filter mean there is great amplification of high-frequency noise. Can you suggest ways to remedy this problem? What are the limitations of your suggestion(s)?

3. Describe spectral extension in terms of its objectives, assumptions, procedure, and limitations.

6.3 Adaptive deconvolution

A source wavelet may change its form and properties with time owing to factors such as attenuation (Q) or multi-dimensional effects. Examples of the multi-dimensional effect include arrivals leaving a seismic source at different angles in the common case of variable radiation, and changes in the relationship between the "ghost" (free surface reflection) and the primary. Box 6.2 shows a graphic comparison between the stationary and non-stationary convolution methods. Here we want to modify the deconvolution algorithms to take the non-stationary changes into account.

6.3.1 Different approaches of adaptive deconvolution

An **adaptive deconvolution** is a method that changes with time to accommodate itself to the change in the time series to be deconvolved. One approach is to divide the data into several segments and separately deconvolve each piece. The outputs of all segments would then be put back together by a weighted addition. These segments should overlap to avoid artificial discontinuities in the output, as shown in Figure 6.11.

A second approach is to adapt the prediction coefficients to account for the apparent changes in the source wavelet. To achieve this, we define a number of weighted data segment windows, each looking like that shown in Figure 6.12. Within each window, we solve for a new prediction operator (source wavelet). However, two issues make this scheme less attractive. The first is the cost; we basically need to solve for a new filter at each output

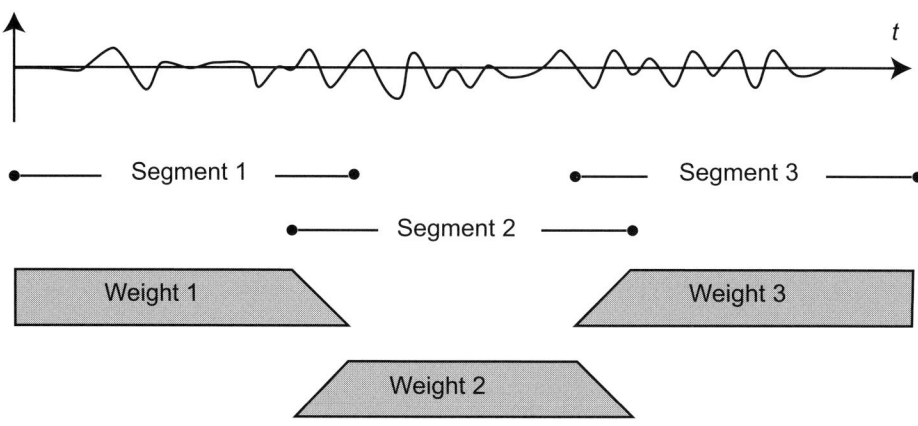

Figure 6.11 A sketch of the idea of adaptive deconvolution.

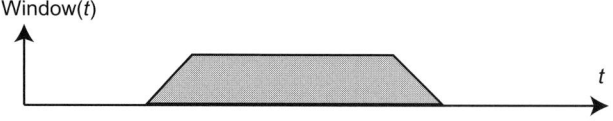

Figure 6.12 A weighted data segment window.

Box 6.2 Graphic illustration of stationary and non-stationary convolutions

One way to make deconvolution adaptive is to develop non-stationary deconvolution. Margrave (1998) gave a nice graphic illustration of the stationary and non-stationary convolutions, and the latter is a direct extension of the conventional stationary convolution to non-stationary processes. This extension provides a way to construct a non-stationary deconvolution. Other possible applications of non-stationary convolution include time-varying filtering, one-way wave propagation, time migration, normal moveout removal, and forward and inverse Q filtering.

Box 6.2 Figure 1 shows the conventional stationary convolution between a minimum-phase band-pass filter $a(t-\tau)$ and a reflectivity series $h(\tau)$ to produce an output seismogram $g(t)$. All columns in the convolution matrix have the same band-pass filter $a(t)$ with different time shifts.

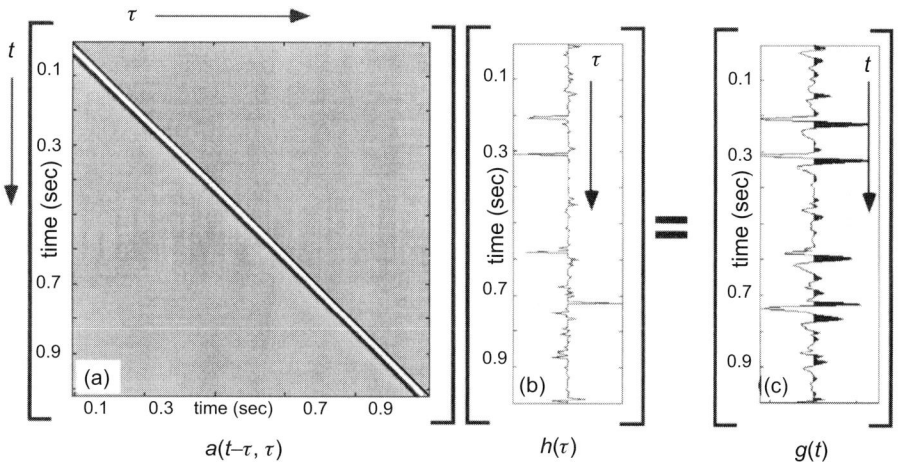

$$a(t-\tau,\ \tau) \qquad\qquad h(\tau) \qquad\qquad g(t)$$

Box 6.2 Figure 1 An illustration of stationary convolution as a time-domain matrix operation. (a) The stationary convolution matrix for a particular minimum-phase band-pass filter. The matrix displays Toeplitz symmetry meaning that each column contains the filter impulse response, each row contains the time reverse of the impulse response, and any diagonal is constant. (b) A reflectivity series in time to which the convolution matrix is applied. (c) The output stationary seismogram. (After Margrave, 1998.)

In contrast, Box 6.2 Figure 2 shows the non-stationary convolution between a non-stationary filter $a'(t-\tau,\ \tau)$ and the same reflectivity series $h(t)$. This time each column of the non-stationary convolution matrix contains the result of convolving the original minimum-phase band-pass filter with the impulse response of a constant Q medium at the corresponding traveltime. In other words, the effect of attenuation due to a constant Q factor is applied to the filter in the non-stationary convolution matrix.

The difference between stationary and non-stationary filters is that the impulse response of the latter varies arbitrarily with time. While the example in Box 6.2 Figure 2 used the impulse response of a constant Q medium, the complete description of a general non-stationary filter requires that its impulse response be known at all times. Following such a definition, it is possible to develop deconvolution schemes to determine the reflectivity series and the non-stationary trend of the filter.

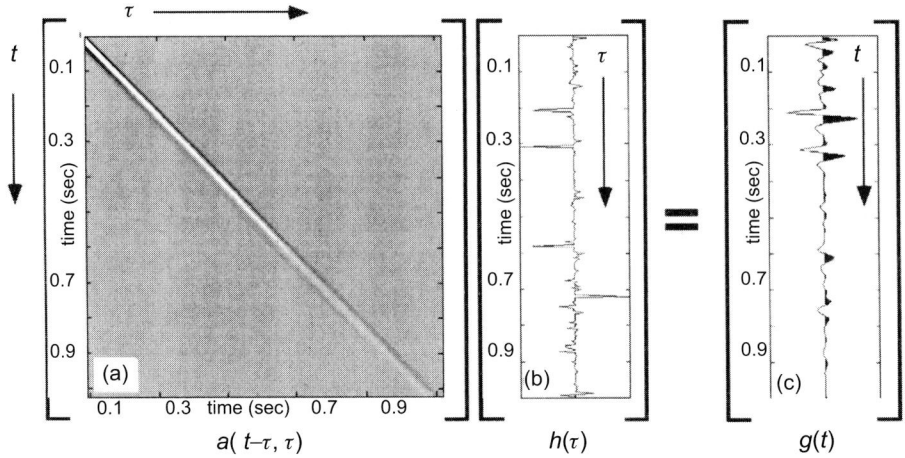

$$a(t-\tau, \tau) \qquad\qquad h(\tau) \qquad\qquad g(t)$$

Box 6.2 Figure 2 An illustration of non-stationary convolution as a time-domain matrix operation. (a) The non-stationary convolution matrix for a particular forward Q filter band limited by the stationary waveform. Each column contains the convolution of the minimum-phase waveform of previous figure and the minimum-phase impulse response of a constant Q medium for a traveltime equal to the column time. (b) The same reflectivity series in time as in the previous figure. (c) The output constant Q seismogram. (After Margrave, 1998.)

point. The second is that the filter can change much too rapidly and hence may remove too much information from the input data.

The third approach is to solve approximately for a new filter. In Burg's algorithm, for example, the error at the last stage in the recursion for the filter is

$$E = \sum_t f_t^2 + \sum_t r_t^2 \qquad\qquad (6\text{--}40)$$

where f_t and r_t are the forward and reverse prediction error, respectively. If we add a new point to a length-1 filter, the error will be

$$E' = E + (x_n + Cx_{n-1})^2 + (x_{n-1} + Cx_n)^2 \qquad\qquad (6\text{--}41)$$

Based on this new point, by setting $\partial E'/\partial C = 0$, the coefficient C will be

$$C = \frac{-2x_{n-1}x_n}{x_{n-1}^2 + x_n^2} \qquad\qquad (6\text{--}42)$$

We can now approximately update C by

$$C_{\text{new}} = C_{\text{old}} + \alpha C \qquad\qquad (6\text{--}43)$$

where $\alpha \in [0, 1]$ is a parameter which controls the rate of the adaptation.

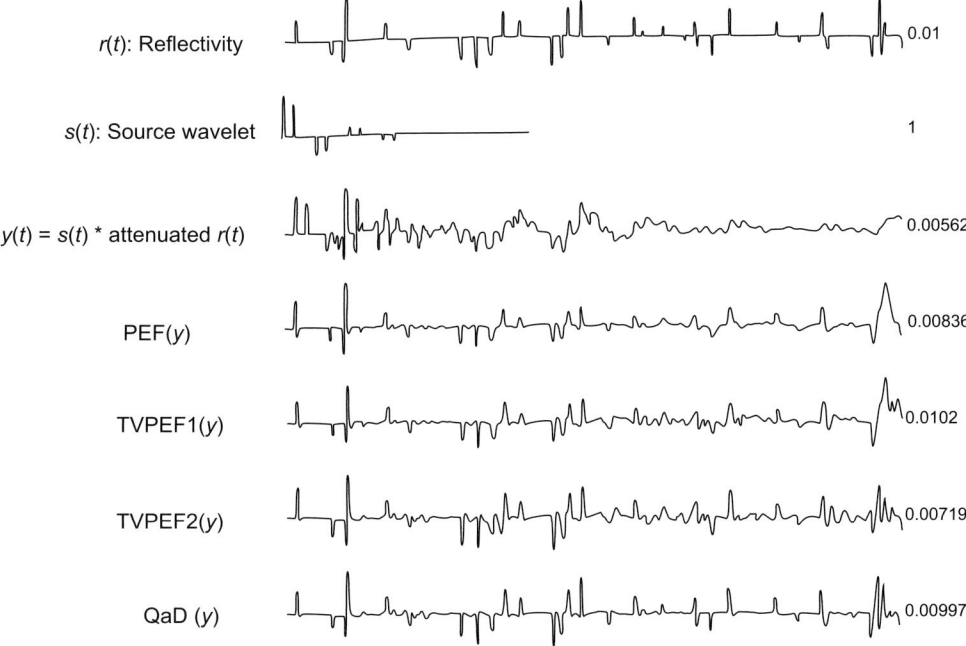

Figure 6.13 A comparison of three conventional and *Q*-adaptive deconvolutions. PEF(*y*) is a time-invariant unit-lag prediction error filtering of *y*(*t*), and TVPEF1(*y*) and TVPEF2(*y*) are the first and second time-varying prediction error filtering of *y*(*t*). PEF(*y*), TVPEF1(*y*), and TVPEF2(*y*) were exponentially gained to amplify events at later times. QaD(*y*) is the result of *Q*-adaptive deconvolution. The numbers shown to the right are maximum amplitudes of each trace.

6.3.2   *Q*-adaptive deconvolution

An example of *Q*-adaptive deconvolution is given in Figure 6.13. The upper three traces show the creation of the synthetic seismogram *y*(*t*) by convolving the source wavelet with an attenuated version of the reflectivity function *r*(*t*), using $Q = 100$. The lower four traces are the results from four different deconvolution methods, in comparison with the true reflectivity function *r*(*t*) as the uppermost trace. PEF(*y*) denotes a time-invariant unit-lag prediction error filtering of *y*(*t*), and TVPEF1(*y*) and TVPEF2(*y*) respectively are the first and second time-varying prediction error filtering of *y*(*t*). Amplitudes of the above three predictive error filtered traces were exponentially gained to amplify events at later times. The lowermost trace is the result of the *Q*-adaptive deconvolution, which gives the best comparison with the reflectivity function among the four deconvolution methods.

Exercise 6.3

1. Following the description of the quality factor *Q* in Section 4.5, compute the expressions of a 20-Hz Ricker wavelet at one-way traveltime of 1 s, 2 s, and 3 s in a medium of a constant velocity of 2 km/s and a constant *Q* value of 50.

2. In the situation of non-stationary convolution shown in Box 6.2 Figure 2, if we know the seismogram $g(t)$, the original filter $a(t-\tau)$, and the fact that a constant Q is a good approximation for the medium, how can we determine the Q value?

3. Comment on the similarities of the four deconvolution results with the reflectivity function $r(t)$ in Figure 6.13. How will you quantify the similarity? How can you quantify the solution quality in real cases for which you know only $y(t)$ but not $r(t)$?

6.4 Minimum entropy deconvolution

6.4.1 The entropy concept

The minimum entropy deconvolution (MED) takes a very different approach from that of predictive error filtering and frequency domain division. **Entropy** here means the degree of uncertainty, which is a statistical measure of the data. The MED relies on the concept of minimizing entropy or maximizing the spikiness of the data (Wiggins, 1977). In order to separate the effects of the source wavelet and the reflectivity series, we have so far assumed that the reflectivity series is white noise. This means that at every time point there is a reflector of random amplitude. This assumption clearly does not match what we see in many real cases.

An alternative assumption is that the reflectivity contains a few spikes of random amplitude. In terms of probability theory, we can cast this assumption in terms of a **probability density function** (PDF) for the reflection amplitude as function of the reflector depth z. Figure 6.14 shows two types of PDF. Panel (a) shows a broad PDF, indicating a situation in which there are many reflectors with significant amplitudes. In contrast, panel (b) shows a slender PDF, indicating there are few reflections with significant amplitude. The case of slender PDF means a spiky reflectivity function, which is the objective of the MED method. In the following, we introduce a measurement of the "spikiness" known as **kurtosis**.

6.4.2 Measurement of kurtosis

Moments of a distribution are the most common statistical measures of a stochastic variable $\{x_t\}$. The **kth moment** of a distribution is defined as

$$m_k(x) = E\left[(x_t - E[x_t])^k\right] \tag{6–44a}$$

where $E[\ ]$ is the expectation operator. For zero-mean distributions, the kth moment is defined as

$$m_k(x) = E\left[x_t^k\right] \tag{6–44b}$$

Using the above formula, we can define several common types of zero-mean moments:

$$\text{Mean:} \quad m_1(x) = \sum_j x_j \tag{6–45}$$

$$\text{Variance:} \quad m_2(x) = \sum_j x_j^2 \tag{6–46}$$

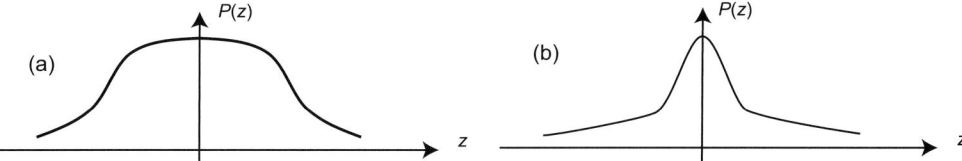

Figure 6.14 Two probability density functions of z, the depth of a reflector. (a) A broad PDF. (b) A slender PDF.

$$\text{Approximated skewness:} \quad m_3(x) = \sum_j x_j^3 \tag{6–47}$$

$$\text{Varimax:} \quad m_4(x) = \sum_j x_j^4 \tag{6–48}$$

Then the kurtosis measurement is defined as:

$$\text{Kurtosis } V(x) = \frac{m_4(x)}{[m_2(x)]^2} = \sum_j x_j^4 \bigg/ \left(\sum_j x_j^2\right)^2 \tag{6–49}$$

which is also known as the **varimax** norm. Now we are ready to use kurtosis to quantify the spikiness, and we assume that this spikiness is reversely proportional to entropy. In other words, maximizing the kurtosis means minimizing the entropy.

In addition, note that we can measure the asymmetry about the mean, called **skewness**, by

$$\text{Skewness } S(x) = \frac{m_3(x)}{[m_2(x)]^{3/2}} \tag{6–50}$$

The two measurements given above are scale-independent; i.e., for a constant a, $\{ax_t\}$ and $\{x_t\}$ will have the same kurtosis and skewness.

Let us consider a simple example of kurtosis for the two spiky time series shown in Figure 6.15.

a. $x_0 = x_1 = 1$, $V = \frac{1+1}{(1+1)^2} = 1/2$

b. $x_0 = \alpha$, $x_1 = 0$, $V = \frac{\alpha^4}{(\alpha^2)^2} = 1$

The comparison of this simple case indicates that in general a distribution with fewer spikes will have a higher kurtosis V.

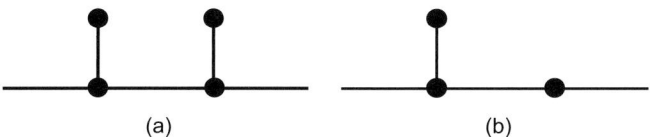

Figure 6.15 Two simple time series.

6.4.3   Derivation of the minimum entropy deconvolution

We assume a convolution model that seismic data g_t are a convolution of the reflectivity function x_t with a source wavelet s_t:

$$g_t = x_t * s_t \tag{6–51}$$

We attempt to recover x_t by constructing a filter f_t so that

$$x_t = f_t * g_t \tag{6–52}$$

Thus, by the convolution formula

$$x_j = \sum_k f_k g_{j-k} \tag{6–53}$$

We solve for the filter f_t by maximizing the kurtosis of x_t and taking its length as arbitrary:

$$V(x) = \sum_j x_j^4 \Big/ \left(\sum_j x_j^2 \right)^2$$

We know that maximizing $V(x)$ is equivalent to minimizing $\ln V(x)$, since the logarithm is monotonic. Then

$$\ln V(x) = \ln \sum_j x_j^4 - 2 \ln \sum_j x_j^2$$

Differentiating $\ln V(x)$ with respect to f_k and then setting the result to zero, we get

$$\frac{\partial}{\partial f_k} \ln V(x) = \left(\sum_j x_j^4 \right)^{-1} \sum_j x_j^3 \frac{\partial x_j}{\partial f_k} - 2 \left(\sum_j x_j^2 \right)^{-1} 2 \sum_j x_j \frac{\partial x_j}{\partial f_k} = 0$$

This means, for $k = 1, 2, \ldots, N$ (length of the filter),

$$\sum_j x_j \frac{\partial x_j}{\partial f_k} = \frac{\sum_j x_j^2}{\sum_j x_j^4} \sum_j x_j^3 \frac{\partial x_j}{\partial f_k}$$

From $x_j = \sum_k f_k g_{j-k}$, we have $\frac{\partial x_j}{\partial f_k} = g_{j-k}$. The above equation then becomes

$$\sum_j x_j g_{j-k} = s \sum_j x_j^3 g_{j-k} \tag{6–54}$$

where the scaling factor is

$$s = \sum_j x_j^2 \Big/ \sum_j x_j^4 \tag{6–55}$$

On the left-hand side of (6–54) we substitute using (6–53)

$$\sum_j x_j g_{j-k} = \sum_j \left(\sum_1 f_1 g_{j-1} \right) g_{j-k} = \sum_1 f_1 \sum_j g_{j-1} g_{j-k} = \sum_1 f_1 r_{1-k} \quad (6\text{–}56)$$

where r_{1-k} is the auto-covariance of g_j.

Equation (6–54) then becomes

$$\sum_1 f_1 r_{1-k} = s \sum_j x_j^3 g_{j-k} \quad (6\text{–}57)$$

where the left side is a filtering, and the right side is a cross-correlation between x_t^3 and g_t. In other words,

$$\mathbf{R}f = s \sum_j x_j^3 g_{j-k} \quad (6\text{–}57')$$

The above equation contains two kinds of unknowns, f and x, and the equation is non-linear with respect to x. It can be solved by an iterative process.

First iteration:

- Let $f_t^{(0)} = \delta(t)$
- Compute r_t, the auto-correlation of g_t.

Further iteration:

- Compute $x_t^{(n)} = f_t^{(n)} * g_t$
- Compute scaling factor $s^{(n)} = \sum_j (x_j^{(n)})^2 / \sum_j (x_j^{(n)})^4$
- Compute the cross-correlation $\sum_j x_j^3 g_{j-k}$
- Update filter by solving (e.g., using least squares) $\mathbf{R}f = \mathbf{d}$
  where $\mathbf{d} = s \sum_j (x_j^{(n)})^3 g_{j-k}$.

The iteration of the above MED process stops when f_t ceases changing with further iterations.

6.4.4 Examples of minimum entropy deconvolution

Two field data examples of the MED are shown here. Figure 6.16 shows a CMP gather before and after the processing of MED. The MED reduces the number of high-amplitude reflectors, particularly in an area denoted by the ellipses. The method is most effective for areas with several well-distinguishable reflectors.

Figure 6.17 shows another comparison for a stacked section. The MED clearly reduces the number of high-amplitude reflectors. However, the multiple reflections are not suppressed much by the MED, because the underlying principle for MED is to reduce the occurrence of high-amplitude events without any preference towards either primary or multiple reflections. If the objective is to suppress multiples, a predictive deconvolution may be more effective.

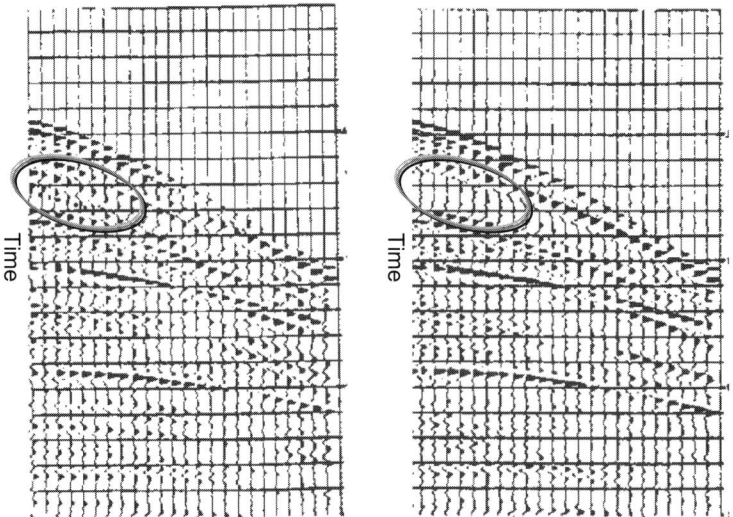

Figure 6.16 Two CMP gathers before (left) and after (right) applying a minimum entropy deconvolution. Ellipses show an area of contrasting differences.

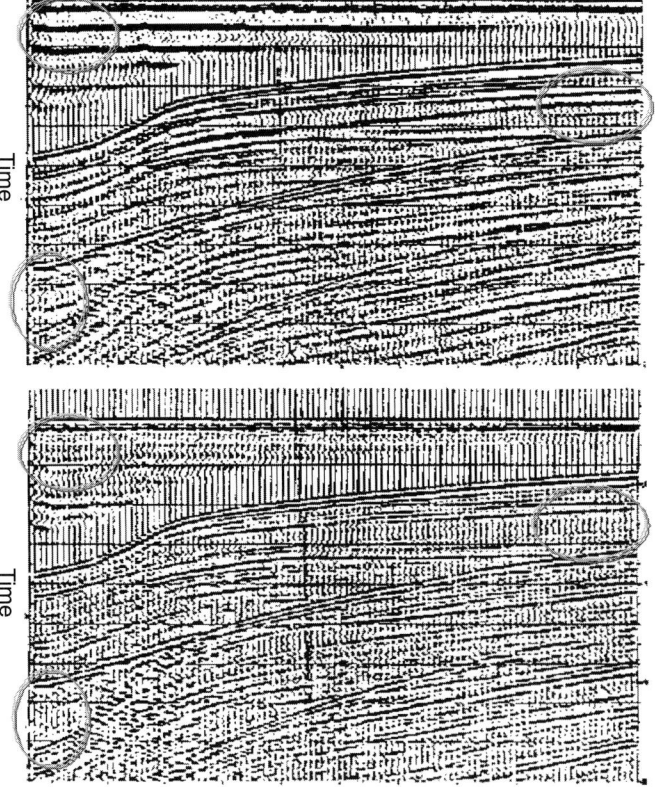

Figure 6.17 An example stack section before (upper) and after (lower) a minimum entropy deconvolution. Ellipses show places of contrasting differences.

Exercise 6.4

1. The four zero-mean moments (mean, variance, skewness, and varimax) have a number of applications in data processing, geo-statistics, and attribute studies. Search geophysics literature for their applications and make a spreadsheet to document these examples.

2. For 2D land pre-stack reflection seismic data, design processing flows to suppress noises specified below. State the type of gather, method, assumptions, and potential pitfalls or cautions.

   (i) Surface wave with source on the receiver line;

   (ii) Surface wave with source off the receiver line;

   (iii) Multiple reverberations.

3. Discuss the effects of minimum phase and white noise in deconvolution. For each factor, illustrate your point using a particular deconvolution method.

6.5 An example of deterministic deconvolution

6.5.1 Motivation

The previous discussion on conventional deconvolution methods has been focused on statistical estimation of two of the three factors involved in a convolution (e.g., Robinson & Treitel, 1980; Ziolkowski, 1984; Yilmaz, 1987). If we know two of the three factors, the determination of the remaining factor is a deterministic process. In this section, we discuss a method called **extrapolation by deterministic deconvolution** (EDD), which is extended from the approach of deconvolution via inversion (Oldenburg *et al.*, 1981; Treitel & Lines, 1982; Claerbout, 1986).

The motivation for EDD comes from the desire for higher resolution from seismic data. For industry and environmental applications, many targeted features such as reservoir sands are at sub-seismic resolution. Individual sands are rarely thicker than a few meters, but the wavelength of conventional seismic data is ten times greater. On the other hand, high-resolution seismic data such as well measurements or hazard surveys sometimes do exist. If high frequencies are missing from the data, they cannot be brought back by processing. However, if high-resolution and low-resolution data are available at the same location, we may be able to determine the relationship between them. If this relationship is stable and does not vary much spatially, we may be able to extrapolate or predict the high frequencies at a neighboring location where only the low-resolution data are available. This is the basic idea of the EDD method.

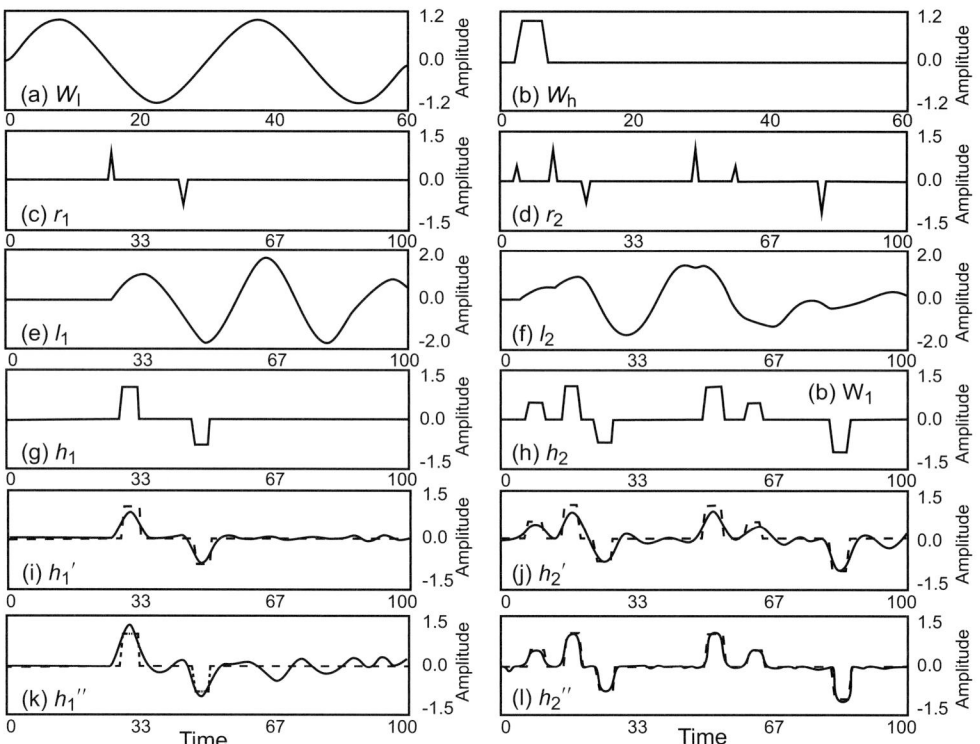

Figure 6.18 A synthetic example of extrapolation by deterministic deconvolution. A low-resolution wavelet (a) and a high-resolution wavelet (b) are convolved with the reflectivity functions at two locations (c) and (d) to produce the low-resolution data in (e) and (f) and high-resolution data in (g) and (h). In the first test, the low-resolution data at two locations (e) and (f) plus the high-resolution data at the first location (g) were used to predict the high-resolution traces in (i) and (j). In the second test, the low-resolution data at the two locations and the high-resolution data at the second location (h) were used to predict the high-resolution traces in (k) and (l). The dashed traces in the lower four panels are the true high-resolution data for comparison.

6.5.2   Derivation

Figure 6.18 shows an example of EDD. Panels (a) and (b), respectively, are a low-resolution wavelet w_l and a high-resolution wavelet w_h. Panels (c) and (d) are reflectivity functions, r_1 and r_2, at two different locations. As shown in panels (e) and (f), the low-resolution seismic responses using wavelet w_l are the convolution of the wavelet with the two reflectivity functions

$$l_1 = r_1 * w_l \qquad\qquad (6\text{–}58a)$$

$$l_2 = r_2 * w_l \qquad\qquad (6\text{–}58b)$$

where the star (*) symbol denotes convolution. Similarly, the high-resolution responses using w_h are

$$h_1 = r_1 * w_h \tag{6-59a}$$

$$h_2 = r_2 * w_h \tag{6-59b}$$

which are shown in panels (g) and (h), respectively.

Based on (6–58) and (6–59), a pair of seismic responses at the same location will be related by

$$l_i * g_i(t) = h_i \tag{6-60}$$

where $i = 1$ or 2, and $g_i(t) = w_h(t)/w_l(t)$ is the relationship filter connecting the two responses.

We now make the important assumption that the filter $g_i(t)$ is invariant over the survey area, i.e.,

$$g(t) \approx g_i(t) = w_h(t)/w_l(t) \tag{6-61}$$

This assumes the ratio between the high- and low-resolution wavelets, $w_h(t)/w_l(t)$, is invariant throughout the survey. When this assumption is valid, the EDD is applicable in two steps.

First, we can determine the filter $g(t)$ at the location where both high- and low-resolution responses are available, by deconvolution of (6–60). To maintain a stable $g_i(t)$, $w_l(t)$ must have a broad bandwidth that includes the bandwidth of $w_h(t)$. In practice, the bandwidth can be broadened by **pre-whitening**. To better handle potential singularities at some frequencies, we solve for $g_i(t)$ in (6–60) in the time domain by using a **minimum-norm inverse** (to be discussed in Section 9.3.2).

Second, we use $g(t)$ to extrapolate the high-resolution data at other locations where only the low-resolution data are known. In the first test shown in Figure 6.18, we determined the filter $g(t)$ in (6–60) using l_1 and r_1 at the first location. We then predicted the high-resolution responses h_j' using $g(t)$ and l_i in

$$h'_j = l_j * g(t) \tag{6-62}$$

The predictions are shown by solid curves in Figure 6.18(i) and (j), along with the true high-resolution responses, shown by dashed curves, for comparison. In this figure the reflectivity function at the first location is simpler than that at the second location.

To see the reverse situation, in the second test, we determined the filter $g(t)$ using the low-resolution and high-resolution data at the second location, and then predicted the high-resolution responses as shown in solid curves in Figure 6.18(k) and (l). The quality of the predictions in the second test is comparable with that in the first test.

6.5.3 Examples

Figure 6.19 shows a test of the EDD algorithm on the well-known **Marmousi model** whose data have formed a benchmark for testing various depth migration algorithms (Versteeg & Grau, 1991; Youn & Zhou, 2001). Here synthetic high- and low-resolution data were created

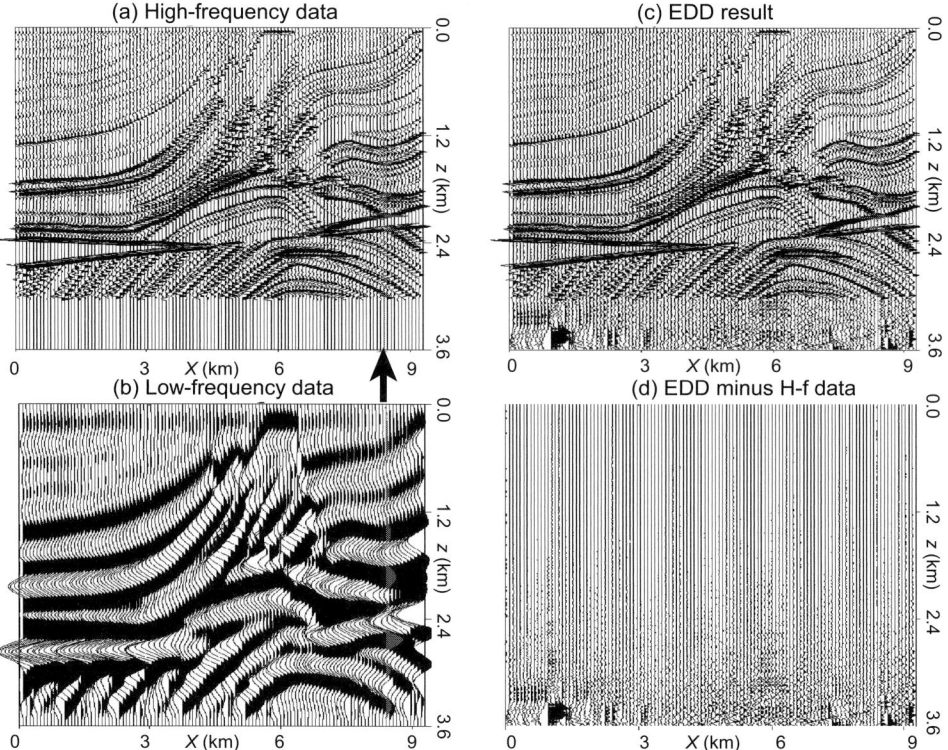

Figure 6.19 (a), (b) High- and low-resolution data obtained by convolving depth changes of the Marmousi model with wavelets of central wavelengths at 50 m and 500 m, respectively. (c) Prediction from the EDD using the low-resolution data (b) and a single high-resolution trace in (a) that is marked by the arrow. (d) Difference between prediction (c) and true model (a). Near the bottom of (c) and (d) there are noises due to edge effects.

by convolution with the Marmousi velocity model. Using the low-resolution data and only a single trace of the high-resolution data, the prediction from EDD as shown in panel (c) is remarkably similar to the true model shown in panel (a), except for some artifacts caused by edge effects near the bottom.

Figure 6.20 shows the result of a test of the EDD on a seismic volume after pre-stack time migration over the Vinton salt dome near Texas/Louisiana border (Constance *et al.*, 1999). In this test, a small time-trace window along Crossline 340 was taken as the high-resolution data $h(x, t)$, shown in panel (c). As shown in panel (a), low-resolution data $l(x, t)$ were created by re-sampling one out of every five points along each trace of the $h(x, t)$, followed by a moving-average smoothing. Shown near the right side of each panel is the flank of the salt dome, with much of the geologic strata dipping away from the dome to the left. The preliminary solution, $h'(x, t)$ in panel (b), was created using the low-resolution data $l(x, t)$ and the first trace of the high-resolution data $h(x, t)$. The result shows that some of the trends of the high-resolution data are restored, though artifacts such as horizontal strips are

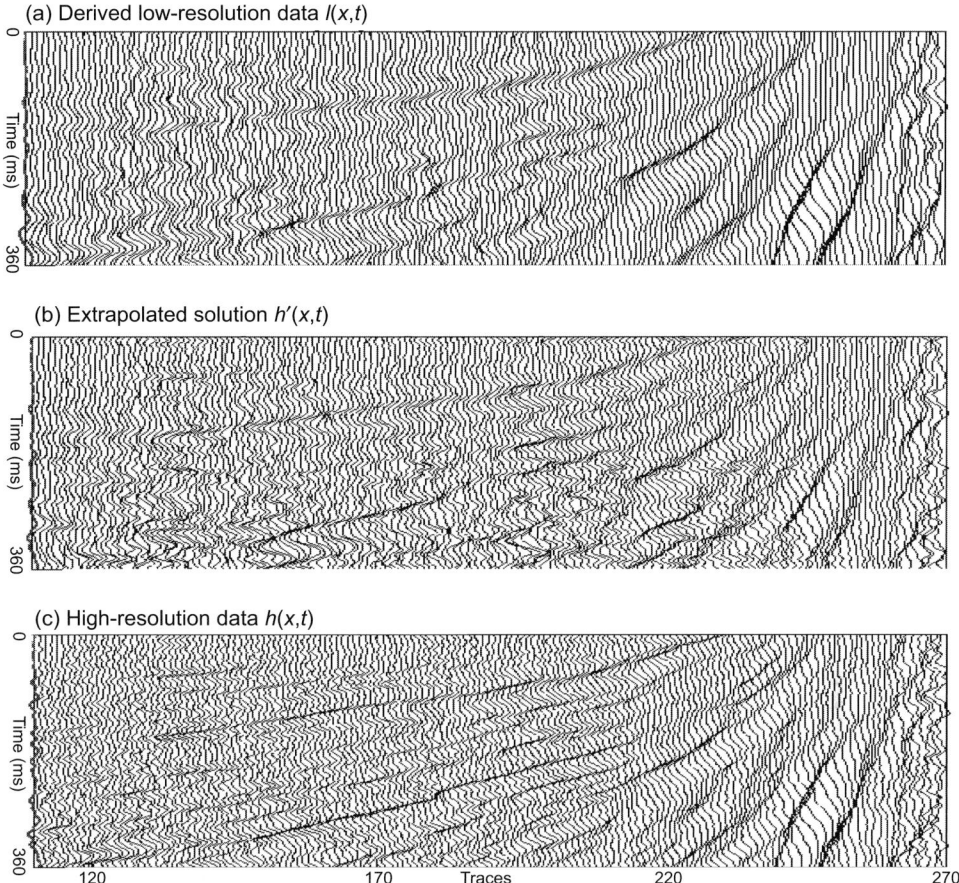

(a) Derived low-resolution data $l(x,t)$

(b) Extrapolated solution $h'(x,t)$

(c) High-resolution data $h(x,t)$

Figure 6.20 Preliminary test of EDD on a field dataset acquired over Vinton salt dome, southwest Louisiana. (c) Part of Crossline 340 of the pre-stack time migrated data. (a) Low-resolution data created by re-sampling (c) and smoothing. (b) Solution of EDD using the low-resolution data in (a) and the first trace of the high-resolution data in (c).

also present, probably due to the fact that the $l(x, t)$ panel was not physically acquired but re-sampled from $h(x, t)$. Considering that only one trace from the high-resolution data was used for such a field data with salt structure, the result shown is quite promising.

Conventional deconvolution has been focused on statistical estimation of two out of the three factors involved in a convolution (e.g., Robinson & Treitel, 1980; Ziolkowski, 1984; Yilmaz, 1987). In contrast, the EDD approach is much more robust because we already know two out of the three factors. This is true in both the first step of inverting for $g_i(t)$ in (6–60) and the second step of forecasting h_j' using (6–62). The EDD method is also applicable to the situation where high-resolution data are available at multiple locations, such as the case of high-resolution hazard surveys. A regional filter $g(t)$ can be determined using part or all of the high-resolution responses and corresponding low-resolution responses. Because EDD is a single-trace operation, it is applicable to 2D and 3D data.

Box 6.3 Making synthetic seismograms for seismic–well tie

For the ultimate goal of identifying the properties of the subsurface reservoirs, a key question is how to tie seismic imagery with wellbore measurements. Well logs achieve high resolution at nearly centimeter scales, but are available only in limited well locations, with extremely high cost. In contrast, seismic imagery typically covers the entire subsurface volume of interest at relatively low cost but very low resolution. For instance, the wavelength of 80-Hz seismic data is 25 m if the average velocity is 2 km/s. Most petroleum-bearing beds are much thinner than the seismic resolution. A major effort of the industry is to make predictions away from the limited well measurements, using seismic imagery.

Well-log-based **synthetic seismograms** are produced traditionally by time-domain convolution of a seismic wavelet with a well-log-based impedance curve. There are two associated assumptions: the convolution model is valid, and the wavelet is invariant in time. While the first assumption is commonly acceptable, the second assumption is invalid in the presence of a moderate amount of attenuation. Modern reservoir characterization has to deal with additional complications such as: (a) the correctness of amplitude and phase of seismic data; (b) scalability of each rock-physical property at different frequencies; (c) validity of the seismic–well tie processing in the presence of fluids; and (d) problems due to the presence of attenuation and anisotropy. More detail will be given in Section 10.5.4.

There are several major obstacles in the making of the seismic–well tie:

- A huge gap exists between the resolutions of the seismic dataset and well dataset. A down-scaling or up-scaling between the two datasets may be impossible for some physical properties that are not scalable, such as viscosity.
- While all well logs are measured along depth, the vertical dimension of seismic data is typically in time. Hence a conversion between the time and depth domains is necessary, requiring an appropriate velocity function.
- In the presence of significant dip in the rock strata or significant deviation of the wellbore from the vertical, it is questionable to make the synthetic seismogram using a convolution model that assumes a layer-cake Earth model.

Box 6.3 Figure 1 compares a well-log-based synthetic seismogram with seismic data in the same field area. The seismic data were acquired using surface shots and receivers and processed through a conventional data processing and migration flow. The well is located at the pilot trace that is shown in a slightly darker color in the figure. The synthetic seismogram was produced in two steps. First, a reflectivity function was built based on sonic log and density log from the well, and it was converted from depth domain to time domain using a smoothed velocity function based on sonic logs. Second, the time-domain reflectivity function was convolved with a Ricker wavelet using frequencies that were comparable with that of the seismic data. To make the comparison more visible, the amplitude of the synthetic seismogram is enlarged by a factor of two in this figure.

To improve the match of the synthetic seismogram with the pilot trace of the seismic data, we can repeat the above two steps of making the synthetic seismograms by altering the frequency and phase of the wavelet, and smoothing the velocity function for the depth-to-time conversion of the reflectivity function. However, we should not arbitrarily stretch

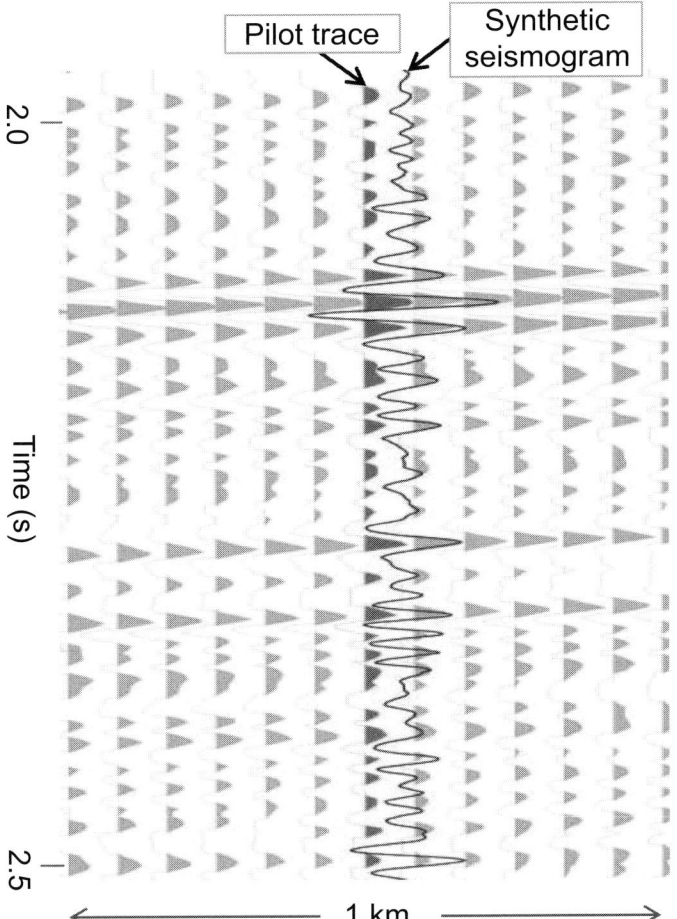

Box 6.3 Fig. 1 Comparison between a synthetic seismogram shown as the black trace and a migrated seismic section shown as the filled gray traces. The well is located at the slightly darker pilot trace. The synthetic seismogram, whose amplitude is enlarged two times, was produced in two steps. First, the reflectivity function based on well log data was converted from the depth domain to time domain using a smoothed velocity function based on sonic logs. Second, the time-domain reflectivity function was convolved with a Ricker wavelet of frequencies comparable to that of the seismic section.

or squeeze the synthetic seismogram to increase the match. A good seismic–well tie can only be proven by a close match between the well-log-based synthetic seismogram and seismic data with a minimum number of constraints. Only a close match can validate the similarities in the physical properties that induced the two types of measurable responses, and hence the possible scalability of the well log prediction at seismic frequencies. Physical limitations often mean that reasonable matches between the synthetic seismograms and the seismic imagery are unachievable.

Exercise 6.5

1. Why is a deterministic processing method usually preferred over a statistical one? Are there exceptions? Please give examples to support your arguments.

2. Will the deconvolution method be applicable to multi-component data? How will you utilize the connection between different data components? Devise an approach and discuss its assumptions, procedure, and potential problems.

3. Make a table describing the deconvolution methods covered in this chapter. Characterize each method using the following entries: the model, the assumptions, requirements (e.g., minimum phase), application procedure, and an example from the literature.

6.6 Summary

- Deconvolution is a direct application of inverse filtering to "undo" a convolution. The original purpose of deconvolution for seismic data processing is to remove the seismic wavelet from the seismic data in order to predict the subsurface seismic reflectivity.

- As a major time processing method, the main benefits of deconvolution include increasing data bandwidth and resolution, suppressing periodicity such as multiples, and removing a known wavelet.

- In practice we may only have the input seismic trace and we need to estimate both the wavelet and the reflectivity. This non-uniqueness problem leads to the approach of predictive deconvolution, which assumes that the predictable components of the input trace belong to the seismic wavelet and the unpredictable components of the input trace belong to the reflectivity.

- To remove a known filter from the data, we may apply a frequency domain deconvolution. However, since the noise will be amplified by the division in frequency domain, we need to take certain measures such as smoothing the data and constraining the deconvolution to a frequency region of a sufficiently high signal-to-noise ratio.

- Because the amplitude and phase of real data vary with wave-propagation time, the deconvolution operator may be applicable only to data within a certain time window. Adaptive deconvolution is a practical way to divide a data trace into small time windows that overlap with each other, to apply deconvolution for each window, and then to integrate the deconvolved results together.

- By quantifying the distribution of seismic wiggles using the concept of entropy, minimum entropy deconvolution seeks to minimize the number of spikes on a seismic trace. The method works well in cases of few major reflectors.

- The method of extrapolation by deterministic deconvolution (EDD) attempts to extrapolate predictions from some locations where joint observations are available to nearby locations that have only a single observation. It provides the possibility of using seismic data to forecast filtered versions of wellbore measurements.

FURTHER READING

Margrave, G. F., 1998, Theory of non-stationary linear filtering in the Fourier domain with application to time-variant filtering, *Geophysics*, 63, 244–259.

Robinson, E. A., 1998, Model-driven predictive deconvolution, *Geophysics*, 63, 713–722.

Schoenberger, M. and Houston, L. M., 1998, Stationary transformation of multiples to improve the performance of predictive deconvolution, *Geophysics*, 63, 723–737.

Ziolkowski, A., 1984, *Deconvolution*, International Human Resources Dev. Cor.

7 Practical seismic migration

As the most widely used subsurface imaging method in petroleum exploration, seismic migration attempts to place seismic reflection data into their correct spatial or temporal reflector positions. Similar to the echo sounding technique to fathom the water bottom from a boat, seismic migration maps the subsurface reflectors in two steps. Step one is to back-project the seismic data measured at the surface downwards using the wave equation and a velocity model, producing an extrapolated wavefield that is a function of space and time. Step two is to use an imaging condition to capture the positions of the subsurface reflectors from the extrapolated wavefield. These two steps are demonstrated by the three common seismic migration methods introduced in this chapter. First, Kirchhoff migration is the most intuitive and flexible migration method, and it uses the ray theory approximation in practice. Second, frequency domain migration is theoretically rigorous and made efficient by taking advantage of the Fourier transform, although it is less effective in the presence of strong lateral velocity variations. Like these two methods, most migrations simplify reality by assuming that the input data contain only primary reflections; hence some pre-processing procedures are necessary to suppress other seismic waves recorded. Third, reverse time migration is a full wave migration method that is capable of using both primary reflections and other waves such as refractions and multiple reflections.

Fundamentally, a velocity model is required for all seismic migration methods. A time migration uses layer-cake models without lateral velocity variations. In contrast, a depth migration may handle a significant level of lateral velocity variations in the velocity model. In the case of gently dipping reflectors, a post-stack migration may be sufficient, using post-NMO stacked traces to approximate zero-offset traces. In the presence of steeply dipping reflectors, a pre-stack migration is usually more suitable but takes many more computational resources. Depending on the complexity of the target structures, we may choose from a suite of migration methods, from the crude but fast post-stack time migration which is not sensitive to velocity variations, to the expensive pre-stack depth migration to handle steep reflector dips and strong lateral variations in the velocity models.

7.1   Seismic imaging via stacking

7.1.1   Modeling, inversion, and migration

Seismic imaging is the process of forming an image of the elastic impedance responses of the subsurface using observed seismic data. There are various types of seismic imaging methods using different types of seismic data such as refractions, reflections, and surface waves; and there are different types of imaging targets and required accuracy. Given a model of the distribution of elastic impedance of the subsurface, we can use forward modeling to calculate traveltimes and waveforms of various types of seismic phases. We may be able to simplify the forward modeling process into a linear operator $\mathbf{A}$ which maps a model vector $\mathbf{m}$ into a predicted data vector $\mathbf{d}$:

$$\mathbf{d} = \mathbf{Am} \tag{7-1}$$

Conversely, an inversion aims to reverse the forward modeling process. It may be possible to use the data vector and the above linear relationship to invert for the model vector. For instance, we can use a generalized linear inversion of the matrix, $\mathbf{A}^{-1}$

$$\mathbf{m} = \mathbf{A}^{-1}\mathbf{d} \tag{7-2}$$

The above two equations show that **seismic modeling** is the process of mapping a given model from the model space into its projection in the data space, and the reverse process of **seismic inversion** is the mapping of a given dataset in the data space into its projection in the model space. There are linear and non-linear operators for both seismic modeling and inversion. Seismic inversion is clearly one type of seismic imaging because it produces subsurface models using seismic data as the input. In current practice, however, it is often unfeasible to apply seismic inversion, owing to limitations in the **data coverage** and effectiveness of the inversion methods. Data coverage for seismic imaging refers to the number of seismic wave or raypaths traversing over the targeted area; we usually prefer a high level of diversity in the traversing angles of the paths over the target area.

The **effectiveness** of seismic processing and imaging methods can be assessed in many ways. Two practical assessments of the effectiveness are **simplicity** and **stability**. If two methods deliver results of similar quality, we prefer the simpler one. A simple method usually has high efficiency in computation. More importantly, it is easier to find the reasons for success and the origin of error in simple methods. Given the common occurrence of noise and insufficient coverage of seismic data, we want to be able to detect the signals as well as artifacts of each seismic imaging method. The assessment of stability is not merely the convergence of computation during an application. A broad definition of stability refers to whether the method works as expected in most situations.

Currently the most effective seismic imaging method is **seismic migration**. This is a process of repositioning seismic energy into its correct spatial or temporal locations. Most seismic migration methods consist of two steps: **downward continuation** and **imaging conditioning**. Downward continuation is an extrapolation process to move seismic data from the recording positions of sources and receivers into a subsurface model space including the target zone, and the result is called the migrated wavefield. Imaging conditioning is the process of producing the migrated images from the migrated wavefield. We may regard the NMO stacking process covered in Chapter 2 as a simple example of seismic migration. In this case, downward continuation means forming the CMP gathers and conducting the NMO processing, and imaging conditioning means stacking the NMO-corrected traces into a post-stack trace for each CMP gather. The quality of the migrated result, the section containing all stacked traces, depends on the validity of the assumptions involved in the NMO and stacking processes. Mathematically, if the forward modeling can be simplified into a linear mapping as shown in (7–1), then seismic migration can be expressed as an approximation of the inversion matrix by a transpose matrix, $\mathbf{A}^T$

$$\tilde{\mathbf{m}} = \mathbf{A}^T \mathbf{d} \qquad (7\text{–}3)$$

where $\tilde{\mathbf{m}}$ is the migrated solution model. This approximation is an **adjoint operator** (matrix transposing) with respect to (7–1). An exemplary list of operators and their adjoints is given in Chapter 5 of Claerbout (1992). For any matrix, it is much easier to find its transpose matrix than its inverse matrix.

7.1.2   Three assumptions of seismic migration

There are three general assumptions for seismic migration. First, all the velocities, in both vertical and lateral sense, are known; this is required for both time and depth migration methods. Second, all of the input signals are primary reflections or diffractions, so that there are no multiples, shear waves, or converted waves. This second assumption has originated from years of practice in exploration geophysics showing that primary reflections are the preferred signal in seismic imaging. This assumption can be removed for some special migration methods that are able to use shear waves, converted waves, and even multiples (e.g., Youn & Zhou, 2001). Finally, all events for a 2D migration come from vertically beneath the seismic line so that there is no **sideswipe** energy. This third assumption spells out the difference between 2D and 3D migration methods. Three-dimensional (3D) migration eliminates the third assumption above, and requires that data be collected in a 3D sense

Box 7.1 An example comparison between time and depth migrations

Box 7.1 Figure 1 shows vertical slices of a pre-stack time migration and a pre-stack depth migration of a 3D field data acquired over a salt dome in southwest Louisiana (Duncan, 2005). Most parts of each panel show imagery of sediments along the flank of a salt dome situated near the right end of each slice.

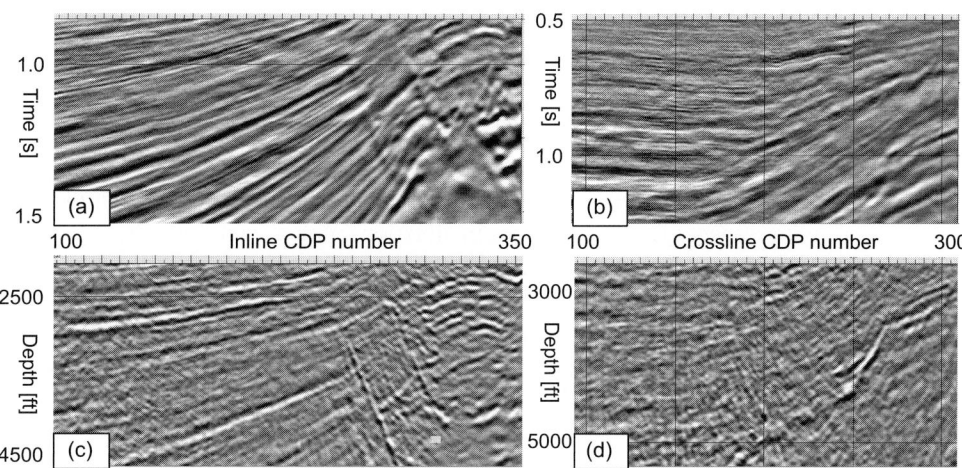

Box 7.1 Figure 1 Vertical slices of 3D pre-stack migrations over a salt dome. (a) and (b) are slices of a pre-stack time migration along south–north and west–east directions. (c) and (d) are slices of a pre-stack depth migration along south–north and west–east directions. The CDP line spacing is about 60 feet.

As shown here, the positions of reflectors often differ considerably, both laterally and vertically, between the time and depth migration results. In this case, the time-migrated results have higher lateral continuity than the depth-migrated results, but the depth-migrated reflectors and faults have steeper dip angles and appear to be more plausible geologically.

with evenly spaced traces in both the x and y directions. Energy is repositioned in both the **inline** and **crossline** directions. After 3D migration, inline images and crossline images will tie in with each other.

In general, seismic migration moves the input energy both laterally from trace to trace and vertically along either time or depth axis. In cases of constant velocity, seismic migration may help in focusing the imaged reflectors and improving their resolution. A **depth migration** maps the input data into the subsurface depth or time spaces, and the mapping process honors the lateral velocity variations. A **time migration** conducts a similar mapping process but assumes there are no lateral variations in the velocity model. Hence, the difference between depth migration and time migration is whether one accounts for lateral velocity variations during the mapping process (see Box 7.1 for a comparison of results).

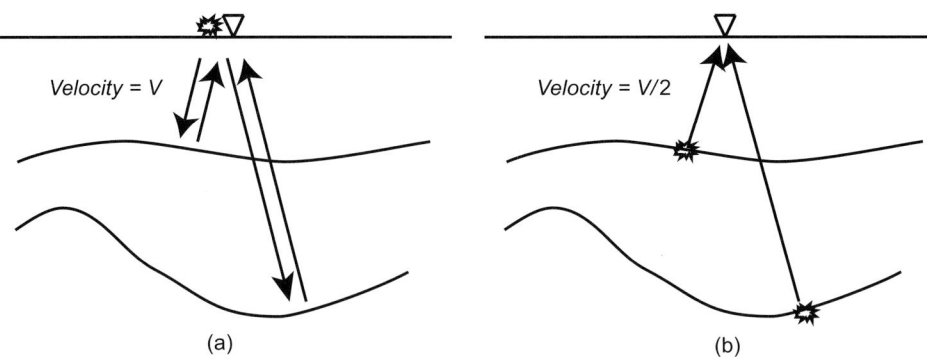

Figure 7.1 (Left) Zero-offset experiment. (Right) Exploding reflector model.

The main advantages of time migration are efficiency in computation and low sensitivity with respect to velocity errors. In situations with a crude velocity model, we can use time migration to get a quick-and-dirty migrated image. However, only depth migration can properly account for lateral velocity variations, and is therefore useful in migration velocity analysis. Although after a time migration we can obtain a depth section by a time to depth conversion using a velocity model, the result will not be considered to be a depth-migrated model.

7.1.3 The exploding reflector model

When the velocity structure can be approximated by local layer-cake models, the NMO stacking process has the advantages of enhancing the SNR and reducing the data volume. The corresponding **post-stack migrations** are much more efficient than the pre-stack migrations. Because of the assumption about the locally layer-cake models, post-stack migrations are less sensitive to velocity variations than the pre-stack migrations. Kinematically, post-stack data can be viewed as equivalent to zero-offset data. For zero-offset data, an **exploding reflector model** can be used to increase the computational efficiency of post-stack migrations. As shown in the right panel of Figure 7.1, this model decomposes all reflectors into a number of point sources. In the case of a constant velocity V, the seismic reflections and diffractions have the same arrival times in both the zero-offset seismic experiment in the left panel and the exploding reflector model, with half the velocity, in the right panel. Except for the difference in the velocity values for the medium, the two panels are kinematically equivalent for the seismic reflections and diffractions.

In terms of the imaging quality, however, there are several problems with the post-stack migrations due to the NMO stacking. First, the NMO stacking produces depth point smear due to both reflection point dispersal and the large size of the horizontal resolution even in case of flat reflectors. Second, the NMO stretch, which is most severe at far offset and shallow depth, results in a reduction in resolution, and in alteration of the wavelet. Third, NMO cannot handle coincident events of different move-outs, such as dipping reflectors and diffractions that have the same zero-offset times. Finally, it is insensitive to velocity variations, particularly lateral velocity variations. In contrast, **pre-stack migration** aims to solve all of the above problems. A major drawback of pre-stack migration is its huge

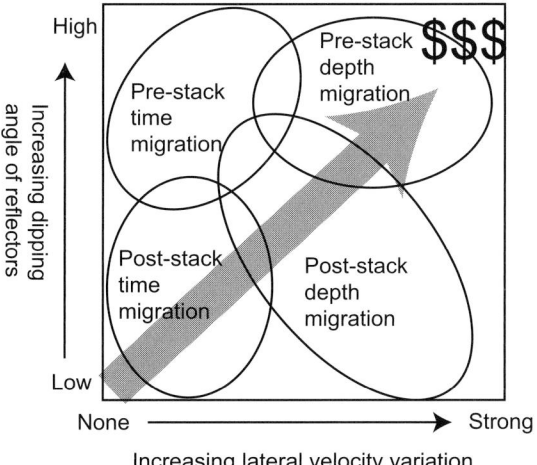

Figure 7.2 Suitability of different migration approaches in terms of dipping angle of main reflectors, level of lateral velocity variation, and cost (modified from Liner, 1999).

appetite for computation power in terms of large memory size, high speed of computing, and fast data transmission.

7.1.4 Classification of migration methods

A practical issue in applying seismic migration is the choice of the appropriate migration method for the given dataset, intended targets, complexity of the geology, and project time span. A practical solution is to quantify the complexity of the geology in two aspects: the level of lateral velocity variations and the dipping angle of main reflectors, especially the targets. Together with the cost in money and time, the suitability of four main migration approaches is depicted in Figure 7.2. Time migrations work well in cases of none or little lateral velocity variations, while the presence of strong lateral velocity variation requires the use of depth migration after **velocity model building** (**VMB**). Post-stack migrations are suitable and efficient in cases of low dipping angle of reflectors, but the presence of steeply dipping reflectors of interest demands the use of pre-stack migrations.

For each approach to seismic migration shown in Figure 7.2, there are a number of migration methods based on various principles. Figure 7.3 shows a hierarchical list of some common types of seismic migration methods (Bednar, 2005). In the rest of this chapter, three of these methods will be introduced. The effect of seismic migration on some simple but common structures can be appreciated intuitively:

- Flat horizons remain unchanged if there are no lateral velocity variations above them;
- Dipping horizons become steeper, shallower in time, and move laterally updip;
- Synclines become broader, with any bow ties eliminated;
- Anticlines become narrower; and
- Diffractions are collapsed to their apex points.

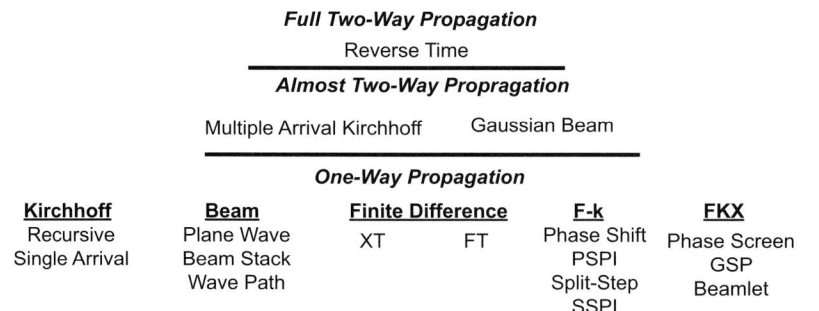

Figure 7.3 A hierarchy of migration methodology (Bednar, 2005).

Exercise 7.1

1. In what situations is the exploding reflector model in Figure 7.1 not applicable?
2. Discuss the differences between a depth migration and a time migration followed by a time to depth conversion.
3. A well log records a physical property at a constant depth increment of 1 ft $(w_j = w(j * \Delta z)$, $\Delta z = 1$ ft). A depth-to-time conversion of the well log gives a variable time interval due to velocity variations. How would you extract the well log with a constant time interval and a minimum loss of information?

7.2   Kirchhoff integral migration

7.2.1   Common-tangent method

Kirchhoff migration is one of the original seismic migration methods and still one of the most popular today. Before exploring it in detail, let us first look at a simple example of using zero-offset data in a constant velocity model. In such case, migration can be carried out by hand using so-called **common-tangent method**. We can demonstrate this method using a case of zero-offset data for a single reflecting horizon in the constant velocity model shown in Figure 7.4. The input trace, denoted by a dashed line, consists of a number of reflectors along the dipping horizon, such as a point A with two-way time T_A and a point B with two-way time T_B. The true reflection positions of the two input points A and B must lie on the circles, with radius T_A and T_B respectively, centered at the corresponding zero-offset shot and receiver locations denoted by the stars. The common tangent line of all such circles defines the correct position of the reflecting horizon, as denoted by the solid line. With v denoting the velocity, the simple geometry leads to a relation between the dipping angles of the input horizon α_{in} and the migrated horizon α_{out}:

$$\tan \alpha_{in} = \frac{2}{v} \sin \alpha_{out} \tag{7–4}$$

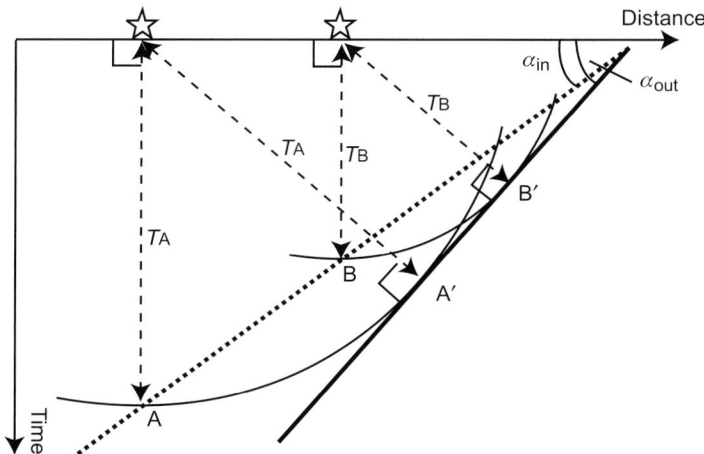

Figure 7.4 Constant velocity migration of a zero-offset dipping horizon (heavy dipping line) from its input position (dashed line) to its output position (solid line). Point A is migrated to point A′, and B is migrated to B′.

Figure 7.5 shows another example of the common-tangent method. Figure 7.5a shows the depth model of a syncline and raypaths of zero-offset reflections. Figure 7.5b shows the reflection time section where each reflection arrival time is plotted right below the corresponding shot-geophone position. The classic "bow-tie" pattern is formed owing to multiple reflections from the synclinal reflector. In Figure 7.5c the common-tangent method is applied with the assistance of a pair of compasses. For a given stacking bin, we measure the arrival time by placing the central pivot of the compass at the stacking bin location at zero time and extending the compass directly beneath this spot so that the pencil point rests on the arrival time curve. With the pivot point fixed, draw an arc of a circle. Repeat this procedure at about every tenth stacking bin and do this for each lobe of the arrival time curve. The curve that is the common tangent to all the arcs is the migrated response, as shown by the thick curve in Figure 7.5c.

Why does the common-tangent method work? One view is to decompose the reflection into a large number of separate pulses. The migration response of each individual pulse is a circular arc in a constant velocity model. The migration of the entire reflector is a superposition or stacking of all these circular arcs. The amplitude along the common tangent of the arcs is cumulated through constructive interference, whereas it cancels at positions away from the common tangent owing to destructive interference.

7.2.2   Gathering approach versus spreading approach

The common-tangent method laid the foundation for the **Kirchhoff migration**, also known as the diffraction summation method. There are two intuitive views of the process of Kirchhoff migration by Schneider (1971). The first view is the **gathering approach**, which loops over each image point in the solution model space and gathers the contributions of all the input traces that are within a **migration aperture**, which is an input zone for migration

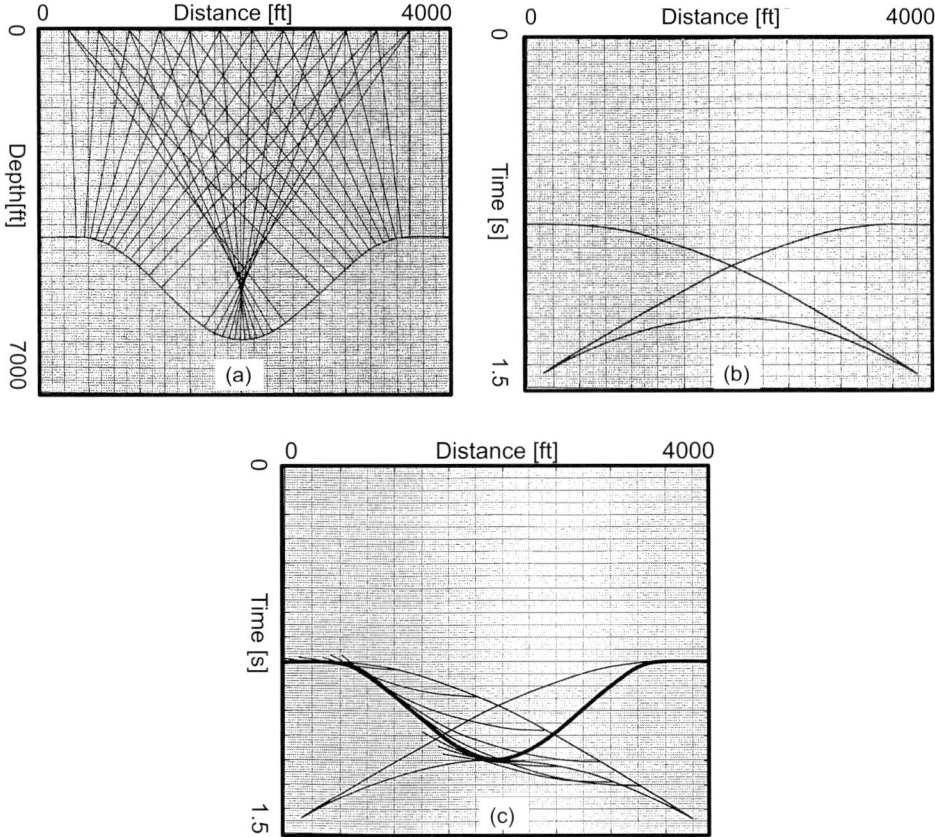

Figure 7.5 (a) A syncline depth model with selected zero-offset reflection rays. (b) Reflection time arrivals as input to migration. (c) The input arrivals together with the migrated arcs using the common-tangent method. The true reflector (solid curve) is along the common-tangent trajectory of all migrated arcs.

defined in the data space. As shown in Figure 7.6, each value of the output trace is a stack of many values of the input traces along the two-way equal-traveltime contour in the data space. In the case of a constant velocity model, the equal-traveltime contour will be a hyperbolic surface centered over the image point. In the case of a variable velocity model, the equal-traveltime contour can be calculated by ray tracing in the velocity model.

The second view is the **spraying approach**, which loops over each input trace in data space and sprays the data over all the image traces within another migration aperture defined in solution model space. As shown in Figure 7.7 using a pre-stack input trace, each value of the input trace is sprayed along the two-way equal-traveltime contour in the solution model space. In the case of a constant velocity model, the equal-traveltime contour will be an elliptical surface with the source and receiver as the two foci. In a variable velocity model, the two-way equal-traveltime contour can also be calculated by ray tracing through the velocity model.

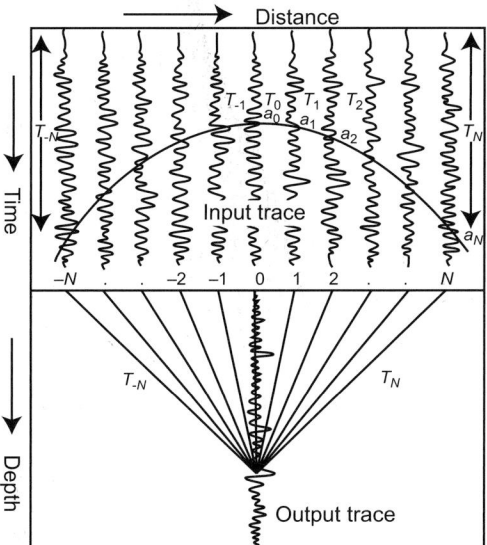

Figure 7.6 The gathering approach, which takes each image point in the solution model space and gathers the contributions it received from all the input traces that are within the migration aperture (Schneider, 1971).

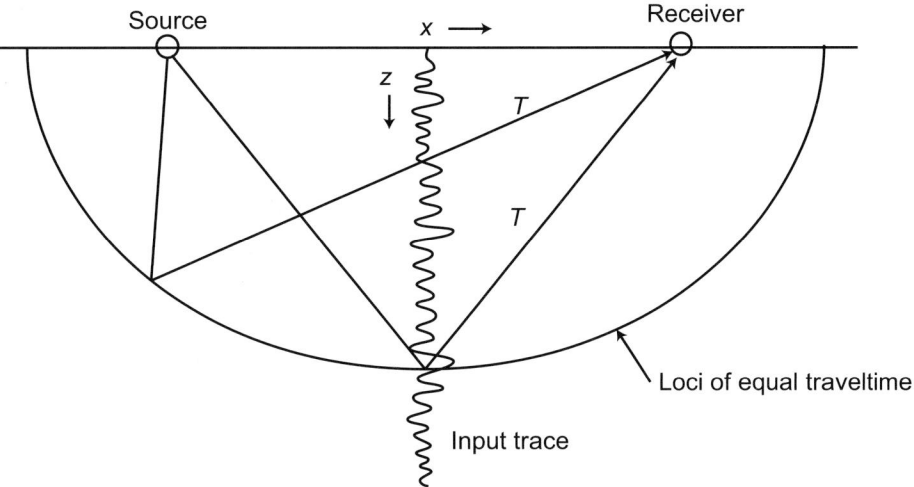

Figure 7.7 The spraying approach, which takes each piece of the input trace (such as the small wiggle shown) and sprays it over all the proper image positions like the elliptic curve shown (Schneider, 1971).

7.2.3   Impulse response and migrated gathers

The previous two views of Kirchhoff migration are rooted in the concept of **impulse response** applied in mapping between the data space and model space. Similar to its usage in filtering theory, here it refers to the response in the mapped space due to an impulse in

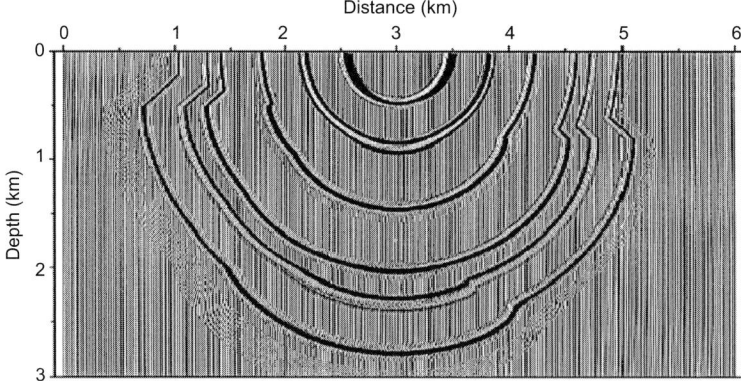

Figure 7.8 An example of the impulse response of a data trace (Zhu & Lines, 1998). The velocity model consists of blocks and the velocity in each block linearly increases with depth.

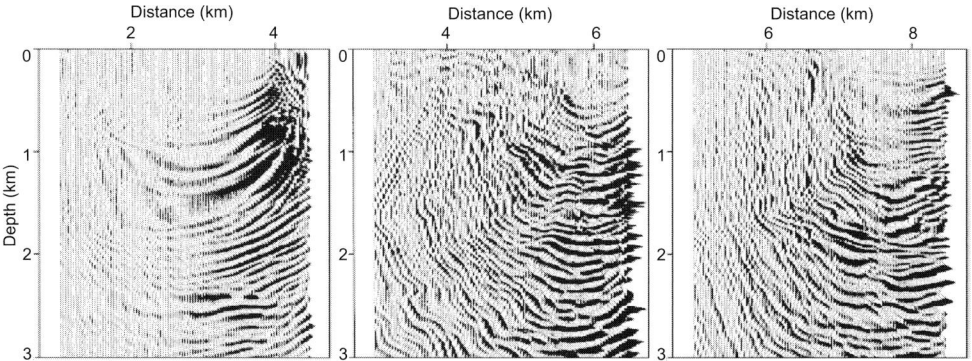

Figure 7.9 Three migrated shot gathers for the Marmousi dataset (Zhu & Lines, 1998).

the mapping space. Using such impulse responses to scan through the spaces involved, one can figure out the entire mapping process. As shown in Figures 7.6 and 7.7, the impulse in the solution model space gives rise to the **gathering approach**, and the impulse in data space gives rise to the **spraying approach**.

The concept of the impulse response can be expanded to use the entire data trace, such as the example shown in Figure 7.8. This is called a **migrated trace gather**, which is the result of apply the spraying approach to an input trace. To extend the concept further, one can apply the spraying approach using a common shot gather of multiple traces, and stack the responses of all trace gathers together to form a **migrated shot gather**. Several examples of migrated shot gathers are shown in Figure 7.9.

The final migrated section can be produced by stacking all the migrated gathers together. In practice, all traces will be migrated first, and the stacking process takes place only in the final stage. Since each input trace will produce many migrated traces over the migrated 2D section or 3D volume, it is a non-trivial matter to design the migration strategy, selecting the proper routine and parameters such as aperture size, sample rate, and weight functions to ensure high efficiency and to minimize problems such as artifacts. The migrated traces can be used to create **partial stacks**, which are stacked sections or volumes involving subsets of the migrated data. The trace gather in Figure 7.8 and shot gathers in Figure 7.9 are examples of such partial stacks. See Box 7.2 for concerns about aliasing.

7.2.4   Mathematical expression of Kirchhoff migration

The mathematical formula for the Kirchhoff migration is one form of the Kirchhoff–Rayleigh integrals. One expression of the Kirchhoff integral is

$$P_{out}(x, z, t) =$$

$$\frac{1}{2\pi} \int dx \left\{ \frac{\cos\theta}{r^2} P_{in}(x_{in},\ z = 0,\ t - r/v) + \frac{\cos\theta}{vr} \frac{\partial}{\partial t} P_{in}(x_{in},\ z = 0,\ t - r/v) \right\} \quad (7\text{–}5)$$

where $P_{out}(x, z, t)$ is the output wave field at a subsurface location (x, z), $P_{in}(x_{in}, z = 0, t)$ is the zero-offset wave field measured at the surface $(z = 0)$, v is the root-mean-square (rms) velocity at the output point (x, z), and $r = [(x_{in} - x)^2 + z^2]^{1/2}$ is the distance between the

Box 7.2 Spatial aliasing concerns

One challenge for migrating and stacking is **spatial aliasing**. Aliasing is a phenomenon of under-sampling of high-frequency events owing to limitations in sample rate. It occurs with multi-trace data, where the mapping process will increase the dip angle of dipping events and therefore create spatial aliasing. Box 7.2 Figure 1 shows an example of migration aliasing, from Zhang *et al.* (2003).

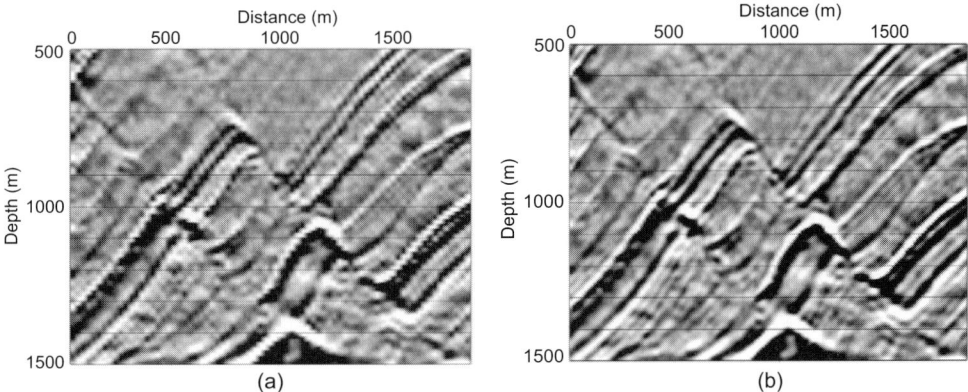

Box 7.2 Figure 1 Common-shot migrated images (Zhang *et al.*, 2003). (a) The imaging condition was applied to uninterpolated wavefields. (b) Imaging-condition aliasing overcome by interpolating the downward-continued wavefields before imaging.

input $(x_{in}, z = 0, t)$ and the output (x, z) points. The above formula is compatible with the migration integral of Yilmaz (1987).

At large distance or r, the first **near-field term** in (7–5), which is proportional to $1/r^2$, is negligible compared with the second **far-field term**, which is proportional to $(1/r)$. As an example, the Kirchhoff migration formula given by Gray (1986) is

$$P(x, z, 0) = \int W(x, z, x_{in}) \left(\frac{\partial}{\partial t} \right)^{1/2} P_{in}(x_{in}, z = 0, t - r/v) dx_{in} \qquad (7\text{–}6)$$

where $W()$ is the weight applied to the input data.

The Kirchhoff integral allows us to appreciate the details of the method. Yilmaz (1987) defines the Kirchhoff migration as the diffraction summation method of migration that incorporates three factors:

- The cosine terms in (7–5) account for the **obliquity factor** or the directivity factor, which describes the angle dependence of amplitudes. It is expressed in terms of cosine of the angle between the direction of propagation and the vertical axis z.
- The denominators of the cosine terms account for the **spherical spreading factor**. Equation (7–5) is formulated for 3D wave propagation. In 2D cases, however, the right-hand side of (7–5) will have $(r)^{-1}$ instead of $(r)^{-2}$, and $(1/vr)^{1/2}$ instead of $(1/vr)$.
- The temporal derivative in the last term of (7–5) and in the right side of (7–6) account for the **wavelet shaping factor**. For instance, **the half-derivative operator** $(\frac{\partial}{\partial t})^{1/2}$ is the temporal expression of the filter $\sqrt{i\omega}$ applied in the frequency domain. It is equivalent to a phase shift of 90° in the 3D case and 45° in the 2D case, aiming to restore the correct phase to the data after migration without greatly altering the locality of the input samples.

7.2.5   A schematic illustration of Kirchhoff migration

Let us see the process of Kirchhoff **pre-stack depth migration** in a schematic illustration. Suppose that a seismic trace d_{ij} is recorded at station R_j from shot S_i. The trace is subjected to a series of pre-processing, such as statics correction to remove near-surface effects, outer mute to remove first arrivals, inner mute to remove ground rolls, and various filtering processes to suppress multiples and other noise. Afterwards, the trace may supposedly contain only primary reflection events. The trace is also corrected for: (1) the obliquity factor which describes the angle dependence of amplitudes; (2) the spherical spreading factor which will make the reflection energy roughly stationary; and (3) the wavelet shaping factor, which is a phase shift of 45° for 2D cases and 90° for 3D cases.

With the processed input trace, d'_{ij}, the mapping process of Kirchhoff pre-stack depth migration is sketched in Figure 7.10. Based on the given velocity model $V(x, z)$, one-way traveltimes from the source and receivers can be calculated, as shown in the first two panels in the figure. Then the two-way traveltime from the source via scatters in any part of the model back to the receiver can be approximated by summing the one-way traveltime fields for the corresponding source and receiver, as shown in panel (c) of the figure. In other words, the two-way traveltime at each pixel in (c) is just the summation of the values of the same pixel positions in the first two panels of the figure. Now following the spraying

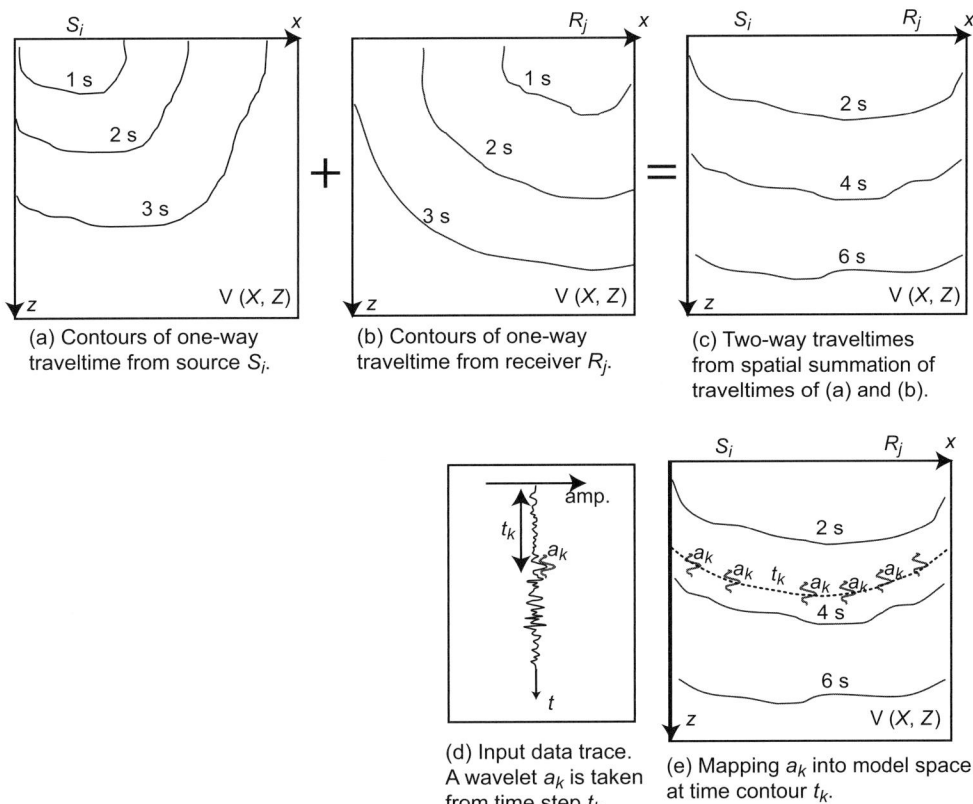

(a) Contours of one-way traveltime from source S_i.

(b) Contours of one-way traveltime from receiver R_j.

(c) Two-way traveltimes from spatial summation of traveltimes of (a) and (b).

(d) Input data trace. A wavelet a_k is taken from time step t_k.

(e) Mapping a_k into model space at time contour t_k.

Figure 7.10 A schematic 2D Kirchhoff pre-stack depth migration. Two-way traveltimes in (c) result from summing one-way traveltimes from source S_i in (a) and from receiver R_j in (b) at every spatial pixel. The Kirchhoff migration maps each piece a_k of input data onto all positions in the output model space along the contour of two-way traveltime t_k.

approach, we can map each point of the input trace (panel (d)) into the model space, to all points along the corresponding traveltime contour as shown in panel (e). Note that such spraying mapping of an input trace will result in a migrated trace gather like that shown in Figure 7.8.

The Kirchhoff migration, based on traveltimes calculated from all shot–receiver pairs to all model grid points, has become a conventional means of pre-stack time and depth migration methods for variable velocity fields. Since it is a ray-theory method, it is cheaper but less rigorous than advanced full-wave migration methods. To carry out the Kirchhoff migration properly, we must have a velocity model of sufficient accuracy and correct for the effects of geometrical spreading, directivity of the shots and geophones, phase shift, and other factors that may affect the pulse shape of reflection energy. A key to improving Kirchhoff migration is the ray tracing methodology, which affects both the accuracy and efficiency of Kirchhoff migration. Advanced ray tracing methods will allow Kirchhoff migration to become a multiple arrival method or to be applicable in the presence of velocity anisotropy. These topics are beyond the scope of this text.

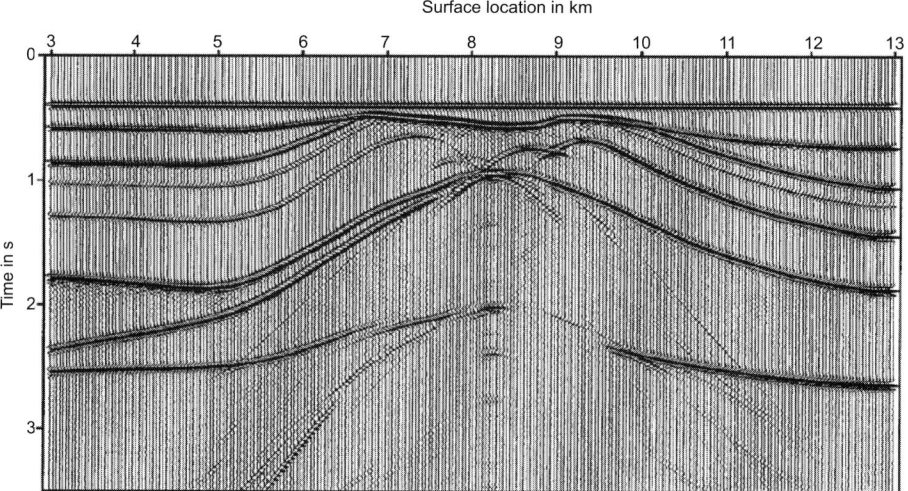

Figure 7.11 A zero-offset seismic section.

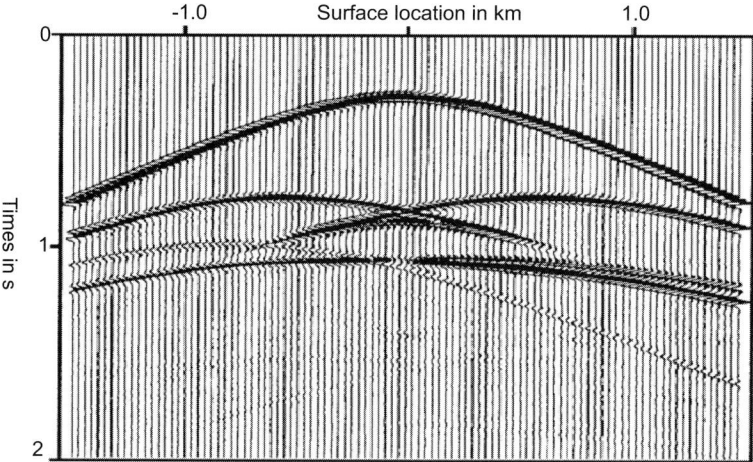

Figure 7.12 A common-shot gather using a line of surface receivers and a single surface shot in the middle.

Exercise 7.2

1. How can you constrain Kirchhoff migration so that it will produce a result similar to that from a NMO stacking? Discuss what is missing from NMO stacking if taking it as a seismic imaging method.

2. For the zero-offset section shown in Figure 7.11, apply the common-tangent method to obtain the migrated section.

3. Figure 7.12 shows a common-shot gather over a five-layer model using a line of surface receivers and a single surface shot in the middle. Write a procedure to migrate these data and show your migration result.

7.3   Frequency domain migration

7.3.1   Development highlights

The methods of carrying out downward continuation and imaging in the frequency domain are known collectively as **frequency domain migration** or **f–k migration**. This type of migration emphasizes two objectives, rigorously following the wave theory and maximizing the computational efficiency. Fourier transform allows seismic waveforms recorded at the surface to be downward-continued to subsurface positions, and the computation is greatly improved by the use of FFT. Ordinary Fourier migration is most suitable for cases of steeply dipping reflectors but relatively low levels of lateral velocity variations. The main advantages of *f–k* migration are:

- A "wave-equation" approach delivering good performance under low SNR conditions;
- Good performance for steep dipping reflectors but smooth lateral velocity variations;
- Computational efficiency.

The main disadvantages of *f–k* migration are:

- Difficulty in handling strong velocity variations, especially lateral abrupt velocity variations, for most *f–k* migration methods;
- Usually requires regular and fully covered shot and receiver geometry.

   The method was developed in the 1970s as people realized the usefulness of expressing downward continuation as a phase shift of monochromatic waves after the creation of the FFT in the 1960s. The ground-breaking publications include that by Stolt (1978) who expressed the post-stack *f–k* migration in constant velocity media as an inverse Fourier transform. This approach of **Stolt migration** will be introduced in detail in this section. Gazdag (1978) showed the phase-shift nature of wave field extrapolation in the frequency domain, leading to the **phase-shift migration** in layer-cake velocity models. An insightful early description of *f–k* migration was given by Chun and Jacewitz (1981). Several years later, in an effort to extend *f–k* migration to cases with lateral velocity variations, Gazdag and Squazzero (1984) refined the phase-shift migration into the **phase-shift plus interpolation** (**PSPI**) method of Fourier migration. A more elegant handling of the lateral velocity variations was given by **split-step Fourier migration** (Pai, 1985; Stoffa *et al.*, 1990). Following this work, many other *f–k* and FKX (frequency, wavenumber, space) migrations (see Figure 7.3) were developed, forming the majority of one-way propagator wave-equation migration methods.

7.3.2   Post-stack *f–k* migration

7.3.2.1   Pre-processing

Let us examine the idea of the **post-stack *f–k* migration** in a 2D constant velocity model. As shown in Figure 7.13, the horizontal distances from the origin to the shot and geophone

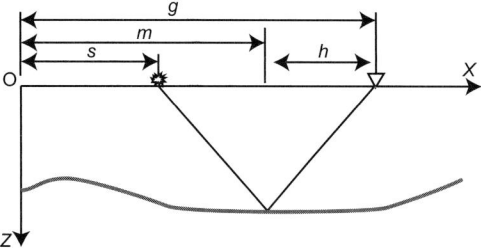

Figure 7.13 Relationship between shot–geophone coordinates and midpoint–offset coordinates on a cross-section.

are denoted as s and g, respectively. The horizontal distance from the origin to the midpoint between shot and geophone is denoted by m, and the offset distance between the shot and geophone is denoted by h. Then the shot–geophone coordinates and the midpoint–offset coordinates are related by

$$m = (g + s)/2 \qquad (7\text{--}7a)$$

$$h = (g - s)/2 \qquad (7\text{--}7b)$$

For a pre-stack 2D dataset $p(s, g, t)$, we can produce a stacked section by taking the following steps:

- Take common-midpoint (CMP) sorting to convert $p(s, g, t)$ into midpoint–offset coordinates $p(m, h, t)$, where t is two-way reflection time;
- Take normal moveout to produce $p(m, h, t_0)$, where t_0 is zero-offset time;
- Stack or sum along the offset axis to produce the stacked section $p(m, t_0)$.

The resulted stacked section is the input to the Stolt migration.

7.3.2.2   Downward continuation via phase shift

The objective of post-stack migration is to produce an image $r(x, z)$ from the stacked section $p(m, t)$, where we have dropped the subscript zero for the zero-offset time. To achieve this we first want to downward-continue the surface data to a depth z:

$$p(m, t, z = 0) - \text{downward continuation} \rightarrow p(m, t, z) \qquad (7\text{--}8)$$

After substituting m by x, the downward continuation is carried out using the acoustic wave equation:

$$p_{zz} + p_{xx} - \frac{1}{v^2} p_{tt} = \left(\frac{\partial^2}{\partial z^2} + \frac{\partial^2}{\partial x^2} - \frac{1}{v^2} \frac{\partial^2}{\partial t^2} \right) p(x, z, t) = 0 \qquad (7\text{--}9)$$

where v is velocity. To solve this **partial differential equation**, we take a 2D Fourier transform (FT) over x and t using:

$$F(k_x, \omega) = \frac{1}{2\pi} \iint f(x, t)e^{-ik_x x + i\omega t} dx dt \qquad (7\text{--}10a)$$

$$f(x, t) = \frac{1}{2\pi} \iint F(k_x, \omega)e^{ik_x x - i\omega t} dk_x d\omega \qquad (7\text{--}10b)$$

Note that differentiation corresponds to multiplication by constants after Fourier transform:

$$\frac{\partial}{\partial x} f(x, t) - FT \rightarrow ik_x \, F(k_x, \omega), \quad \frac{\partial^2}{\partial x^2} f(x, t) - FT \rightarrow -k_{x^2} \, F(k_x, \omega); \quad (7\text{--}11a)$$

$$\frac{\partial}{\partial t} f(x, t) - FT \rightarrow i\omega \, F(k_x, \omega), \quad \frac{\partial^2}{\partial t^2} f(x, t) - FT \rightarrow -\omega^2 F(k_x, \omega) \quad (7\text{--}11b)$$

Therefore, after the Fourier transform, (7–9) is transferred into an **ordinary differential equation**

$$\left(\frac{\partial^2}{\partial z^2} - k_x^2 + \frac{\omega^2}{v^2} \right) P(kx, z, \omega) = 0 \qquad (7\text{--}12)$$

or

$$\frac{\partial^2}{\partial z^2} P = -\left(\frac{\omega^2}{v^2} - k_x^2 \right) P \qquad (7\text{--}12')$$

Solving the above equation as an initial value problem in z, we obtain

$$p(k_x, z, \omega) = P(k_x, z = 0, \omega)e^{\pm i \sqrt{\omega^2/v^2 - k_x^2} \, z} \qquad (7\text{--}13)$$

where the left-hand side is the downward-continued data, the first term on the right-hand side is the surface data, and the second term on the right is the **phase-shift term**. An inverse Fourier transform of (7–13) over k_x and ω leads to

$$p(x, z, t) = \iint P(k_x, z = 0, \omega)e^{\pm i \sqrt{\omega^2/v^2 - k_x^2} \, z} \, e^{ik_x x - i\omega t} dk_x d\omega \qquad (7\text{--}14)$$

What is the physics behind the above procedure? The 2D Fourier transform decomposes the input waveform data into many monochromatic wavefields each with a single variable z, the depth. We can call this **plane-wave decomposition**. For each monochromatic wavefield, its downward or upward movement can be realized by a phase shift as denoted by (7–13). While the example in this equation moves the data wavefields from depth 0 to depth z, you can use the form of this equation to move wavefields from any depth z_1 to another depth z_2 with the corresponding phase shift. After phase shifts of all monochromatic wavefields, the inverse Fourier transform (7–14) is effectively a **plane-wave superposition**.

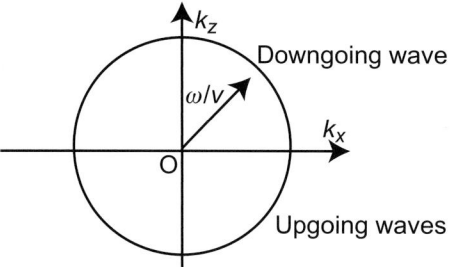

Figure 7.14 Relationship between Fourier components k_x, k_z and ω^2/v^2 for downgoing and upgoing waves.

7.3.2.3   Choosing the one-way operators

We now need to decide the sign in the phase-shift exponent, which determines the type of waves that we use in the migration. To do so, we take another single Fourier transform of (7–12) over z:

$$\left(-k_z^2 - k_x^2 + \frac{\omega^2}{v^2}\right) P(k_x, k_z, \omega) = 0 \tag{7-15}$$

This equation shows the dispersion relation

$$k_z^2 + k_x^2 - \frac{\omega^2}{v^2} = 0 \tag{7-16}$$

which is a circle in the (k_x, k_z) space shown in Figure 7.14

Now we can separate the solutions of (7–10) into **upgoing** and **downgoing** waves.

$$\text{Downgoing waves:} \quad k_z = +\sqrt{\omega^2/v^2 - k_x^2} > 0 \tag{7-17a}$$

$$\text{Upgoing waves:} \quad k_z = -\sqrt{\omega^2/v^2 - k_x^2} < 0 \tag{7-17b}$$

As shown in Figure 7.15, in the case of downgoing waves, z increases as t increases; in the case of upgoing waves, z increases as t decreases. While the pressure field data recorded at $z = 0$ are the same for both cases, the choice of sign for k_z determines where we think the source or the exploding reflector is.

To map subsurface reflectors using upgoing waves, we must use a minus (−) sign for the exponent of the phase-shift term. Then, the downward continuation operator becomes

$$P(k_x, z, \omega) = P(k_x, z = 0, \omega)e^{-i\sqrt{\omega^2/v^2 - k_x^2}\, z} \tag{7-18}$$

or

$$P(x, z, t) = \iint P(k_x, z = 0, \omega)e^{-i\sqrt{\omega^2/v^2 - k_x^2}\, z} e^{ik_x x - i\omega t} dk_x d\omega \tag{7-19}$$

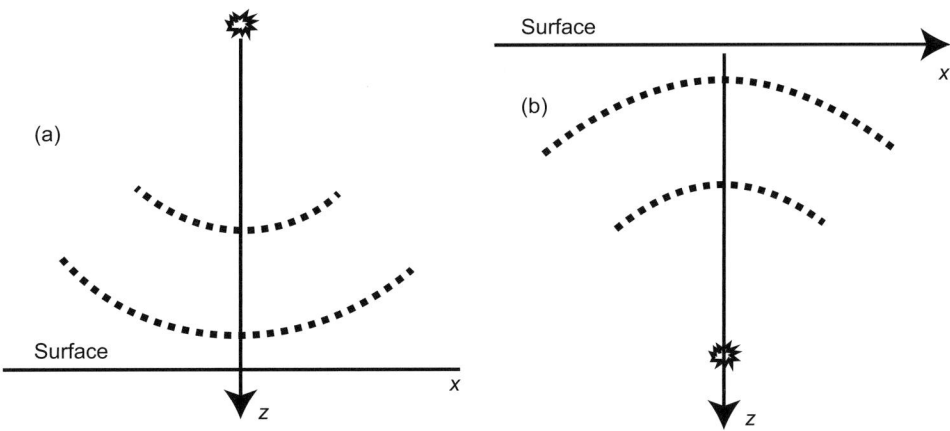

Figure 7.15 Relationship between time and depth for (a) downgoing waves and (b) upgoing waves.

7.3.2.4   Imaging condition for post-stack data

Now let us turn to the second step of migration, the imaging condition. Consider a medium with two reflecting points A and B as shown in Figure 7.16. A line of zero-offset shots and receivers are placed at four depths. At the surface $z = 0$, the record shows two diffractions from the two reflecting points. As the recording line gets closer to the reflection points, the corresponding diffractions become narrower at shorter reflection times. At an arbitrary depth z, the recording line is on top of reflecting point A, and the recorded data show a very narrow diffraction with its apex on the true position of point A. As the recording depth goes deeper, the recorded diffraction from point A appears on the minus side of time.

The above experiment let us conclude that, for the exploding reflector model, the imaging condition is: ***Reflections exist at $t = 0$***. Putting the above together, we have the result of downward continuation as

$$P(x, z, t) = \iint P(k_x, z = 0, \omega)e^{-i\sqrt{\omega^2/v^2 - k_x^2}\,z}e^{ik_x x - i\omega t}\,dk_x\,d\omega \qquad (7\text{--}19')$$

Applying the imaging condition, we have

$$r(x, z) = \lim_{t \to 0} p(x, z, t) \qquad (7\text{--}20)$$

This leads to the 2D post-stack Fourier migration formula:

$$r(x, z) = \iint P(k_x, z = 0, \omega)e^{-i\sqrt{\omega^2/v^2 - k_x^2}\,z}e^{ik_x x}\,dk_x\,d\omega \qquad (7\text{--}21)$$

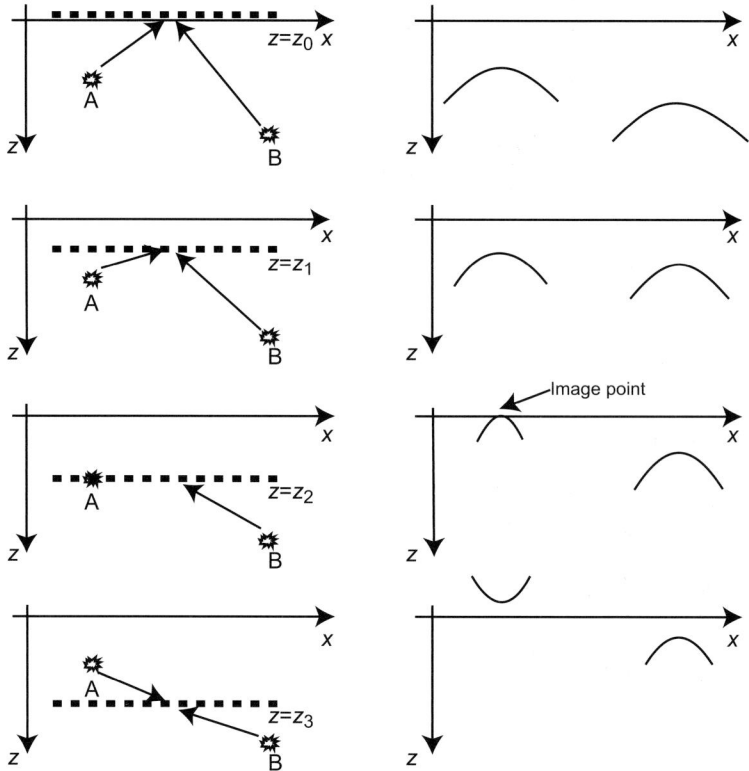

Figure 7.16 The left panels show four steps of an experiment in model space with two reflectors A and B, with the zero-offset shot and receiver lines at four depths. The right panels are the corresponding data recorded at the four depths.

7.3.2.5   Stolt migration

Stolt (1978) noticed that the previous expression for $r(x, z)$ looks almost like an inverse 2D Fourier transform, except for the $\sqrt{\omega^2/v^2 - k_x^2}$ term in the exponent. He suggested a substitution to realize the inverse Fourier transform. Let

$$k_z = -\sqrt{\omega^2/v^2 - k_x^2}$$

Hence

$$\omega = v\sqrt{k_x^2 + k_z^2}$$

and

$$\frac{\partial \omega}{\partial k_z} = \frac{v\,|k_z|}{\sqrt{k_x^2 + k_z^2}} \tag{7–22}$$

which is the obliquity factor. Taking this into equation (7–21) leads to the Stolt migration

$$r(x, z) = \iint Q(k_x, k_z)e^{ik_x x}e^{ik_z z}\,dk_x\,dk_z \tag{7–23}$$

Practical seismic migration

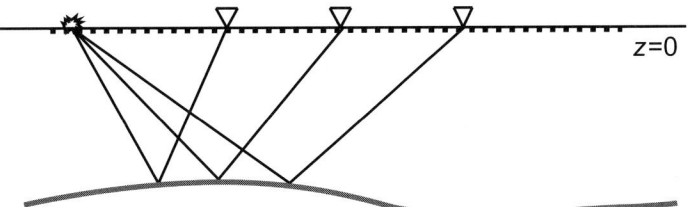

Figure 7.17 A cross-section view of pre-stack shot gather data.

where the Fourier transform of the image $r(x, z)$ is

$$Q(k_x, k_z) = \frac{v\,|k_z|}{\sqrt{k_x^2 + k_z^2}} P\left[k_x, z = 0, \omega = -\mathrm{sgn}(k_z)v\sqrt{k_x^2 + k_z^2}\right] \qquad (7\text{–}24)$$

In the right-hand side of the above equation, the first term is the **obliquity factor**, and the second term is a **mapping factor** or **stretching factor** in ω. The application of the Stolt (f–k) migration has three steps:

- Double Fourier transform data from $p(x, z = 0, t)$ to $P(k_x, z = 0, \omega)$.
- Interpolate $P()$ onto a new mesh so that it is a function of k_x and k_z. Multiply $P()$ by the obliquity factor.
- Inverse Fourier transform back to the (x, z)-space.

A synthetic example of the Stolt migration is shown in Box 7.3.

7.3.3 Pre-stack f–k migration

Let us now examine the process of **pre-stack f–k migration**. As being discussed previously, pre-stack migration avoids one or more of the following problems associated with the NMO stacking:

- Reflection point smear, especially for steep dip;
- NMO stretch of far offset traces;
- Inability to simultaneously NMO-correct for flat reflectors, dipping reflectors, and diffractors that have the same zero-offset times.

On the other hand, pre-stack migration consumes large amounts of computer resource, and the method is very sensitive to errors in the velocity model. As was the case for Stolt migration, the following illustrations use a constant-velocity 2D case to illustrate the multi-offset f–k migration.

7.3.3.1 Survey sink

To image using multiple offset data, a procedure known as **survey-sink** is applied. We start in the (s, g, t) space as shown in Figure 7.17. The pre-stack 2D data have five dimensions, $P(z_s, z_g, g, s, t)$, where the depths of shot and geophones are z_s and z_g, and the distances of

Box 7.3 The original 2D synthetic example of Stolt migration

Box 7.3 Figure 1 shows an example of the Stolt migration using 2D synthetic data (Stolt, 1978). The three synclinal reflectors are imaged well. However, there are "ringing" artifacts associated with the true reflectors in the solution panel (c). These artifacts are due to error in the Fourier transform as well as the edge effect in the numerical computation. Nevertheless, in cases of little lateral velocity variation, Stolt migration is the most efficient wave equation method for post-stack time migration. Its migration velocity is an integral of the overburden velocities that is similar to stacking velocity.

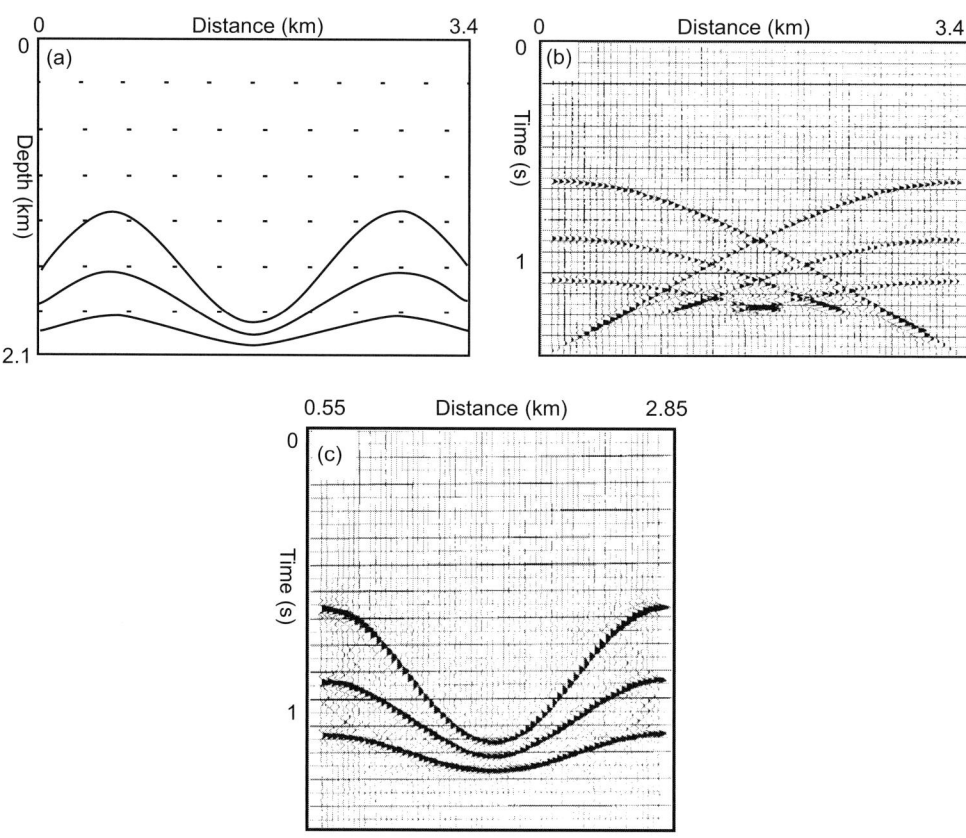

Box 7.3 Figure 1 (a) 2D model with three reflectors in a constant velocity of 2926 m/s. (b) Synthetic seismic section. Trace spacing is 36.6 m. (c) Result of Fourier migration of the synthetic seismic section.

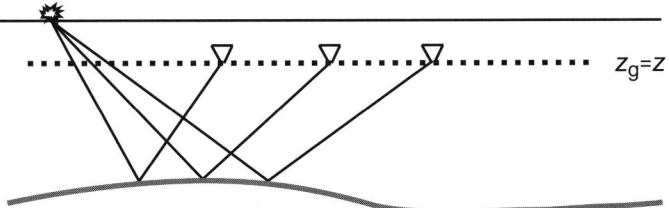

Figure 7.18 A cross-section view of pre-stack common shot gather data after downward continuation of receivers to z.

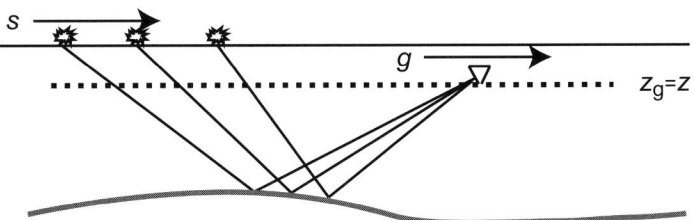

Figure 7.19 A cross-section view of pre-stack common receiver gather data after downward continuation of receivers to z.

shot and geophone are g and s. From the surface-recorded data, we first downward-continue the receivers using the wave equation.

$$p_{zz} + p_{gg} - \frac{1}{v^2} p_{tt} = 0 \qquad (7\text{--}25)$$

The situation is illustrated in Figure 7.18. The downward-continued field is

$$P(z_s = 0, z_g = z, k_g, k_s, \omega) = P(z_s = 0, z_g = 0, k_g, k_s, \omega)e^{-i\sqrt{\omega^2/v^2 - k_g^2}\,z} \quad (7\text{--}26)$$

where k_g and k_s are the Fourier duals of g and s.

After downward-continuing the geophones of all shots, we can sort data into common receiver gathers, as shown in Figure 7.19. Then we can downward-continue the shots of each common receiver gather to depth z. Based on the wave equation

$$p_{zz} + p_{ss} - \frac{1}{v^2} p_{tt} = 0 \qquad (7\text{--}27)$$

Like that shown in Figure 7.20, the corresponding phase-shift equation is

$$P(z_s = 0, z_g = z, k_g, k_s, \omega) = P(z_s = 0, z_g = 0, k_g, k_s, \omega)e^{-i\sqrt{\omega^2/v^2 - k_s^2}\,z} \quad (7\text{--}28)$$

We can switch from common receiver gathers to common shot gathers as shown in Figure 7.21. Now the entire downward-continued data are related with the surface data by

$$P(z_s = z, z_g = z, k_g, k_s, \omega) = P(z_s = 0, z_g = 0, k_g, k_s, \omega)e^{-i\sqrt{\omega^2/v^2 - k_g^2}\,z}e^{-i\sqrt{\omega^2/v^2 - k_s^2}\,z}$$

$$(7\text{--}29)$$

We can drop the distinction between z_s and z_g using $z_s = z_g = z$, or

$$P(z, k_g, k_s, \omega) = P(z = 0, k_g, k_s, \omega)e^{-i\psi(k_g,k_s,\omega)z} \qquad (7\text{--}30)$$

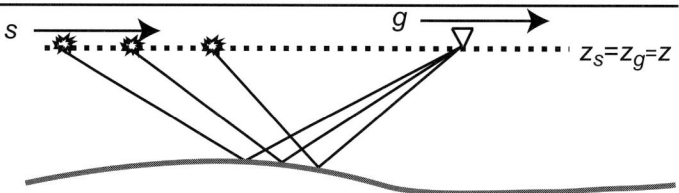

Figure 7.20 A cross-section view of pre-stack common receiver gather data after downward continuation of receiver and shots to z.

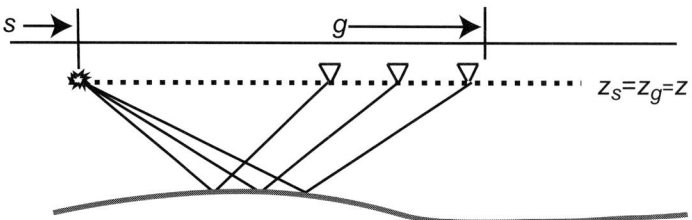

Figure 7.21 A cross-section view of a pre-stack common shot gather data after downward continuation of receivers and shots to z.

where $\psi(k_g, k_s, \omega) = \sqrt{\omega^2/v^2 - k_g^2} + \sqrt{\omega^2/v^2 - k_s^2}$. We now have a procedure for downward-continuing the multi-offset data in physical coordinates (s, g, t) or equivalently (k_g, k_s, ω).

7.3.3.2   Double square root (DSR) equation

In practice midpoint–offset coordinates are preferred over the shot–receiver coordinates. To convert the above phase-shift equation to midpoint–offset coordinates, we derive the following based on equation (7–7)

$$g = m + h \tag{7–31a}$$

$$s = m - h \tag{7–31b}$$

To realize this coordinate change in the wavenumber domain, apply chain differentiation:

$$\frac{\partial}{\partial m} = \frac{\partial g}{\partial m}\frac{\partial}{\partial g} + \frac{\partial s}{\partial m}\frac{\partial}{\partial s} = 1\frac{\partial}{\partial g} + 1\frac{\partial}{\partial s} \tag{7–32}$$

This implies that in Fourier space

$$ik_m = ik_g + ik_s$$

or

$$k_m = k_g + k_s \tag{7–33a}$$

Similarly, we find that

$$k_h = k_g - k_s \tag{7–33b}$$

We therefore have

$$k_g = (k_m + k_h)/2 \tag{7-34a}$$

$$k_s = (k_m - k_h)/2 \tag{7-34b}$$

Now we can convert the data to midpoint–offset coordinates by changing the phase,

$$p(z, g, s, t) \rightarrow p(z, m, h, t)$$
$$P(z, k_g, k_s, \omega) \rightarrow P(z, k_m, k_h, \omega)$$
$$\psi(k_g, k_s, \omega) \rightarrow \psi(k_m, k_h, \omega) \tag{7-35}$$

where

$$\psi(k_m, k_h, \omega) = \sqrt{\frac{\omega^2}{v^2} - \left(\frac{k_m + k_h}{2}\right)^2} + \sqrt{\frac{\omega^2}{v^2} - \left(\frac{k_m - k_h}{2}\right)^2} \tag{7-36}$$

The above is known as the **double square root (DSR) equation** (e.g., Sections 3.3 and 3.4 of Claerbout, 1985b). Note that it is not separable in (k_m, k_h, ω)-coordinates as it was in (k_g, k_s, ω)-coordinates. This is the reason that downward continuation and stacking have to be done together.

To put the downward-continued experiment into (m, h, t) space, we apply a 3D inverse Fourier transform

$$p(z, m, h, t) = \iiint P(z = 0, k_m, k_h, \omega) \, e^{-i\psi(k_m, k_h, \omega)z} \, e^{i k_m m + k_h h - i \omega t} \, dk_m \, dk_h \, d\omega \tag{7-37}$$

This equation facilitates the downward continuation for the pre-stack data.

7.3.3.3   Imaging condition for pre-stack data

To image the pre-stack data, we use a slightly modified imaging condition based on the following general principle:

Reflectors exist where up- and downgoing waves are coincident in time and space.

In our downward-continued experiment, the waves are coincident in time when $t \rightarrow 0$, and are coincident in space when $h \rightarrow 0$ (i.e., no offset between source and receiver). Hence, the formula of our imaging condition is:

$$r(z, m) = \lim_{\substack{t \to 0 \\ h \to 0}} p(z, m, h, t) \tag{7-38}$$

Therefore,

$$r(z, m) = \iiint P(z = 0, k_m, k_h, \omega) \, e^{-i\psi z} \, e^{i k_m m} \, dk_m \, dk_h \, d\omega \tag{7-39}$$

To convert the above to a workable migration method, substitute

$$k_z = -\psi(k_m, k_h, \omega) = -\sqrt{\frac{\omega^2}{v^2} - \left(\frac{k_m + k_h}{2}\right)^2} - \sqrt{\frac{\omega^2}{v^2} - \left(\frac{k_m - k_h}{2}\right)^2} \tag{7-40}$$

This leads to

$$\omega(k_m, k_h, k_z) \equiv \omega = \frac{vk_z}{2}\sqrt{\left(1 + k_m^2/k_z^2\right)\left(1 + k_h^2/k_z^2\right)} \tag{7-41}$$

this is the ω-stretching for pre-stack cases. Therefore,

$$\frac{\partial \omega}{\partial k_z} = \frac{v}{2}\frac{1 - k_m^2 k_h^2/k_z^2}{\sqrt{\left(1 + k_m^2/k_z^2\right)\left(1 + k_h^2/k_z^2\right)}} \tag{7-42}$$

The Jacobian $\frac{\partial \omega}{\partial k_z}$ is known as the obliquity factor.
   Now following (7–39),

$$r(z, m) = \int\int Q(k_m, k_h)e^{ik_z z}e^{ik_m m}dk_z dk_m \tag{7-43}$$

where

$$Q(k_m, k_h) = \int \left|\frac{\partial \omega}{\partial k_z}\right| P(z = 0, k_m, k_h, \omega = \omega(k_m, k_h, k_z))dk_h \tag{7-44}$$

7.3.3.4   Evanescent waves

In making the substitution we have to guarantee that the Jacobian of the transformation, $\left|\frac{\partial \omega}{\partial k_z}\right|$, never goes to zero. This implies the restriction

$$|k_m k_h| < k_z^2 \tag{7-45}$$

This restriction keeps the square roots from going to zero. In terms of the original downward continuation, it keeps **evanescent waves** out of the problem. These are standing waves forming at an interface with their amplitude decaying exponentially away from the interface. In other words, expression (7–45) means that

$$k_g^2 < (\omega/v)^2 \tag{7-46a}$$

$$k_s^2 < (\omega/v)^2 \tag{7-46b}$$

   In practice, an important issue for pre-stack *f–k* migration is to choose adequate range in the offset dimension. Larger offset means wider illumination angle and better solution for steeply dipping events, but demands more computation resources and more stringent requirement for small and smooth lateral velocity variations. The *f–k* migration requires that the input data be sampled regularly, and it is most suitable for cases of variable reflector dipping angles but no or smooth lateral velocity variations.

7.3.4   Two *f–k* depth migration methods

As mentioned previously, in order to handle lateral velocity variations, the pioneers of Fourier migration developed several approaches. One approach, known as the **phase-shift plus interpolation** (**PSPI**) method (Gazdag & Squazzero, 1984), applies several phase-shift migrations using a number of 1D reference velocity models based on the given 2D or 3D velocity models. The final image of the PSPI method is an interpolation of the multitude

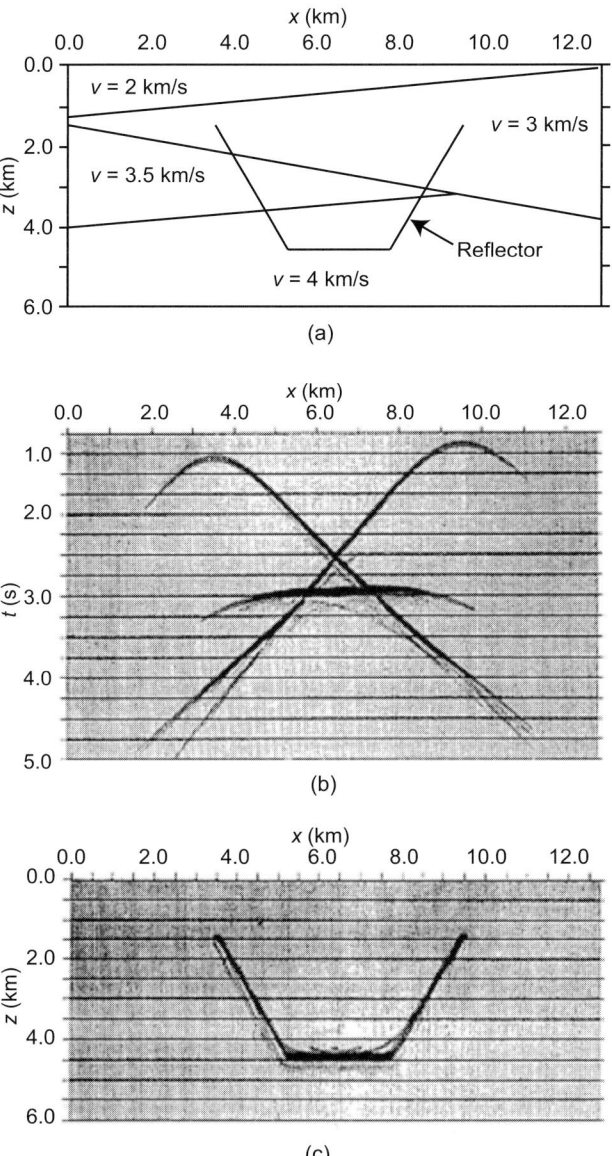

Figure 7.22 PSPI migration of a synthetic data (Gazdag & Squazzero, 1984). (a) Model with a ditch-shaped reflector in a four-layer velocity model. (b) Zero-offset data. (c) Result of PSPI migration.

of 1D-model solutions based on the 2D or 3D velocity models. A synthetic example of PSPI is shown in Figure 7.22. The synthetic "**Earth model**" of the velocity field and reflectors is shown in panel (a). As is traditional in seismic migration, the velocity field is totally decoupled from the reflector distribution. The zero-offset data shown in panel (b) indicate the influence of lateral velocity variations.

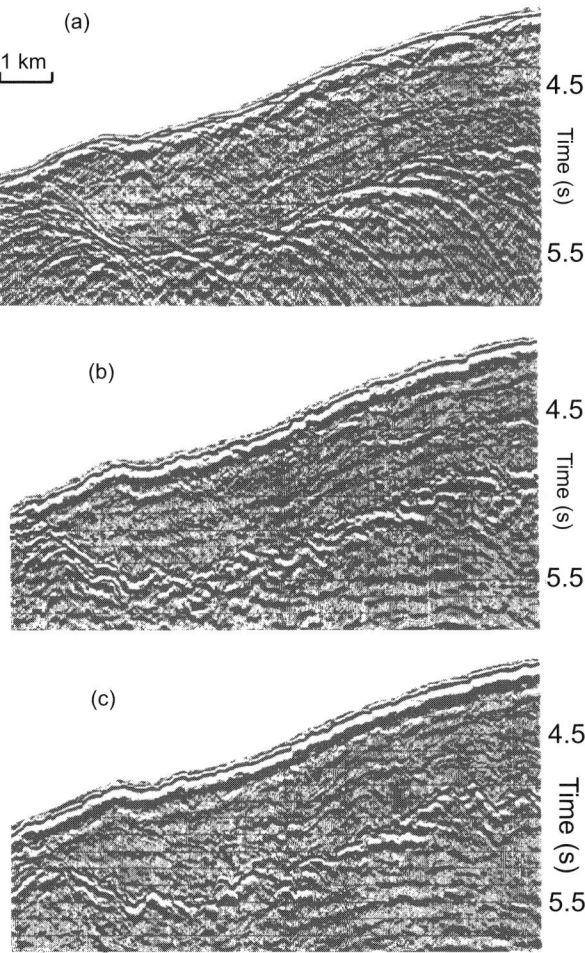

Figure 7.23 Dip-line slices of migration results of field data (Stoffa *et al.*, 1990). (a) Stacked 3D data. (b) 2D split-step migration. (c) 3D split-step migration.

A more elegant approach to the lateral velocity variations was provided by **split-step Fourier migration** (Pai, 1985; Stoffa *et al.*, 1990). This approach decomposes the total velocity model into two portions, a layer-cake average reference velocity field plus the lateral velocity variations. For each depth interval the phase-shift downward continuation of monochromatic wavefields in the wavenumber domain is split into two steps. The first step is laterally homogeneous downward continuation using the layer-cake reference velocity for the depth interval. The second step is another downward continuation using the laterally varying velocity variations to adjust the waveform data within the depth interval.

Figure 7.23 shows the application of 2D and 3D split-step Fourier migration of field zero-offset data acquired over the accretionary wedge off the shore of Costa Rica (Stoffa *et al.*, 1990). Thirty lines of a 3D data volume were used to compare 2D and 3D split-step migrations. Each line has 512 traces with a spacing of 33 m. The line spacing was 50 m. The slice shown is a portion of dip line in the middle of the data volume. The diffractions in the stacked data are gone in both of the migration results. There are considerable differences between the 2D and 3D migration images.

Exercise 7.3

1. Explain the use of Fourier transform in Stolt migration. What are the benefits of using FT? What is the assumption involved in the use of FT?
2. Read the paper by Stoffa *et al.* (1990) on split-step Fourier migration, and create a summary covering the procedure, assumptions, benefits, and limitations of the method.
3. Compare and contrast between the PSPI and split-step *f*–*k* migration methods. How does each method treat the lateral velocity variations?

7.4　Reverse time migration

7.4.1　The concept of reverse time migration

As shown previously in Figure 7.3, **reverse time migration (RTM)** is the only seismic migration method that is able to employ full two-way propagation in the downward-continuation process (e.g., Gray *et al.*, 2001; Bednar, 2005). In laypeople's words, a two-way propagation of seismic waves means consideration of multiple reflections or scatterings. Such a consideration for a layer-cake model means an exponential growth in the number of multiple reflections involving both upgoing and downgoing waves with time. In practice the computation of complicated multiples is feasible only with a forward modeling approach. The RTM takes the real shot and receiver positions as the computation "sources" in forward modeling processes.

The schematic on the left side of Figure 7.24 illustrates the concept of reverse time, by reversing the time sequence of the letters in an echo of "Hello" with respect to the original word from the mouth of the speaker. In other words, if a shot is placed at the speaker's position and a line of receivers are placed along the wall, then to re-create the correct sequence backwards from the receivers to the shot using the signal received by the receivers, we must reverse the sequence of the recorded signal. This situation is shown in the four pairs of panels on the right side of Figure 7.24. Each pair consists of two views in data space and model space at a time step. At the first three time steps with a forward wavefield from the shot to a line of receivers, the data space shows the expansion of a hyperbolic first arrival wavefield recorded by the line receivers. At the last time step, to send the recorded signal from the receivers back to the source, the signal must be time reversed before sending it back.

Reverse time migration is generically applied in the common shot gather domain. For each common shot gather, the downward continuation is realized using two forward modeling processes: one creates a forward wavefield from the shot using an estimated wavelet, and the other creates a backward wavefield from all receivers with the time-reversed recorded data

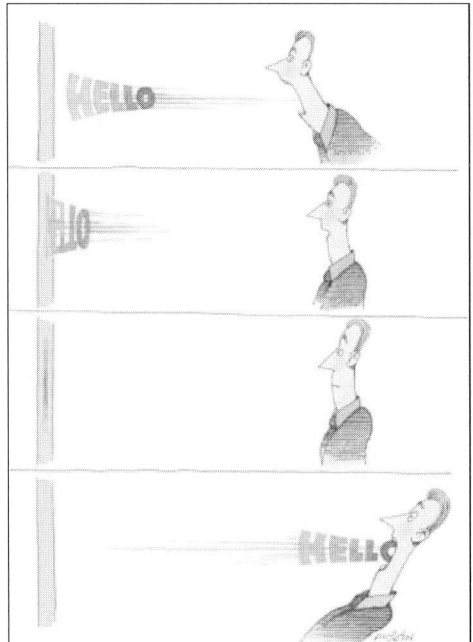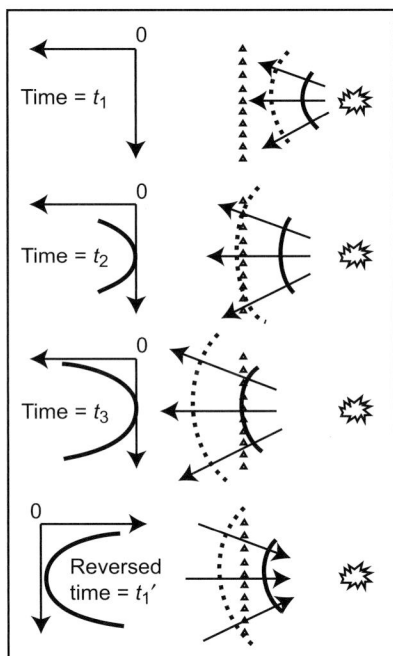

Figure 7.24 (Left) Cartoon by Dusan Petricic on time-reversed acoustics (Fink, 1999). (Right) Illustration of reverse-time migration with four pairs of data and model panels. Receivers and shot are shown by triangles and star. The top three pairs are the wavefield at three forward time steps, and the lowest pair is at a reversed time step.

as source functions. Let us see these processes in Figure 7.25, after Biondi (2004). Panel (a) is a synthetic model with two reflectors in gray color, and panel (b) is a common shot gather consisting of only primary reflections recorded by a line of surface receivers and a surface shot in the middle. Using an estimated wavelet, the forward wavefield from the shot position is shown in panels (c), (f) and (i) at traveltimes of 1.2, 0.75, and 0.3 seconds, respectively. The recorded data shown in panel (b) were used to create the backward wavefield. At each receiver, the recorded trace is time-reversed first, and then sent out as the source function from a source placed at this receiver position. The backward wavefield shown in panels (d), (g) and (j) are superpositions of the modeled waves from all receivers using the reverse time data, at traveltimes of 1.2, 0.75, and 0.3 seconds, respectively. Note that the diffractions along the shallow reflection are due to the small steps used in creating the dipping reflector shown in panel (c).

7.4.2   The imaging condition for RTM

The imaging condition for RTM is:

Reflectors exist where the forward wavefield from the source and backward wavefield from the receivers using the reverse time data are coincident in time and space.

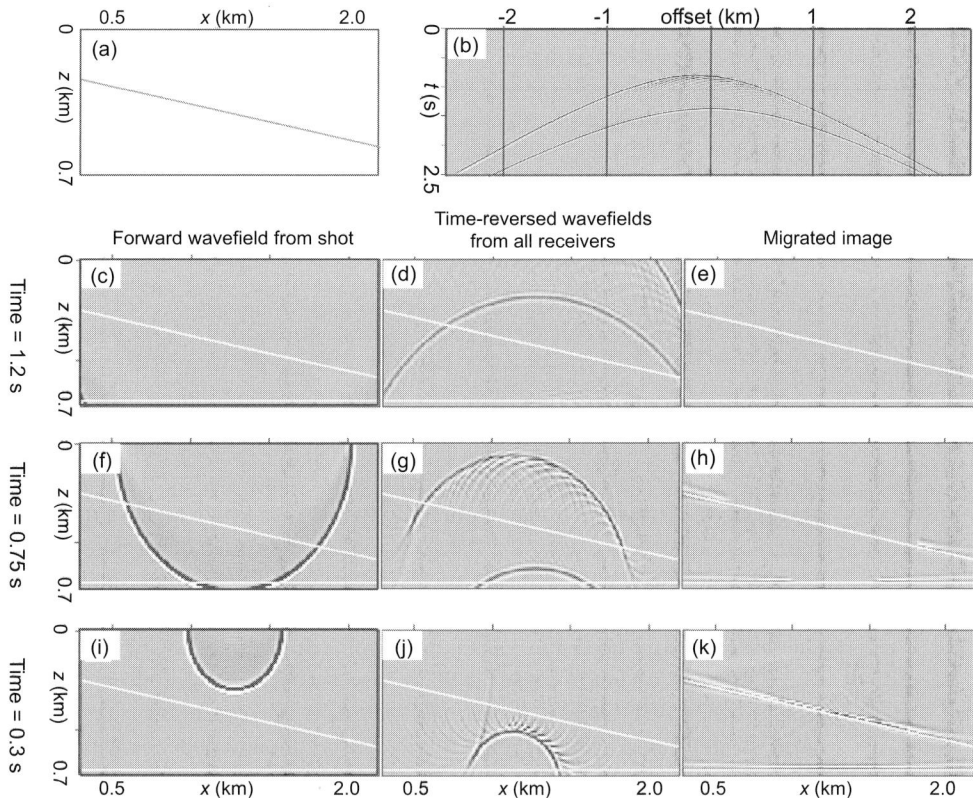

Figure 7.25 Illustration of reverse-time migration (Biondi, 2004). (a) A model with two plane reflectors. (b) A synthetic shot gather with a line of surface receivers and a surface shot in the middle. The remaining nine panels show, at three time steps of 1.2, 0.75 and 0.3 s, the forward wavefield from a shot in the left column; the wavefields from all receivers using time-reversed data in the middle column; and the migrated image in the right column.

This imaging condition can be realized at each pixel in the model space by a time cross-correlation between the forward and backward wavefields at the pixel position, and then taking the amplitude of the cross-correlation function at time zero as the image value for the pixel. At spatial position (x, y, z) in the model space, Claerbout (1971) gave the following expression for the cross-correlation:

$$S(x, y, z) = \sum_t F(x, y, z, t)B(x, y, z, t) \qquad (7\text{--}47)$$

where $S(x, y, x)$ is a shot record image, $\mathbf{F}(x, y, z, t)$ is the forward wavefield at time index t, and $\mathbf{B}(x, y, z, t)$ is the backward wavefield at the corresponding time index. This process is a zero-lag correlation, i.e., it takes the zero-time amplitude of the correlation as the image value.

The imaging condition originally defined in Claerbout's unified mapping theory (1971) was primarily formulated for a one-way wave equation to map reflection geometries. However, the two-way wave equation propagates waves in all directions, and the cross-correlation produces the strength of the wavefield at a particular time and space location. Thus, high correlation values will be obtained at all points of homogeneous velocity and density. At boundaries, the correlation values will be smaller than those for the homogeneous zones. Therefore, the correlation image is like a photographic negative. It needs to zero-out all high-valued homogeneous transmission areas and to map only heterogeneous boundaries. To do this, a Laplacian image reconstruction operator is devised:

$$I(x, y, z) = \left(\frac{\partial^2}{\partial x^2} + \frac{\partial^2}{\partial y^2} + \frac{\partial^2}{\partial z^2} \right) S(x, y, z) \tag{7–48}$$

where $I(x, y, z)$ is the reconstructed image frame for a shot record.

The resulting image represents boundaries in the velocity–depth model by the rate of change in correlation amplitudes between forward and backward propagation wavefields. A divergence operator for image reconstruction depicts boundaries in one direction (positive coordinate direction), which is the nature of a first-order differential. However, a Laplacian operator depicts boundaries in all directions (both positive and negative coordinate directions), which makes it suitable to represent natural wave propagation in all directions. One associated issue is the phase and amplitude changes in the output spatial wavelets. The Laplacian operator causes a 90° phase shift and amplitude changes because it is a second-order differential. A phase-shift filter can correct the phase problem. However, the issue of amplitude changes, particularly the frequency-dependent behavior of Laplacian amplitude, needs to be investigated further.

Since reverse time migration makes fewer approximations to the equations governing seismic wave propagation than any other migration method, it is the most accurate migration method currently available. Unfortunately, it is usually the slowest in terms of computation. Nevertheless, it has served us well, both as a conceptual model to calibrate other migration techniques and as a method of last resort in extremely complex areas with good velocity control.

7.4.3   Highlights and examples of RTM

Many people have contributed to the development of the reverse time migration method. Baysal et al. (1983) and McMechan (1983) separately introduced RTM as they noticed that the numerical finite-difference modeling can be carried out in forward and reverse directions. Loewenthal and Mufti (1983) conducted RTM in the spatial frequency domain. Although some synthetic examples of RTM were shown by Chang and McMechan (1986, 1987), 2D field data examples of RTM were first reported by Rajaskaran and McMechan (1995) and Lines et al. (1996). Mufti et al. (1996) applied RTM to 3D field data using relatively large horizontal grid steps to save computation resources. Zhu and Lines (1998) compared RTM with Kirchhoff migration in pre-stack depth imaging. While nearly all early RTM applications used only primary reflections, Youn and Zhou (2001) showed the benefit of using multiple reflections in pre-stack depth migration that is facilitated by the two-way propagating RTM.

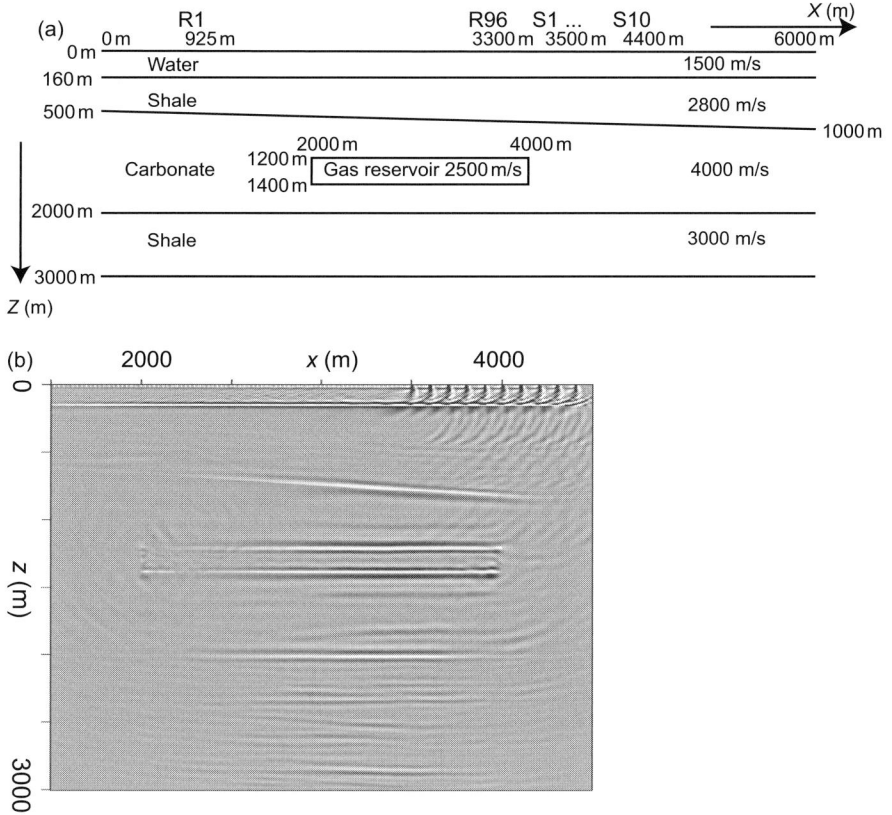

Figure 7.26 (a) A multiple-generating model and shooting geometry. (b) The RTM result of 10 synthetic shot gathers (Youn & Zhou, 2001).

Two examples of RTM from Youn and Zhou (2001) are shown in the following. The process goes through four main steps:

1. Forward modeling with an estimated source wavelet using the full two-way scalar wave equation from a source location to all parts of the model space.
2. Backward modeling of the time-reversed recorded traces in a shot gather using the full two-way scalar wave equation from all receiver positions in the gather to all parts of the model space.
3. Cross-correlating the forward- and backward-propagated wavefields and summing for all time indices.
4. Applying a Laplacian image reconstruction operator to the correlated image.

As shown in Figure 7.26, a line of 10 synthetic shot gathers was acquired over a simple multiple-generating model simulating a carbonate reservoir. The model has a flat water-bottom of shale at a depth of 160 m, with a relatively high normal incidence reflectivity of 0.3, which generates abundant water-bottom multiples. Below the shale, we placed a tight carbonate formation which contains a porous reservoir zone in the middle. The boundary between the carbonate and shale has about a 5° dip. Another shale formation is placed

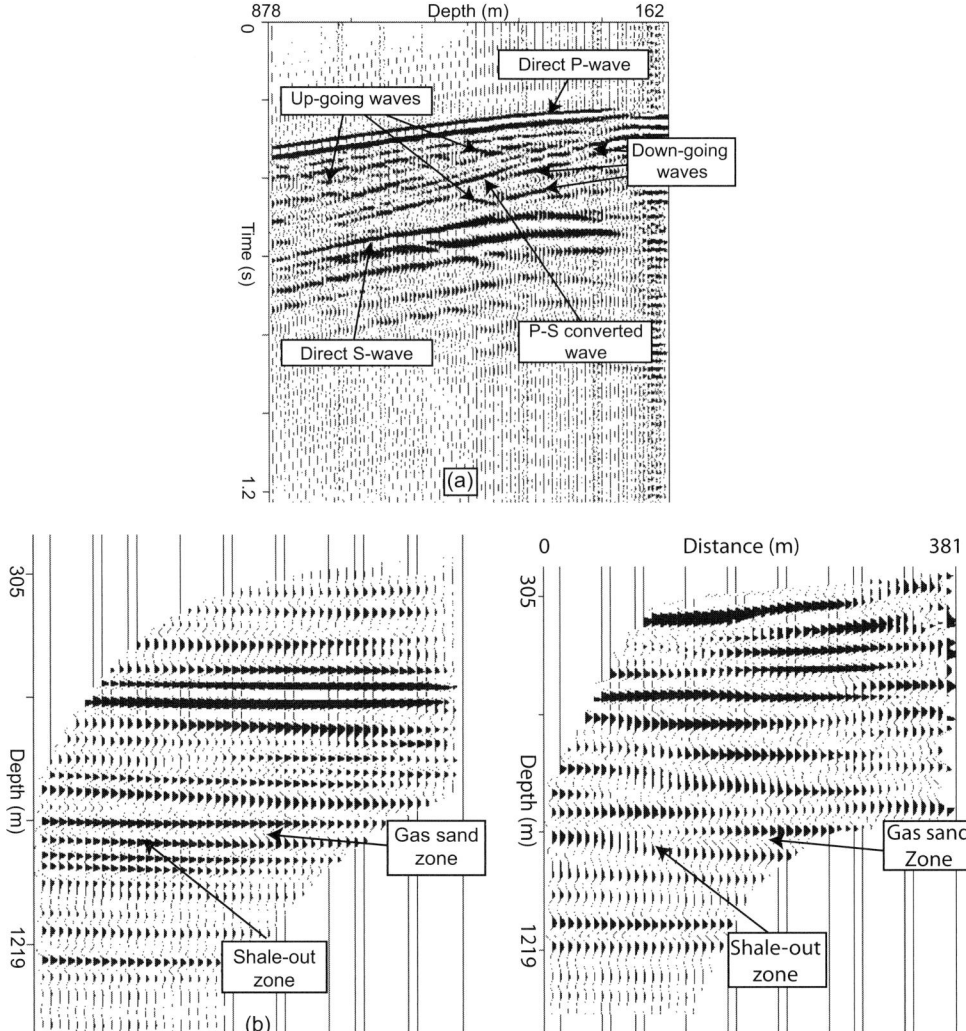

Figure 7.27 (a) A raw offset VSP gather. The shot offset from the borehole is 722.7 m. There are 95 receivers with a 7.62 m interval from 162 to 878 m depth. (b) Result from a depth migration using processed upgoing wavefield data. (c) Result from a two-way RTM using raw shot gathers and check-shot velocities (Youn & Zhou, 2001).

below the carbonate layer. The velocity contrasts between layers were intentionally set at high values within realistic bounds to generate a significant number of multiples. The lower panel in the figure is the RTM solution using all 10 shot gathers. In this case the two-way wave equation for forward and backward directions of wave propagation was solved using a finite-difference technique. The method is able to use all types of acoustic waves such as primary waves and reflections, refraction, diffraction, transmission, and any combination of these waves.

The RTM method can be applied to all types of seismic data such as surface seismic, vertical seismic profile (VSP), crosswell seismic, vertical cable seismic, and ocean bottom cable (OBC) seismic. Because all wave types can be migrated, the input data should be raw shot gathers without any processing. Hence, it is only a one-step process from the raw field gathers to a final depth image. External noise in the raw data will not correlate with the forward wavefield except for some coincidental matching; therefore, it is usually unnecessary to do signal enhancement processing before the RTM using multiple reflections. After migration of an entire 2D line or 3D cube, all the image frames are registered at their correct spatial locations, and summed to produce a final depth image dataset. In the stacking process, it may be desirable to mute marginal portions of the image frames and to apply amplitude normalization for uneven fold or coverage distribution.

Figure 7.27 compares a two-way RTM with a one-way depth imaging using a field VSP dataset. Like that shown in panel (a), VSP data typically show many downgoing multiples with a much weaker upgoing wavefield. The upgoing wavefield also has built-in multiple trains within it, so a downgoing-wavefield deconvolution before common depth point (CDP) transform or depth imaging is usually done. Chang and McMechan (1986) and Whitmore and Lines (1986) have investigated the VSP depth imaging using RTM techniques.

The main objective of this offset VSP study was to discover the extent of the sand reservoir and shale-out zones that were apparently encountered during horizontal drilling. As shown in Figure 7.27a, the data consist of upgoing and downgoing P and S waves as well as P to S converted waves. Figure 7.27b shows a depth migration by a contracting company, which has gone through the standard separation of upgoing wavefield and downgoing wavefield, and only the upgoing reflection wavefield was used for this depth migration. The output CDP interval is 7.62 m and there are 48 CDPs. In comparison, Figure 7.27c shows the result of the two-way RTM using the raw gather without any processing. As indicated on the figure, the shale-out zone can be seen clearly on the two-way RTM result, but is undetectable on the one-way depth migration result.

Exercise 7.4

1. Describe the assumptions and procedure of reverse time migration.
2. If only first arrivals can be recognized in our data, will the RTM still be applicable? What kinds of target may be imaged by such RTM? What may be the problems?
3. The imaging condition for the RTM described in Section 7.4.3 involves cross-correlating the forward- and backward-propagated wavefields. Can you suggest other types of imaging condition? For the described imaging condition and for your imaging conditions, what situations may limit their effectiveness?

7.5   Practical issues in seismic migration

As the final processing step to turn seismic data into images of subsurface distribution of elastic impedance variations, seismic migration is among the most important seismic imaging methods. Various seismic migration methods, from simple stacking to pre-stack depth migrations, have served as leading seismic imaging methods for the petroleum industry for at least the past 50 years. However, errors and artifacts always exist in the imaging results because of practical limitations. For example, the discovery rate for deep water (>1000 feet) exploratory wells in the Gulf of Mexico only reached around 17% for 2008, according to the US Minerals Management Service. Considering that each such well costs more than $100 million, it is critical for the petroleum industry to improve the quality of seismic data and seismic imaging methods, and to improve the practical procedure of seismic migration and its associated methodology.

To most users of seismic migration in the petroleum industry or other physical science disciplines, it is critical to comprehend the practical issues of seismic migration. The major issues include:

- Selection of the appropriate migration methods for the scientific and business objectives;
- Velocity model building, detection of velocity errors, and assessment of the consequences;
- Appreciation of various artifacts from seismic migration and associate processing methods;
- Choices of migration parameters, such as migration aperture and sample rate.

7.5.1   Selection of methods

In order to select a suitable migration method, we need to consider several questions. (1) What are the target features according to the scientific and business objectives? Examples include pinnacles and fractures in carbonate reservoirs or major discontinuities in crustal seismology. (2) Will the targeted features be detectable by the given data coverage and SNR? This question may be addressed by conducting resolution tests and/or checking similar case studies. (3) Which migration methods might potentially be able to address the first two questions within the given time span? Since all real-life projects have deadlines, we must give the best answer within those time constraints. (4) What are the limitations and artifacts from the candidate methods? And will you be able to assess the limitations and artifacts? While the answer to this question belongs to the third issue, the limitations and artifacts are among the key criteria for choosing the right method(s).

The above questions may be addressed by running some preliminary tests. Although most commercial seismic processing software contains all common seismic migration methods, a method given the same name by two different vendors may have considerable differences in implementation and emphasis. Hence, given any unfamiliar migration software, one must conduct a series of standard tests, such as checking the **impulse responses** in both data and model spaces using constant and variable velocity models. It is also helpful to produce some migrated gathers or partial stacks to assess all the parameters of the code. Any unfamiliar migration software should be tested using standard synthetic model datasets such as the Marmousi dataset for 2D pre-stack migration or various versions of the SEG/EAGE 3D datasets for different imaging purposes.

Some general properties of common migration approaches hold true in most cases, such as the insensitivity of time migration to lateral velocity variations, and the effectiveness of Fourier migration to situations with steeply dipping reflectors but smooth lateral velocity variation. Another common situation is the need to consider the relative effectiveness of Kirchhoff migration and wave equation migration. As shown with a synthetic 2D salt-body example in Figure 7.28, wave equation migration handles the top salt **rugosity** (the rough undulation of the interface) much more effectively than Kirchhoff migration. The rugosity

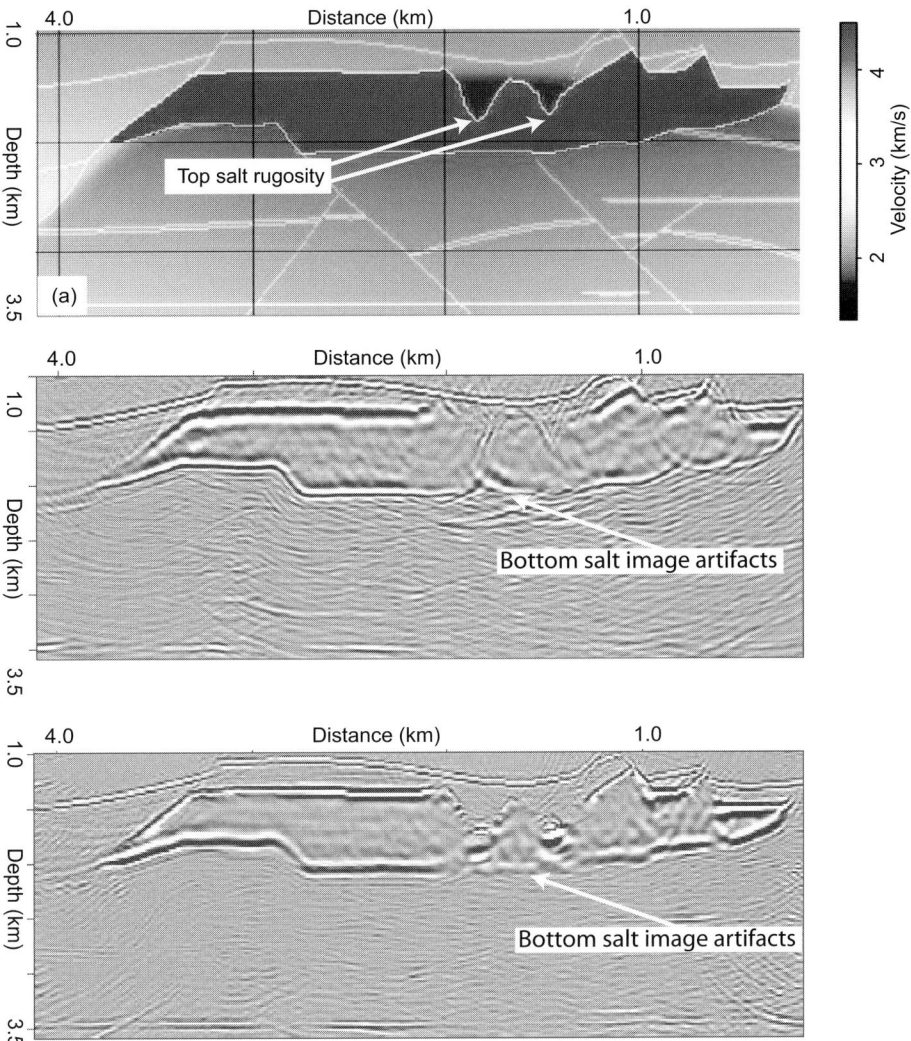

Figure 7.28 (a) Crossline section of the interval velocities for a SEG-EAGE model. (b) Kirchhoff migration result. (c) Wave equation migration result (Biondi, 2004).

of the interface between the low-velocity sediments and high-velocity salt body produces multiple paths of traversing waves that make it difficult to image this top salt interface and cause imaging artifacts for all interfaces beneath it. In this case, the artifacts in the bottom salt image are more apparent in the Kirchhoff migration solution than in the wave equation migration solution. On the other hand, most wave equation migration methods require regularly sampled data and an even distribution of data quality over the model space. In the presence of velocity model errors and uneven distribution of data over the model space, Kirchhoff migration may be easily customized to the situation and therefore produce a more reliable result than the wave equation migration.

One way to help in selecting the proper migration method is to run and compare all available methods using a subset of the data. In the above example, the difference between the Kirchhoff migration and wave equation migration results for the bottom salt interface will alert us to the possible presence of imaging artifacts. In general, a careful comparative analysis of the results from different migration methods may allow us to evaluate many important questions such as: (1) reliability of chosen signals; (2) accuracy of velocity model; (3) effectiveness of chosen migration and processing methods for the geologic targets; and (4) processing artifacts.

7.5.2 Velocity model building

The second practical issue of seismic imaging is that of velocity model building, details of which will be given in the next chapter. Because a velocity model is a precondition for all seismic migration methods, a practical issue is how to build the migration velocity models and how to assess their quality. Many migration methods use interval velocities. These may differ from the Dix interval velocities (see Section 8.2.2) derived from the stacking velocities, especially in cases of dipping reflectors. Using too fast a velocity may cause **over-migration**, where the reflected energy is moved too far. Using too slow a velocity may cause **under-migration**, where the reflected energy is not moved far enough. If we are uncertain about the given velocities, using a smoothed velocity model usually gives a more stable result, so it has become common practice to smooth the migration velocity model in order to minimize the impacts of incorrect velocities. To choose between a fast and a slow velocity model, it is usually better to use the slower velocity model so that the reflection energy will be more separated in most places. Owing to the sensitivity of depth migration to velocities, migration velocity analysis has been devised as the most powerful way of velocity determination.

7.5.3 Assessing migration artifacts

The third practical issue is that of migration artifacts. The common types of migration artifacts include:

- **Acquisition footprints**. These are image artifacts resembling the survey geometry but caused by insufficient fold at shallow depths. Recall from Chapter 1 that "fold" is the number of midpoints within each CMP bin. An example of a time slice in the Permian Basin is shown in Figure 7.29. The intensity of the footprints typically decreases with depth owing to a combination of decreasing frequency and more even fold distribution with increasing depth.
- **Smearing**. This refers to artificial stretching of imaged anomalies along isochrons owing to insufficient or uneven illumination. Figure 7.30 shows a cross-section of migrated VSP data across the San Andreas fault in California using surface shots and wellbore receivers. In such cases, reliable image solution is only defined within the "**VSP corridor**", which is a portion of the model space with sufficient **seismic illumination** as defined by a sufficient level of crisscrossing rays or wavepaths.
- **Spatial aliasing**. This is spatial alignment of artificial events due to under-sampling as discussed in Chapter 1. Because migration moves input dipping events in the up-dipping direction so that dipping events become steeper on the output image, the vertical sample rate for the input may not be sufficient to prevent spatial aliasing of the output. In addition, the time-to-depth mapping in depth migration may increase the

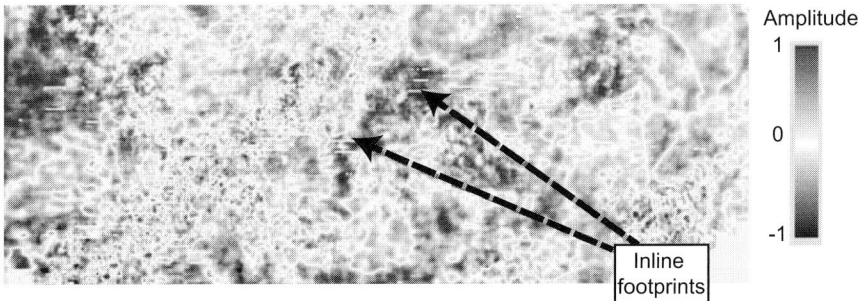

Figure 7.29 A time slice of seismic data from East Ranch, Permian Basin, showing acquisition footprints (short horizontal lines) along the inline direction.

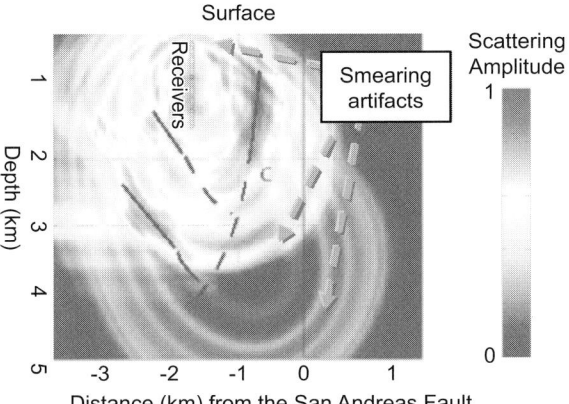

Figure 7.30 A cross-section of VSP Kirchhoff migration near the San Andreas Fault using receivers in wellbore and surface shots. Smearing artifacts are evident along isochrons, or contours of equal traveltime between the shot and receiver via the reflector.

apparent vertical frequencies at some portions of the output image. Consequently, we need to estimate the maximum dipping angle and frequency of the output image and see if the output is in danger of spatial aliasing. The spatial aliasing artifact can be suppressed either by sampling the result using a fine enough sample rate, or by smoothing the result so that its frequency is properly sampled by the given sample rate.

- **Local velocity-induced artifacts**. These are anomalous pull-ups or pull-downs of otherwise flat reflection events beneath a localized velocity anomaly. The so-called **velocity pull-ups** are upward displacements of events beneath a high-velocity anomaly such as a salt body, a carbonate pinnacle, or a patch reef. The **velocity pull-downs** are downward displacements of events beneath a low-velocity anomaly such as a **gas chimney**, a gas hydrate body, or a carbonate sink hole. As illustrated in Figure 7.31, the pull-ups and pull-downs can be created by a fault offsetting a layered velocity structure. In reality, a fault may consist of several branches and cut through a number of velocity layers, hence producing a set of time sags and pull-ups that are collectively called a **fault shadow**, like that shown in Figure 7.32. Velocity-induced artifacts occur much more often for time-migrated results because such results ignore lateral velocity variations.

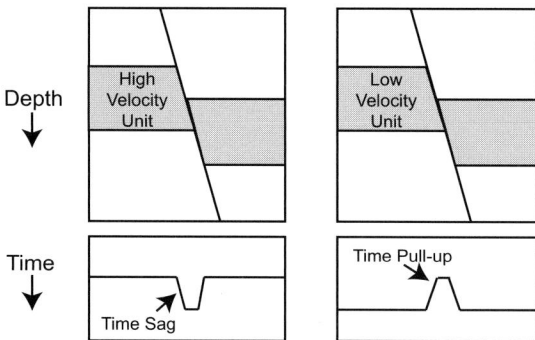

Figure 7.31 Two pairs of schematics showing the traveltime effects that structural extension creates on reflectors below high-velocity and low-velocity bodies (Fagin, 1996).

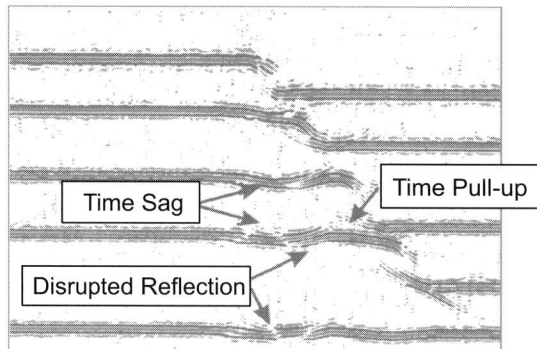

Figure 7.32 A synthetic time-migrated section created from CMP gathers after NMO velocity analysis and stacking, showing fault shadow effect with time sags and time pull-ups (Fagin, 1996).

- **Fake events**. When the input contains coherent events that are unacceptable in the migration method, fake events will be produced on the migrated results which may mislead the interpretations. For instance, many migration software only accept primary reflections as the signals in the input, so it is necessary for pre-migration data processing to suppress all events other than primary reflections. In this case, the "colored noises" may include first arrivals, surface waves, multiple reflections, and converted waves. As another example, a 2D migration will assume that all input signals are from the vertical plane containing the survey line, or that there is no sideswipe energy from scatters or reflectors away from the vertical plane. As shown in Figure 7.33, taking sideswipe events as input to 2D migration results in fake events like the fuzzy channel-like anomalies (right panel of Figure 7.33) which disappear from the result of a proper 3D migration.

7.5.4 Setting migration parameters

The last practical issue to be addressed here is the choice of parameters for a migration method. Depending on the chosen method, there are a number of key parameters for seismic

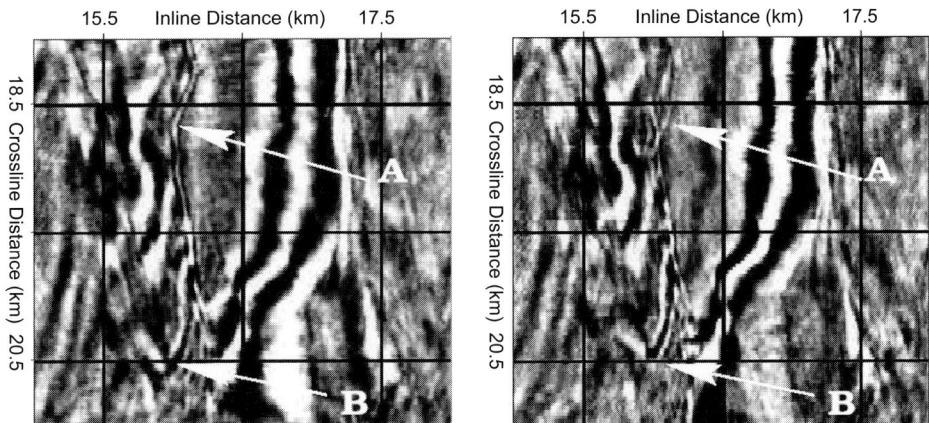

Figure 7.33 Depth slices of 3D (left) and 2D (right) migration results (Biondi, 2004). Arrows labeled A and B denote fake events in the 2D migration result that disappear in the 3D migration result.

migration, such as migration aperture or its equivalents such as cable length, sample rate in the solution space with consideration on spatial aliasing, weighting coefficients for Kirchhoff migration, and concerns about the preservation of amplitude and phase of seismic wavelets. To become an expert user of a seismic migration method, one needs to satisfy a set of minimum requirements:

- Comprehension of the underlying principles, assumptions and application domains of the method, so that you know why the method is chosen for the given situation.
- For a chosen commercial migration software, the capability to quantify its effectiveness by running standard tests to produce impulse responses and partial stacks, and the ability to learn how to use the software.
- For the chosen migration method, the ability to understand or find out the impact of the velocity model and data parameters on the migration result.
- Knowledge of the common artifacts such as smearing and spatial aliasing for the method, and ability to find ways to QC the validity of all migrated features.

Exercise 7.5

1. Describe the general assumptions and procedure of seismic migration. What are the specifics for Kirchhoff, Stolt, and reverse time migration methods? For seismic data in an area with nearly flat reflectors, will migration be necessary?

2. How would you check the quality of a seismic migration method? Make a list of tests that you would conduct in order to quantify the quality.

3. Someone has said that smoothing the velocity model is a practical way to improve the migration result. Discuss the possible effects of smoothing velocity model on seismic migration. What are the potential benefits and drawbacks from the smoothing?

7.6   Summary

- Seismic migration is a mapping method to place seismic reflection data into their correct spatial or temporal reflector positions in two steps. Step one is to back-project the measured seismic data into subsurface using the wave equation and a velocity model, producing an extrapolated wavefield. Step two is to apply an imaging condition to capture the image of subsurface reflectors from the extrapolated wavefield.
- All migration methods assume, first, that the given velocity model is sufficiently accurate. Second, many traditional migration methods also assume that the input data contain only primary reflections. Third, each 2D migration method assumes that the reflections in the input data are confined within a vertical plane containing the 2D profile. The second and third assumptions are the reasons for extensive pre-processing before applying migration.
- A velocity model is a precondition for seismic migration. Time migration uses a layer-cake velocity model without lateral velocity variation. In contrast, depth migration is able to use a velocity model containing a significant level of lateral velocity variation.
- In the case of gently dipping reflectors, a post-stack migration may be sufficient using post-NMO stacked data traces to approximate zero-offset data traces. In the presence of steeply dipping reflectors, a pre-stack migration is usually more suitable but takes much more computational time and resource than post-stack migration.
- Kirchhoff migration uses two-way reflection traveltime tables calculated based on wave or ray theories to map the amplitude- and phase-calibrated waveform records to the subsurface, and then stack all mapped data together to capture the image of subsurface reflectors. It is intuitive and flexible, although it uses the ray theory approximation.
- f–k migration is efficient and thorough because it takes advantage of Fourier transform. It is very effective in migrating steeply dipping reflectors in smoothly varying velocity models, but usually less effective in the presence of strong lateral velocity variations.
- Reverse time migration is a full wave migration method that is capable of using both primary reflections and other waves such as refractions and multiple reflections. It maps time-reversed data from receivers into a backward wavefield, propagates a forward wavefield from each source, and finally applies an imaging condition to take the common denominators from the backward and forward wavefields as the image of the subsurface structure.

FURTHER READING

Biondi, B. L., 2004, *3-D Seismic Imaging*, SEG.

Chun, J. H. and Jacewitz, C., 1981, Fundamentals of frequency-domain migration, *Geophysics*, 46, 717–732.

Hatton, L., Worthington, M. H. and Makin, J., 1986, *Seismic Data Processing: Theory and Practice*, Chapter 4, Blackwell Science.

Sheriff, R. E. and L. P. Geldart, 1995, *Exploration Seismology*, 2nd edn, Section 9.12, Cambridge University Press.

Yilmaz, O., 1987, *Seismic Data Processing*, SEG Series on Investigations in Geophysics, Vol. 2, Chapter 4, SEG.

8 Practical seismic velocity analysis

Chapter contents

Velocity analysis is synonymous with **velocity model building (VMB)** because the goal is to produce a velocity model for the subsurface. VMB is among the most common practices in seismology for two reasons. First, for any study area, its seismic velocity model is one of the main measurable results from geoscience. Second, a velocity model is a precondition for seismic migration and other seismic imaging methods to map subsurface reflectors and scatters using reflected or scattered waves. Since seismic velocity is inferred from traveltimes of seismic waves, the resolution of each velocity model is limited by the frequency bandwidth and spatial coverage of seismic data. This chapter starts with definitions and measurements of different types of seismic velocities. The observations reveal the trends of seismic velocities as functions of pressure, temperature, and other physical parameters. The dominance of the 1D or $V(z)$ velocity variation at large scale in the Earth leads to the classic refraction velocity analysis based on seismic ray tracing in a layer-cake velocity model.

The three common seismic velocity analysis methods based on NMO semblance, seismic migration, and seismic tomography are discussed in three sections. Assuming gentle to no changes in the reflector dip and lateral velocity variation, semblance velocity analysis provides stable estimates of the stacking velocity. The stacking velocity at each depth, marked by the corresponding two-way traveltime, is the average velocity of all layers above the depth. By taking advantage of the dependency of depth migration on velocity variations, migration velocity analysis enables the velocity model to be refined using horizons defined by data to accommodate lateral velocity variations. Nowadays most migration velocity analyses are conducted on **common image gathers** (**CIG**s). To constrain the velocity variation using all data together, tomographic velocity analysis derives or refines the velocity model through an inversion approach to update the velocity perturbations iteratively. Some practical issues in tomographic VMB are discussed in the final section, mostly on inversion artifacts and deformable layer tomography.

8.1   Velocity measurements and refraction velocity analysis

8.1.1   Introduction to seismic velocities

Seismic velocity is the speed at which a **seismic phase** travels through a medium and hence is a primary property of the medium. A **seismic phase** is a packet of traveling wave energy representing a particular wave type, such as a reflected P-wave, a P-to-S converted wave, a multiple reflection, or a Rayleigh wave. A seismic phase may have different velocities as it traverses through different media, referred to as **velocity inhomogeneity**, and may have different velocities in different directions at the same medium position, referred to as **velocity anisotropy**. Detecting the properties of the media is a major motivation for studying seismic velocities of different types, because they may help us decipher the lithology and structure of the media. The task of detecting lithology from seismic velocity is very challenging because seismic velocity depends on many factors in addition to lithology. Some factors are properties of the medium, such as its density, porosity, or their structure variation, other factors are physical conditions such as temperature and pressure, and yet others are wave properties such as frequency and wave type. The traveling speed of a single-frequency component is called the **phase velocity**. For a broadband wave, the traveling speed of its main energy is called the **group velocity**. Group velocity is the rate at which the energy in a wave train travels. Even for a homogeneous medium, phase velocity may differ from group velocity in the presence of seismic attenuation and/or anisotropy.

Why do we need seismic velocity? First, we need to have seismic velocity in order to obtain information for interpreting subsurface properties in terms of structure and lithology. A good example is the case of **over-pressure prediction** in petroleum exploration. Over-pressured areas of subsurface have abnormally high porosity and therefore abnormally low seismic velocity. Second, a pre-condition for applying seismic migration is the availability of a proper velocity model. Even for time migration, we still need to have a velocity model. In order to tie seismic data with well-log measurements, we need a velocity model to conduct time-to-depth conversion since seismic records are acquired as a function of time, and well logs are acquired as a function of depth.

In the practice of seismic migration, people often decouple the variation in velocity field from the variation of reflector geometry. How can they do this? The answer lies in the general

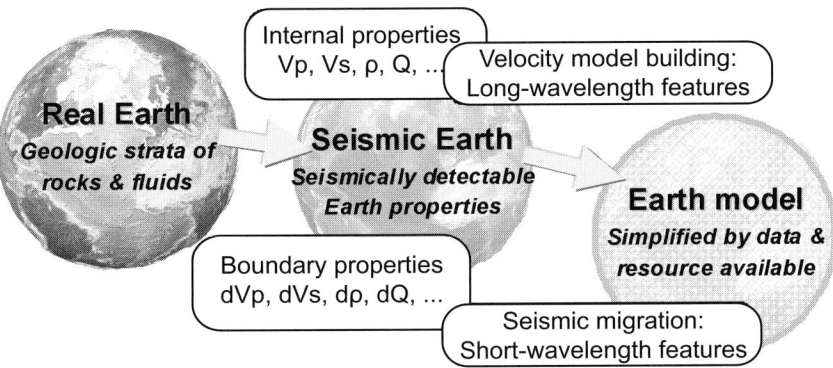

Figure 8.1 A view of seismic imaging – from real Earth via seismic Earth to the Earth model.

view of seismic imaging as shown in Figure 8.1. In this view, we decompose the spatial distribution of the elastic impedance field into two components. One component consists of long-wavelength features that vary slowly over the **model space**, which is the velocity model that is targeted by **velocity model building (VMB)**. The parameters in the velocity model may include compression velocity, shear velocity, density, and the quality (Q) factor. The other component consists of short-wavelength features that quickly vary over the model space, which are interfaces or the reflector geometry targeted by seismic migration. We may regard the interfaces as spatial derivatives of the velocity model. Mapping the long-wavelength features is the goal of VMB, while mapping the short-wavelength features is the goal of seismic migration. Hence velocity model building and seismic migration are two complementary processes of subsurface imaging.

In theory, seismic velocity V is a function of density ρ and **incompressibility** K for an acoustic medium such as a fluid:

$$V = (K/\rho)^{1/2} \tag{8–1}$$

For general elastic media, K can be regarded as the **effective elastic parameter**. For solid media, K is equivalent to the **bulk modulus**, and the velocities of compression wave V_P and shear wave V_S are

$$V_P = [(\lambda + 2\mu)/\rho]^{1/2} \tag{8–2a}$$

$$V_S = [(\mu)/\rho]^{1/2} \tag{8–2b}$$

where λ and μ are the two Lamé constants of the medium. The parameter μ is also known as **shear modulus** or **rigidity**, and λ is the bulk modulus less two-thirds of the shear modulus:

$$\lambda = K - 2/3\mu \tag{8–3}$$

For fluid media, the shear modulus vanishes, hence we have

$$V_P = [(\lambda)/\rho]^{1/2} \tag{8–4a}$$

$$V_S = 0 \tag{8–4b}$$

Measurements of seismic velocities

Seismic velocity can be obtained either through direct measurements, such as well logs, or through indirect inferences based on seismic data. Since the direct measurements and indirect inferences can only be conducted in discrete forms, we always need to know more about the physical and geologic tendencies of seismic velocity in order to populate its values in space based on a limited set of measurements. The difference in seismic velocity of various rocks is a primary reason that seismology is a useful tool for deciphering rocks. The problem, however, is that different rocks may have similar velocities, while the same rock type may have a wide range of velocities. This can be seen in Figure 8.2, from Sheriff and Geldart (1995, p. 101).

Because the level of bending of seismic wavepaths is proportional to the velocity variation rather than the absolute velocity difference, it is useful to quantify the velocity variation of rock types in terms of a percentage between the minimum and maximum velocities. As an example of comparing the velocity variation versus the absolute velocity difference, Grant and West (1965) cited V_P for alluvium as 0.3–1.6 km/s, and V_P for limestone as 2.75–6.3 km/s. As shown in the two lines ending in filled circles in Figure 8.2, alluvium has a narrower range of absolute velocity than limestone. However, in terms of velocity variation measured as a percentage, that of alluvium is 1.6/0.3 = 533%, which is much wider than that of limestone at 6.3/2.75 = 229%. Since the level of velocity variation of a lithology depends mainly on its minimum velocity, places of low velocities such as near surface and low velocity zones require special attention in velocity analysis.

The primary causes for the wide range of velocity for the same rock type are variations in temperature, pressure and **rock texture**. We may use rock texture to describe the consequence of different geologic histories, resulting in differences in age, tectonic and solidification level due to compaction and re-crystallization. The difference in rock texture can be genetic, due to variations in porosity, grain distribution, and depositional environment at the original formation stage; or diagenetic, caused by alterations after the lithification of the original rock. The wide range of rock velocities means their lithological interpretations can be non-unique. On the other hand, if we understand the effects on seismic velocity of various factors at genetic and evolutionary stages, we may have a chance to decipher the depositional environment and evolution history through analyzing seismic velocity in conjunction with all geologic and geophysical information.

The near-surface depth range, roughly from the surface down to several hundred meters in depth, has the largest variation in velocities owing to the extremely low velocities of the **weathering layer**. Table 2.1 showed the P-wave velocities of some typical rocks in the near-surface depth range. We have discussed corrections for the effect of near-surface velocities in Section 2.5, a topic that is important in exploration seismology and geo-engineering studies.

The near-surface depth range is a phase boundary layer. The largest variations in seismic velocities exist in all phase boundary layers, such as the air–solid, air–water, water–solid, and core–mantle boundary (D″) layers of the Earth. As an example, the weathering layer is usually a **low velocity layer** (**LVL**) whose bottom is marked by the water table where velocity increases significantly. In arid areas the water table can be very deep (100 m) and marked by a strong change in velocity (0.6–3.0 km/s for the P-wave). Eroded rocks such as limestone caverns could be buried, with significant velocity variations. Cox (1999)

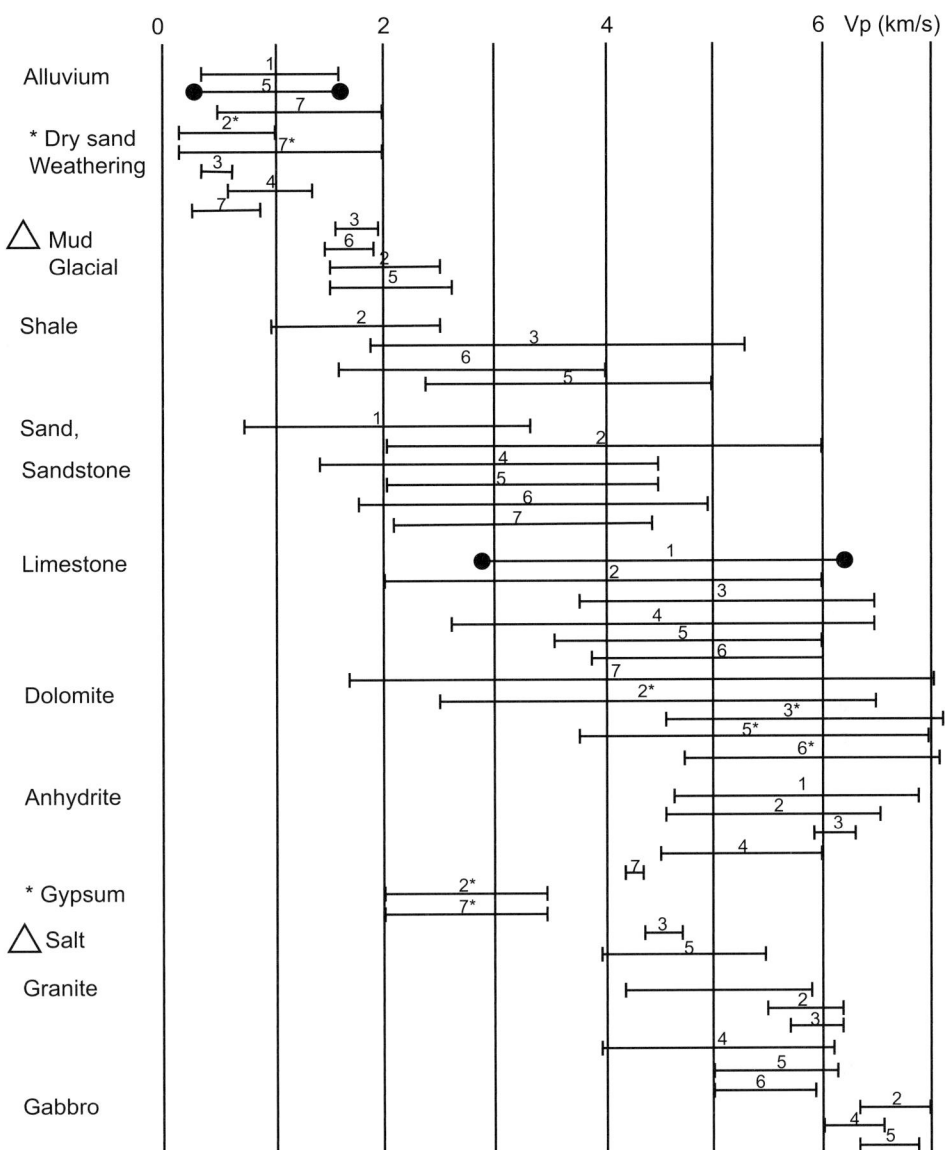

Figure 8.2 P-wave velocities for common rocks, based on: (1) Grant and West (1965); (2) Kearey and Brooks (1984); (3) Lindseth (1979); (4) Mares (1984); (5) Sharma (1976); (6) Sheriff and Geldart (1995); and (7) Waters (1987). From Sheriff and Geldart (1995); see that paper for the original reference details.

summarized the velocity characteristics of some near-surface layers:

- Sand dunes (height up to 200 m) have dips on the windward slopes of 15–20° and slip slope of 30°. Owing to differences in height and compaction, velocities near the base of a dune will be higher than near the edge. The base of a sand dune is usually of higher velocity and gentler relief (Sabkhas gravel plains). Velocities could be from 0.15 km/s to 1.5 km/s depending on thickness and compaction, and vary from dune to dune.

- A highly irregular weathered layer has rapid changes in near-surface velocity and thickness associated with recent unconsolidated sediments. Examples would be swamps, deltas, or river mouths with high rate of sedimentation (in the Mississippi Delta the top 30 m has 0.3 km/s velocity). Mud with shallow gas could have very low velocities, less than 100 m/s. Loess topography could be very thick, as much as 300 m, with velocities from 0.6 km/s above the water table to 1.2 km/s at the water table. Karst limestone, alluvial fans, stream channels, mature topography with irregular bedrock, etc. could also be examples of irregular weathered layers.
- Permafrost is defined as permanently frozen soil or rock (Sheriff, 1991). It is characteristic of areas like the Arctic, Antarctic, and similar regions. Because of its high velocity (~3.6 km/s) relative to surrounding lower velocities, it is important to know its base and extent. It may change thickness, decreasing from land to offshore.
- A combination of old topography, such as river channels or glacial valleys, and young sediments can disguise the signs of large changes below the surface. These sediments could be very thick and cause high lateral velocity variations which depend on channel depth and width.
- Near-surface velocities also have time-variant changes due to daily and seasonal variations. This factor needs attention in seismic acquisition and 4D seismic surveys.

8.1.3 Trends of seismic velocities

It is very useful in practice to understand the general behavior of seismic velocity under various conditions. Such knowledge is largely supported by laboratory measurements. In most VMB projects we will never be able to obtain all possible measurements to constrain all aspects of seismic velocity. Thus, knowledge of the general behavior of seismic velocity can be used to make predictions in areas of poor or no data constraints. More importantly, knowledge of the general trend of seismic velocity may assist VMB projects in designing data acquisition and model parameterization.

Figure 8.3 from Tatham and McCormack (1991) summarizes the effects of various rock properties on seismic velocities and V_p/V_s ratio. Some of the effects are monotonic, such as the effects of porosity, pore shape, pressure, and temperature. Others are somewhat non-linear, such as the effects of lithology, pore fluid type, and anisotropy. It is good practice to compile a list of such "rules of thumb" in each VMB project.

Seismic velocities usually follow the trend of density, which generally increases with burial depth and/or increase in the confining pressure. Seismic velocities usually decrease with the increase of porosity, because pore space typically has much lower velocity than that of the rock matrix. Figure 8.4 shows some measurements of velocity versus porosity for sandstone and limestone samples (Wyllie *et al.*, 1958). Porosity in a clastic rock decreases with depth of burial or the extent of compaction. At comparable depths, porosity increases with increasing level of sorting, but decreases with increasing level of cementation. Porosity is usually unchanged by uplifting of the rocks after their formation.

On the other hand, velocity decreases with an increase in the number of microcracks, as shown in Figure 8.5. Age is typically proportional to the extent of consolidation of the rocks. Hence the effect of age is similar to the effect of burial depth and increase of pressure.

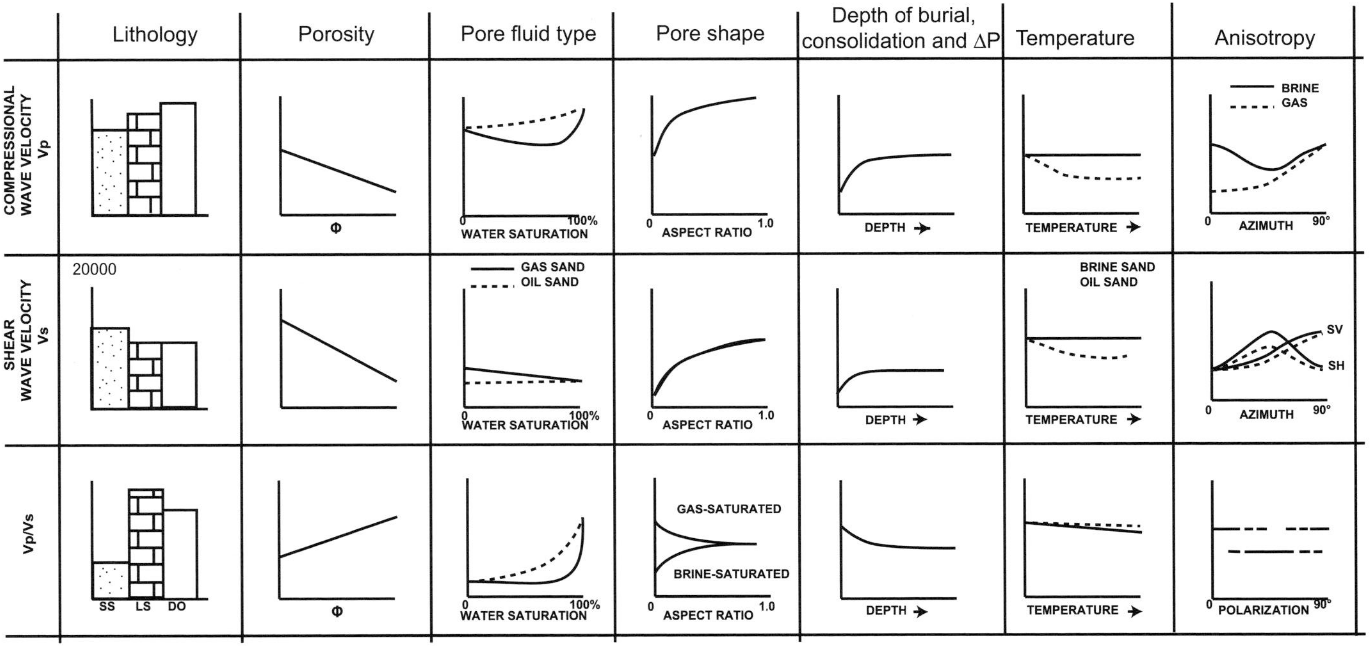

Figure 8.3 Summary of effects of different rock properties on V_p and V_s and their ratios. (After Tatham & McCormack, 1991.)

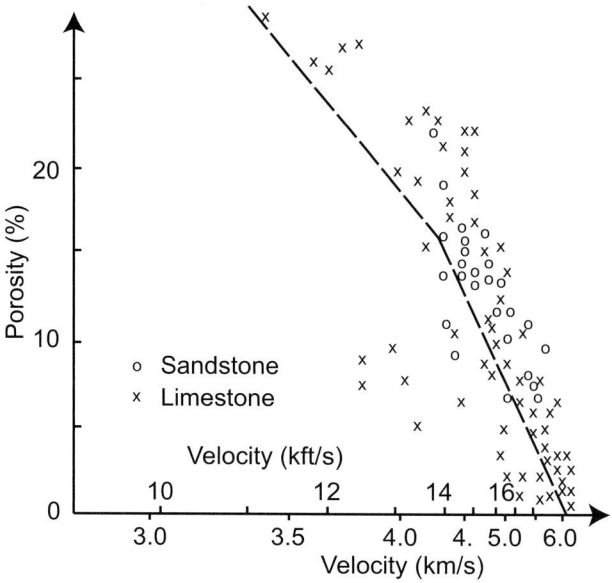

Figure 8.4 Measurements of velocity and porosity relationship for some sandstone and limestone samples. The dashed line is the time-average for V_p at 5.94 km/s and V_s at 1.62 km/s. (After Wyllie *et al.*, 1958.)

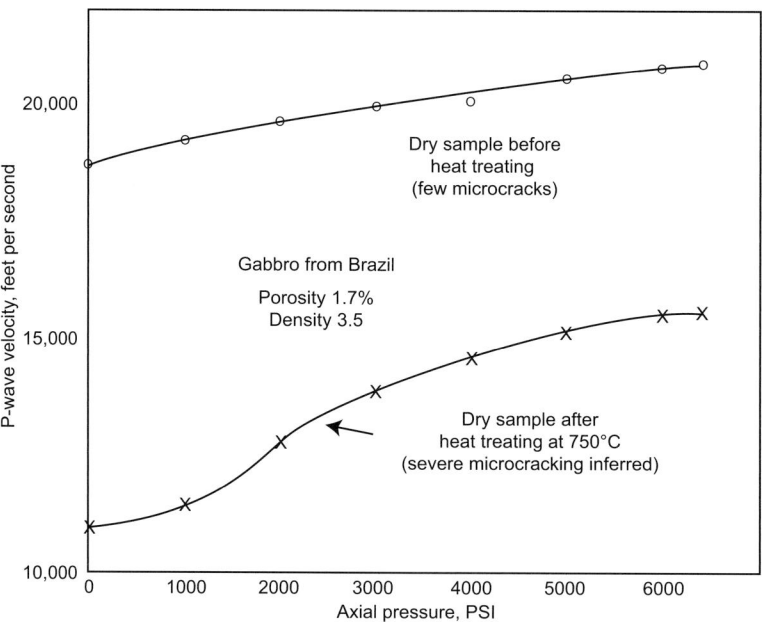

Figure 8.5 Effect of microcracks on velocity of gabbros. The upper curve is for dry samples with few microcracks. The lower curve is for dry samples after heat treating at 750 °C, causing severe microcracking. (From Gardner *et al.*, 1974.)

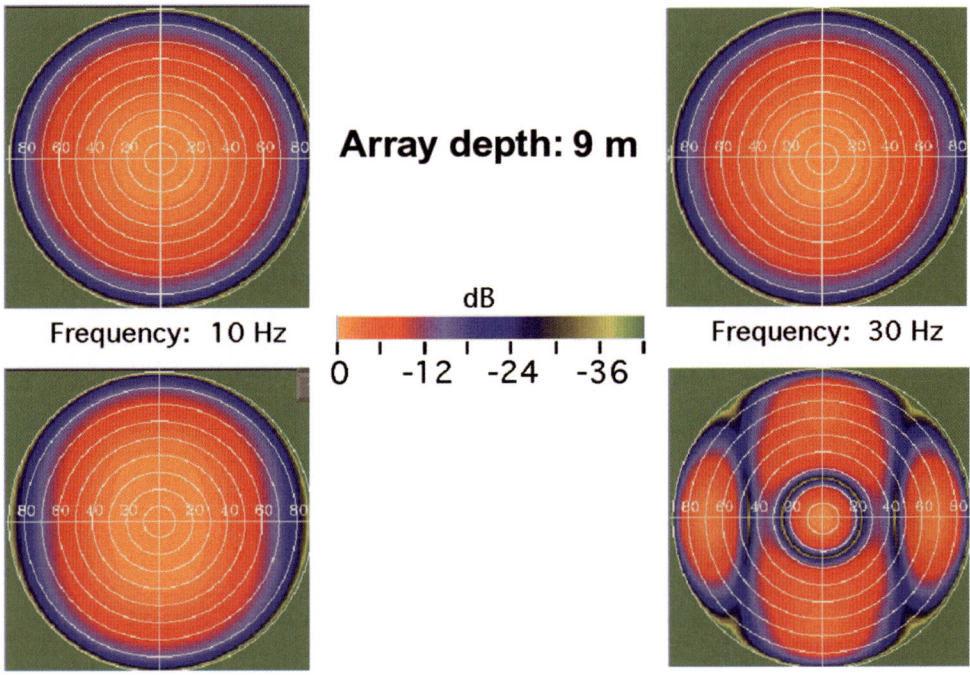

Figure 1.14 Radiation pattern of an airgun array at a tow depth of 9 m. Each panel is a lower-hemisphere projection of the wave amplitude at a particular frequency as a function of the azimuth and dip angles (Caldwell & Dragoset, 2000).

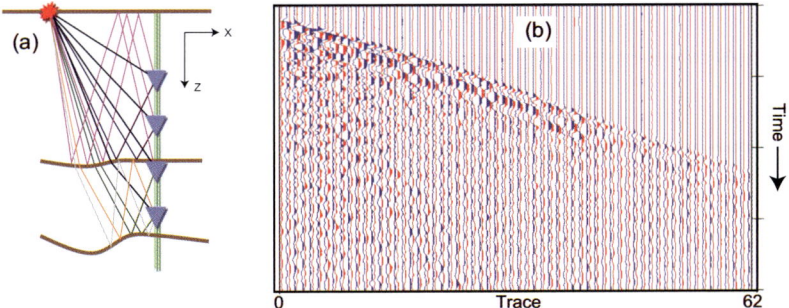

Figure 2.2 (a) A sketched cross-section of offset VSP, where raypaths show various waves from a shot (star) to the receivers (triangles) along the well bore. (b) A common shot gather of the two horizontal components from an offset VSP.

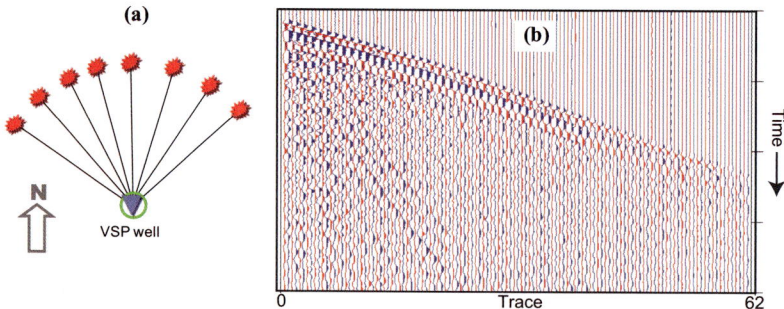

Figure 2.4 (a) A map-view sketch showing a selected group of shots (stars) of similar source-to-receiver offsets but different azimuths from the VSP well (triangle). (b) The common shot gather after correction for orientation errors of the geophones.

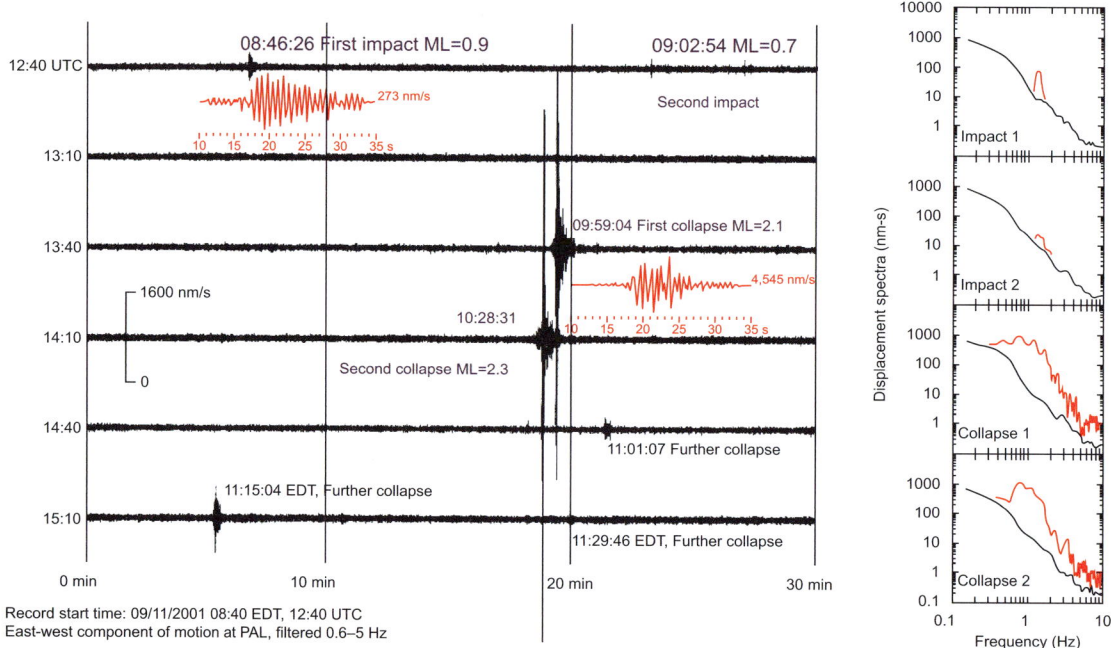

Box 3.3 Figure 1 Seismic record at Palisades, NY, 34 km north of the World Trade Center during the 9/11 disaster. (Left) East–west component of time record started at 8:40 EDT, or 13:40 WTC, on 9/11/2001. Two inserted seismograms are zoom-in plots of the first impact and the first collapse. (Right) Displacement spectra [nm s]. In each panel the upper curve is the signal spectrum, and the lower curve is the noise spectrum (from Kim *et al.*, 2001).

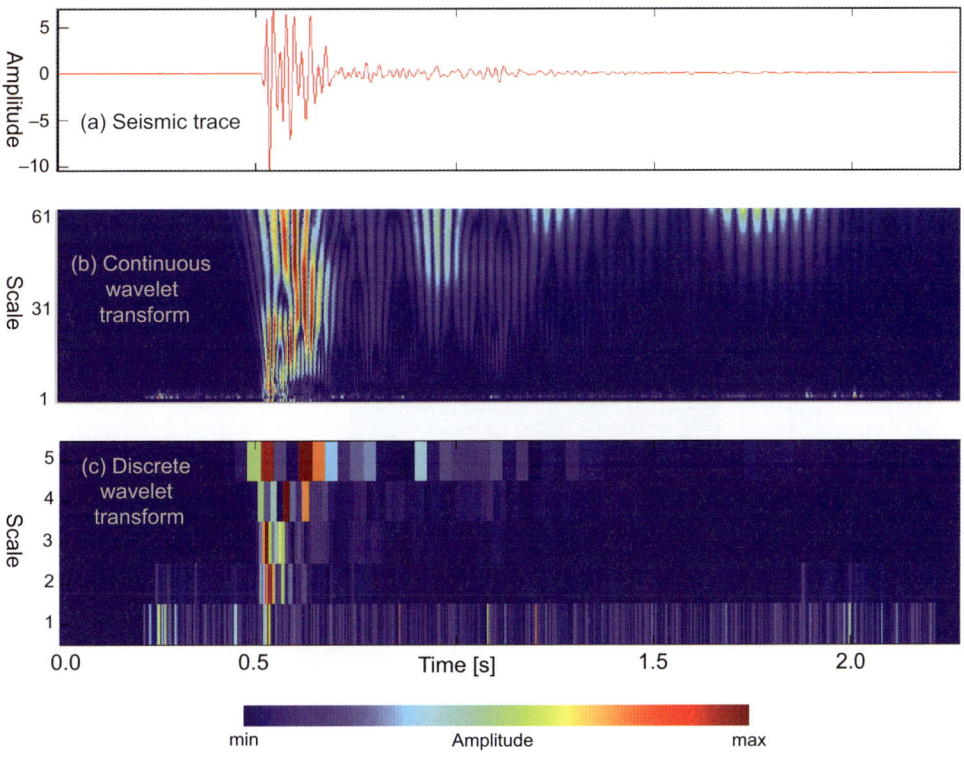

Figure 3.21 (a) A seismic trace. (b) Continuous wavelet transform. (c) Discrete wavelet transform.

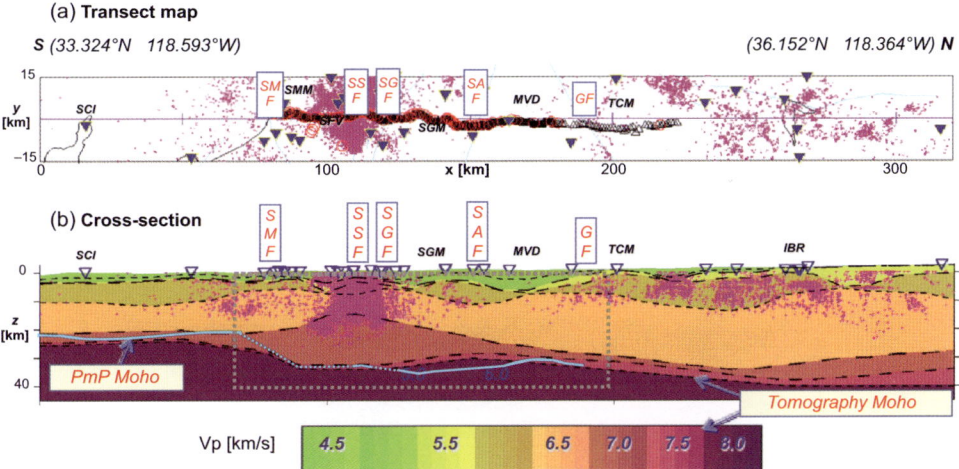

Figure 4.14 (a) Map of the profile in southern California showing earthquake foci (small crosses), seismic stations (triangles), faults (lines), shots (circles), and receivers (triangles). (b) Cross-section of tomographic velocity model. The tomography Moho at the top of the layer with 8 km/s velocity is compared with the PmP Moho in light color. Locations: Santa Catalina Island (SCI), Santa Monica Mountains (SMM), San Fernando Valley (SFV), San Gabriel Mountains (SGM), Mojave Desert (MVD), Techchapi Mountains (TCM) and Isabella Reservoir (IBR). Faults in boxes: Santa Monica (SMF), Santa Susana (SSF), San Gabriel (SGF), San Andreas (SAF), and Garlock (GF).

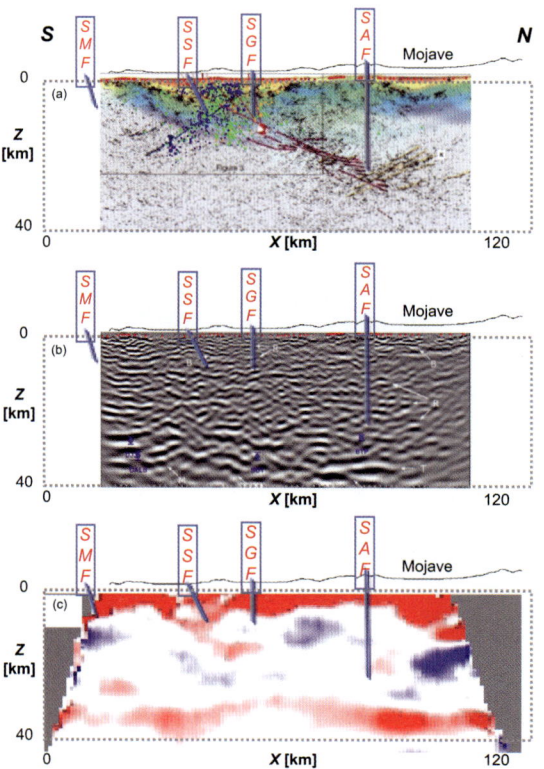

Figure 4.15 Comparison of three seismic cross-sections enclosed by the dashed box that is also shown in Figure 4.14. (a) Reflection stack (Fuis *et al.*, 2003). (b) Pre-stack depth migration (Thornton & Zhou, 2008). (c) Receiver functions (Zhu, 2002). Faults denoted: Santa Monica (SMF), Santa Susana (SSF), San Gabriel (SGF), and San Andreas (SAF).

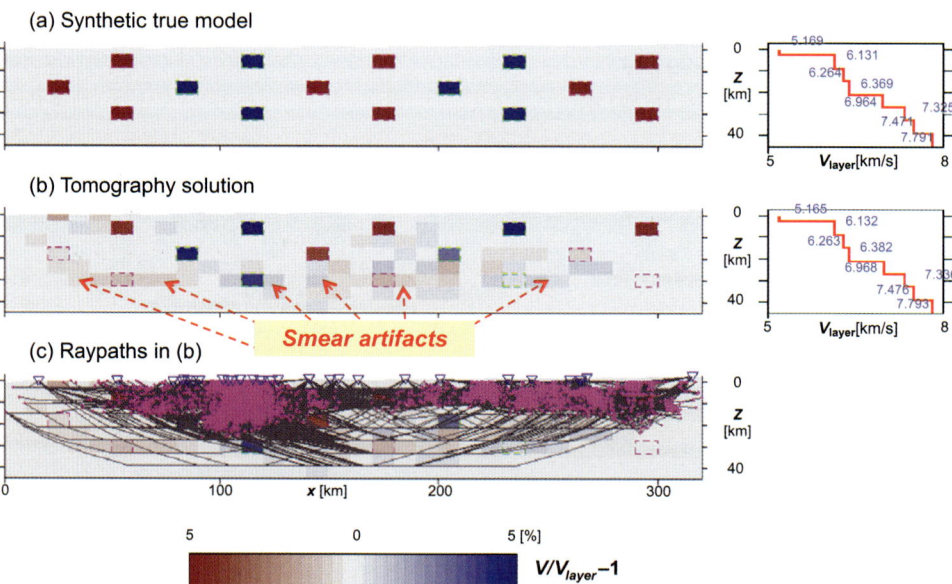

Figure 4.17 Checkerboard resolution test for a traveltime tomography. The layer velocity averages are shown in the right panels and the numbers are in km/s. The left panels are the lateral velocity variations after removal of layer velocity averages. (a) The synthetic true model. (b) Tomography solution. The small dashed boxes outline the correct positions of the checkerboard anomalies. (c) Raypaths for model (b).

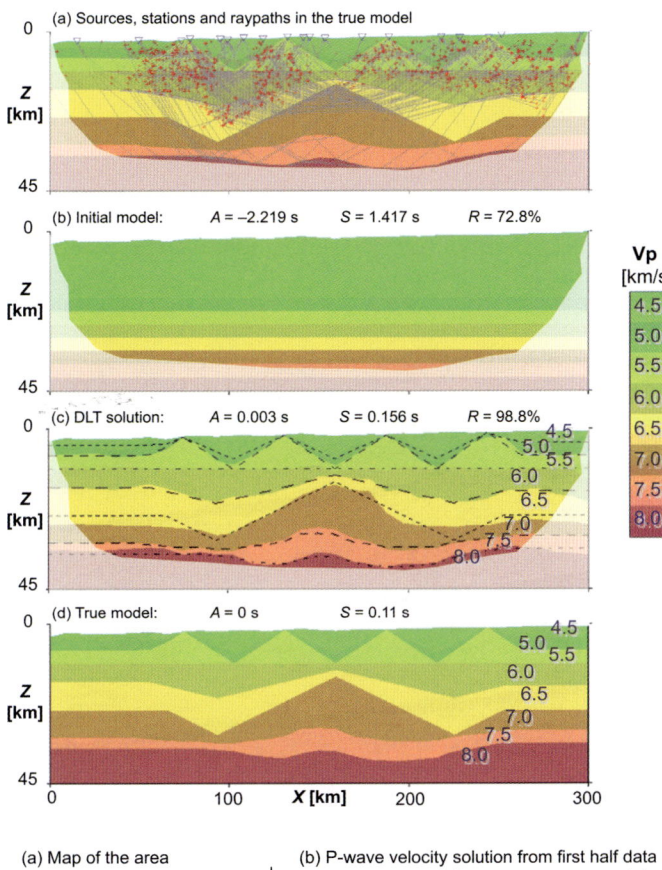

Figure 4.18 Restoration test of a deformable layer tomography. (a) Stations (triangles), sources (crosses), and rays (curves) in the synthetic true model. (b) The initial reference model. (c) Solution velocities in comparison with true model interfaces denoted by dashed curves. (d) True model. In (c) and (d) the numbers denote layer velocities in km/s. *A* and *S* are the average and standard deviation of traveltime residuals, and *R* is the correlation with the true model. Areas without ray coverage are shadowed.

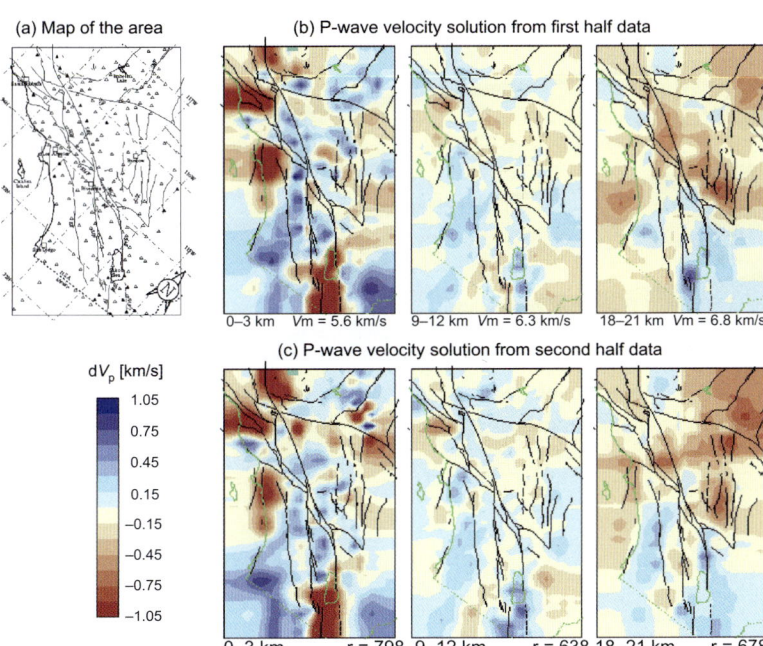

Box 4.3 Figure 1 A delete-half jackknife test for crustal P-wave tomography in southern California (Zhou, 2004a). (a) Map view of major faults and seismologic stations (triangles). (b) Horizontal slices of the velocity model from the first half of the data. (c) Horizontal slices of the velocity model from the second half of the data. The dataset consists of more than 1 million P-wave arrivals from local earthquakes to local stations. V_m is the average velocity of the layer, and *r* is the correlation coefficient between two model slices in the same depth range.

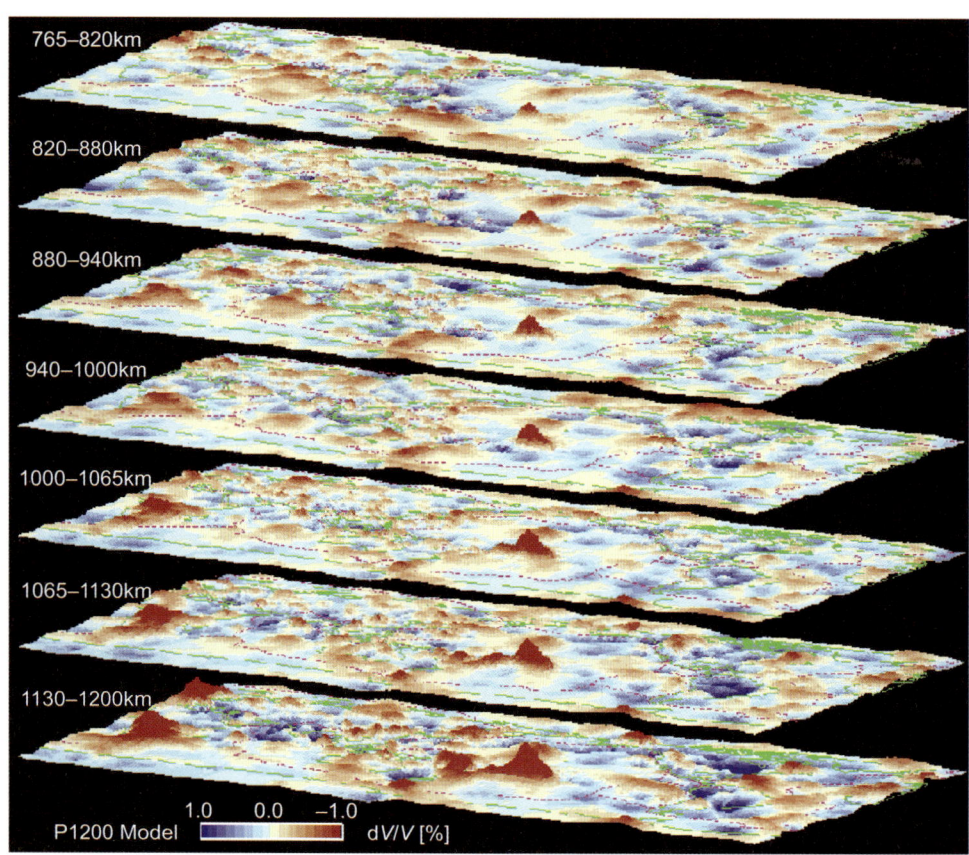

Figure 8.7 3D view of seven layers of lateral variations of V_p in the lower mantle, where the slow and fast velocity anomalies are displayed as red hills and blue basins, respectively.

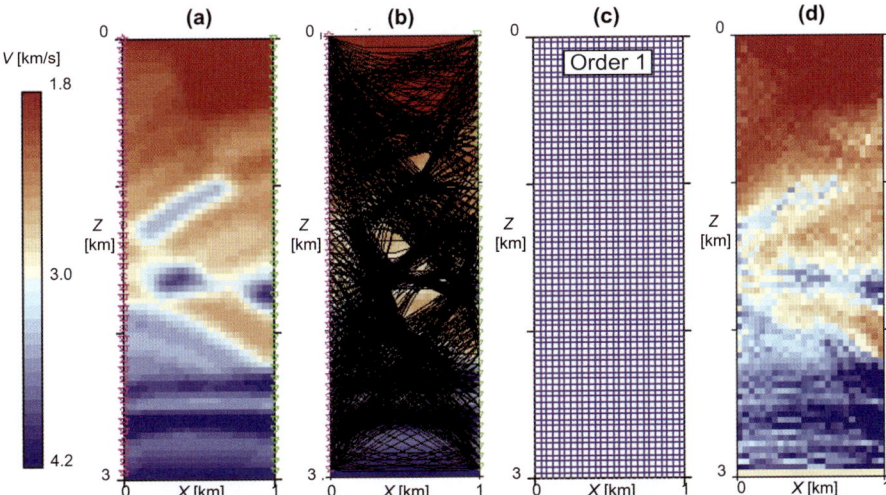

Figure 8.25 (a) A synthetic true velocity model in a cross-well survey using 43 shots (purple stars) and 43 receivers (green triangles). (b) Raypaths in the true model. (c) A cell tomography grid. (d) Solution of single-cell tomography using first-arrival traveltimes in the true model.

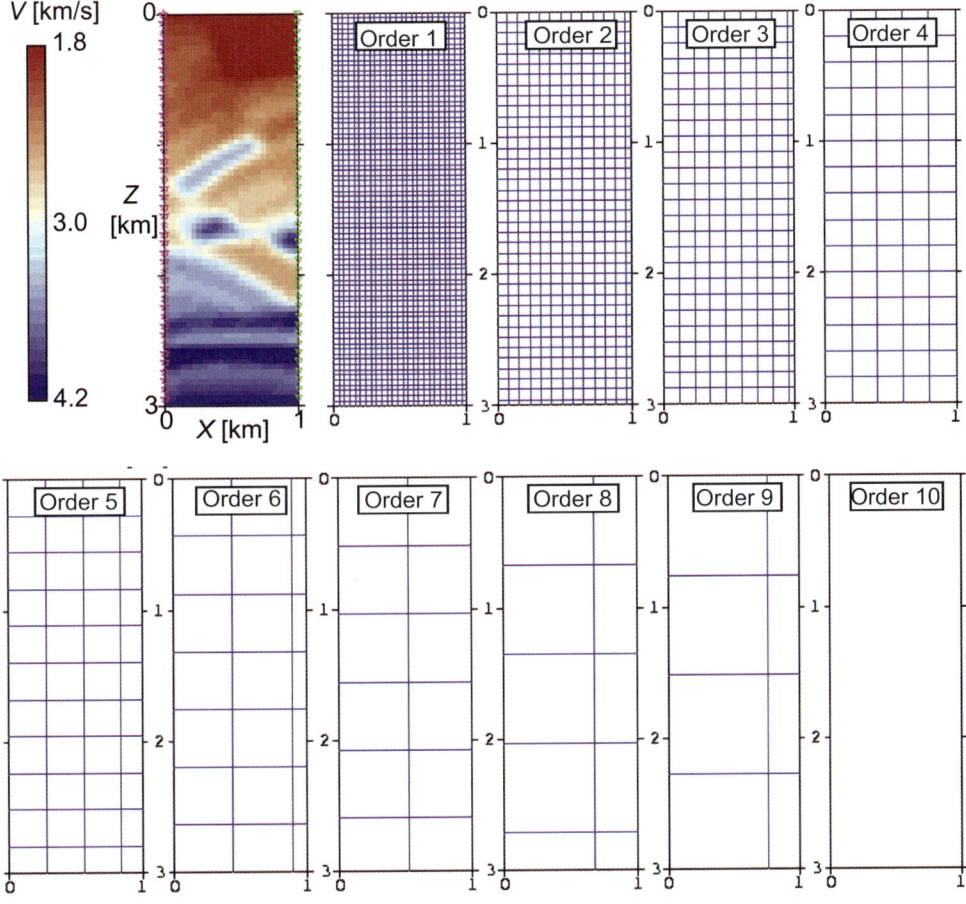

Figure 8.26 The upper-left panel is the same synthetic true model as in Figure 8.25 in a cross-well survey. The other panels show 10 MST sub-models. The Order 1 sub-model has the same geometry as the SST model in Figure 8.25.

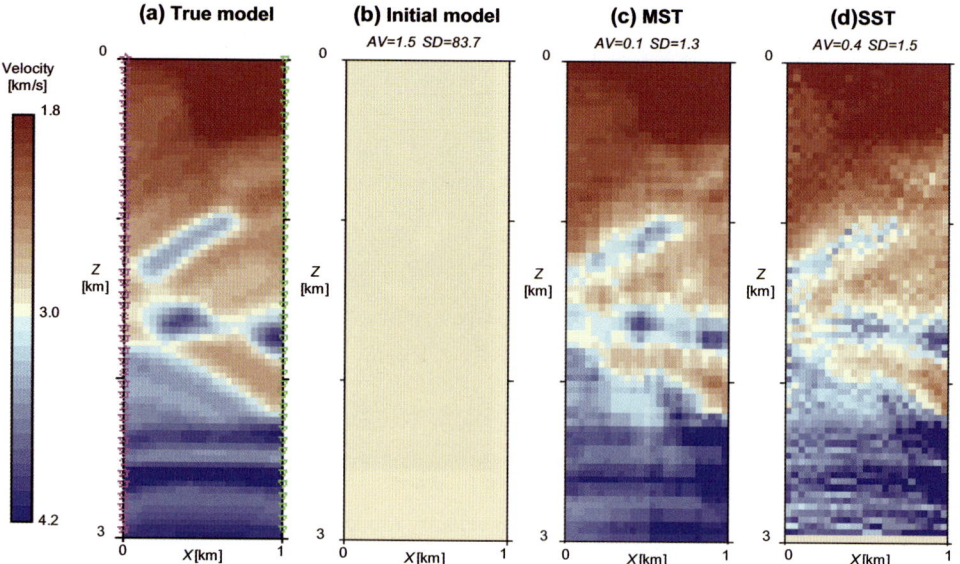

Figure 8.27 (a) A synthetic true velocity model. (b) The initial reference velocity model. (c) The fourth iteration MST solution. (d) The fourth iteration SST solution. The average (AV) and standard deviation (SD) of traveltime residuals in milliseconds are indicated on top of the solution models.

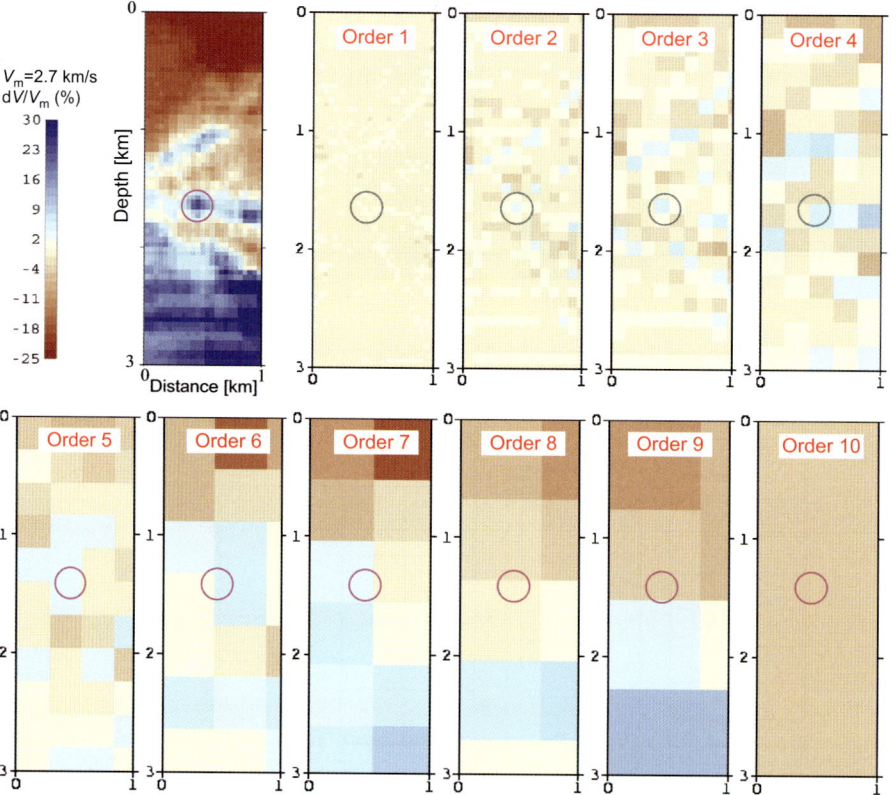

Figure 8.28 The upper-left panel is the MST solution, which is a superposition of the solutions of 10 sub-models shown in the rest of the panels. Plotted are the ratios in per cent between the velocity perturbations and the mean velocity $V_m = 2.7$ km/s.

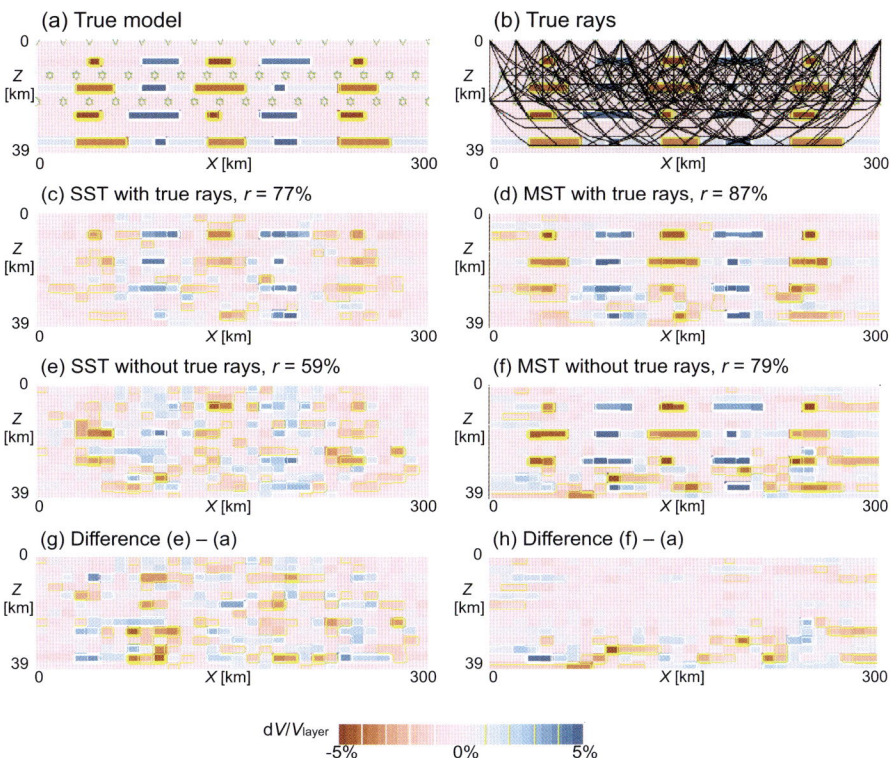

Figure 8.29 Velocity perturbations are contoured in the unit of percentage of the layer averages. (a) True model. (b) First-arriving rays in true model. (c) and (d) are SST and MST solution using true rays. (e) and (f) are solutions without the knowledge of true rays. (g) and (h) are difference between model pairs. r is correlation between the solution and the true model.

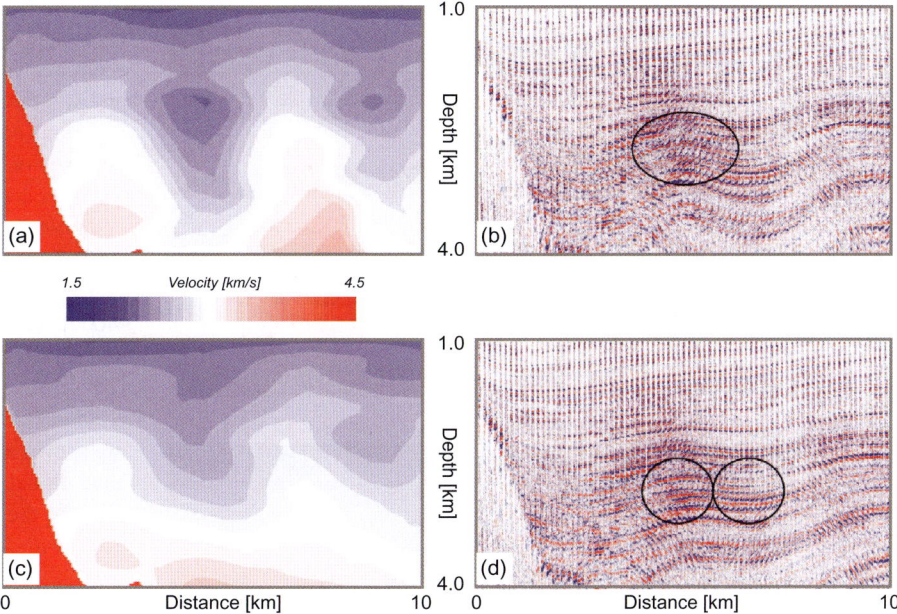

Box 8.3 Figure 1 Comparison between SST and MST for a Gulf of Mexico dataset (Cao *et al.*, 2008). (a) and (c) are velocity models derived using SST and MST, respectively. (b) and (d) are the corresponding common image gathers.

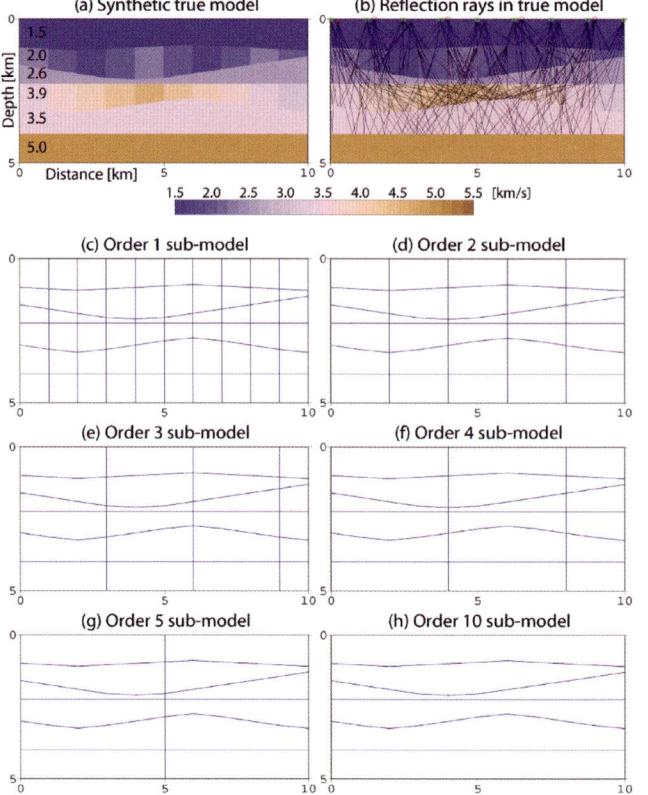

Figure 8.30 (a) 2D synthetic model for a synthetic reflection tomographic test. The values in this panel denote average velocities of each model layer in km/s. (b) Raypaths in the true model from eight shots (stars) to nine receivers (triangles). (c)–(h) Examples of the multi-cell sub-models.

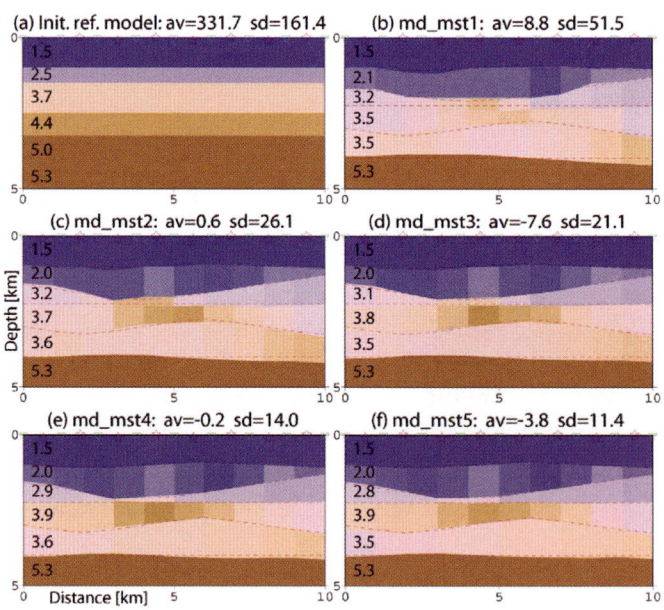

Figure 8.31 The initial reference velocity model and solutions of five iterations of the MST. The numbers on the models are layer average velocities in km/s. The dashed curves show the interfaces of the true model. The average (av) and standard deviation (sd) of traveltime residuals in milliseconds are shown at the top of each model. See Figure 8.30 for the velocity scale.

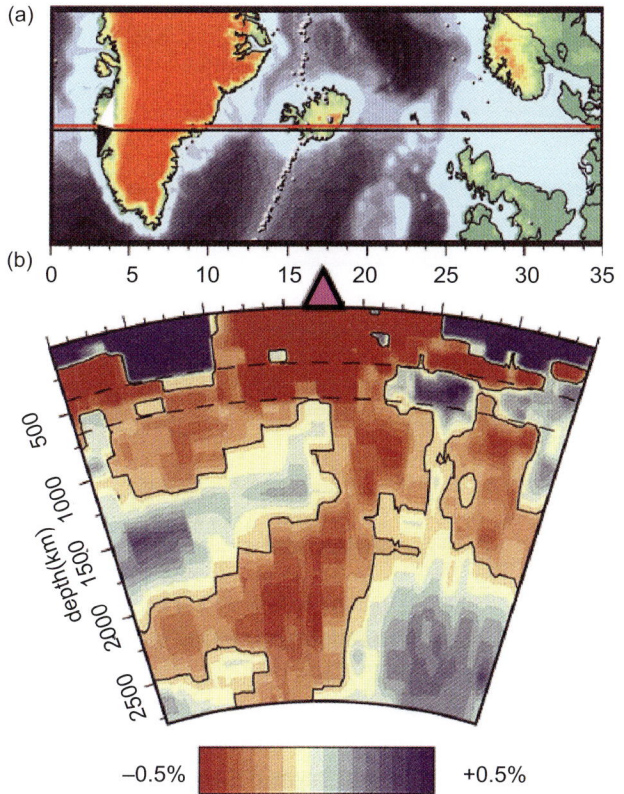

(a)

(b)

0   5   10   15   20   25   30   35

depth(km)

500
1000
1500
2000
2500

−0.5%   +0.5%

Figure 8.33 (a) Map-view of a cross-section through the Iceland. (b) Cross-section of lateral velocity variations from the tomographic inversion (Bijwaard & Spakman, 1999). Can you tell the signature of the Iceland plume from artifacts?

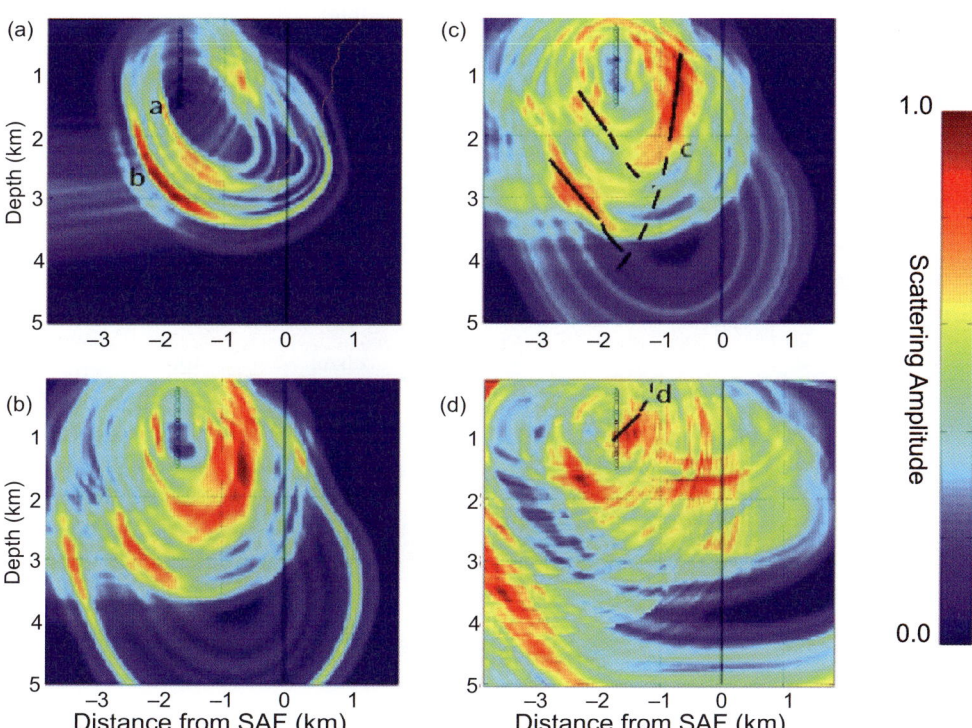

(a)

Depth (km)

1
2
3
4
5

a
b

−3   −2   −1   0   1

(c)

1
2
3
4
5

c

−3   −2   −1   0   1

(b)

Depth (km)

1
2
3
4
5

−3   −2   −1   0   1
Distance from SAF (km)

(d)

1
2
3
4
5

d

−3   −2   −1   0   1
Distance from SAF (km)

1.0

Scattering Amplitude

0.0

Figure 8.34 Images from a VSP Kirchhoff migration near the San Andreas Fault (SAF) illustrating the along-isochronal smear artifacts.

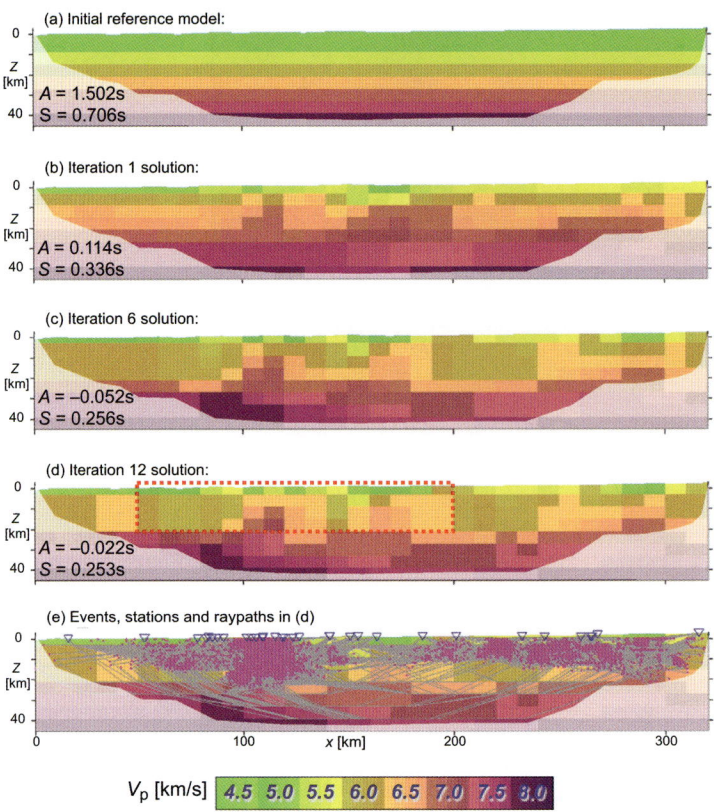

Figure 8.38 The initial reference model (a) and three DLT solutions (b–d). Panel (e) shows raypaths in model (d). Small triangles are stations. Small crosses are earthquake foci. A and S are the average and standard deviation of traveltime residuals in seconds. Areas without ray coverage are shadowed. Area in dashed box in (d) is the next figure.

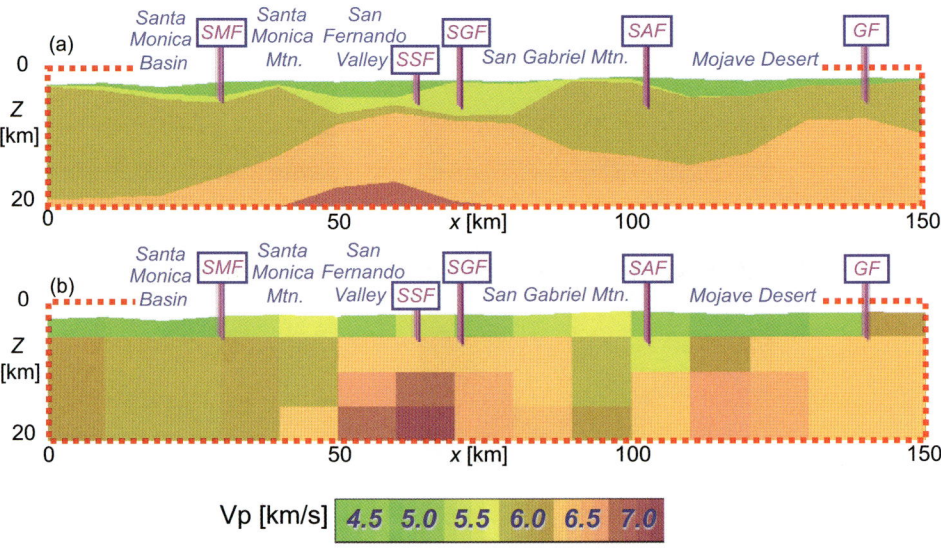

Figure 8.39 Comparison between a DLT model and a cell tomography model in California using the same data. Faults: Santa Monica (SMF), Santa Susana (SSF), San Gabriel (SGF), San Andreas (SAF) and Garlock (GF).

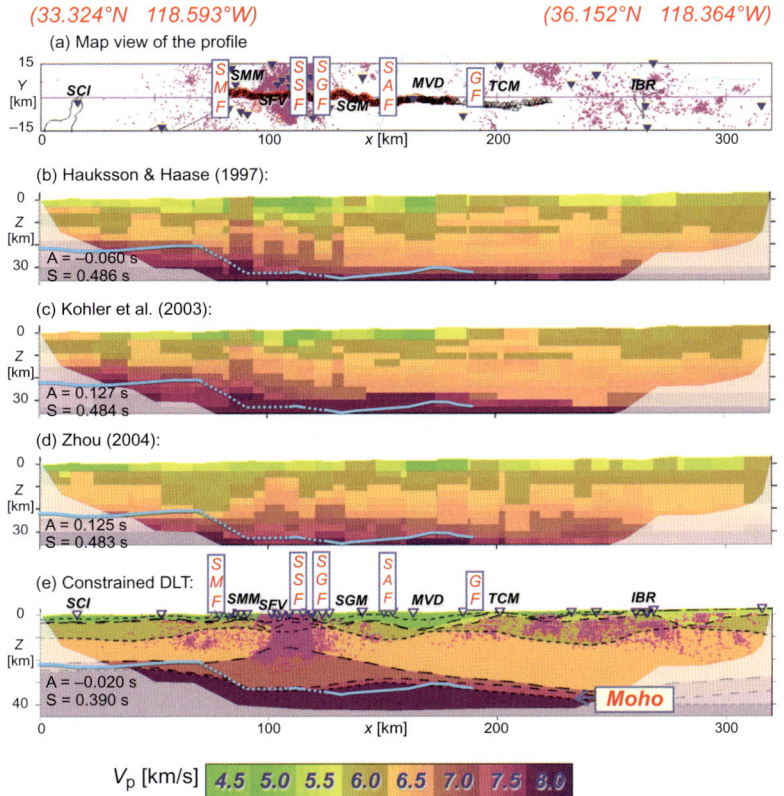

Figure 8.40 Comparison of constrained DLT model (e) with three cell tomography models: (b) Hauksson and Haase (1997), (c) Kohler *et al.* (2003), (d) Zhou (2004a). Purple crosses in (e) are earthquake foci. Faults (pink) are: Sierra Madre (SMF), Mission Hills (MHF), San Gabriel (SGF), San Andreas (SAF), and Garlock (GF). *A* and *S* are the average and standard deviation of traveltime residuals. Solid and dotted curves of light-blue in (b) to (f) denote the PmP Moho interpreted by Fuis *et al.* (2007).

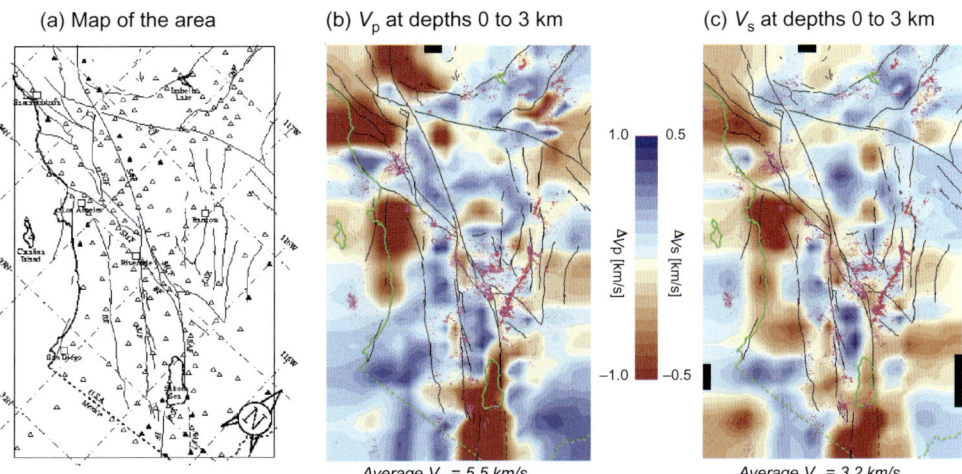

Figure 9.25 (a) Map view of major faults (black curves) and seismologic stations (triangles) in southern California. (b), (c) P- and S-wave velocities in the depth range 0–3 km from inversions using a LSQR algorithm (Zhou, 1996). Areas of insufficient data coverage are blacked out. The correlation coefficient is 66% between the lateral variations of the P- and S-velocities of this layer. Pink dots are earthquake foci.

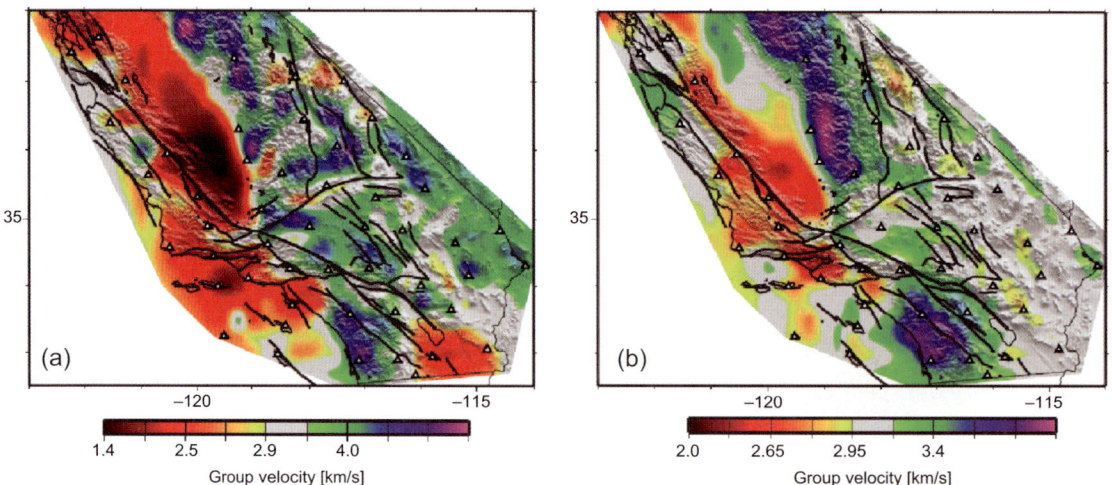

Box 10.1 Figure 1 Maps of Rayleigh-wave group velocities at (a) 7.5 s period and (b) 15 s period in southern California, based on cross-correlating 30 days of ambient noise between USArray stations denoted by triangles (Shapiro *et al.*, 2005). Black curves are major faults.

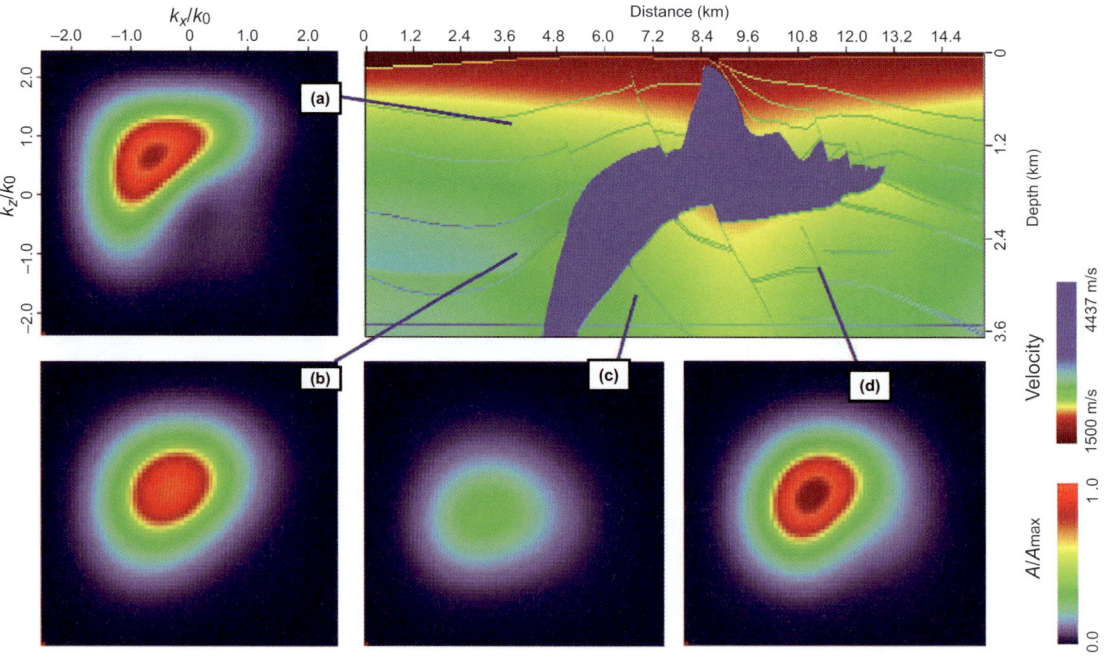

Figure 10.11 Local illumination matrices at four locations in the 2D SEG/EAGE salt model shown in the upper right panel (Xie *et al.*, 2006).

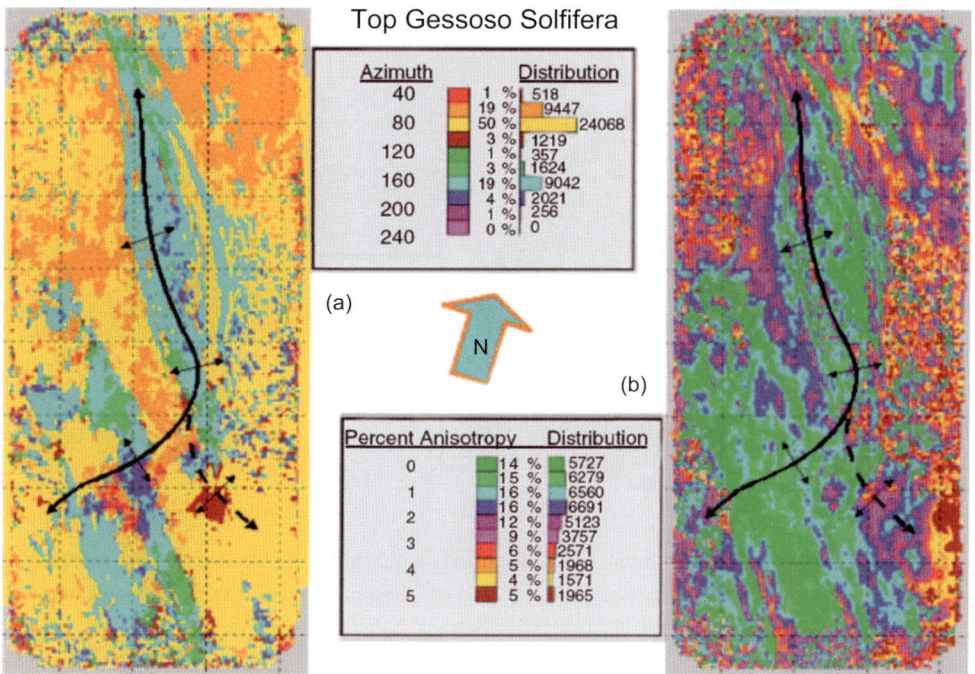

Figure 10.39 (a) Polarization azimuth of the PS$_1$-wave and (b) percentage of shear-wave splitting coefficient above the Gessoso Solfifera formation of the Emilio field in the Adriatic Sea (after Gaiser *et al.*, 2002). The PS$_1$-wave is polarized parallel to the crest of a doubly plunging anticline (thick black arrows), where anisotropy is generally higher.

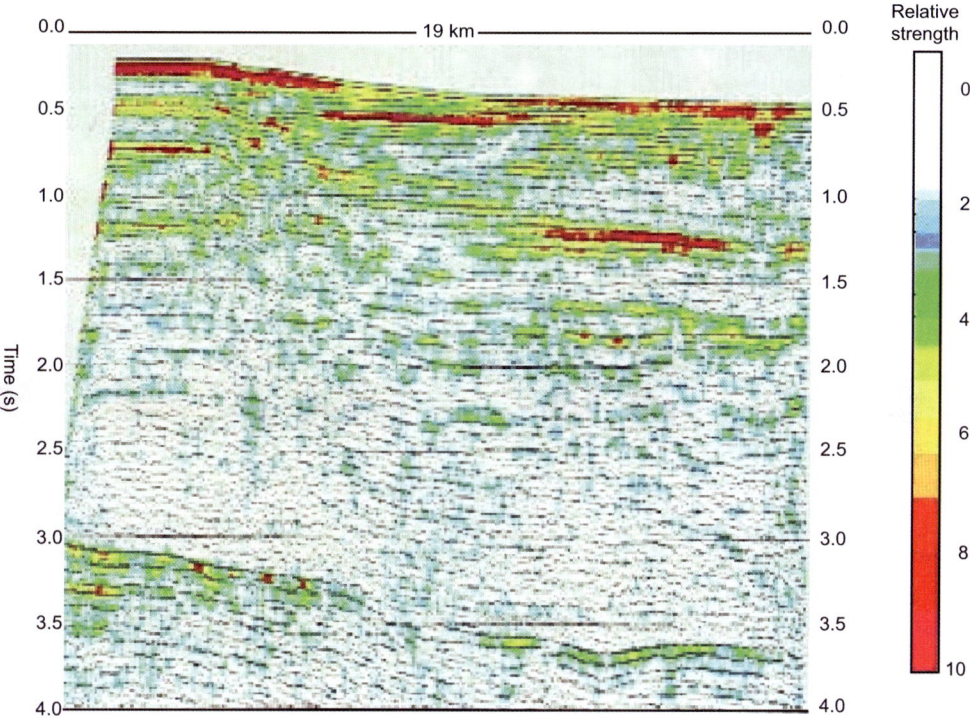

Figure 10.55 A section from the North Sea with bright spots (areas of high seismic reflectivity) identified. (After Anstey, 2005.)

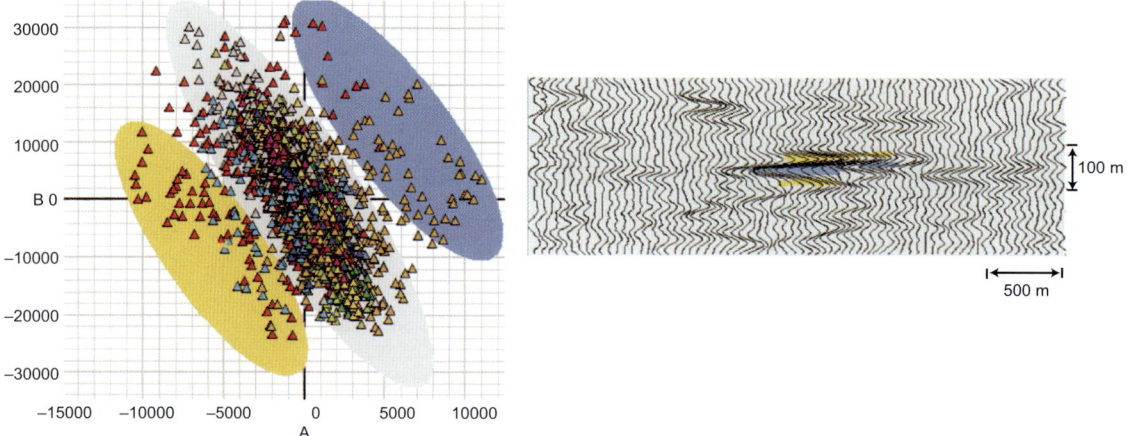

Figure 10.59 (a) Intercept-gradient cross plot using a 200 ms window centered on a bright spot anomaly from a gas sand in the Gulf of Mexico. (b) Color-coded seismic section using the cross-plotted zones in (a). (After Ross, 2000.)

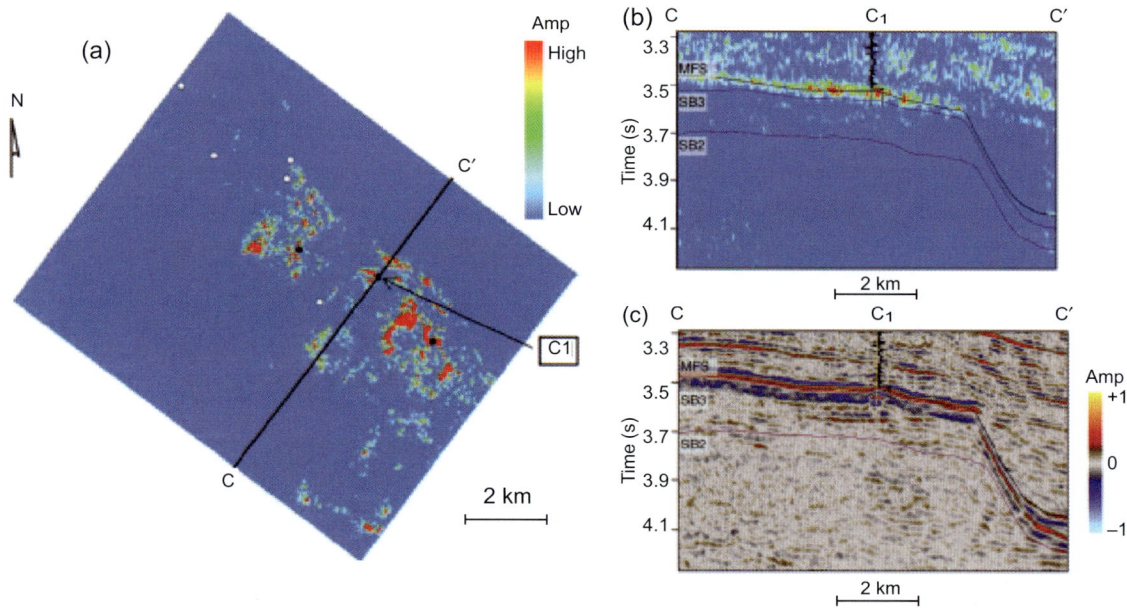

Figure 10.70 Two slices of a 40 Hz iso-frequency volume from the Central Tarim Basin in NW China as (a) horizon slice along the oil and gas zone between the maximum flooding surface (MFS) and sequence boundary 3 (SB3), and (b) vertical slice C–C′ whose location is shown in the mapview (a). In (a) the three solid black circles are oil- and gas-producing wells, and the five open circles are wells without a good reservoir. (c) Original seismic amplitude on vertical slice C–C′ across an oil and gas well C1. (After Li *et al.*, 2011.)

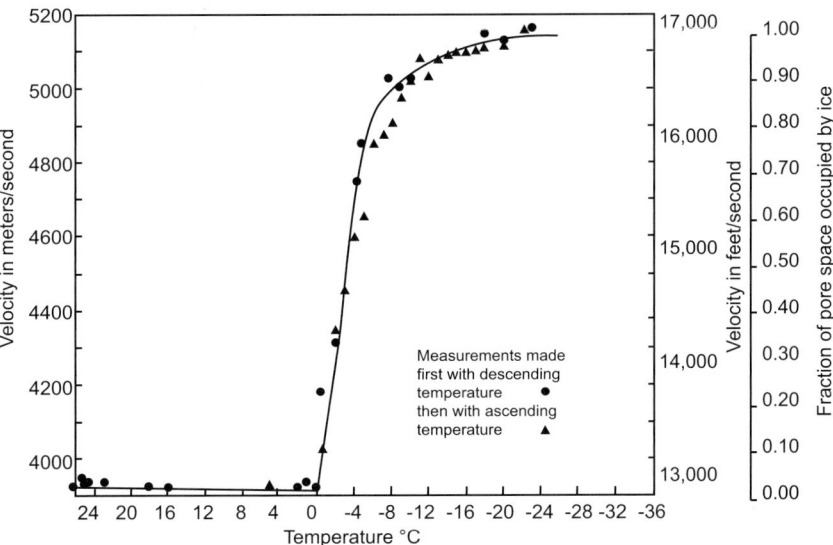

Figure 8.6 P-wave velocity in Berea sandstone as the temperature passes through the freezing point. (From Timur, 1968.)

The effect of temperature on seismic velocities can be viewed in two aspects. First, for near-surface rocks, there is a large effect when the temperature passes through the freezing point, as shown in Figure 8.6. This figure shows a velocity jump of more than 20% when the rock is frozen. In many cold regions, seismic acquisitions are conducted during the frozen winter season to take advantage of the temperature effect on seismic velocity.

Second, in crustal and mantle seismology, velocity anomalies are usually expressed in terms of **lateral velocity variation**, or $(dV/V_{layer} - 1)$, where V_{layer} denotes average velocity of the layer. The **de-mean process**, or removal of the layer average, takes away the effect of pressure or depth. Hence the lateral variation in velocity can be interpreted mostly as the consequence of lateral temperature variation. owing to thermal expansion, the slow lateral velocity anomalies indicate areas of higher or hotter temperature, and fast lateral velocity anomalies indicate areas of lower or cooler temperature. An example of such an interpretation for mantle P-wave tomography is shown in Figure 8.7.

8.1.4   Refraction velocity analysis

8.1.4.1   Refraction and turning rays

Our initial understanding of large-scale trends of seismic velocities in the real Earth comes largely from refraction surveys, which were the first type of seismologic studies. Such studies were motivated by the fact that the velocity variation at large scale in the Earth is dominated by the 1D or $V(z)$ trend. The data for refraction study are primarily the first arrivals, or the first breaks. As indicated by the name, a refraction study implies that the seismic waves travel through a depth-varying or layer-cake velocity model so that the seismic rays are making refraction-type turns. The paths of all seismic waves depend on the velocity variations. For first arrivals, their raypaths can be pure refraction rays in a

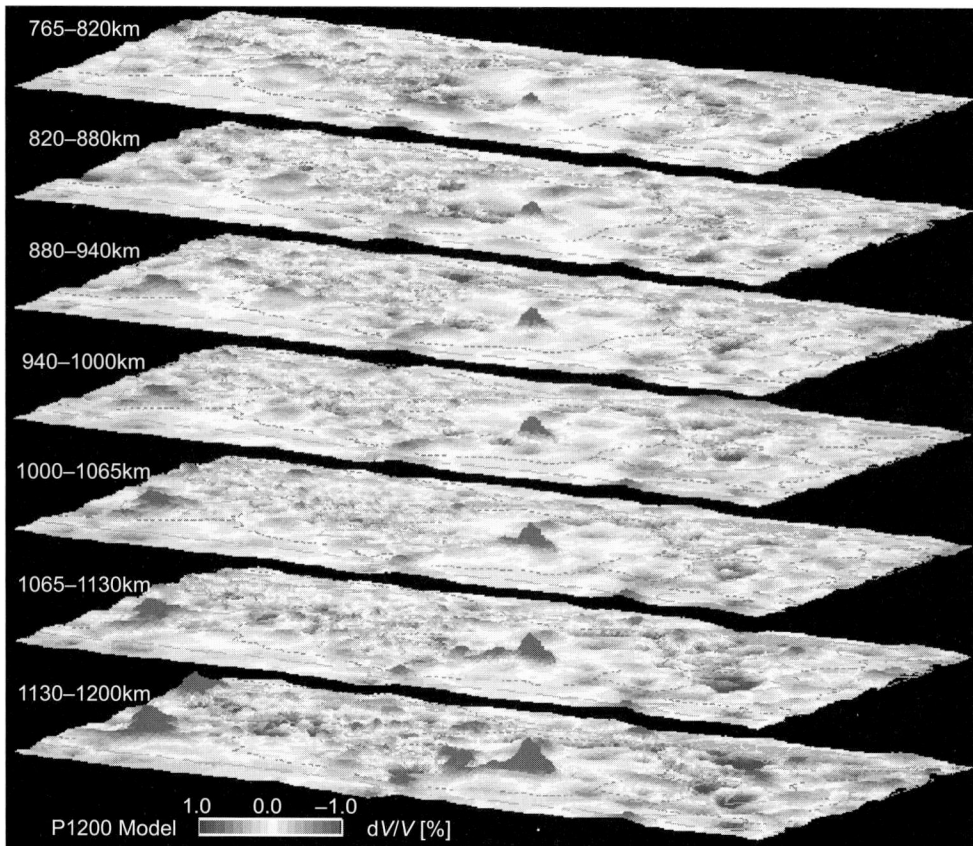

765–820km

820–880km

880–940km

940–1000km

1000–1065km

1065–1130km

1130–1200km

1.0   0.0   −1.0
P1200 Model ▨▨▨▨▨▨ dV/V [%]

Figure 8.7 3D view of seven layers of lateral variations of V_p in the lower mantle, where the slow and fast velocity anomalies are displayed as red hills and blue basins, respectively. For color versions see plate section.

layer-cake model, or **turning rays** in a gradient model where velocity varies gradually with depth, as shown in Figure 8.8.

Since refraction rays and turning rays will not graze a greater depth without an increase in velocity with depth, a refraction study will assume that velocity increases with depth either monotonically or throughout most of the depth range of the study. The existence of a low-velocity layer (LVL) below a higher velocity layer (HVL) is sometimes called **velocity inversion** (please be sure to distinguish this term from the other meaning of inversion – which means determining model parameters from data). The appearance of velocity inversion is troublesome for refraction studies.

8.1.4.2   Global 1D velocity models

Why should we care about the 1D velocity model? While it is true that it is simple to use a 1D velocity model, the main reason is that at large scale the Earth's velocity field is predominantly 1D. This is evident from the good match between observed traveltimes and predictions from the **JB model** (Jeffreys & Bullen, 1940) in Figure 8.9. Taking the Earth

Practical seismic velocity analysis

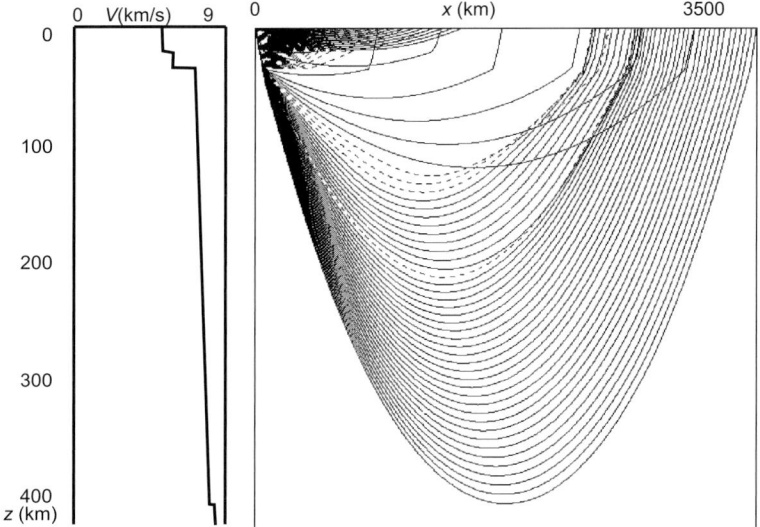

Figure 8.8 (Left) A layer-cake velocity model of the crust and upper mantle of the Earth. (Right) Refracted rays in the model. On decreasing the take-off angle from the source, rays shown as solid curves increase the offset, and rays shown as dashed curves decrease the offset.

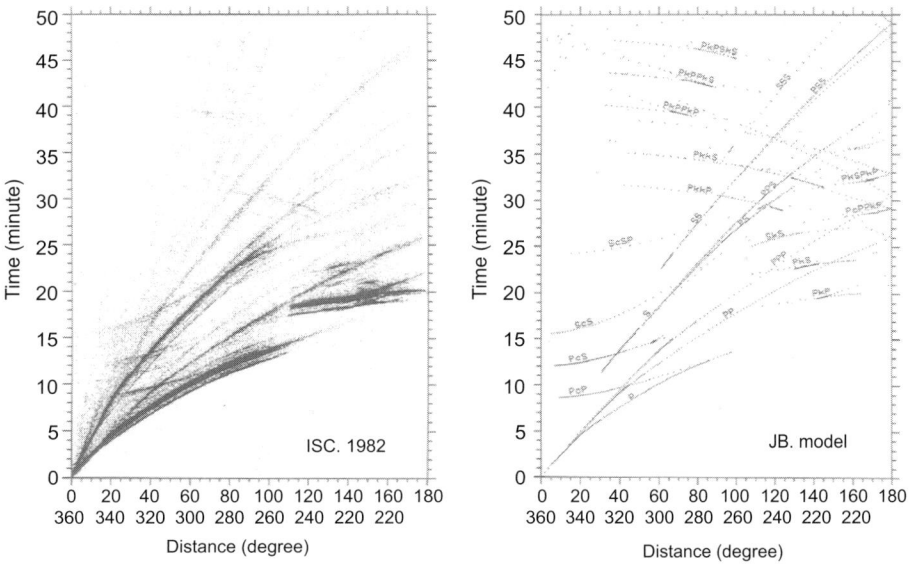

Figure 8.9 (Left) Observed traveltimes of earthquakes that occurred in 1992, compiled by the International Seismologic Centre (ISC). (Right) Predicted traveltimes based on the JB model. Major body wave phases can be identified by comparing two panels. Different focal depths cause widening of data phases such as P, S, and PkP. (After Zhou, 1989.)

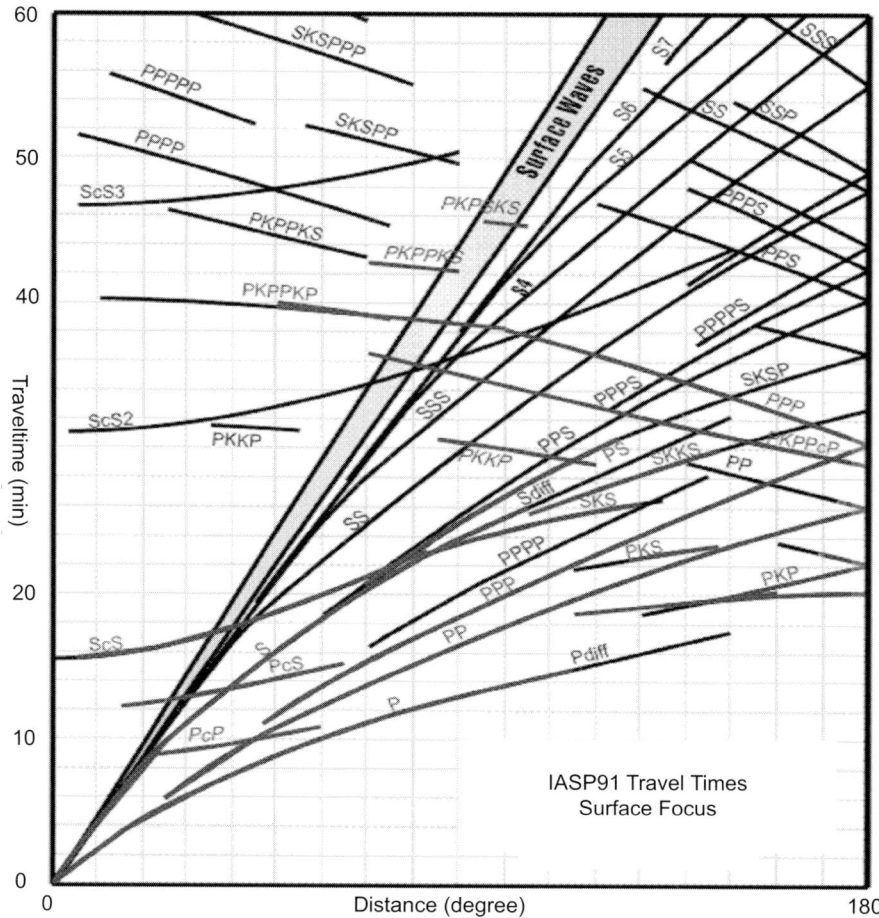

Figure 8.10 Traveltime versus distance graph in the IASP91 velocity model.

as a whole, because of the combined effects of billions of years of tectonic recycling of Earth's interior materials and predominately depth-varying nature of the pressure and temperature fields, the Earth's velocity field is mostly depth-varying.

Consequently, over 98% of the traveltimes of major seismic phases, such as P, pP, sP, PP, S, SS, PcP, PKP, SKS, and PKIKP, can be explained by a standard 1D Earth model. The most famous seismologic 1D Earth models include the **JB model**, the **Herrin model** (Herrin, 1968), the **PREM model** (Dziewonski & Anderson, 1981), and most recently the **IASP91 model** (Kennett & Engdahl, 1991). Figure 8.10 shows the predicted traveltimes from the IASP91 model of the global 1D P-wave and S-wave velocities.

Except in the case of a constant-velocity model, obtaining the **1D velocity model** is often a useful first step in velocity model building. A 1D velocity model is a depth-varying and laterally homogeneous velocity model. A further simplification is the layer-cake model, which consists of a set of flat and constant-velocity layers. The layer-cake velocity model is most useful in refraction and reflection studies, where the velocity profile is often a step function on a graph of velocity versus depth graph or velocity versus zero-offset time. The

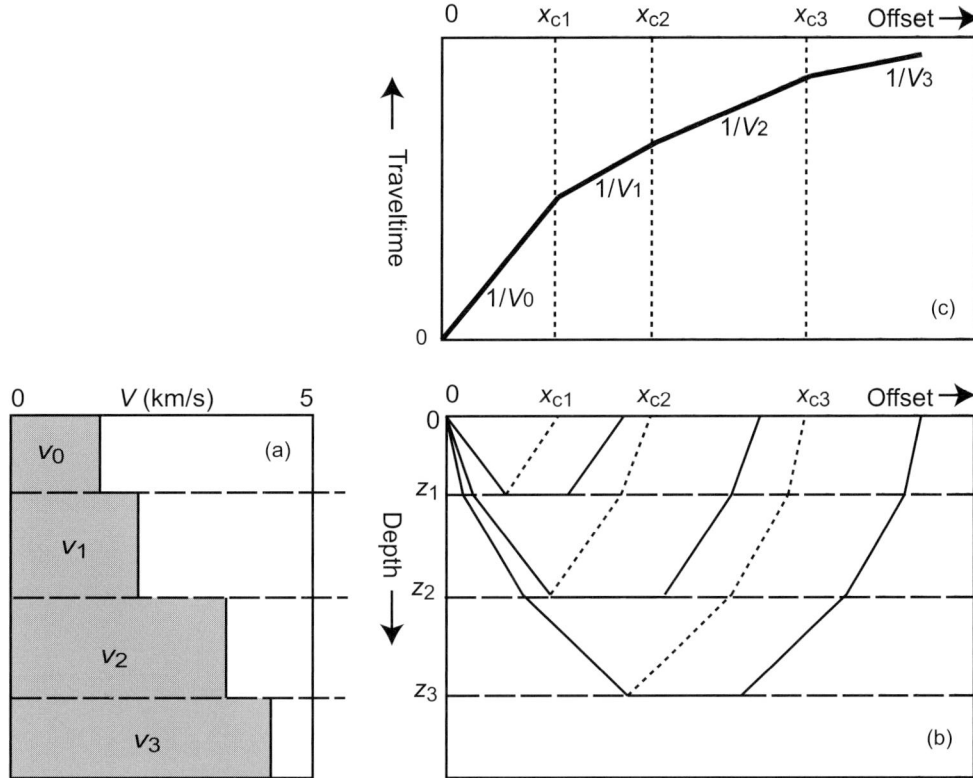

Figure 8.11 (a) A four-layer velocity model; (b) refraction rays in the model. Dashed curves denote critical refraction rays for the three interfaces; (c) first arrival versus offset, whose slope corresponds to the interval velocity at the bottom portion of the rays.

simplification from a continuous velocity function into a layer-cake function focuses our attention on the velocity interfaces, or reflectors. Some prominent velocity interfaces are called **seismic discontinuities**, such as the Moho discontinuity which separates the crust from the upper mantle, and the 670-km discontinuity which separates the upper mantle from the lower mantle.

8.1.4.3   Use of traveltime versus distance graphs

A traveltime versus distance graph is the most basic way to analyze seismic data. Here the distance means source-to-receiver offset, and traveltime can be one-way or two-way times for refraction data, or two-way traveltime for reflection data.

A traveltime versus distance graph can be used to solve some common problems:

- QC signals and noises
- Distinguishing the types of wave modes
- Estimating and extracting velocity information (apparent velocity, etc.)

In a layer-cake velocity model, we may use traveltime versus source-to-receiver offset plots to estimate velocities traversed by the bottoming portions of refraction rays. Figure 8.11

Box 8.1 Estimate velocities using reduced traveltime versus offset plot

Here a real data example demonstrates the estimation of interval velocities using a plot of reduced traveltime versus offset. Box 8.1 Figure 1a shows the geometry of a vertical seismic profile (VSP) study (Zhou, 2006). Panel b shows first arrivals from two common receiver gathers of seismic records from the surface shots to the receivers placed along a nearly vertical wellbore. The thin lines in the figure show the theoretical arrival times for several reduction velocities. If the bottoming portion of a first-arrival ray travels at the speed of a reduction velocity, then it will fall on the corresponding line on the plot of reduced traveltime versus offset.

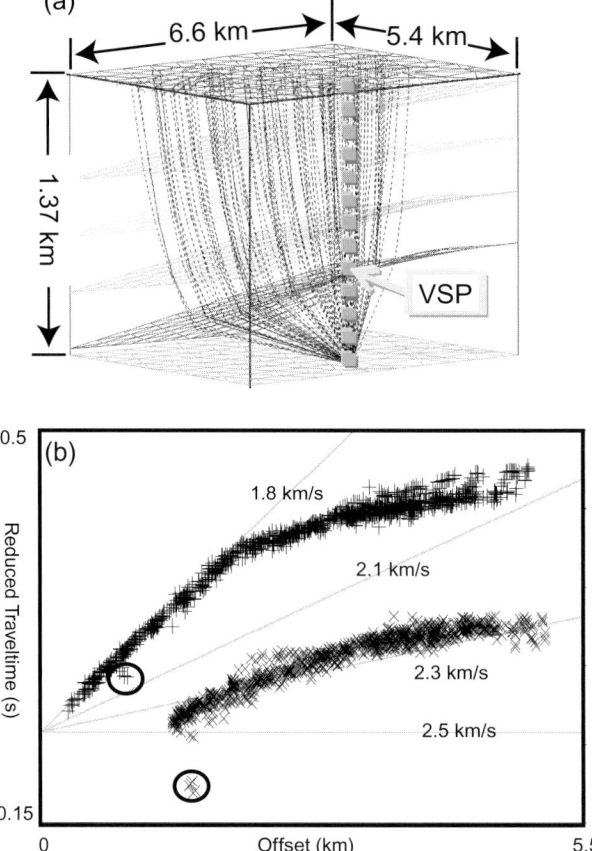

Box 8.1 Figure 1 (a) 3D view of a VSP survey with dashed curves denoting first arrival raypaths from surface shots to receivers along the wellbore. (b) First arrivals with a reduction velocity of 2.5 km/s versus shot–receiver distance for the shallowest receiver (+ symbols) and the deepest receiver (×). Lines show the theoretical arrival times for the reduction velocities indicated. Picks inside the small circles probably traversed along the salt flank. (After Zhou, 2006.)

In panel (b) the first arrivals of the shallowest receivers show at least three segments in their slope trend. The slope at the near offset follows the trend of 1.8 km/s in reduction velocity. At intermediate offset the slope is parallel with the trend of 2.3 km/s in reduction velocity. At the far offset the slope is between 2.3 and 2.5 km/s. Hence the first arrivals of this common receiver gather indicate the general pattern of the velocity model: three layers, with interval velocities of 1.8, 2.3, and 2.4 km/s, respectively. This pattern is consistent with the first arrivals for the deepest common receiver gathers.

gives a schematic illustration using a four-layer model in which the interval velocity increases with depth. Along each model interface, a critical distance x_c denotes the minimum offset of the refracted head wave traversing along the interface. The corresponding critical refraction rays for the three model interfaces are denoted by dotted curves in Figure 8.11b. Between zero offset and the critical distance of the first interface, x_{c1}, the first arrivals are direct waves; hence their slope on the traveltime–offset plot corresponds to the slowness of the first layer, $1/V_0$. Between x_{c1} and the critical distance of the second interface, x_{c2}, the first arrivals are head waves traversing along the first interface; hence their slope on the traveltime–offset plot corresponds to the slowness of the second layer, $1/V_1$. Similarly the slopes of the first arrivals in the next two distance intervals correspond to the slowness of the other two layers. The slope on the traveltime–offset plot can be more easily measured using the reduced traveltime τ as defined in Chapter 2:

$$\tau = t - x/v \qquad (2\text{–}23')$$

where t is traveltime, x is source to receiver offset, and v is the reduction velocity.

A field example of estimating interval velocities using a reduced traveltime–offset plot is demonstrated in Box 8.1.

8.1.5   Basics of seismic ray tracing

8.1.5.1   Overview

The predicted traveltimes from a velocity model are calculated using ray tracing, which is one of the oldest topics in applied seismology but still a useful one. **Kinematic ray tracing** involves determining the traveltime and raypath from point A to point B in a velocity model. **Dynamic ray tracing** may involve additional parameters that are frequency-dependent and/or amplitude-dependent. In addition to seismology, ray tracing has been applied widely in many other fields. For instance, modern computer graphic design often uses rendering techniques that are based on ray tracing of light. Kinematic ray tracing is a fundamental step for traveltime tomography, because it connects the traveltimes with velocity anomalies. Seismologists also use ray tracing to study illumination of seismic data, and Kirchhoff depth migration is based on traveltime tables generated by ray tracing.

One of the early 3D ray tracing techniques (Jacob, 1970) is designed to incorporate the effect of a subducted slab. Based on their different approaches to the problem there are four types of 3D ray tracing technique. The two traditional methods include the shooting method (Julian, 1970) and the bending method (Wesson, 1971). There are many hybrid versions of these two methods (e.g., Julian & Gubbins, 1977; Pereyra *et al.*, 1980; Thurber & Ellsworth, 1980). The third method is the finite-difference eikonal solver (Vidale, 1990; Qin *et al.*, 1992), and the fourth method is the shortest-path method (Moser, 1991) which follows network theory. Practically, however, it is more convenient to classify different ray tracing methods based on the number of dots to be connected. By this classification, there are usually three ray tracing classes: (1) one-point ray tracing, which simulates shooting of rays; (2) two-point ray tracing, which simulates transmission or refraction rays; and (3) three-point ray tracing, which simulates reflection rays.

8.1.5.2   Ray tracing in V(z) models: the Moho velocity distribution

Ray tracing in 1D or $V(z)$ models has been of interest to seismologists since the birth of seismology. The main reason is that on a global scale, the Earth's velocity structure is predominately 1D. For many large-scale studies, a 1D model or a 2D model (assuming that the raypath stays in the great-circle plane through the source and receiver) is often sufficient. The main benefit of using the $V(z)$ model is its computational efficiency. For example, in 1985 I was able to trace around 100 rays per second in a $V(z)$ mantle model using a SUN2 computer. Another benefit of using the $V(z)$ model is the stability of the solutions. Since the reference velocity model usually differs from the reality, such stability is very helpful.

One particularly interesting way of ray tracing in global $V(z)$ models is based on a classic approximation due to Mohorovicic (Chapter 7 of Bullen, 1963). By this **Moho velocity distribution**, the Earth's velocity field is composed of many shells or layers, and in the *i*th layer, the velocity is

$$v_i(r) = a_i \exp(r, b_i) \tag{8–5}$$

where r is the distance to the center of the Earth, and a_i and b_i are two constants for the *i*th layer. Both Figures 8.8b and 8.9b were produced following this approach.

Two factors make the Moho velocity distribution very useful. First, the 1D velocity profiles in many areas can be closely modeled by the Moho velocity distribution. Second, when the Moho velocity distribution becomes valid, analytical solutions exist for source–receiver traveltime and distance. For a refraction ray with ray parameter p, the distance $\Delta(p)$ and traveltime $T(p)$ of first arrivals are given by

$$\Delta(p) = 2p \int_{\gamma_p}^{\gamma_0} \frac{dr}{r\sqrt{\eta^2 - p^2}} \tag{8–6}$$

$$T(p) = 2 \int_{\gamma_p}^{\gamma_0} \frac{\eta^2 dr}{r\sqrt{\eta^2 - p^2}} \tag{8–7}$$

With the above Moho velocity distribution, we will have

$$\eta = r/v = \exp(r, 1 - b)/a \tag{8–8}$$

Table 8.1 Format of a velocity model using the Moho approximation.

Order	Bottom radius	1st constant	2nd constant
1	r_1	a_1	b_1
2	r_2	a_2	b_2
3	r_3	a_3	b_3
...	...	...	...

Then the solutions of the distance and time terms are

$$\Delta(p \text{ or } i_0) = \frac{2}{1-b} \cos^{-1}\left(a \cdot p \cdot r_0^{b-1}\right) = \frac{2}{1-b} \cos^{-1}(\sin i_0) \qquad (8\text{--}9)$$

$$T(p \text{ or } i_0) = \frac{2}{(1-b)} \sqrt{a^{-2}r^{2-2b} - p^2} = \frac{2}{(1-b)} \frac{r_0^{1-b}}{a} \cos i_0 \qquad (8\text{--}10)$$

where i_0 is the take-off angle at the surface. In a $V(z)$ model described by the Moho velocity distribution with a single pair of constants a and b, the above solutions can be used to compute distance and traveltime from the surface ray take-off angle. Since the equations are analytical, one can also obtain the take-off angle i_0 from given distance Δ.

A much more useful $V(z)$ model is composed of multiple shell layers, each following the Moho approximation. The model may be expressed like that in Table 8.1. In this case, for a given **ray parameter** p, we can use Equations (8–9) and (8–10) to compute changes in distance and traveltime between any two levels within each velocity layer. For example, suppose that we have two radius levels r_{upper} and r_{lower} within the ith layer, then for ray parameter p,

$$\Delta(r_{\text{upper}}) - \Delta(r_{\text{lower}}) = \frac{2}{1-b_i}\left[\cos^{-1}\left(a_i \cdot p \cdot r_{\text{upper}}^{b_i-1}\right) - \cos^{-1}\left(a_i \cdot p \cdot r_{\text{lower}}^{b_i-1}\right)\right] \qquad (8\text{--}11)$$

$$T(r_{\text{upper}}) - T(r_{\text{lower}}) = \frac{2}{(1-b_i)}\left[\sqrt{a_i^{-2}r_{\text{upper}}^{2-2b_i} - p^2} - \sqrt{a_i^{-2}r_{\text{lower}}^{2-2b_i} - p^2}\right] \qquad (8\text{--}12)$$

Repeated use of the above two equations for each velocity layer will lead to total distance and traveltime of any ray in the multi-layered $V(z)$ model.

8.1.5.3   Ray tracing using reference tables

One particularly efficient way to do ray tracing in the above type of $V(z)$ velocity model is to use a reference ray table. We construct the reference ray table using the fact that all seismic stations are usually located at the surface. To make the table from a surface source a series of rays in the given $V(z)$ model is computed for source–receiver distances increased by a small step Δx, as shown in Figure 8.12.

As shown in Table 8.2, each row of the reference ray table contains information for one ray. The first part of the ith row is the "header" part that includes ray parameter p_i, total traveltime T_i, and bottoming depth of the ray zb_i. The remaining part of the row contains

Table 8.2 An example reference ray table.

Ray	Header			Ray depth and traveltime						
1	p_1	zb_1	T_1	$z_{11}{:}t_{11}$	$z_{12}{:}t_{12}$	$z_{13}{:}t_{13}$	...	...	0:0	0:0
2	p_2	zb_2	T_2	$z_{21}{:}t_{21}$	$z_{22}{:}t_{22}$	$z_{23}{:}t_{23}$	...	...	0:0	0:0
3	p_3	zb_3	T_3	$z_{31}{:}t_{31}$	$z_{32}{:}t_{32}$	$z_{33}{:}t_{33}$	...	...	0:0	0:0
4	p_4	zb_4	T_4	$z_{41}{:}t_{41}$	$z_{42}{:}t_{42}$	$z_{43}{:}t_{43}$	...	...	0:0	0:0

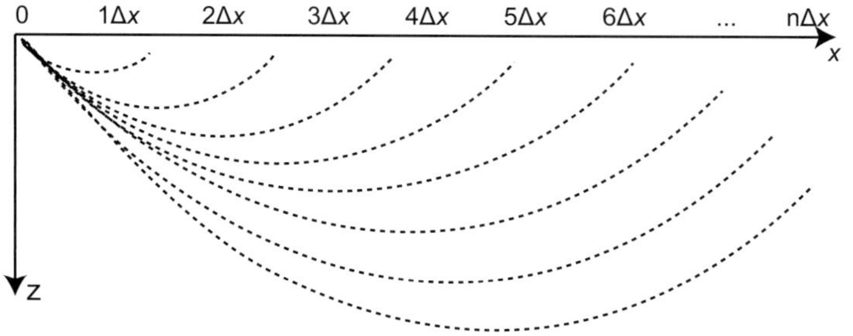

Figure 8.12 A series of rays of incremental distance Δx.

the ray depth z_{ij} and traveltime t_{ij} at the intersection point between the ith ray and the jth distance step.

To use the reference ray table, we start with the given source–receiver distance and source depth. We will locate the column of the table that has a distance nearest to the given source–receiver distance. We will then search through the column to find the row with a depth nearest to the given focal depth. A close approximation of the given ray is thus yielded, represented by the ray parameter of the row. If a better approximation is desired, one can either use a finer reference ray table, or trace the raypath further from the current ray information.

Exercise 8.1

1. Using a spreadsheet to compile observed 1D, or $V(z)$, profiles from the literature, make a list of entries to include location, geologic setting, wave type, frequency, depth range, velocity profile, reference (authors, year, and publication), etc.

2. If velocity increases linearly with depth, what is the maximum depth of the raypath of a first arrival at 10-km offset between a source and a receiver located on the surface?

3. The traveltimes and raypaths of transmitted and reflected waves are dependent on velocity variation but independent of density variation. The amplitudes of transmitted and reflected waves depend on the variations of both velocity and density. Can you devise a way to extract density variation using transmission and reflection data jointly? As a first-order approach, you may simplify the model and acquisition geometry.

Semblance velocity analysis

NMO semblance velocity analysis

In Section 2.2 we discussed normal moveout (NMO) stacking and associated semblance velocity analysis that deliver two products for each CMP gather: the stacking velocity profile and a stack trace defined at the midpoint in lateral position. Here the usefulness and limitations of **NMO semblance velocity analysis** are examined. In most seismic data processing projects, NMO semblance velocity analysis is the first type of velocity analysis. At each CMP location, after identifying some major primary reflection events, we may approximate the velocity function above each chosen event with an average velocity that is the **stacking velocity**, and perform NMO with a profile of stacking velocities. The stacking velocity profile should be able to flatten most primary reflections after NMO, so that all data traces can be stacked across the offset axis to form a single stacked trace. Therefore, the stacking velocity at each depth, marked by the corresponding two-way traveltime, is a medium average of all the overburden velocities above that depth.

Semblance velocity analysis has the following advantages:

1. Simplicity – the underlying model, assumptions, and equations involved are all very straightforward.
2. Robustness – the processes of NMO and stacking are computationally robust.
3. Objectivity – the semblance can be computed regardless of the data quality and validity of the assumptions. It is up to the user to interpret the validity of the analysis and pick the stacking velocity.

The semblance velocity analysis also has a number of pitfalls in terms of:

1. Assumptions – a number of assumptions are involved to compute the semblance:
   - All events are primary reflections (this requires the elimination of multiples and converted waves when possible);
   - The velocity field is locally layer-cake (this requires the use of CMP gathers);
   - Reflection raypaths are straight due to the layer-cake assumption, but real reflection raypaths are curved like the letter U (this is one of the reasons behind the hockey-stick at far offset which will be discussed in the next section).
2. Non-uniqueness – it is non-unique to determine interval velocities from rms velocities. This issue will be examined in a moment.
3. Challenges – the process has to tolerate a number of problems and associated noises, such as:
   - Narrow offset range of reflections at shallow times;
   - NMO stretch at farther offset and shallower times;
   - Static problem at short and long wavelengths;
   - Increase of data frequency with depth;
   - Cross-over events, or events with conflicting dipping angles.

offset

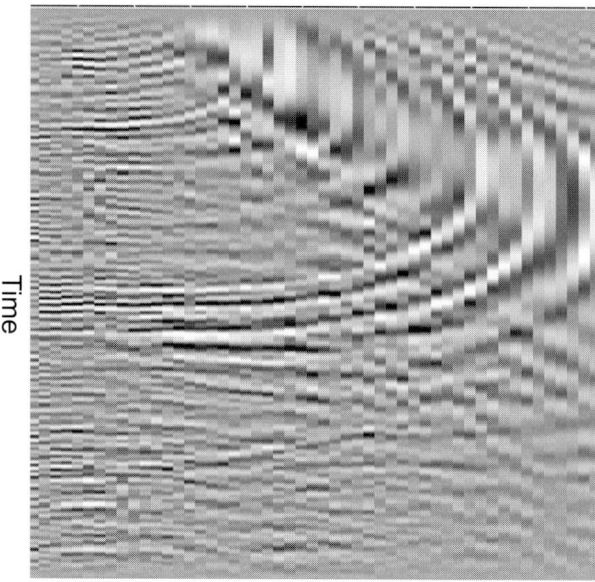

Time

Figure 8.13 A migrated gather illustrating the hockey-stick pattern at far offset.

An example of the consequence of some of the challenges is the "hockey-stick" pattern after NMO, or on a time-migrated gather such as that shown in Figure 8.13. Since both processes contain far-offset stretch, one explanation for the cause of the hockey-stick is the stretch effect at far offset. However, there are a number of other factors that may cause or at least affect the pattern. These include the existence of velocity anisotropy and the effect of ray bending. It will be illuminating to compile all factors that may contribute to the hockey-stick pattern and then quantify their impact using both synthetic and field datasets.

In practice, semblance velocity analysis can be affected by a number of factors. Box 8.2 illustrates the impact of data filtering on the result of semblance velocity analysis.

The result of the semblance analysis for each CMP gather is a stacking velocity profile for the midpoint position of the CMP. For 2D or 3D seismic data, we can construct a 2D or 3D model of stacking velocities by combining the semblance velocity solutions along the 2D profile or 3D data volume. The stacking velocities are functions of zero-offset two-way reflection time, and they can be used in post-stack migration.

8.2.2 Interval velocities and Dix formula

For many seismic imaging processes, we need to have a model of **interval velocity**, which is the velocity of a local interval either in spatial or temporal scales. Hence, there is a need to relate the stacking velocity profile to either the temporal interval velocity profile or the depth interval velocity profile. Dix (1955) formulated the relationship between the temporal velocity profile and its **rms velocity**, V_{rms}, which may approximate the stacking velocity.

Box 8.2 Effects of filtering on semblance velocity analysis

The accuracy of picking stacking velocity via the semblance velocity analysis can be affected by various pre-processing methods and parameters such as fold and offset range. Here the effects of filtering on the semblance analysis are illustrated in Box 8.2 Figure 1

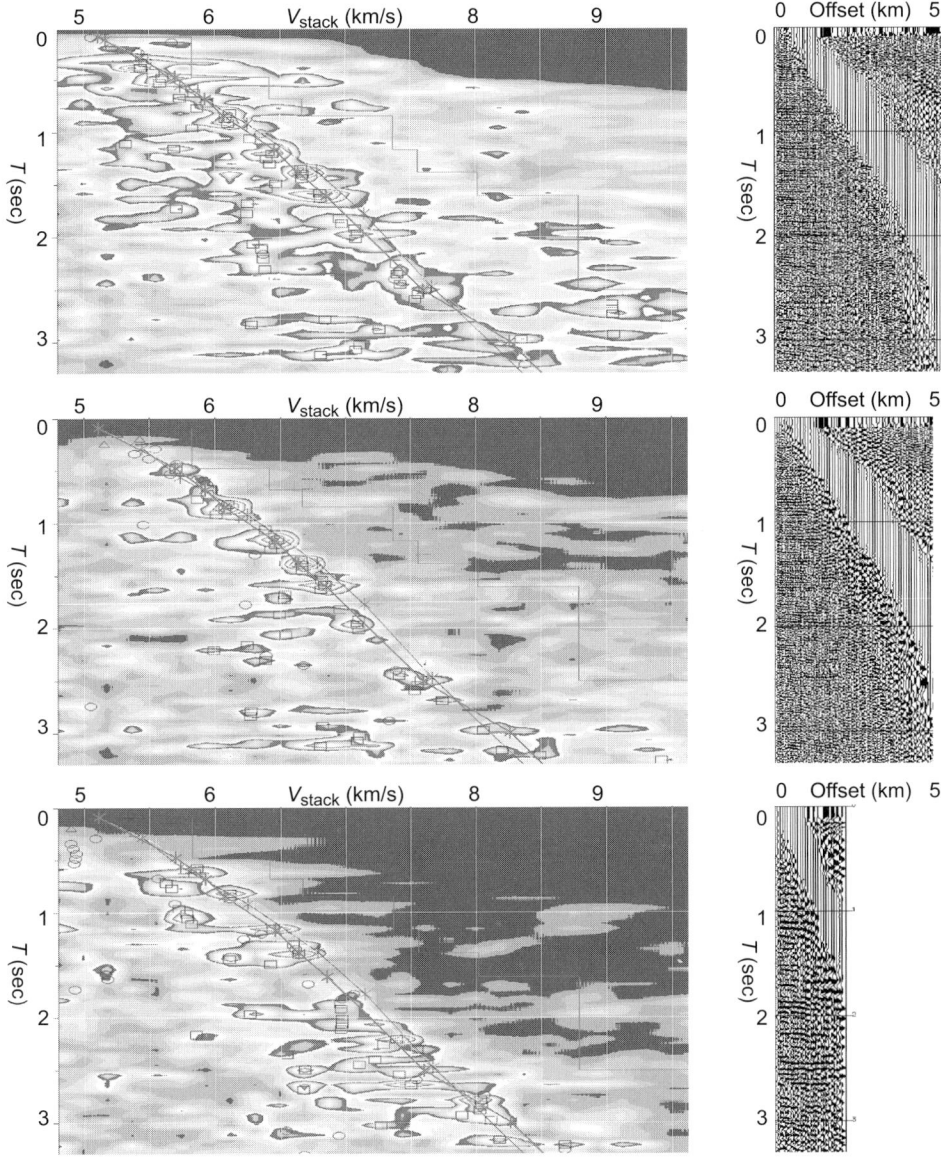

Box 8.2 Figure 1 The left panels are the semblance plots, and the curves with stars denote some of the velocity profiles picked. The right panels are the CMP gathers after NMO using the picked velocity profiles in the left panels. (Upper) The original data. (Middle) After a low-cut (25–30–55–65 Hz Ormsby) filter. (Lower) After a high-cut (3–6–25–30 Hz Ormsby) filter.

(D. Reynolds, personal communication, 2007). The data were acquired in the Gulf of Mexico using an airgun source and ocean bottom node receivers that were tethered to a buoy with antenna sending the signal to the recording vessel. Shot spacing is 33.5 m, receiver spacing is 100.6 m, and the maximum offset 10 737 m. A surface-consistent deconvolution was applied prior to the semblance analysis.

The upper panels of the figure show the semblance and NMO plots of the original data. In comparison, the data after a low-cut filter in the middle-left panel have more focused semblance but lose shallow events; the data after a high-cut filter in the lower-left panel maintain more slow events, losing the high-velocity "clouds" and shallow events. Overall, the low-cut filter has improved the focused trend in the semblance plot, hence helped the process of picking the correct stacking velocity. Cutting out the low frequencies may remove much of the average values of the input data and hence increase the distinction between events of different dips. In contrast, cutting out the high frequencies will reduce the distinction between events of different dips.

For a layer-cake model, let Δt_i and v_i be the time interval and interval velocity of the ith layer, respectively. Then from the first to the kth model layers the square power of the temporal rms velocity will be

$$V_{rms}^2(k) = \sum_{i=1}^{k} v_i^2 \Delta t_i / T(k) \tag{8–13}$$

where $V_{rms}(k)$ is the temporal rms velocity of the top k layer, and $T(k)$ is the total two-way reflection time from the surface down to the bottom of the kth layer:

$$T(k) = \sum_{i=1}^{k} \Delta t_i \tag{8–14}$$

Notice in (8–13) that rms velocity is an average of the interval velocity of the k layers weighted by their time intervals. A layer of greater time interval will therefore have a greater impact on the rms velocity.

Similar to the above temporal rms velocity, if Δz_i is the depth interval of the ith layer, we can obtain the squared power of the depth rms velocity of the top k layers $U_{rms}{}^k$:

$$U_{rms}^2(k) = \sum_{i=1}^{k} v_i^2 \Delta z_i / Z(k) \tag{8–15}$$

where $Z(k)$ is the total thickness from the surface down to the bottom of the kth layer:

$$Z(k) = \sum_{i=1}^{k} \Delta z_i \tag{8–16}$$

The major motivation for Dix (1955) is to derive the reverse process, i.e., from the rms velocity back to the interval velocities. From (8–13) we know that

$$V_{rms}^2(k)T(k) - V_{rms}^2(k-1)(k-1) = v_k^2 \Delta t_k$$

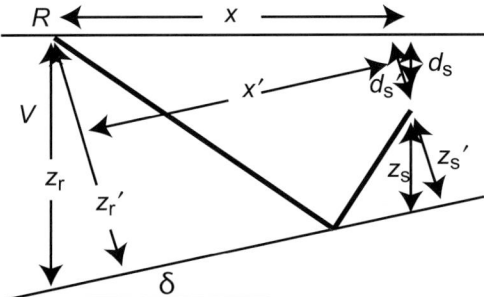

Figure 8.14 A cross-section showing the model of normal moveout with a dipping reflector.

Hence the squared power of the interval velocity of the kth layer is

$$v_k^2 = \left[V_{rms}^2(k)T(k) - V_{rms}^2(k-1)T(k-1) \right] / \Delta t_k \tag{8–17}$$

From the top layer downwards, we can use the above **Dix formula** repeatedly to obtain the temporal interval velocity profile from the temporal rms velocity profile. Notice that for the ith layer its depth interval and temporal interval must satisfy

$$v_i = \Delta z_i / \Delta t_i \tag{8–18}$$

Thus one can easily derive the depth intervals from a temporal interval velocity profile.

From a stacking velocity profile, it is common practice in reflection seismology to approximate the stacking velocity as the rms velocity and use the Dix formula (8–17) to calculate the interval velocities of a layer-cake model. However, it is a non-unique process to obtain the interval velocities from the rms velocities, although the reverse process based on (8–13) is unique. When the interval velocities are given, the corresponding time intervals of all model layers will also be given; thus the process to calculate the rms velocity is unique. In contrast, when an rms velocity profile is given, the number of model layers and their traveltime intervals are not given. It is well known (e.g., Aki & Richards, 1980) that a number of different combinations of time intervals and interval velocities may give exactly the same rms velocity. Hence, the solution of the interval velocities is often determined with a number of constraints, such as determining the number and time intervals of model layers based on data quality and geology, and constraining the range or trend of interval velocities using other information.

8.2.3   Dipping reflectors and point scatters

From the viewpoint of velocity analysis, the essential idea behind the semblance velocity analysis is to create a situation in which the value of the velocity will affect the semblance or stacked amplitude of a specific seismic gather; hence we can derive the velocity based on the intensity of the semblance. We may extend the above idea to gathers other than the NMO process for CMP gathers. Let us examine a couple of cases in the following.

In the case of a dipping reflector like that shown in Figure 8.14, the reflection traveltime can be derived from a coordinate rotation. The original coordinate system (x, z_r), is rotated

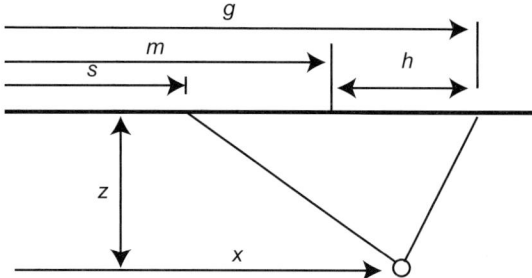

Figure 8.15 A cross-section showing the geometry of a point scatter.

to a new coordinate system $(x', z_{r'})$. In the rotated coordinates with dipping axes, we have

$$(vt)^2 = (x')^2 + \left(z'_r + z'_s\right)^2 \tag{8-19}$$

We can transform the above back to the original coordinates with non-dipping axes to yield

$$(vt/\cos\delta)^2 = (x - d_s\tan\delta)^2 + (z_r + z_s)^2 \tag{8-20}$$

The existence of dip reflectors argues the need for **dip move-out** (**DMO**). Aiming to preserve conflicting dips with different stacking velocities, DMO is a mapping process that is applied to pre-stack data after NMO correction. A tutorial discussion on NMO and DMO is given by Liner (1999).

Another case is that of point scatter, which is fundamental to applications such as Kirchhoff migration. Figure 8.15 sketches a cross-section with a point scatter at location (x, z), and the shot and geophone are located at $(s, 0)$ and $(g, 0)$ respectively.

The total raypath length from the shot to scatter and back to the geophone is

$$tv = \sqrt{z^2 + (x - s)^2} + \sqrt{z^2 + (x - g)^2} \tag{8-21}$$

We can convert from the shot and geophone coordinates to the midpoint (m) and half-offset (h) coordinates using

$$m = (g + s)/2 \tag{7-7a'}$$

$$h = (g - s)/2 \tag{7-7b'}$$

In other words,

$$g = m + h \tag{7-31a'}$$

$$s = m - h \tag{7-31b'}$$

Hence from (8–21) we have

$$tv = \sqrt{z^2 + (x - m + h)^2} + \sqrt{z^2 + (x - m - h)^2} \tag{8-22}$$

The contours of traveltimes of different values of either (m, h) or (g, s) coordinate systems, as portrayed by either (8–21) or (8–22), respectively, form a graph similar to the **Cheops**

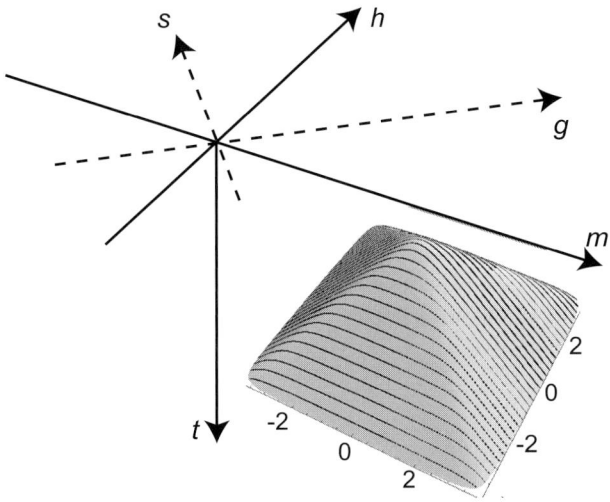

Figure 8.16 A 3D view of the Cheops Pyramid for a single scatter (modified from Biondi (2004).

Pyramid. Figure 8.16 shows an example of the pyramid, modified from Biondi (2004). Would you be able to draw a cross-section of the Cheops Pyramid using either Equation (8–21) or (8–22) and using a spreadsheet?

Exercise 8.2

1. In semblance velocity analysis we usually stack all traces with equal weight. Should we apply variable weights as a function of offset and intersection time? Devise a way to conduct such a weighted semblance stack process.

2. Explain why the root-mean-square velocity is not the exact stacking velocity. Which of the two velocities is usually faster than the other and why?

3. Use an Excel spreadsheet to illustrate the Cheops Pyramid in imaging a point diffractor.

8.3   Migration velocity analysis

8.3.1   Motivation

Migration velocity analysis refines the velocity model by utilizing the dependency of depth migration on velocity variations. It is an extension of semblance velocity analysis for migrated gathers. While a velocity model is a pre-condition for seismic migration, we often do not have a sufficiently good velocity model and we still need to deliver the best possible migration result. This motivated the creation of time migration methods which attempt

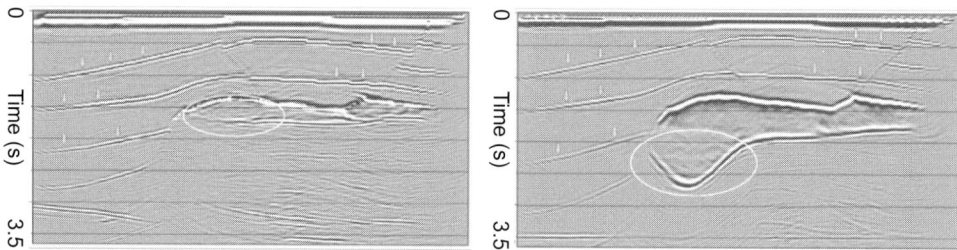

Figure 8.17 Comparison between cross-sections of: (left) Kirchhoff curved ray pre-stack time migration; and (right) Kirchhoff pre-stack depth migration for a synthetic salt model (Lazarevic, 2004). The pre-stack depth migration produces a much better image of the sub-salt areas.

to image reflectors of different dipping angles without the presence of lateral velocity variation. Specifically, a post-stack time migration will take either a constant velocity model or a semblance velocity model, while a pre-stack time migration will use an interval velocity model created based on semblance velocity analysis plus a conversion using the Dix formula. In general, time migrations are not sensitive to changes in velocity model.

The high sensitivity of depth migration methods toward velocity variations is the main motivation for migration velocity analysis. In the presence of lateral velocity variations the quality of depth migration is much superior to time migration, as shown in Figure 8.17. Post-stack depth migration may be a relatively inexpensive way to image in the presence of strong lateral velocity variations such as salt flank and highly faulted and folded areas. Pre-stack depth migration is currently the state-of-the-art imaging method to deal with cases that have high structural complexity as well as strong lateral velocity variations.

8.3.2   Velocity–depth ambiguity

One challenge to reflection velocity analysis is the **velocity–depth ambiguity**, which refers to the notion that many different velocity and depth combinations are able to satisfy the same traveltime of seismic reflections. This non-uniqueness problem has been studied by Bickel (1990), Lines (1993) and Tieman (1994) for the flat reflector case. Rathor (1997) extended the analysis to the dipping reflector case. Let us examine the velocity–depth ambiguity with the simple case shown in Figure 8.18.

In Figure 8.18 the left panel shows a simple post-stack case, and the right panel shows a pre-stack case. Since velocity is constant, we can calculate the traveltime with small changes in velocity and reflector depth. As shown in Table 8.3, the zero-offset traces will have the same traveltime of 1 s in three different combinations of reflector depth and velocity. However, the traveltime varies for the three combinations when offset is 1 km. Hence, we need to use pre-stack depth in order to determine the correct velocity and reflector depth.

8.3.3   The impact of velocity errors on migration solutions

Migration frowns and **migration smiles**, respectively, are often used as indicators of **under-migration** (migration velocity is too slow) or **over-migration** (migration velocity

Table 8.3 Comparison between zero-offset and non-zero-offset cases.

	Zero-offset			Offset = 1 km		
Reflector depth (km)	0.9	1	1.1	0.9	1	1.1
Velocity (km/s)	1.8	2	2.2	1.8	2	2.2
Traveltime (s)	1	1	1	1.144	1.118	1.098

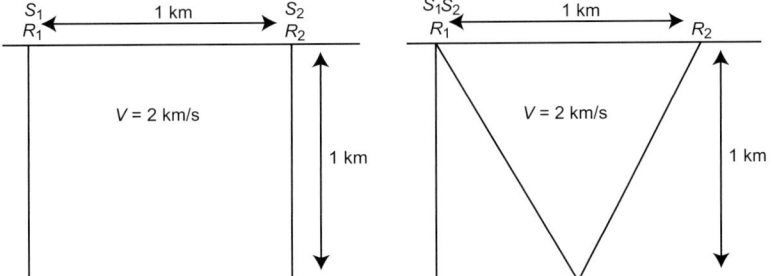

Figure 8.18 Reflection rays in a one-layer model. (Left) Two zero-offset reflection rays. (Right) A zero-offset and a non-zero offset reflection ray.

is too fast). Figure 8.19 shows an example by Lazarevic (2004) illustrating the influence of velocity on Kirchhoff pre-stack depth migration of a synthetic SEG/EAGE salt model. As shown in panel (a), when the migration velocity is too fast, the observed reflection time is longer than the predicted time, resulting in deeper reflector positions than the correct ones. A number of migration smiles also appear near the edges of the reflection features in panel (a). Beddings near the fault traces show signs of migration smiles above discontinuous points along interfaces and faults. In contrast, when the migration velocity is too slow, the observed reflection time is shorter than the predicted time, resulting in shallower reflector positions than the correct ones. Events tend be of lower dip angles, and the synclinal features tend to have cross-over events.

A careful comparison between the three migrated images in the figure reveals that the image has the best focus when the correct velocity model is used. However, such evidence may not be easy to capture in real cases where we may not have much clue about what is the correct image. There are also data noise and processing artifacts in practice. Nevertheless, the better the velocity model, the less appearance of offsetting events at kinked corners of synclines, anticlines, and faults.

Another question is about whether it is better to use slower or faster velocity models when we are not sure about their correctness. Since a faster velocity model will result in shallower reflection events, there will be more reflection events when such a model is used, as shown in the first panel of Figure 8.19. Hence, from the perspective of capturing as much reflection energy as possible, we would be better off trying faster migration velocity models until we reach to the most reasonable velocity model. Without knowledge of the correct velocity model, people often smooth the most reasonable velocity model in order to reduce the short-wavelength errors for the velocity field.

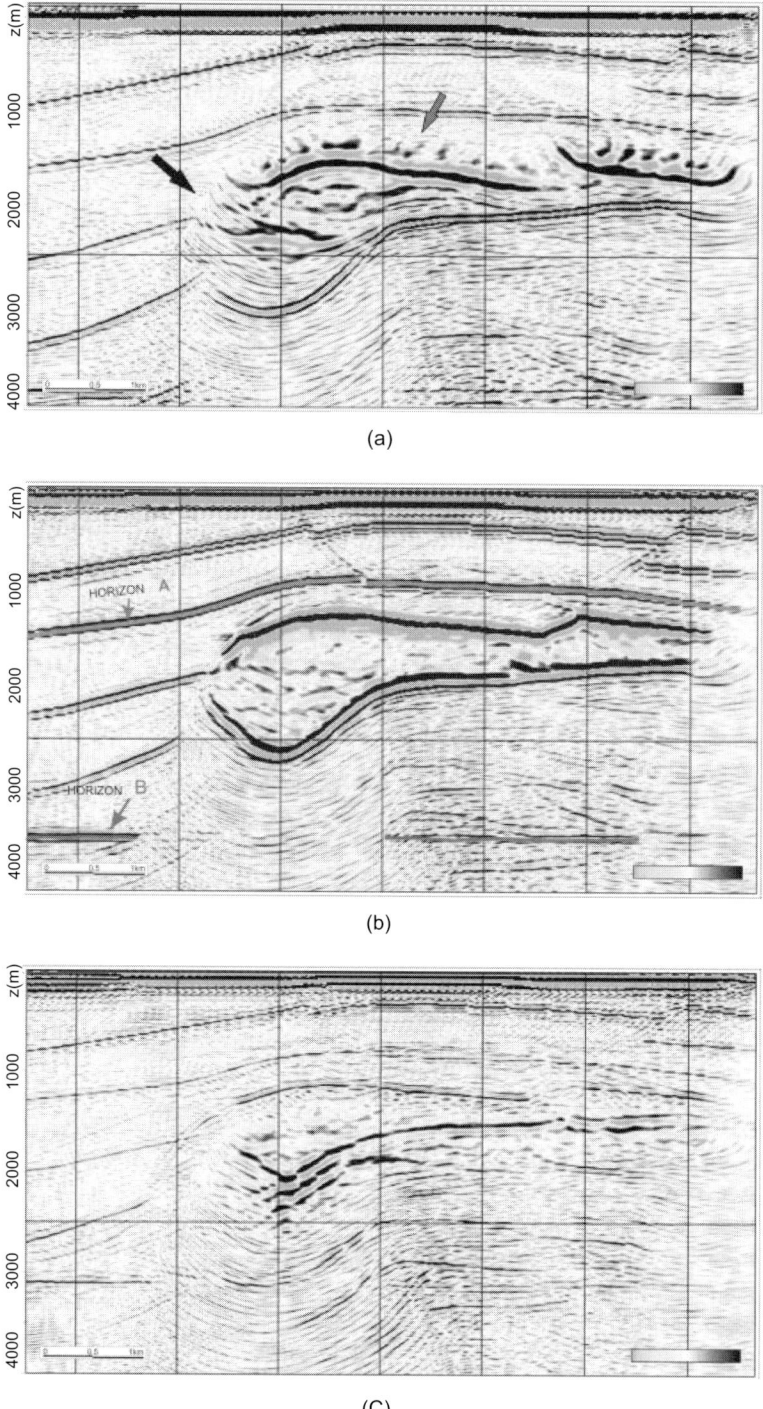

Figure 8.19 Comparison of cross-sections of Kirchhoff pre-stack depth migration using: (a) 10% faster velocity, (b) correct velocity, and (c) 10% slower velocity (Lazarevic, 2004).

8.3.4   The common image gather (CIG) flattening principle

Following the discussion in Section 7.2, let us re-visit the general process of the single-arrival Kirchhoff pre-stack depth migration as shown in Figure 7.10. Each piece of the input data trace is mapped to all model points that satisfy the corresponding two-way traveltime. As shown in Figure 7.10, the migration process produces extra dimension(s) of traces. In other words, each input trace will produce many migrated traces over the migrated section in 2D or volume in 3D. While the final migration image can be produced by stacking all migrated traces, we can make **partial stacks** by using a subset of migrated traces. We can collect some of the migrated traces to form different types of **migrated gathers**, such as taking migrated traces from each common shot gather to produce migrated shot gathers.

For migration velocity analysis, we want to select such migrated gathers that will be most sensitive to velocity variations. One of the best choices is the **common image gather (CIG)**, which is a collection of migrated traces at a common image location. A CIG is defined for a fixed lateral position in the solution space: we may call it the image position, image location, or **image bin**. The CIG for each image position is simply a collection of all migrated traces within the image bin – a stack of these traces forms a single trace of the final image at the image position.

Let us see the CIG using schematic plots shown in Figure 8.20. Panel (a) shows the model space with one image bin highlighted. The final migrated image will be divided into many such image bins. Following the concept of the exploding reflector model, the portions of the two reflectors inside the given image bin can be regarded as two diffractors denoted by open and filled stars, respectively. For each of the three shot–receiver pairs, the two-way reflection raypaths from the shot to each of the diffractors are represented by the dashed curves connecting the mid points with the diffractors. Panel (b) shows the data traces recorded by the three receivers. Each data trace contains reflected events from all parts of the two reflectors, including the portions from the two diffractors inside the image bin that are highlighted by the open and filled stars in this panel.

In Kirchhoff migration, each data trace will be mapped into the entire solution space based on two-way traveltime tables. Out of the many migrated traces from a single input data trace, there may be one migrated trace that falls in the given image bin, and we have plotted this migrated trace at the corresponding midpoint in panel (c) of Figure 8.20. The **offset CIG**, as shown in panel (d), is a graph of all migrated traces of this image bin according to the offset between the image bin and the midpoint of each trace.

In the final migrated section, the single trace in the given image bin of this example will be a stack of all migrated traces shown in panel (c). If the velocity model used agrees well with the true velocity field, then the diffractors of each of the migrated traces in this CIG will be at the same depths as the two reflectors in the image bin. In other words, with a correct velocity model, events on the CIG tend to be flattened. If the velocity is in error, however, the diffractors of some of the migrated traces may deviate from being flat on the CIG. In fact, when the velocity field follows a layer-cake model, the offset CIG is nearly the same as the NMO-corrected CMP gather.

The **CIG flattening principle** describes the notion that the correct velocity model should result in flat reflectors on all CIGs. Figure 8.21 shows an example of the CIGs with three

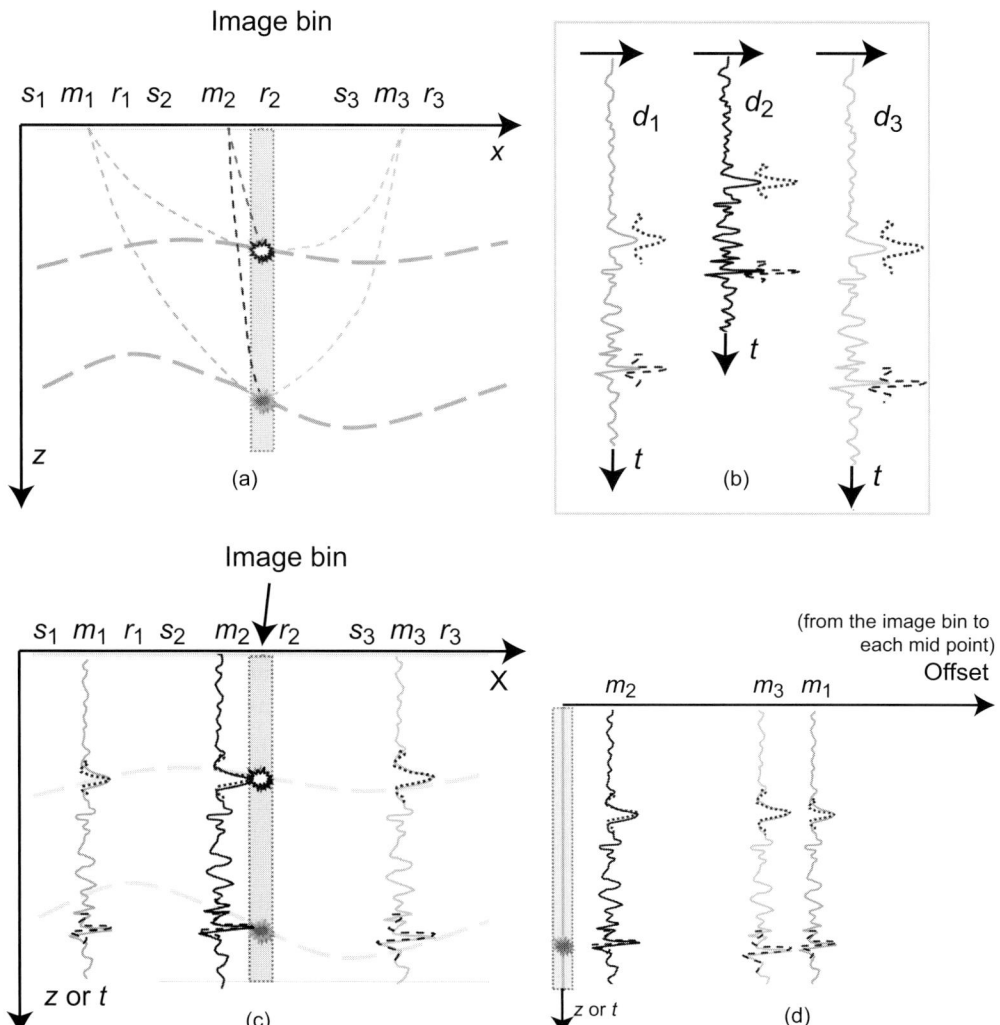

Figure 8.20 A schematic offset CIG in 2D imaging. (a) The portions of reflectors within the image bin are viewed as diffractors. Dashed curves follow diffraction rays to three midpoints, m_i. (b) Three data traces, d_i. The two dotted wavelets denote events from the two diffractors. (c) Migrated traces for the image bin plotted at midpoints, with diffractions from the image bin. (d) Offset CIG; the offset is between the image bin and each midpoint.

different velocity models. Only the correct velocity model in the middle panel produces a CIG with most events flat. Notice the appearance of hockey sticks at far offset in all panels. Can you explain why there are non-flat events even when the correct velocity model is used?

The CIG flattening principle is among the foundations of migration velocity analysis (Al-Yahya, 1989). Such analyses are based on exploitation of the sensitivity of migration

Practical seismic velocity analysis

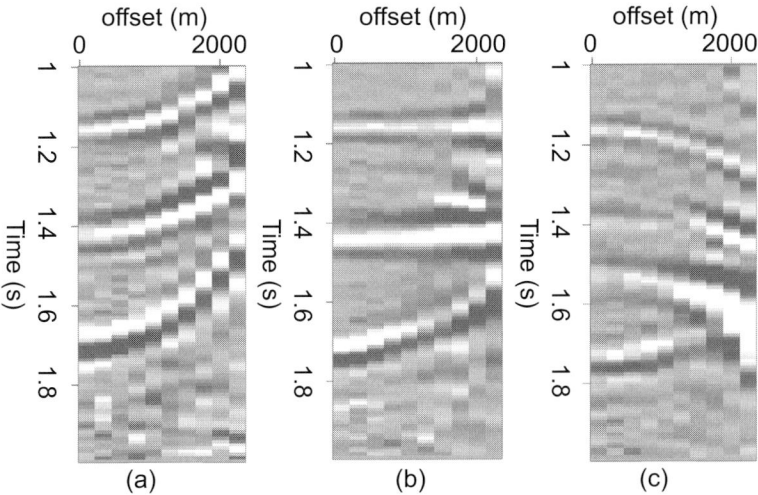

Figure 8.21 CIGs after migration with the total velocity field scaled by a factor of: (a) 0.92; (b) 1; (c) 1.08 (Biondi, 2004).

to velocity variations. An incorrect velocity model may not only lead to wrong reflector positions, but also blur the images in the forms of over-migration, under-migration, or a mixture of the two effects. The migration velocity analysis basically performs migration using different velocity models and checks the focusing level of the images. The level of focusing may be measured in terms of spatial alignment, particularly on CIGs. It may also be measured in terms of the coherency of events in the stack of individual depth-migrated sections.

In a generalized sense, we can see that semblance velocity analysis exploits the sensitivity of some particular seismic data gathers to velocity variations and measures the velocity error using semblance or other criteria, such as flatness of the reflection events on some particular gathers. The most common examples are the CMP gathers for NMO velocity analysis and the CIGs for migration velocity analysis. A remaining task is to convert the semblance velocities into interval velocities.

Exercise 8.3

1. Some people say that a correct velocity model will result in a more focused migration image. Explain the possible reasons for this statement, and its limitations.

2. Explain why, even when using the correct velocity model, some reflection events on a CIG may not be flat (e.g. in Figure 8.21b).

3. Explain the notion that the NMO plot is a special case of the CIG plot. Compare and contrast the two plots. Can you construct a semblance graph for a CIG?

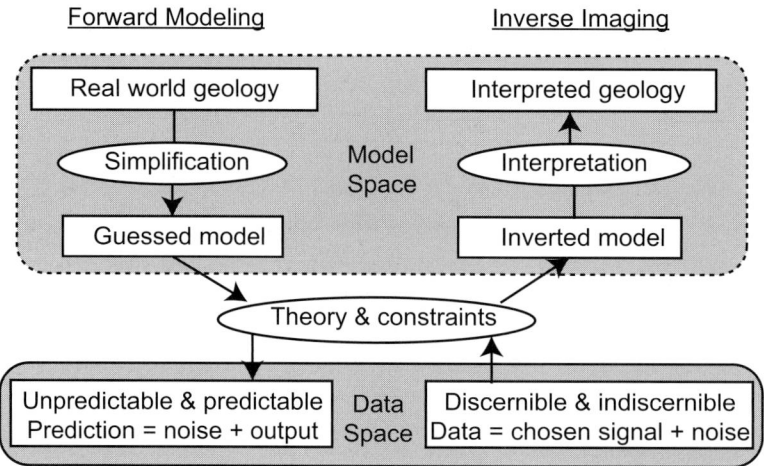

Figure 8.22 Comparison between forward modeling and inverse imaging processes.

8.4   Tomographic velocity analysis

8.4.1   Forward modeling versus inverse imaging

8.4.1.1   Comparison and contrast

Seismic **tomographic velocity analysis** is an application of tomographic inversion for VMB. It is based on the relationship between the velocity model and the traveltimes and waveforms of seismic data. Typically a linear approximation of the relationship is taken to allow a mathematical inversion for the updates of the velocity model based on misfits between the observed data and model predictions. The non-linearity of the relationship and poor data coverage are the major sources of non-uniqueness in the tomographic inversion. Therefore, solving non-unique inverse problems in seismology is a major goal of seismic tomography.

As shown in Figure 8.22, **forward modeling** and **inverse imaging** are two opposite mapping processes between the model space and **data space**. Although the sense of "forward" and "inverse" is only relative in terms of mathematics, geoscientists usually refer to a forward problem as a process that uses values of intrinsic variables (model) to predict the values of observational parameters (data), and inverse imaging as the process of finding the images of intrinsic parameters (model) from the observed values (data). For both forward and inverse approaches, usually the first step is to formulate a theoretical relationship between the data and model spaces and the underlying physics.

Consider, for instance, the problem of inferring subsurface density distribution from the surface or aerial gravity measurements. We first use Newton's gravitational law to relate the observed gravity field to a density model. Then we can use geological information to build an initial reference model as well as a range of model variations. If we choose to take the forward modeling approach, we can use the initial reference model and the range of model variations to make a set of guessed models, and use the theoretical formula to compute their predictions. The fit of each prediction with the observed data quantifies the likelihood of the corresponding guessed model. As shown in Figure 8.22, the forward modeling produces only predictable output that is constrained by the model simplification and theoretical constraints. Consequently, the forward modeling can be as simple as a trial-and-error process, or as extensive as an exhaustive search through the model space.

In contrast, inverse imaging attempts to constrain some properties of the Earth into an inverted model using a chosen signal that is the most identifiable (discernible) portion of the observed data. Inversion is desirable in geophysics because a major goal of geophysics is to detect the properties of the Earth's interior using measurements made mostly along the surface of the Earth. In addition, within a given initial reference model and its variation range, the inverse imaging is usually more objective than the forward modeling. Often the inverse is more efficient than forward modeling when we look for the best model possible for a given set of data and model constraints. In other words, the inverse approach is more attractive as more data become available. The basic tasks of inverse imaging are:

- To establish the inverse problem, usually based on a forward problem;
- To invert for, or to determine, the model parameters such as seismic velocity, Q-values, density, and seismic source locations, from observations;
- Within the inevitable limitations of data, to evaluate the reliability and other aspects of those obtained parameters.

8.4.1.2   Linearization using perturbations

Since most observational data are collected in discrete form and are processed in digital computers, it is sometimes inevitable but generally desirable to discretize a continuous forward problem directly to formulate the corresponding discrete inverse problem. We may express the forward problem as

$$\mathbf{d} = f(\mathbf{m}) \tag{8–23}$$

where $\mathbf{d}$ is a vector of observations (data), $\mathbf{m}$ is the unknown model vector, and $f()$ is the functional describing the forward problem. The functional is usually assumed to be known, and is generally non-linear. The model vector $\mathbf{m}$, an assorted collection of parameters, is a discrete description of the continuum model.

We can linearize the forward system by mathematically expanding it into a Taylor series about a starting reference model $\mathbf{m}_o$:

$$\mathbf{d} = \mathbf{d}_o + \frac{\partial f}{\partial \mathbf{m}}(\mathbf{m} - \mathbf{m}_o) + O(\mathbf{m} - \mathbf{m}_o)^2 \tag{8–24}$$

or, in a linearized form by omitting the higher-order terms

$$\Delta \mathbf{d} \approx \frac{\partial f}{\partial \mathbf{m}} \Delta \mathbf{m} \tag{8–25}$$

where $\Delta \mathbf{d} = \mathbf{d} - \mathbf{d}_o = \mathbf{d} - f(\mathbf{m}_o)$, $\Delta \mathbf{m} = \mathbf{m} - \mathbf{m}_o$, and $\frac{\partial f}{\partial \mathbf{m}}$ is the **Frechet differential kernel** matrix $\mathbf{A}$.

The linear system (8–25) can therefore be denoted by conventional matrix notation

$$\Delta \mathbf{d} = \mathbf{A} \, \Delta \mathbf{m} \tag{8–26}$$

using $\mathbf{A} = \{a_{ij}\}$ with $a_{ij} = \frac{\partial f_i}{\partial m_j}$. $\mathbf{A}$ is a matrix of M columns (number of model parameters) and N rows (number of observations, or number of equations). Consequently, the discrete linear inverse is

$$\Delta\mathbf{m} = \mathbf{A}^{-1}\Delta\mathbf{d} \tag{8–27}$$

The above formula almost never works in a deterministic sense, because $\mathbf{A}^{-1}$ generally does not exist. This is because $\mathbf{A}$ is usually rank defective owing to:

- **Data inaccuracy**: Two or more measurements under the same conditions give different data values. In other words, the number of equations is more than the number of unknowns in (8–27).
- **Model non-uniqueness**: Different models give the same predictions in the data space. In other words, the number of equations is less than the number of unknowns in (8–27).

The above two challenges can be dealt with by the approaches of either a generalized inverse or some types of approximated inverse. A greater challenge, however, is that the kernel matrix $\mathbf{A}$ is unknown in most real applications. This means that we know only the data vector $\mathbf{b}$ in the forward system (8–26) or the inverse system (8–27).

8.4.2   Iterative tomographic inversion

Practically, we often use the kernel matrix in the reference model to approximate the kernel matrix of the true model. This gives the dependency of the solution on the initial reference model. As a result, most tomographic inversions are done in an iterative fashion. At the kth iteration, we use the reference model $\mathbf{m}^{(k-1)}$ to compute the kernel matrix $\mathbf{A}^{(k-1)}$ as well as the traveltime residual vector $\Delta\mathbf{d}^{(k-1)} = \mathbf{d} - \mathbf{A}\ \mathbf{m}^{(k-1)}$. We then try to invert the following forward system for the inverse solution $\Delta\mathbf{m}^{(k)}$:

$$\Delta\mathbf{d}^{(k-1)} = \mathbf{A}^{(k-1)}\Delta\mathbf{m}^{(k)} \tag{8–28}$$

The solution perturbations can be used to update the reference model as

$$\mathbf{m}^{(k)} = \mathbf{m}^{(k-1)} + \Delta\mathbf{m}^{(k)} \tag{8–29}$$

The general processing flow of an iterative traveltime tomography using first arrivals is shown in Figure 8.23. The first three steps include data input and preparation for the inversion; Step 4 is the tomographic inversion iteration; and Step 5 is the QC and output of the final model. The kth tomographic iteration in Step 4 will continue until pre-defined stopping criteria are reached. The details will be illustrated using several examples in the following.

Figure 8.24 shows a synthetic test for a 2D VSP case. The true model has four layers of constant interval velocities and with lateral variations in the interface geometry. The test used nine surface shots and three wellbore receivers. The objective of this VSP simulation is to determine the geometry of model interfaces using known layer velocities from well data that were used to establish the initial velocity model, as shown in the top-right panel of this figure. Ray tracing of the first arrivals in the reference model produces traveltime residual and raypath of each source–receiver pair, which are taken into the data vector and the kernel matrix in the forward system (8–28). For example, for the ith ray, we calculate

Practical seismic velocity analysis

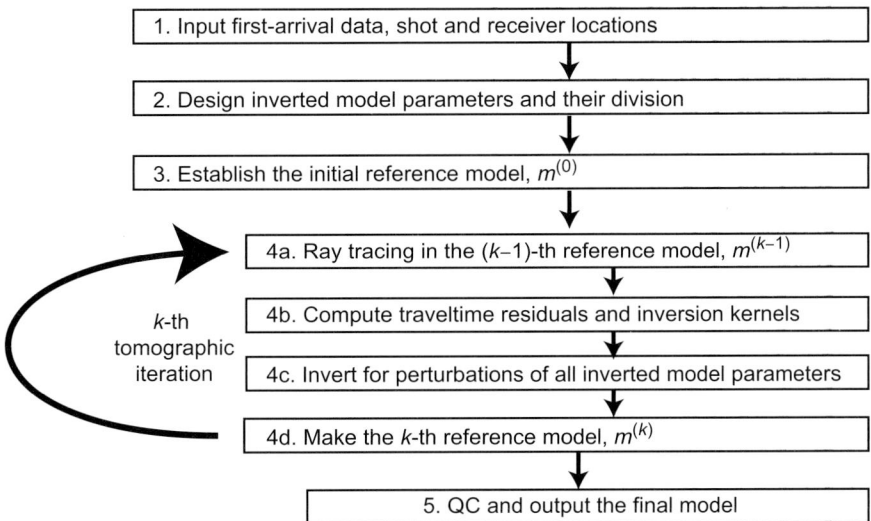

Figure 8.23 Processing flow of an iterative tomographic inversion using first arrivals.

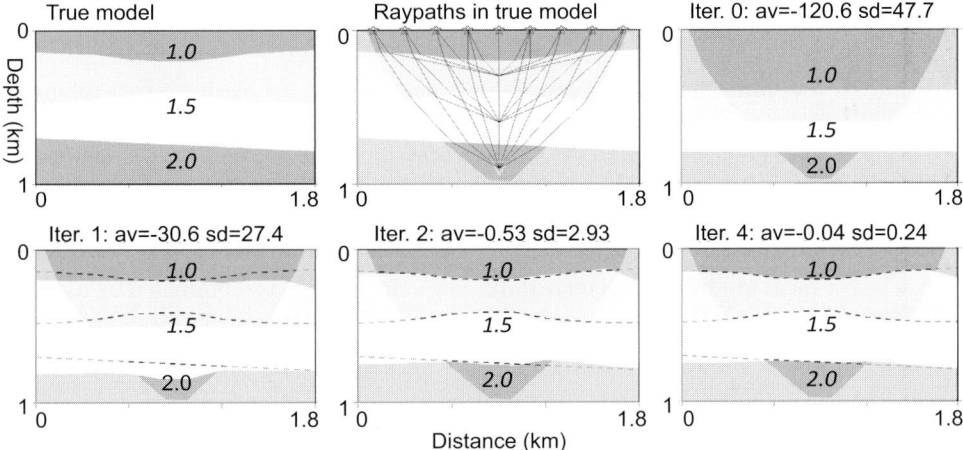

Figure 8.24 The upper-left panel shows the true model with four layers, and the upper-middle panel shows first-arrival rays from nine surface shots to three VSP receivers in the true model. The other panels show the initial reference model and the solutions of the first, second, and fourth tomographic iterations. The values on the layers are the velocities in km/s. The dashed curves denote the true interface positions. The average (av) and standard deviation (sd) of traveltime misfits in milliseconds are shown.

traveltime residual δd_i and a raypath which traverses near many model nodes defining the interface geometry. The position of the jth model node is (x_j, z_j). Thus, the ith row of the forward system becomes

$$\delta d_i = \sum_j^J c_{ij} \delta m_j \qquad (8\text{–}30)$$

where J is the number of model perturbations, one of which is δm_j, and c_{ij} is the corresponding kernel. In this particular case, δd_i is the ith traveltime residual, δm_j is the spatial perturbation of the jth interface node, and c_{ij} is an element of the kernel matrix at the ith row and jth column.

The **stopping criteria** for the tomographic iteration may be based on observations in both the data and model spaces. In the data space, the statistics on data misfit level explicitly reflect the fitness of model prediction as function of tomographic iteration. In the case shown in Figure 8.24, the average and standard deviation of traveltime residuals show a nice reduction with further iterations, but the reduction stops around iteration four. In the model space, the inverted models do not change much beyond the fourth iteration. The combined use of the stopping criteria in both data and model space usually works when the change in data fitness matches well with change model updates. A significant help is if we know the trend or some other characteristics of the model from *a priori* geological and geophysical information. It is also helpful if we have a good estimate of the data noise level. One concern is that the inversion may converge to a wrong solution in cases of erratic data and poor ray coverage. Ray coverage is a measure of the hit count and angular coverage of all traversing raypaths in a given model. Concern about various imaging artifacts will be addressed in Section 8.5.

8.4.3 Three groups of model parameterization

Seismic tomography seeks a model that best fits the observed data for given parameterizations in the data and model spaces. Parameterization in the data space is done through the data acquisition process. At the processing stage, we may have a number of different ways of model parameterization. Model parameterization is an important issue because the data are erratic and incomplete, and portions of the model space are poorly covered in nearly all seismic applications. Most of the existing tomographic velocity analyses employ the **local model parameterization**. The velocity field is represented by its values on a discrete mesh of grids or cells, and the tomographic inversion seeks the velocity values on the grids or cells that will best fit the observed traveltimes. Figure 8.25 shows an example of cell tomography.

Another model representation is **global model parameterization**, which expresses the velocity field using some base functions, such as the spherical harmonics used in some global mantle tomography studies (e.g., Dziewonski, 1984; Dziewonski & Woodhouse, 1987). As an application of the law of decomposition and superposition (see Section 3.1.1), global model parameterization is commonly used to retrieve long-wavelength velocity anomalies, and it requires that all the model areas have at least a minimum level of ray coverage. In contrast, local parameterization seeks to achieve locally high resolution and leaves uncovered areas untouched. Some comparative studies in mantle tomography show that the two types of parameterization can achieve solutions of comparable quality (Pulliam *et al.*, 1993; Wang & Zhou, 1993).

Most traveltime tomography studies took the local model parameterization using cells or grids. Figure 8.25 shows a synthetic example of traveltime cell tomography using first arrivals in a cross-well survey. The velocity field is represented by its values on a set of non-overlapping square-shaped cells. This is a **single-scale tomography (SST)**, which represent the velocities by non-overlapping and regularly spaced cells. Once specified, the

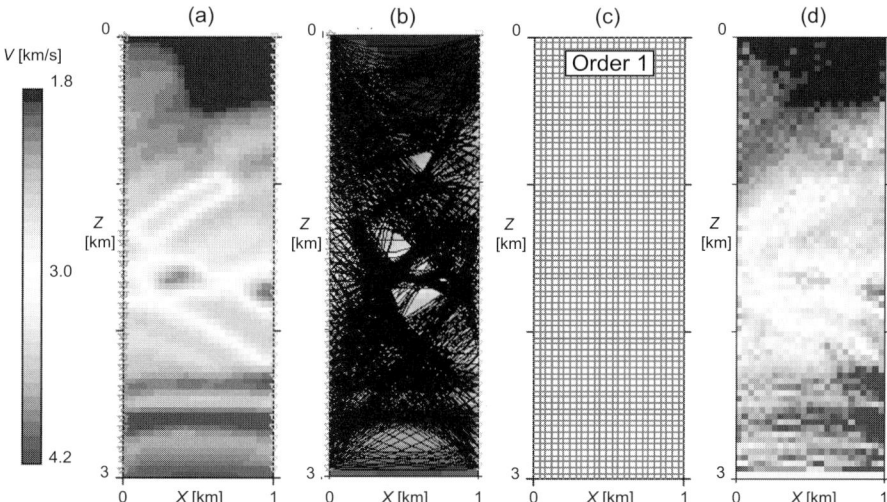

Figure 8.25 (a) A synthetic true velocity model in a cross-well survey using 43 shots (purple stars) and 43 receivers (green triangles). (b) Raypaths in the true model. (c) A cell tomography grid. (d) Solution of single-cell tomography using first-arrival traveltimes in the true model. For color versions see plate section.

regular spacing of model parameters defines a Nyquist wavenumber to which the inversion is especially sensitive. For given raypaths, an increase in the cell size will degrade the resolution but will also enhance the robustness of the slowness solution because the inverted result will be more unique and less sensitive to errors in raypaths and residuals. For real data, the ray coverage depends not only on the positions of shots and receivers, but also on the velocity gradient of the reference velocity model. Even with regularly spaced shots and receivers, unevenness in ray coverage will coexist with velocity heterogeneities. In Figure 8.25b, for instance, the paths of the first arrival rays tend to stay along fast velocities and avoid slow velocities. This is a main cause of non-linearity in traveltime tomography, and it is difficult to accommodate all parts of a model with a single cell size.

A third model representation is the **wavelet model parameterization**, which represent the model by a set of overlapping wavelets that are defined locally. This is another application of the law of superposition and decomposition. This representation is more complex than the first two as it attempts to take advantage of both the basis functions and local decomposition.

8.4.4 Multi-scale tomography

8.4.4.1 Methodology

There are many measures proposed to overcome the non-uniqueness in tomographic velocity analysis, such as the use of penalty functions to constrain the inversion. Because poor and uneven ray coverage is a major source of non-uniqueness in the practice of tomography, a **multi-scale tomography** (**MST**) method (Zhou, 1996) has been devised. As a simple form of wavelet model parameterization, the MST aims to cope with **mixed-determinacy** due to unevenness in ray coverage. Like a combined use of many SST using different cell sizes, the MST consists of three steps. The first step of MST is to define a set of **sub-models** with different cell sizes; each sub-model covers the whole model region. The sub-models allow a decomposition of velocity anomalies into components of different cell sizes.

The second step of MST is to determine all sub-model values simultaneously rather than progressively. Many previous tomographic studies have attempted to use multiple sets of cell sizes progressively (Fukoa *et al.*, 1992; Bijwaard *et al.*, 1998). In contrast, a simultaneous determination of the parameters of all sub-models in the MST will minimize the spread of model values between different model cells. Each spatial location belongs to all sub-models, while at the same location the consistency between the data contributions, such as the apparent slowness of different rays, may vary over different sub-models. The sub-model with a higher level of consistency between data contributions will gain more from the inversion. Owing to the weight of their sizes, larger cells will take more model values when they match well with the geometry of the true model anomalies. The sum of velocity values of all sub-models at each position should approximate the true velocity value there.

The final step of MST model is a superposition of all sub-model values. We may choose K sub-models, and let $M_j^{(k)}$ be the model value of the jth cell in the kth MST sub-model. Then the final value of the jth model variable is

$$m_j = \sum_k^K w^{(k)} M_j^{(k)} \tag{8-31}$$

where $w^{(k)}$ is a weighting factor which can be defined *a priori*. The **unbiasedness condition** in statistics requires that the sum of all weighting factors is equal to one, i.e.

$$\sum_k^K w^{(k)} = 1 \tag{8-32}$$

The default value of $w^{(k)}$ is $1/K$, or a constant value for all sub-models. Differentiating (8–31) on both sides and inserting the result into (8–30), we obtain the general forward equation for the MST

$$\delta d_i = \sum_k^K w^{(k)} \sum_j^{J_K} c_{ij}^{(k)} \delta M_j^{(k)} \tag{8-33}$$

where J_k is the number of model parameters in the kth sub-model. The inversion variables for the MST are the function values $\{\delta M_j^{(k)}\}$.

As an example using notation specified for traveltime tomography, the forward multi-scale equation is

$$\delta t_i = \sum_k^K w^{(k)} \sum_j^{J_k} l_{ij}^{(k)} \delta s_j^{(k)} \tag{8-34}$$

where $\delta s_j^{(k)}$ is the slowness perturbation of the jth cell of the kth sub-model; and $l_{ij}^{(k)}$ is the length of the ith ray in the jth cell of the kth sub-model. At each model location the cells containing a higher level of consistency between data contributions will stand to gain more

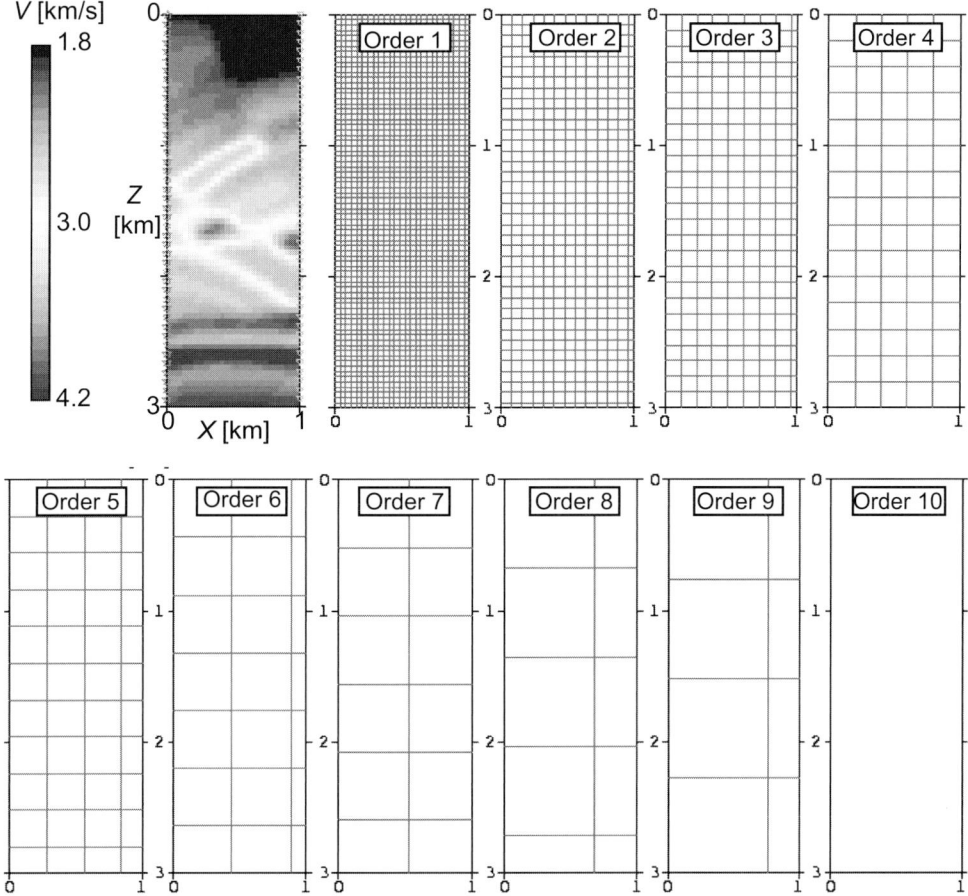

Figure 8.26 The upper-left panel is the same synthetic true model as in Figure 8.25 in a cross-well survey. The other panels show 10 MST sub-models. The Order 1 sub-model has the same geometry as the SST model in Figure 8.25. For color versions see plate section.

from the inversion. After inverting the system (8–34) for $\{\delta s_j^{(k)}\}$, the final MST solution is a superposition of solutions of all multi-scale sub-models

$$\delta s_j = \sum_k^K w^{(k)} \delta s_j^{(k)} \qquad (8\text{–}35)$$

Owing to the above superposition, the number of model variables and cell size of the final solution of the MST is identical to that of the SST with the dimension of the smallest MST sub-model. Notice that the superposition in the above equation will not alter the traveltime residues because it is consistent with the decomposition in (8–34).

8.4.4.2   A synthetic example

Figure 8.26 shows some MST sub-models for the cross-well synthetic case shown previously in Figure 8.25 using the SST. The cell size of the SST model is 40×40 m², giving a total

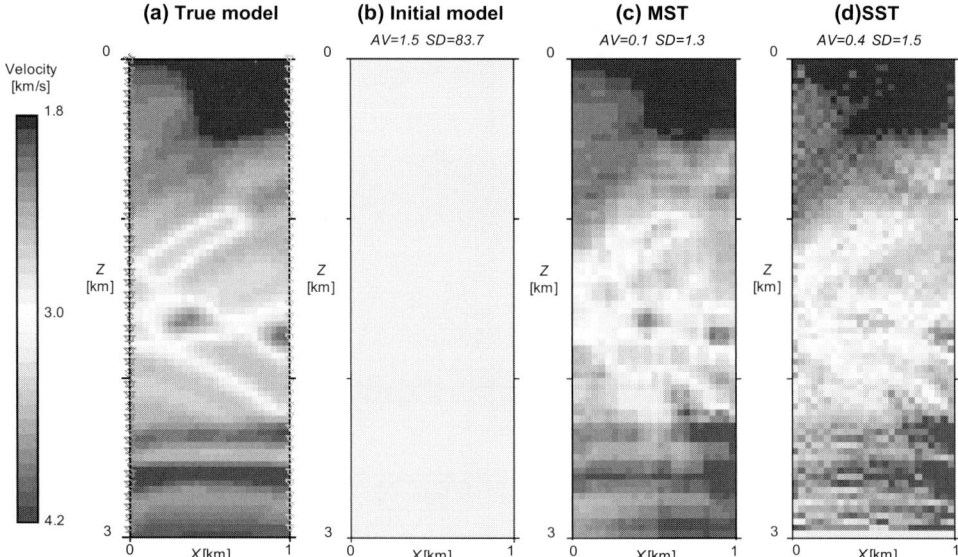

Figure 8.27 (a) A synthetic true velocity model. (b) The initial reference velocity model. (c) The fourth iteration MST solution. (d) The fourth iteration SST solution. The average (AV) and standard deviation (SD) of traveltime residuals in milliseconds are indicated on top of the solution models. For color versions see plate section.

of $25 \times 75 = 1875$ inversion variables for the SST. Taking the SST model as the first-order MST sub-model, the cell sizes of the first nine MST sub-models are multiples of the cell size of the SST model. The last MST sub-model takes the whole model area as a single cell. Combining all sub-model cells, there are 2765 inversion variables for the MST model. Depending on the ray coverage, the actual number of inversion variables might be lower. In contrast, the 43 shots in the left well and 43 receivers in the right well give a total of 1849 first arrivals. After the post-inversion superposition of all MST sub-models, the final MST solution has the same 1875 cells as the SST solution.

Figure 8.27 shows the true model, the initial reference model with constant velocity, and the solution of the MST and SST. Using the same data and initial reference model, both the MST and SST solutions were derived after four iterations of ray tracing in the reference model, inverting for slowness perturbations, and updating of the reference velocity model. The mean and standard deviation of traveltime residuals are 0.1 ms and 1.3 ms for the MST solution, and 0.4 ms and 1.5 ms for the SST solution. The similar level of data fitting for these two solutions manifests the non-uniqueness in tomographic inversion. However, the MST solution is smoother than the SST solution. By the "principle of parsimony", the simpler (or smoother here) of the two equally fitting solutions is the best solution. Comparing with the true solution in this synthetic example, the MST solution outperforms the SST solution.

Figure 8.28 provides a detailed breakdown of the contributions from the MST sub-models. The solutions are plotted as percentage ratio of the velocity perturbations to the mean velocity of 2.7 km/s. The final MST model shown in the upper left panel is a spatial superposition of the 10 sub-model solutions that were inverted simultaneously. Clearly, the

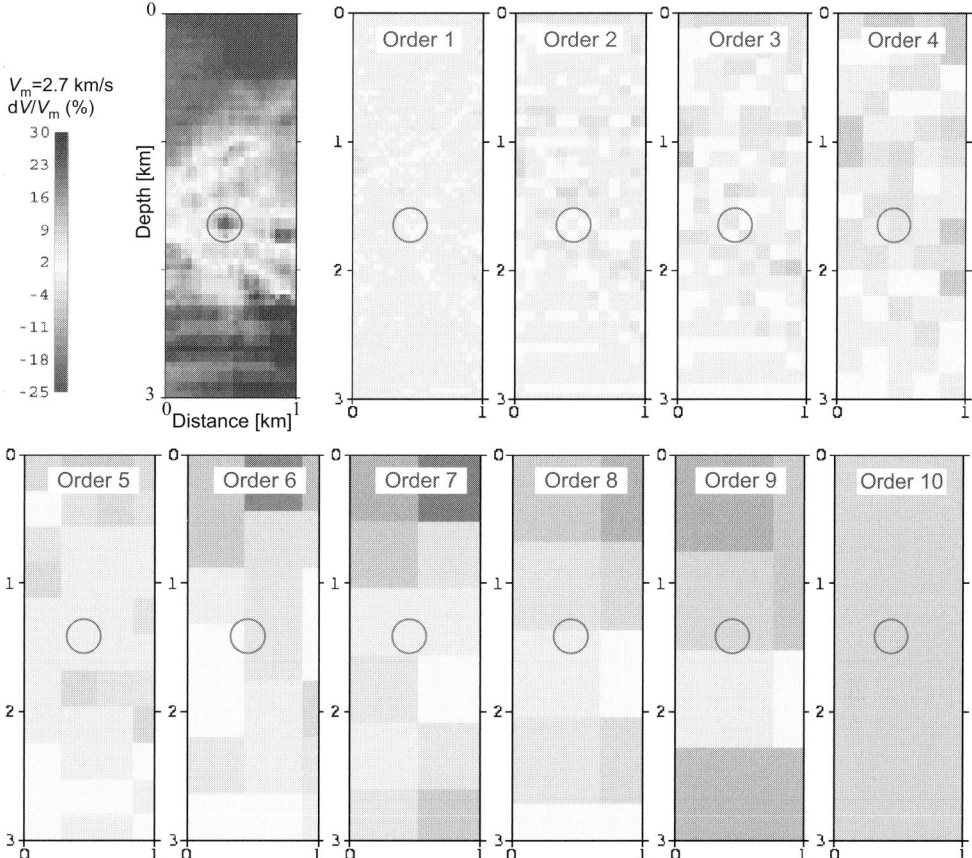

Figure 8.28 The upper-left panel is the MST solution, which is a superposition of the solutions of 10 sub-models shown in the rest of the panels. Plotted here are the ratios in per cent between the velocity perturbations and the mean velocity $V_m = 2.7$ km/s. For color versions see plate section.

sub-models of different wavelengths all contributed significantly to the final model. We may visualize that each velocity anomaly, such as the fast velocity anomaly enclosed by a dotted circle in the middle part of the true model, is composed of contributions from the portions of solutions enclosed by the purple circles in different sub-models in this figure. Interestingly, patterns of the X-shaped stretches in the first-order sub-model in this figure are similar to that of the SST model in Figure 8.27d. These stretches are along-raypath smearing artifacts due to lack of crisscrossing rays for small cell sizes.

8.4.4.3   Impact of the ray dependency on velocity variations

A major reason for the non-linearity of traveltime tomography lies in the dependency of raypaths on the velocity variations. Each raypath tends to bend away from areas of slower velocities and toward areas of faster velocities. Consequently, the ray coverage in the reference model may differ from that in the true model. Figure 8.29 shows a set of synthetic

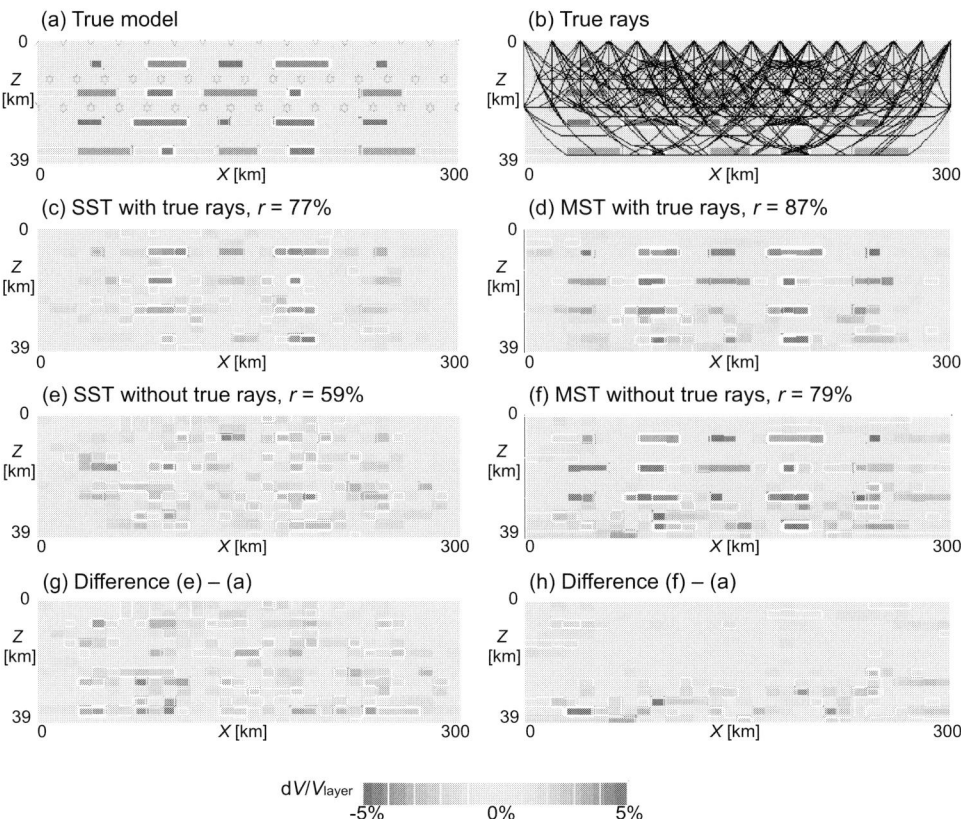

Figure 8.29 Velocity perturbations are contoured in the unit of percentage of the layer averages. (a) True model. (b) First-arriving rays in true model. (c) and (d) are SST and MST solution using true rays. (e) and (f) are solutions without the knowledge of true rays. (g) and (h) are difference between model pairs. r is correlation between the solution and the true model. For color versions see plate section.

tests over a 300 km wide and 39 km deep cross-section simulating a crustal tomography. The true model consists of a series of lateral inhomogeneities that are superposed over a depth-varying 1D velocity profile which is not shown in this figure. There are 16 surface receivers and 31 sources at two focal depths. The true rays in panel (b) indicate that the ray coverage is good in the top portion of the model and poor in the lower portion.

Here each layer (rather than the cross-section) is decomposed into MST sub-models. The lateral size of the single cells is 10 km, while the lateral sizes of the nine multi-cells used are 10, 20, 30, 45, 60, 75, 100, 150, and 300 km. The applications of the SST and MST differ only in their model representation, and all other processing parameters are identical. Panels c and d show solutions of SST and MST, respectively, using true rays to obtain the best possible answers. In real situations, we will not know the true raypaths and have to derive them with the velocities. Panels e and f show SST and MST solutions without *a priori* knowledge of true rays, following the procedure outlined in Figure 8.23. Panels g and h display the difference between the solutions and the true model. As expected, more difference occurs at places of poor raypath coverage. The superiority of the MST over the

Box 8.3 CIG-based multi-scale tomography

Based on the CIG flattening principle described in Section 8.3.4, we can pick the depth deviations of major reflectors on the CIGs and then use these depth deviations in a tomographic inversion to determine the velocity perturbations required to improve the velocity model. The levels of flatness of the reflectors in the CIGs are good indications of the correctness of the corresponding velocity models.

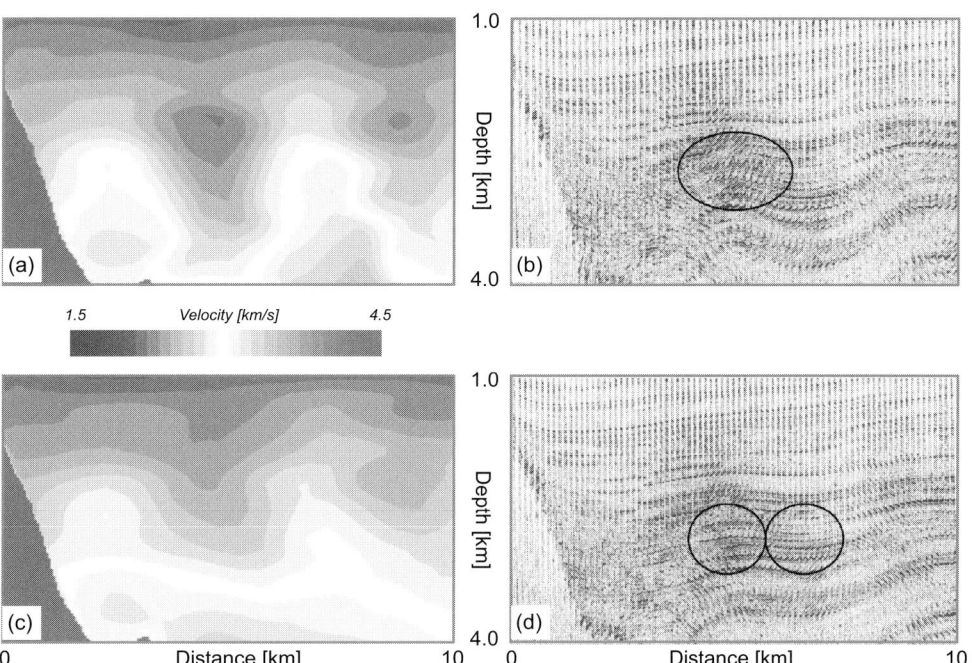

Box 8.3 Figure 1 Comparison between SST and MST for a Gulf of Mexico dataset (Cao *et al.*, 2008). (a) and (c) are velocity models derived using SST and MST, respectively. (b) and (d) are the corresponding common image gathers. For color versions see plate section.

Box 8.3 Figure 1 compares the inverted velocity models and CIGs from the SST and MST methods for a 10 km long seismic profile in the Gulf of Mexico (Cao *et al.*, 2008). Note that the velocity model derived from the MST in panel (c) has smoother lateral variations than the velocity model from the SST in panel (a). The reflectors shown in the CIGs of the MST model in panel (d) are flatter than in the CIGs of the SST in panel (b). The ellipses in these panels highlight the areas of high contrast between the two CIG panels. Further improvement in the model could be made by applying more tomographic iterations.

SST is evident from their correlation coefficients with the true model. Box 8.3 shows a field data example of multi-scale tomography which uses common image gather to measure velocity errors.

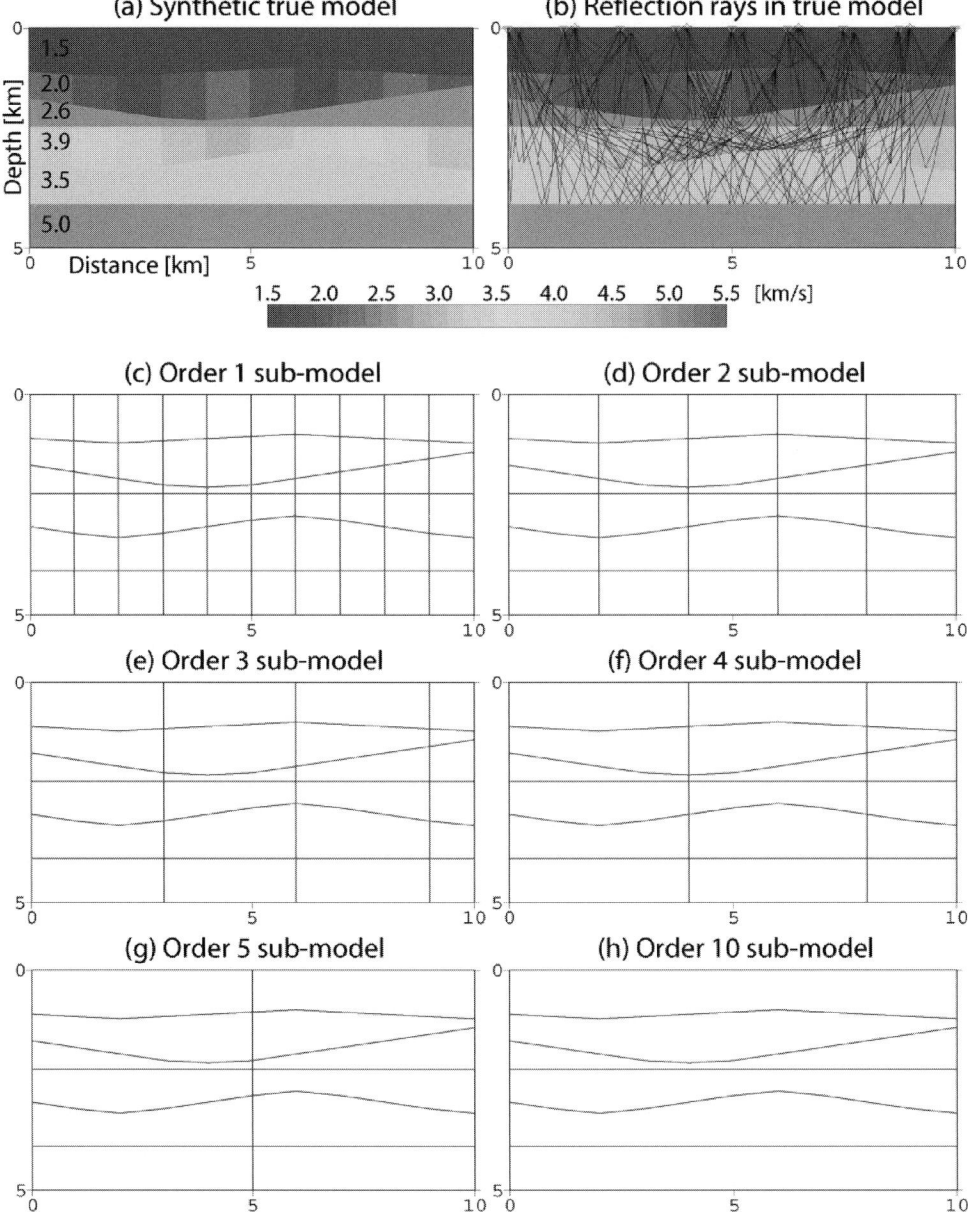

Figure 8.30 (a) 2D synthetic model for a synthetic reflection tomographic test. The values in this panel denote average velocities of each model layer in km/s. (b) Raypaths in the true model from eight shots (stars) to nine receivers (triangles). (c)–(h) Examples of the multi-cell sub-models. For color versions see plate section.

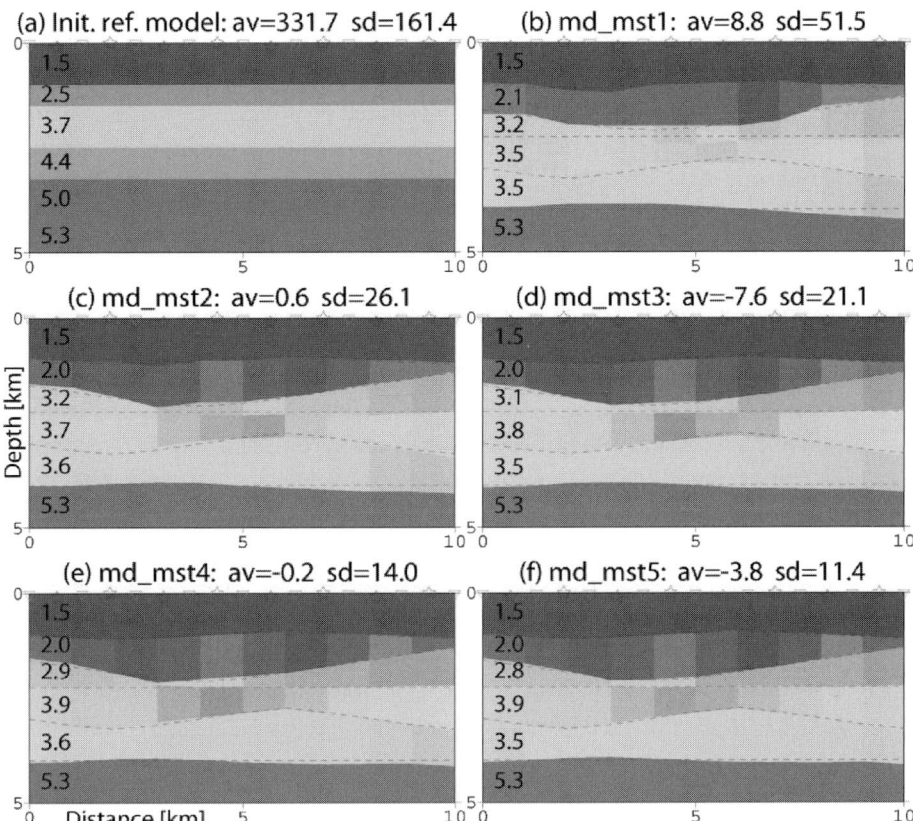

Figure 8.31 The initial reference velocity model and solutions of five iterations of the MST. The numbers on the models are layer average velocities in km/s. The dashed curves show the interfaces of the true model. The average (av) and standard deviation (sd) of traveltime residuals in milliseconds are shown at the top of each model. See Figure 8.30 for the velocity scale. For color versions see plate section.

8.4.5   Reflection traveltime tomography

In reflection seismology, angle-limited ray coverage results in velocity–depth ambiguity (Bickel, 1990; Lines, 1993; Tieman, 1994), owing to the competing effects of medium velocity and reflector depth on traveltimes. One way to cope with the ambiguity is to apply tomographic inversion to account for the effect of velocity and reflector depth properly and to assess the non-uniqueness.

In the synthetic example in Figure 8.30, we use a reflection tomography to recover the slowness and interface geometry for a 2D model simulating a marine survey. Panels (a) and (b) show the synthetic true model and reflection rays from eight shots to nine receivers. The true model consists of a water column over five layers of sediment, with velocities ranging from 1.5 km/s near the surface to 5 km/s in the bottom layer. Lateral velocity variation exists in the second and fourth layers, and in terms of variable interface geometry. The MST sub-models are applied for each layer to describe both the slowness cells and the interface nodes. Panels (c) to (h) show six examples of the sub-models. The sub-model with the largest cell size has the layer average values. Since there are 10 columns of cells in each layer, we used 10 velocity sub-models and 11 interface sub-models (Zhou, 2003).

In the synthetic test, reflection traveltime data were created by ray tracing in the true model. A set of Gaussian random noise of 20 ms in standard deviation is added to the synthetic traveltime data. These data are used by the reflection traveltime tomography following the procedure of Figure 8.23.

Figure 8.31 shows the initial reference model and solutions of five iterations of the MST. The initial model has six flat and constant-velocity layers whose interval velocities are similar but slightly differ from that of the true velocity model. The standard deviation of traveltime residuals decreases monotonically with the MST iterations, reaching a level that is comparable to the added noise in the synthetic data. The MST solution from the fifth iteration matches the true model very well in both velocity values and interface geometry. While the lateral velocity variations in the second and fourth layers are recovered well, there are small amounts of erroneous lateral velocity variation in the third and fifth layers.

Exercise 8.4

1. While traveltime tomography derives model updates by minimizing the traveltime misfit, are there other constraints and information that we can use to help the process? Make a list of such extra information and give some details on how to use each constraint.

2. Devise a residual velocity tomography method for common image gathers based on the Kirchhoff migration. Please provide the main formula and a list of procedures of your method.

3. What are the factors that we should consider in selecting the multi-scale elements in multi-scale tomography? Make a checklist for best-practice procedure in conducting multi-scale tomography.

8.5   Practical issues in tomographic velocity model building

8.5.1   Origins and symptoms of artifacts in seismic tomography

Various artifacts produced from velocity model building and seismic imaging processes are of major concern in practice. Nearly all geophysical imaging methods produce both signals and artifacts, which refer to a range of undesirable anomalies in the solution images due to inadequacy in data quality, coverage, and processing algorithms. Artifacts in tomography images are especially harmful because they are often regarded as real features. Much of this section is devoted to examining pitfalls due to artifacts in tomographic velocity modeling building. Near the end, a tomography method is designed to cope with the artifacts in velocity model building. In some ways, advances in seismic tomography might be measured by our ability to detect and appreciate the presence of artifacts.

The main causes of artifacts in seismic tomography include the following factors, which often overlap with each other:

- Non-linearity of the tomography system due to flaws in the assumptions and theory that describe each inverse problem, as well as the non-linear nature of most geophysical inverse problems themselves;
- Poor and biased ray coverage, which lead to an ill-conditioned kernel matrix;
- Dependency of the kernel matrix on the solution's pattern and magnitude – this means that the resolution may depend on the solution model as well as ray or wave coverage;
- Under-determinacy, or too many unknowns for given data, in global and local scales;
- Data error that is not recognized and appreciated.

The above and other factors are the causes of non-uniqueness of the tomographic solution. Such non-uniqueness may produce artifacts, or features in the solution that are incorrect and often unrecognizable as errors. Table 8.4 lists 15 common types of tomography artifacts and their symptoms.

These artifacts are detailed in the following:

1. Poor resolution is the inability to resolve the morphology and amplitude of targeted features, usually caused by insufficient data ray coverage or erroneous ray coverage due to inadequate processing.
2. Along-raypath smearing, or smearing, is elongation of imaged anomalies along raypaths due to lack of angular variation of raypaths.
3. Near-edge artifacts are erroneous amplitudes and patterns near the edges of tomographic solutions due to lack of ray coverage or inadequate processing.
4. Illumination shadow is a footprint artifact due to uneven and insufficient ray coverage or inadequate processing.
5. Low-velocity holes are low velocity anomalies without ray coverage, usually due to the rays' avoidance of low velocities during tomographic iterations.
6. Depth–velocity ambiguity is the tradeoff between the depth extent of a raypath and the velocities it traversed.
7. Out-of-model contribution can produce artifacts that are indistinguishable from real features; it is due to use of data rays that traverse outside the model volume.
8. Misidentification of event can cause undetectable and severe artifacts.
9. Source and receiver position error can cause spiky pattern near sources and receivers.
10. Erroneous ray tracing is a common error in traveltime tomography which can produce anomalies of wrong location, shape, and amplitude.
11. De-mean is the removal of either the traveltime means before inversion, or the layer velocity means after inversion. This common practice in tomography can result in under-estimation of amplitude of sub-horizontal features and various artifacts.
12. Inadequate model parameterization is a mismatch between model grid and morphology of the real. This will limit the resolution of tomography and cause various artifacts.
13. Inadequate initial reference model can result in wrong images due to the fact that traveltime tomography uses raypaths and traveltime residuals in a reference model to update the velocity iteratively.
14. Over- or under-damping in tomographic inversion can produce overly smoothed or spiky solution models.
15. Inconsistent display is a major factor of many poor interpretations of tomographic solutions.

Table 8.4 Typical artifacts in seismic tomography.

	Name of the artifact	Mainly due to poor data/ray coverage	Mainly due to errors in interpretation	Mainly due to errors in processing	Symptoms / verification of the artifact
1	Poor resolution	mostly		partly	Blurred images, random appearance, and systematic variation in image amplitude / Resolution tests
2	Along-raypath smearing	mostly		partly	Elongation of imaged anomalies along raypaths / Resolution tests
3	Near-edge artifact	mostly		partly	Erroneous amplitude and pattern at edges / Resolution tests, changing model size
4	Illumination shadow	mostly		partly	Casting shallow patterns in deeper parts of the image / Resolution test
5	Low-velocity holes	partly		largely	Persistence of low-velocity anomalies through tomographic iterations / Study ray coverage
6	Depth–velocity ambiguity	largely	partly	partly	Tradeoff between the depth extent of a ray and the velocities it traversed
7	Out-of-model contribution	partly	largely	partly	Use rays partly traversing outside model volume / Compare with solutions using only inside data
8	Misidentification of event		mostly	partly	May cause undetectable artifacts / Careful modeling and QC with other info
9	Source and receiver position error		mostly	partly	Somewhat spiky pattern near source and receiver / Checking raw data geometry
10	Erroneous ray tracing		partly	largely	May produce anomalies of wrong location and pattern / Testing and comparing ray tracing codes
11	De-mean, or removal of data or layer means		largely	partly	Lowering amplitude of horizontal anomalies and reversed polarities around the main features
12	Improper model parameterization		partly	largely	Mismatch between model grid and geology for available data coverage
13	Inadequate initial reference model		partly	largely	Verifiable by tests using different initial reference models
14	Over- or under-damping			mostly	Appearance of overly smoothed or spiky images / Resolution tests
15	Inconsistent display		mostly	partly	Study the original tomographic model with different display options

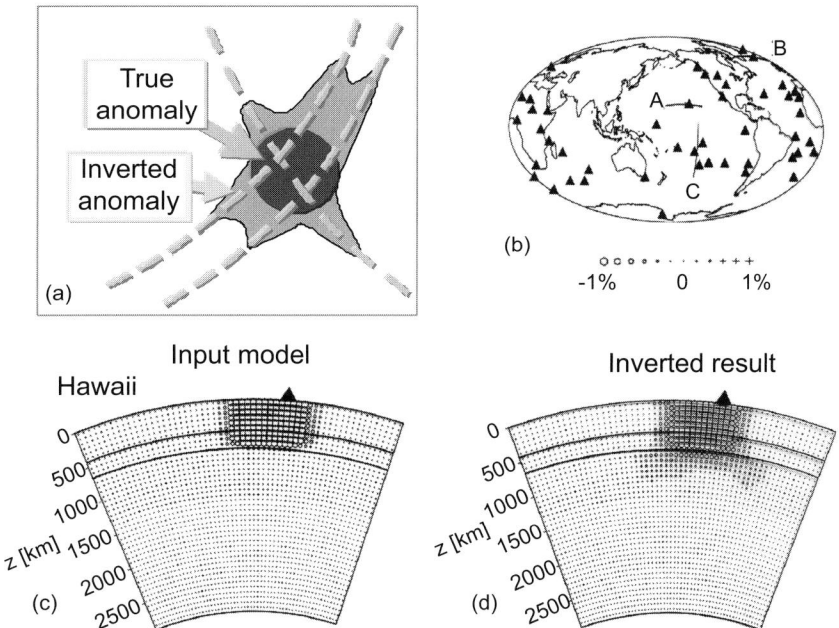

Figure 8.32 (a) A cartoon illustrating the along-raypath smear artifact. Owing to uneven raypaths (dashed curves), the inverted anomaly differs from the true anomaly (circle). (b) A map showing many hotspots (triangles) and cross-section A across Hawaii. (c) Input synthetic true velocity model on cross-section A. (d) Inverted result using traveltimes from the synthetic true model and real source and station locations. The inverted result shows some along-raypath smears from the base of the slow-velocity anomaly. (Panels (b)–(d) are from D. Zhao, personal communication.)

8.5.2   Smear artifacts

Let us examine several types of tomography artifacts graphically. Figure 8.32 shows a synthetic restoration resolution test to check the ability of using real earthquake raypaths to resolve a slow velocity anomaly beneath Hawaii (D. Zhao, personal communication). In all restoration resolution tests, a synthetic true model and real source and receiver locations are used to see how well the synthetic true model can be restored. Such tests check the resolution of source and receiver coverage and the tomographic algorithm. The checkerboard resolution tests shown in the previous section are restoration resolution tests using checkerboard type synthetic models. Here in Figure 8.32 the test is on a slow velocity anomaly that assembles the high-temperature and therefore low-velocity plume anomaly in the mantle beneath the Hawaii. The inverted result shows two along-raypath smears from the base of the slow-velocity anomaly in the lower right panel of this figure. The appearance of the smear artifact here is due to the lack of seismologic stations and earthquake events in the Pacific Ocean.

Figure 8.33 shows the lateral velocity variations on a west–east cross-section beneath Iceland. This figure is from Bijwaard and Spakman (1999) who attempted to map the low-velocity signature of the mantle plume beneath the Iceland hotspot. As was the case for

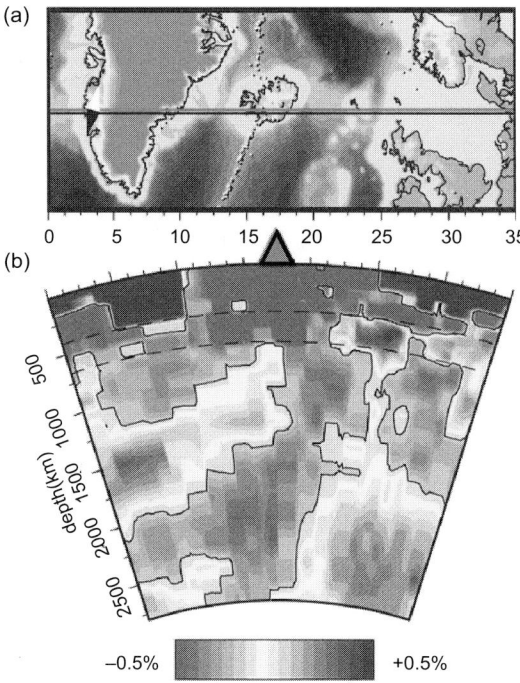

Figure 8.33 (a) Map-view of a cross-section through the Iceland. (b) Cross-section of lateral velocity variations from the tomographic inversion (Bijwaard & Spakman, 1999). Can you tell the signature of the Iceland plume from artifacts? For color versions see plate section.

Hawaii, the raypaths are very biased because of the lack of stations on the ocean floor and lack of earthquakes at different distances and azimuths from the target areas. Consequently, most of the raypaths are within a conical fan centered on the island. In fact, one can almost see the trajectories of raypaths from the cross-section solution in this figure. A challenge to all interpreters of such images is: can you really tell the signature of the Iceland plume from tomographic artifacts?

The smear artifact occurs not only in traveltime tomography, but also in all imaging procedures that involve back-projection of scattered data to form image of the scatter sources. Kirchhoff migration, for instance, can produce along-isochronal smear artifacts. Because the Kirchhoff migration maps the reflected or scattered waves along the corresponding isochrons, or equal-traveltime contours, the lack of crisscrossing isochrons due to poor source and receiver coverage will result in such an along-isochronal smear. Figure 8.34 displays an example of such an artifact from an attempt to map faults near the San Andreas fault using a VSP dataset. In this case the acquisition of the VSP data employed a line of surface sources and several receivers in a wellbore down to a depth of nearly 1.6 km from the surface. The spatial distribution of the sources and receiver allows seismic mapping reflectors and scatters within a limited space called a VSP corridor. Unfortunately, the authors in this study attempted to map subsurface anomalies widely outside the VSP corridor, resulting in the spherical smear pattern in this figure that is known as the "cabbage artifact" in VSP imaging.

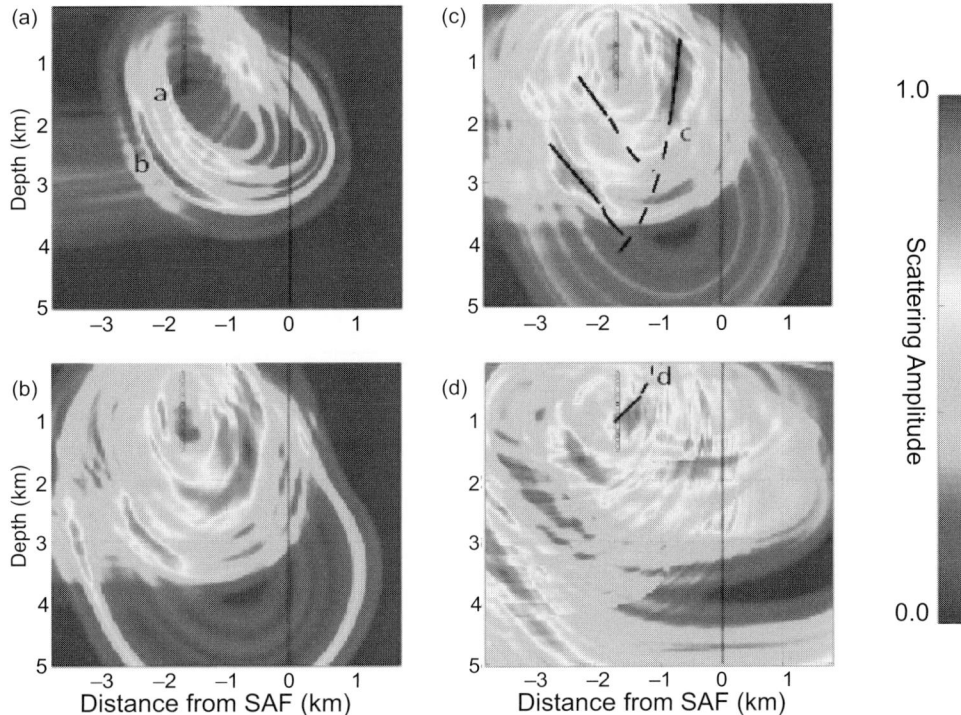

Figure 8.34 Images from a VSP Kirchhoff migration near the San Andreas Fault (SAF) illustrating the along-isochronal smear artifacts. For color versions see plate section.

8.5.3   De-mean artifacts

Another common artifact is due to the de-mean process, or removal of the average value. There are three types of de-mean processes in traveltime tomography. The first type is removal of traveltime average, such as that using traveltime residual rather than total traveltime. Many tomography papers use relative time between two seismic phases, which is an implicit de-mean process. The second type of de-mean is due to the assumption in most inverse operators that the statistical mean of the misfit between data and model prediction is zero. The third type of de-mean is the removal of some averages from the tomographic solution. For instance, the average velocities of all model layers are often removed in velocity tomography because many interpretations are based on the lateral variations of the velocity field rather than the absolute velocity field.

Figure 8.35 shows a cartoon illustrating an artifact due to the third de-mean process, or removal of the layer averages of the velocity model after tomographic inversion. As we do not know the true 1D velocity model in a specific region, we often use the layer averages of the tomographic model to approximate the 1D velocity model. Then if the true velocity anomaly is biased toward either slow or fast polarity, as in the case of slow velocity due to a mantle plume as shown in this figure, the layer averages of the velocity model will be biased, as shown in panel (e). Consequently, the de-mean or removal of such a biased 1D velocity model will produce artifacts in both the background portion and the anomaly portion of the lateral velocity variation model (e.g., panel (f)).

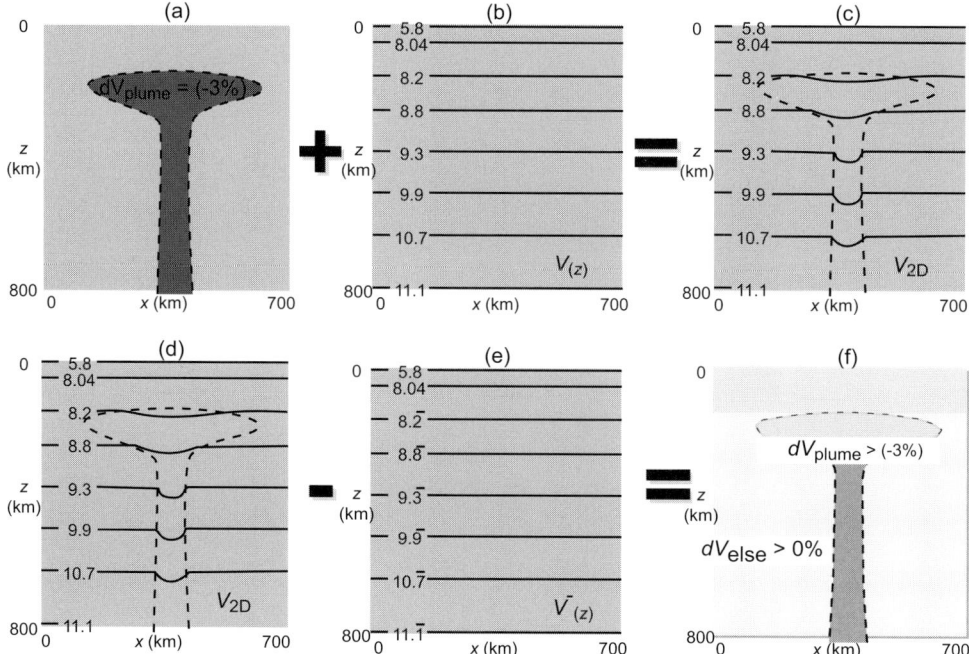

Figure 8.35 Schematic cross-sections showing the de-mean artifact. (a) Lateral velocity variation of a low-velocity plume anomaly. (b) 1D velocity model where the unit is km/s. (c) The absolute velocity field as a summation of the lateral velocity variation and the 1D velocity model. (d) Solution of the absolution velocity field. (e) Estimates of 1D velocities based on (d). Superscript bars (‾) indicate estimates are slower owing to the plume anomaly. (f) The model of lateral velocity variations from the de-mean process, or subtracting (e) from (d).

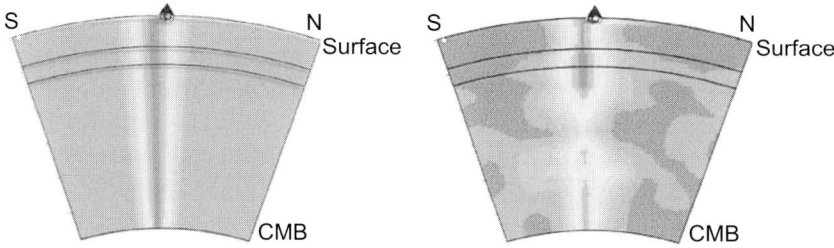

Figure 8.36 A restoration resolution test along a cross-section beneath Hawaii (back triangle) from surface down to the core–mantle boundary (CMB). (Left) Synthetic true model of a low-velocity plume. (Right) Tomographic solution using real sources and station locations and traveltimes in the synthetic true model.

Figure 8.36 shows another restoration resolution test (from D. Zhao, personal communication) to check the ability to resolve a slow-velocity cylinder resembling a mantle plume beneath Hawaii from the surface all the way down to the core–mantle boundary (CMB). Can you identify on this north–south cross-section the artifacts due to the de-mean process and due to along-raypath smearing?

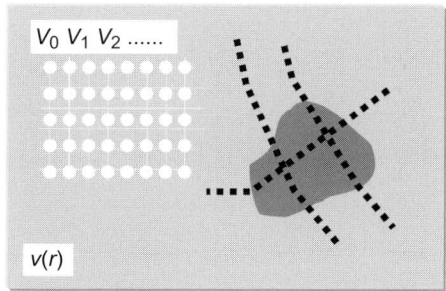 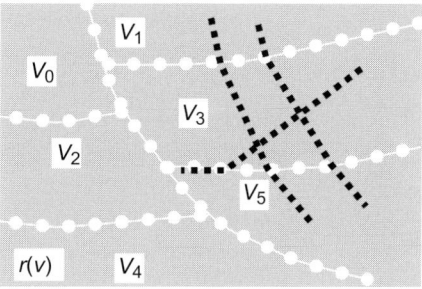

Figure 8.37 (Left) Conventional tomography determines velocity as a function of location, $v(r)$. Crisscrossing raypaths (dark dashed line) are required. (Right) Deformable layer tomography (DLT) determines geometry of velocity interfaces, $r(v)$.

8.5.4   Deformable layer tomography

8.5.4.1   Motivation and method

As an example of how to cope with the artifacts in traveltime inversion, let us examine a **deformable-layer tomography (DLT)** approach (Zhou, 2004b). Seismic tomography traditionally inverts for velocity field on a regularly spaced and fixed-in-space model grid, as shown in the left panel of Figure 8.37. Such an approach cannot adequately describe pinchouts or wedge-shaped velocities as commonly seen at basin boundaries, faulted rock beds, and back-arc mantle wedges. A regularly spaced model grid also requires the use of a large number of model variables. The DLT attempts to invert for velocity interfaces, as shown in the right panel of this figure. The use of thickness-varying layers allows a much smaller number of model variables for the DLT than the regularly spaced model grid. In the DLT the geologic framework and known velocity range are adopted into the initial reference model, and the best-data-fitting geometry of the velocity interfaces is determined.

Conceptually, the traditional cell tomography (left panel of Figure 8.37) regards the velocity field v as a function of the space r, or $v(r)$. Then the objective is to determine the velocity values of all model grids or cells. In practice, however, we know the velocity ranges of most Earth materials and velocity range of most study areas. What we do not know is the location of these materials and velocity values. Hence, the concept of DLT is to regard the velocity field as a series of velocity contours or interfaces, and the objective is to determine the spatial location of these velocity contours. In other words, DLT seeks to determine the depth of velocity interfaces using thickness-varying velocity layers.

Figure 8.38 shows the initial reference model and results of iterations 1, 6, and 12 of a DLT using earthquake data along a crustal profile in southern California. The depth variations of the seven model interfaces have been updated through each of the DLT iterations of ray tracing in the current velocity model and inversion. The averages and standard deviations of the traveltime residuals, as measure of data misfit level, show a systematic reduction through the DLT iterations. The misfits of the final solution at iteration 12 are −8 ms for the average and 257 ms for the standard deviation.

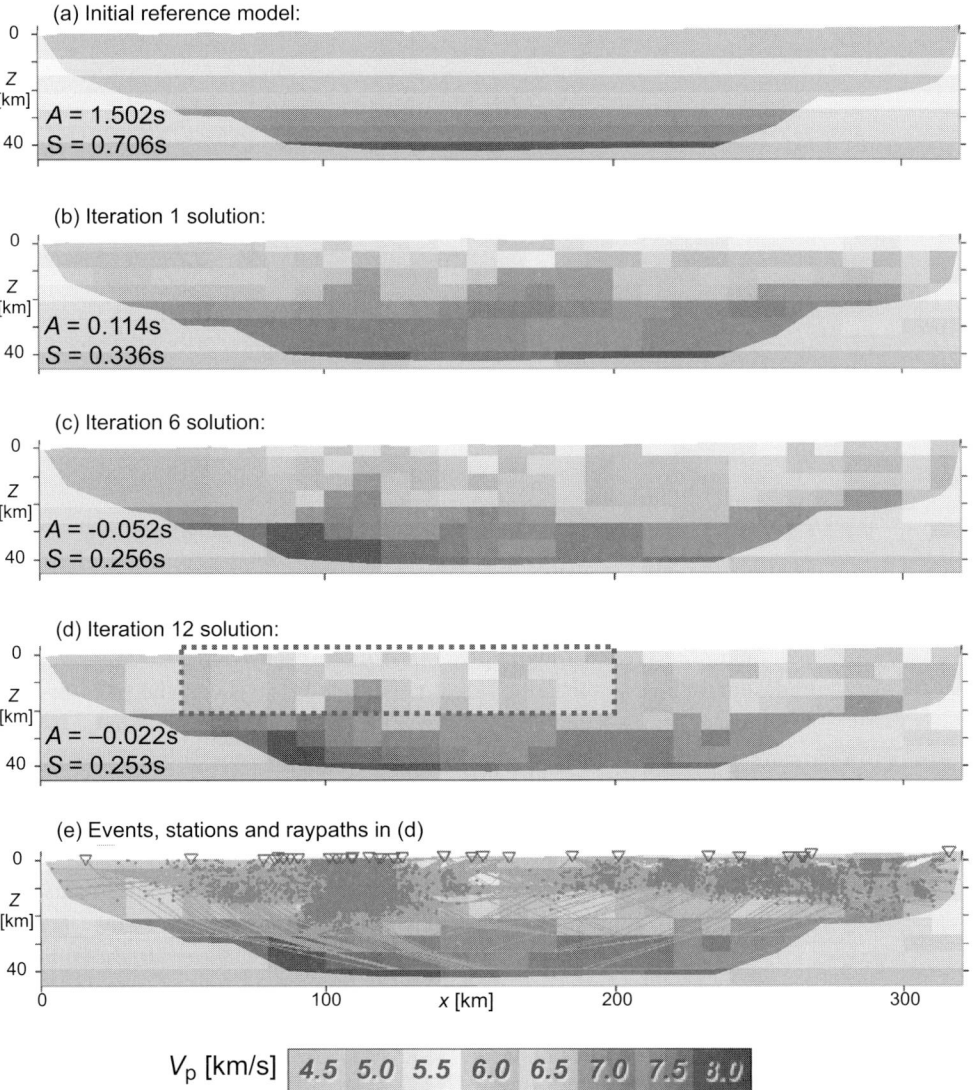

Figure 8.38 The initial reference model (a) and three DLT solutions (b–d). Panel (e) shows raypaths in model (d). Small triangles are stations. Small crosses are earthquake foci. A and S are the average and standard deviation of traveltime residuals in seconds. Areas without ray coverage are shadowed. Area in dashed box in (d) is the next figure. For color versions see plate section.

8.5.4.2 Comparison between cell tomography and DLT

In Figure 8.39, the solutions from DLT and cell tomography are compared along the same 150-km-long and 20-km-deep transect. These two models were derived from the same earthquake dataset, same initial reference model, and roughly the same level of data misfit.

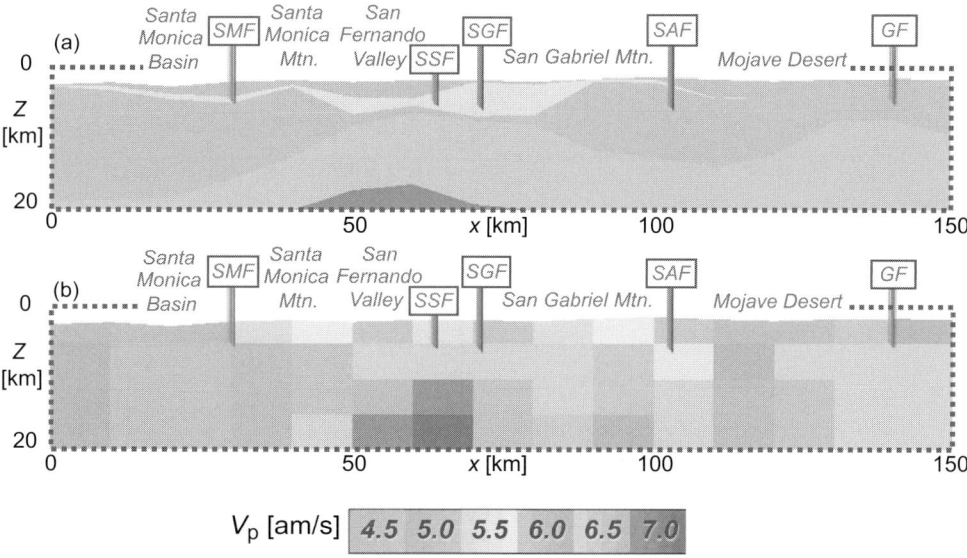

Figure 8.39 Comparison between a DLT model and a cell tomography model in California using the same data. Faults: Santa Monica (SMF), Santa Susana (SSF), San Gabriel (SGF), San Andreas (SAF) and Garlock (GF). For color versions see plate section.

While these two solutions are an example of non-uniqueness of tomography, the model features can be compared directly with many geologic features such as basins and mountains denoted on the figure. The DLT profile exhibits much better vertical resolution than cell tomography because DLT directly inverts for the depths of velocity interfaces and is not limited by cell dimensions. The DLT solution in panel (a) seems to be more geologically plausible than the cell tomography solution in panel (b). At shallow depths in the DLT solution, the Santa Monica Basin, San Fernando Valley and Mojave Desert are lens-shaped basins and correspond to low-velocity anomalies near the surface. High-velocity anomalies at shallow depths correlate with the Santa Monica Mountains and San Gabriel Mountains. The block-shaped, pixelated images of the cell tomography model do not clearly depict the lens-shaped geometries of many crustal structures. A decrease of cell size to improve cell tomography resolution increases the number of unknown parameters and is limited by the density of ray coverage. The locations of pinchouts or wedge-shaped features as shown in the DLT solution simply cannot be replicated by the block-shaped images of cell tomography. In the vicinity of pinchouts, cell tomography velocity values are the averages of velocities of different lithological units and not representative of wedge-shaped lithological units.

Cell tomography models are usually represented on profiles and maps in publications by velocity contours, which are interpolations of grid point values. The contoured representations do not represent traveltime best-fit solutions of the original data and are not unique solutions of the grid point array. The differences between the original inverse solution model and the contoured representation of the model are potentially significant, especially in the areas of wedge-outs or pinchouts of layers. The blocky model parameterization of cell tomography truncates pinchout features that are thinner than the cell thickness, and

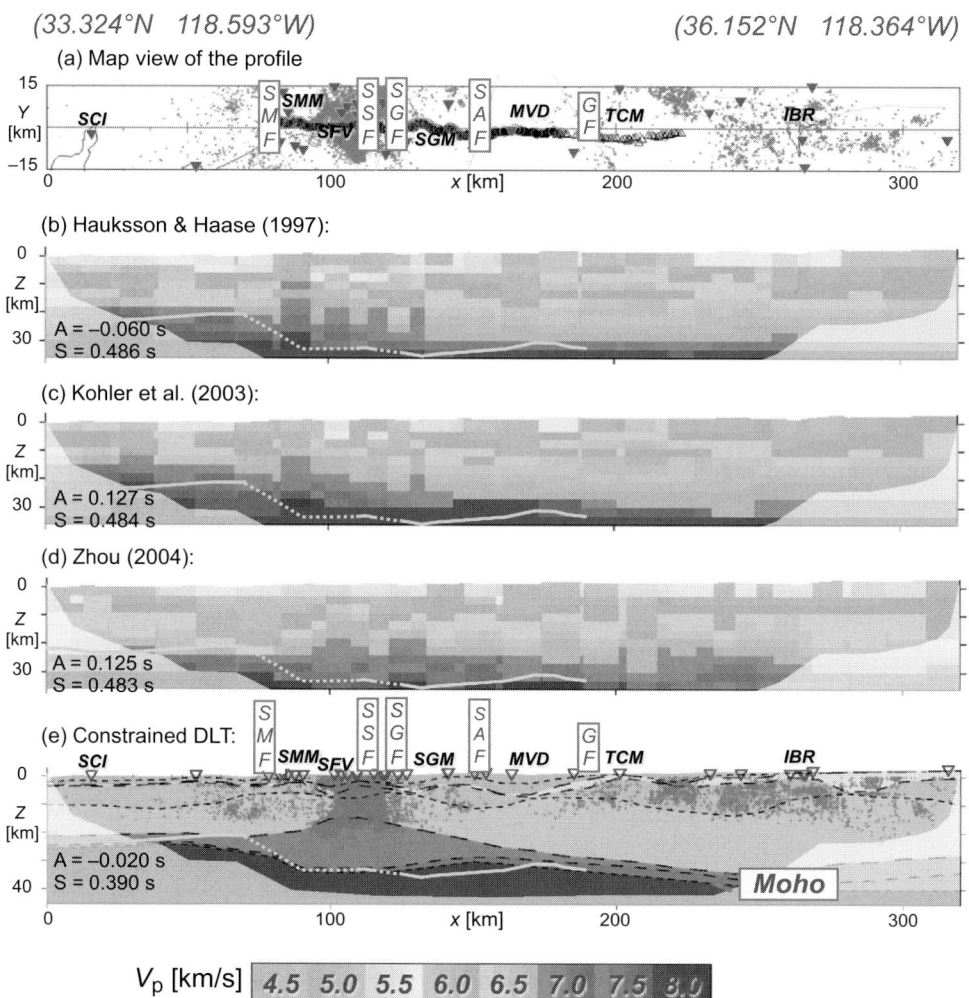

Figure 8.40 Comparison of constrained DLT model (e) with three cell tomography models: (b) Hauksson and Haase (1997), (c) Kohler *et al.* (2003), (d) Zhou (2004a). Purple crosses in (e) are earthquake foci. Faults (pink) are: Sierra Madre (SMF), Mission Hills (MHF), San Gabriel (SGF), San Andreas (SAF), and Garlock (GF). *A* and *S* are the average and standard deviation of traveltime residuals. Solid and dotted curves of light-blue in (b) to (f) denote the PmP Moho interpreted by Fuis *et al.* (2007). For color versions see plate section.

contour representations cannot recover pinchout features not preserved by the inversion process. Because of the problems introduced by contour representations of a grid of data points, we have not used post-inversion smoothing or contouring here.

8.5.5   Constrained traveltime tomography

An important measure to help increase the geologic plausibility and reduce the non-uniqueness is to incorporate constraints in the tomographic inversion process using existing knowledge from other studies. Some of the constraints can be inserted into the design of model parameterization, such as the use of deformable layers in DLT in contrast to the fixed-in-space model grid in cell tomography. We can implement hard constraints that a tomographic inversion must satisfy. We can also implement soft constraints that a tomography inversion needs to satisfy within a given tolerance range.

As an example, Figure 8.40 shows the use of soft constraints, the ranges of crustal thickness from receiver functions, to constrain a DLT crustal velocity profile. Here a constrained DLT model (panel (e)) is compared with three published cell-based crustal P-wave velocity models of southern California (Hauksson & Haase, 1997; Kohler *et al.*, 2003; Zhou, 2004b). The DLT profile fits the data significantly better than the cross-sections of three published cell tomography models. The DLT model has been constrained by the crustal thicknesses estimates based on analyses of receiver functions at seven locations along the profile (Zhu & Kanamori, 2000). In addition, the tomographic solutions are compared with the PmP Moho interpreted from the Moho reflections (Fuis *et al.*, 2007). The DLT Moho correlates well with the PmP Moho along this profile. The difference between these two independently derived Moho depths is less than 3 km at most places.

The velocity distribution patterns of the DLT solution and the cell tomography solutions in Figure 8.40 show some similarities. The constrained DLT solution supports an upward trend of the Moho toward the continental margin. The blocky patterns of cell-based models severely limit realistic representations of crustal features, especially where raypath smearing artifacts are present. Post-inversion contouring of blocky models will differ markedly in regions of artifacts from DLT models, which do not require post-inversion smoothing or contouring, and will not necessarily delineate the shapes of real crustal features. The better data fitness and the absence of raypath smearing artifacts suggest that DLT methodology has the potential to significantly improve 3D crustal velocity models.

Exercise 8.5

1. Compile a list of the "rule of thumb" ways to QC at least five types of common tomographic artifacts. Specify the tests that can be conducted to verify the artifact and quantify its extent.

2. Make a computer algorithm to illustrate the de-mean artifact in traveltime tomography.

3. How should we deal with localized velocity anomalies, such as a small salt body, in deformable-layer tomography? Make a 2D computer algorithm to illustrate your idea.

8.6 Summary

- Seismic velocity is the speed at which a seismic phase travels through a medium. The velocity structure is indicative of the properties of the medium. Velocity inhomogeneity refers to variation of velocity as a function of spatial location, and velocity anisotropy refers to variation of velocity in different directions at a fixed location.

- Velocity model building (VMB) is among the most common practices in seismology, owing to the value of the velocity model for geologic interpretation and seismic imaging. The resolution of each velocity model is limited by the frequency bandwidth of seismic data.

- On a global scale, over 98% of the traveltimes of major types of seismic waves can be explained by one of the standard $V(z)$ 1D Earth models. The dominance of the 1D velocity variation at large scales in the Earth leads to the classic refraction velocity analysis based on seismic ray tracing in a layer-cake velocity model.

- Semblance velocity analysis estimates the stacking velocity by assuming gentle to no changes in the reflector dip and lateral velocity variation. The stacking velocity at each depth is the average velocity of all layers above the depth. The stacking velocity can be approximated by the rms velocity, which is linked with the interval velocity via the Dix formula. The derivation of the interval velocity from the rms velocity is non-unique.

- The reflection traveltimes from a single scatter in a constant velocity field, if expressed as a function of midpoint and offset, form a geometrical shape resembling the Cheops Pyramid. Such expression is useful in studies associated with 3D time migration.

- Migration velocity analysis refines the velocity model by utilizing the dependency of depth migration on velocity variations. Migration velocity analyses are usually conducted using common image gathers (CIGs); each CIG is a collection of migrated traces at an image location. Using the CIG flattening principle that the correct velocity model will make reflectors flat on the CIGs, the depth variations of major reflectors on the CIGs are used to refine the lateral velocity variations.

- Tomographic velocity analysis takes an inversion approach based on relationship between the velocity model and the traveltimes and waveforms of seismic data. Typically a linear approximation of the relationship is taken to allow a mathematical inversion for the updates of the velocity model based on misfits between the observed data and model predictions.

- The non-linearity of the relationship and poor data coverage are the major sources of non-uniqueness in the tomographic inversion. As an illustration of the means to overcome the non-uniqueness, multi-scale tomography (MST) decomposes the model space into overlapping sub-models of different cell sizes, inverts for the solutions of all sub-models simultaneously, and stacks all sub-model solutions into the final model. It is an application of the law of decomposition and superposition.

- Artifacts are often produced in velocity model building and seismic imaging processes owing to inadequacy in data quality, coverage, and processing methods. It is important to know the characteristics of artifacts because they may be misinterpreted as real features.

FURTHER READING

Al-Yahya, K., 1989, Velocity analysis by iterative profile migration, *Geophysics*, 54, 718–729.

Lines, L., 1993, Ambiguity in analysis of velocity and depth, *Geophysics*, 58, 596–597.

Sheriff, R. E. and L. P. Geldart, 1995, *Exploration Seismology*, 2nd edn, Chapter 5, Cambridge University Press.

Zhou, H., 2003, Multiscale traveltime tomography, *Geophysics*, 68, 1639–1649.

Zhou, H., 2006, Deformable-layer tomography, *Geophysics*, 71, R11–R19.

9 Data fitting and model inversion

A common objective of geophysics is to probe the properties of the Earth's interior based on data from observations. Geoscientists often use seismic data to build a model of the subsurface as a representation of various assessments of some simplified key aspects of the real world. The validity of each model depends on its consistency with observations. All observable datasets constitute a data space, and all possible models constitute a model space. Data fitting and model inversion are two complementary approaches in geophysics to relate the data space to the model space. Data fitting uses forward modeling to search for models that fit well with the observed data and satisfy our scientific intuition. Model inversion uses our scientific intuition to set up rules about how the models should behave and then determines the model variations that fit best with the available data. The usefulness of data fitting and model inversion is evident in many applications illustrated in this chapter.

The basic theories of seismic modeling and inverse theory are reviewed here. Data fitting is introduced in the first two sections via several seismic forward modeling methods and a simple example of regression. The basic theories on inverting a system of linear equations are given in the next three sections, in conjunction with the tomographic velocity analysis in Section 8.4. The least squares method as a classic linear inversion is widely applicable in geophysical data analysis and beyond. Some mathematic insights on inversion of linear equations are illustrated via several common ways of matrix decomposition. The common causes of non-uniqueness in geophysical inversion include insufficient constraining power of the data, the non-linear relationship between data and model, and dependency of that relationship on the solution. Several practical inverse solutions discussed here include the Backus–Gilbert method and the LSQR algorithm. Practically, seismic inversion is synonymous with the inverse imaging in Section 8.4. The inverse approach has the advantage of subjectively determining the values of model properties based on the given model parameterization and data. For some applications such as inverse filtering and tomographic velocity analysis, inversion is the preferred method because of its objectiveness in obtaining the solutions.

9.1 Seismic forward modeling

9.1.1 Introduction to seismic modeling

A common objective of seismic data processing is to come up with models quantifying the subsurface structures and properties. To achieve this objective, we need to consider two common approaches to seismic model building: forward modeling and inversion. **Seismic modeling** is the process of verifying the likelihood of different models by comparing the observed seismic data with model predictions. The process is guided by our scientific intuition as to what extent each model is geologically plausible. **Seismic inversion** is the process of determining certain model values as representations of the subsurface properties by directly inverting seismic data. Seismic modeling has three general steps: selecting the data and models, making predictions, and validating the models. In practice these three steps are often repeated until the most satisfactory set of models has been obtained.

Modeling can be carried out mentally, physically, and numerically. By mentally, we may propose hypothetical models of the subsurface, and then predict the consequences of each hypothetical model. A high level of agreement between the predictions and observations means there is a high likelihood that the corresponding model represents the reality. However, we have to consider uniqueness, because there are usually a number of different models that fit equally well with available observations. In practice, seismic modeling is most often carried out using numerical algorithms in computers. There are occasionally physical modeling studies using selected materials.

One useful notion is to visualize a **data space** including all datasets possible, and a **model space** consisting of all models possible. The data space constitutes all the observables, while the model space contains all possible causes of the observables. Typically, the data space in geophysics consists of observations made at the surface of the Earth, while the model space covers a targeted volume in the Earth's interior. As was discussed previously in Section 8.4.1.1, forward modeling is the process of making predictions using the given seismic

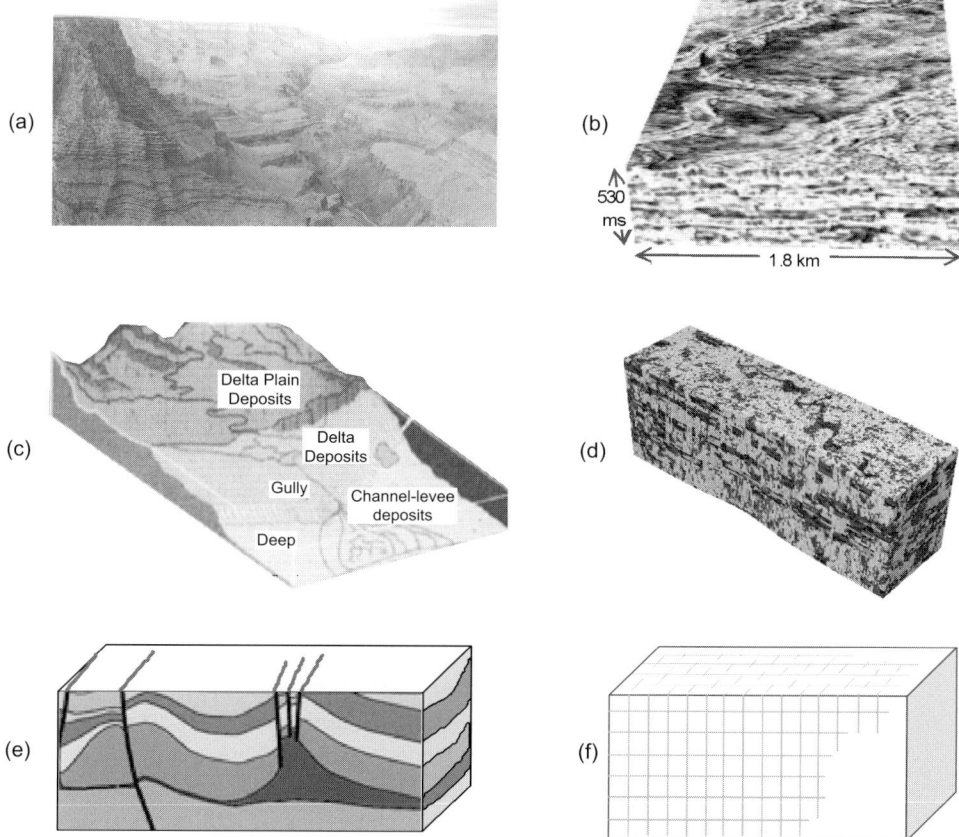

Figure 9.1 (a) A photograph of sediments in Grand Canyon, USA. (b) A seismic facies "cube" (Gao, 2009). (c) A schematic geologic model illustrating the depositional environment of a turbidite reservoir (Soroka *et al.*, 2002). (d) A statistical model in which many variables behave randomly. (e) A layered model. (f) A grid model.

acquisition setup in each given model, and inverse imaging is the process of determining images of the subsurface based on the given data.

In the first step of seismic modeling, we need to choose the datasets that are most suitable for our scientific or business objectives, and come up with hypothetical models. All seismic models are simplified scenarios of the subsurface seismic properties, such as the spatial distributions of seismic velocities, density, and their derivatives such as reflectivity, that are interpreted from the geologic and geophysical information. Making models requires the use of a certain format, called **model parameterization**, or model partition. Let us see some examples.

Figure 9.1a shows a photograph of the Grand Canyon, exhibiting sediments in layered geometry that are exposed on a highly irregular surface topography. Figure 9.1b is a seismic facies "cube" exhibiting the differences between sediments, fluvial channels, and interfaces in terms of geometry and fabrics of rocks. These observations may lead to geologic models like the one shown in Figure 9.1c, as well as seismic models. A seismic model can be statistical, such as a fixed level of porosity but a random pore distribution (Figure 9.1d). In practice, most seismic models are deterministic, with geologically meaningful and fixed values at all spatial positions in each model. Two of the most popular types of seismic models are the layered model (Figure 9.1e) and the grid model (Figure 9.1f). The layered model emphasizes the lateral continuity of rock strata and depth dependency of physical properties such as pressure and temperature in the Earth's interior. The grid model provides a simple and equally spaced representation of spatial and/or temporal distributions of physical properties.

In the second step of seismic modeling, we use a given seismic model and the spatial distributions of sources and receivers to produce measurable seismic responses in terms of traveltime, amplitude, and phase angle of various wave types as functions of space, time and frequency. The responses are produced based on scientific principles of seismic wave and ray theories, which will be elaborated in the next three subsections. The process allows us to map each model's responses in the data space. Conversely, mapping from the data space back to the model space is called an **inverse** with respect to the forward modeling. Compared with forward modeling, an inverse approach is often considered as a more objective, but challenging, task.

In the third step of seismic modeling, we quantify the fitness between the model predictions and observed data, and evaluate the validity or resolvability of the model parameters for the given data and data coverage. As was discussed in Section 4.4, the resolvability is the constraining power of the given data to define the model parameters. The validity of a model requires a good fit between its predictions and the observed data as well as sufficient data coverage to constrain its uniqueness. Though there are always non-unique models for each dataset in geophysical practice, a tighter connection between the data space and model space of a given application means a better chance that our best data-fitting model may agree with reality in some key aspects.

As an example, Figure 9.2 shows a seismic modeling study by Dreger *et al.* (2008) on the source mechanisms of a magnitude 3.9 seismic event associated with the tragic collapse of a coal mine in the Crandall Canyon in central Utah on August 6, 2007. Here the predicted seismograms were computed using a synthetic model of the source that is dominated by a horizontal closing crack to resemble the collapse of the coal mine shafts. A reasonable level of fit between the observed data (dark curves) and predicted seismograms (light curves) is shown in this figure, suggesting that the source model of a horizontal closing crack is a reasonable explanation of what has happened there. The wide azimuthal coverage of the six seismologic stations as shown in the left panel of this figure strengthens the likelihood of the source model. The quality of the source model can be quantified further via resolution tests.

Modeling is a useful way to solve many practical problems. It is the most straightforward way to verify our intuitive concepts and to scan through many hypothetical scenarios. Modeling is a direct way to illuminate the relationship between observations and the models and, most importantly, to quantify the relationship between the variation of the observations and variation of the models. In addition, one can easily adopt *a priori* information to confine the likely models to within a subset of the model space. One can use the modeling approach even without a complete understanding of the mapping functions between the model space and the data space.

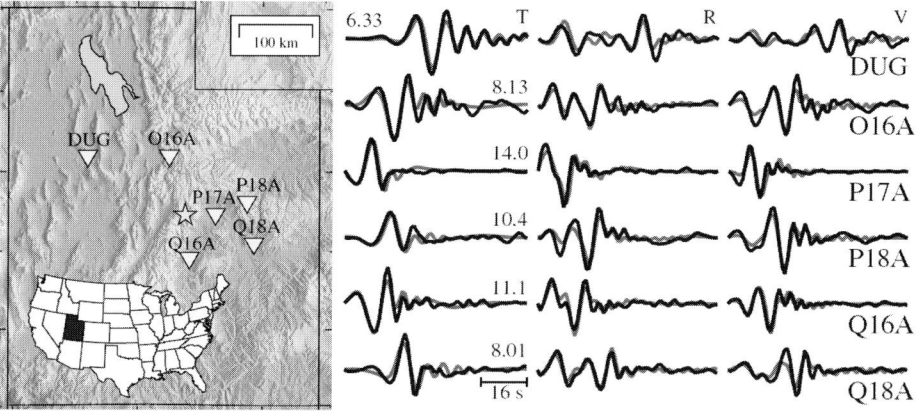

Figure 9.2 (Left) A map of Utah showing the location of the magnitude 3.9 seismic event (star) associated with the tragic collapse of a Utah coal mine on August 6, 2007, and six nearby seismologic stations (triangles). (Right) Fitting observed seismograms (dark curves) with waveform predictions (light curves) for a source dominated by a horizontal closing crack (Dreger *et al.*, 2008). Each row shows the transverse (T), radial (R), and vertical (V) component of a seismologic station whose name is denoted at the right.

9.1.2   Seismic ray modeling

As a simple model method, **seismic ray modeling** aims at verifying the raypaths and traveltime of seismic events under investigation in order to understand the origin of the events, or to verify the velocity model containing the spatial distributions of seismic velocities and density. Using modern computers, numerical modeling has become the most popular means of seismic modeling. In applied seismology, the numerical modeling algorithms follow either ray theory or wave theory. The **ray theory** approach assumes that the spatial dimension of the velocity and density anomalies is much greater than the wavelength of seismic waves, and the theory is based on a series of principles such as Snell's law. The ray theory enables many simple, intuitive, and often efficient approximations. In contrast, the **wave theory** approach takes a rigorous path to quantify the variations in amplitude and frequency during seismic wave propagation.

Most of the ray theory methods focus on computing traveltimes and/or raypaths of distinctive waves such as first arrivals, primary reflections, and converted waves. In particular, **ray tracing** methods, which compute traveltimes and raypaths of distinctive waves, are useful in many applications such as ray modeling, Kirchhoff migration, and traveltime tomography. As was discussed in Section 8.1.5, numerical ray theory methods include **kinematic ray tracing** and **dynamic ray tracing**. The former computes raypaths and traveltimes only, while the later also computes the pulse shape with amplitude and phase information. Kinematic ray tracing is intuitive and efficient, but cannot handle many frequency-dependent wave phenomena. Dynamic ray tracing usually solves a system of dynamic ray equations. Examples of dynamic ray tracing include Gaussian beam and paraxial ray tracing.

There are many practical ways to conduct dynamic ray tracing, or ray-based waveform modeling. A popular and simplified approach in exploration seismology is **convolutional**

modeling which computes the seismic wave events of interest in two steps. First, it traces the raypaths and traveltimes of the interested wave events in the velocity model, and then convolves impulses at the corresponding arrival times of the events with a calibrated source wavelet. The convolutional modeling approach is intuitive and practical, despite its over-simplification of many wave propagation phenomena. Specifically, we can single out events of interest using the ray model approach without worrying about other events that may complicate the picture on the seismograms.

Figure 9.3 shows a ray-based convolutional modeling to verify the seismic structure of a gently dipping subduction zone near Hokkaido, Japan (Iwasaki *et al.*, 1989) using ocean bottom seismographs (OBS). The map in Figure 9.3a shows a NW–SE oriented seismic profile with seven OBSs (denoted by dots with labels such as S6 and SD), and a line indicating the shooting positions of the airgun sources. The bathometry is portrayed by contours of a 1 km interval. Figure 9.3b is a cross-section from SE to NW of the 2D velocity model structure. The lines are velocity interfaces, and the numbers are P-wave velocities in km/s. The P-wave and S-wave 1D velocities beneath the S5 OBS site are shown in two inserts in the lower-left corner of this panel. Two reduced traveltime versus distance plots in Figure 9.3c and Figure 9.3d shows the observed waveform data at OBS S3 and the modeled predictions from the convolutional modeling, respectively. As explained in Section 8.1.4.3, the reduced time helps the identification of refraction events at different bottoming depths. In this case, nearly flat first arrival events may be Moho refractions because the reduced velocity of 8 km/s is close to the upper mantle velocity directly beneath the Moho at 7.9 km/s. The modeled raypaths shown in Figure 9.3e are indicative of the origins of various events on the plot of predicted traveltime versus distance in Figure 9.3d.

OBS data such as shown in Figure 9.3c are typically of poor quality owing to their narrow bandwidth and the presence of multiple reflections. Understanding of the OBS data is further challenged by the small number of receivers and difficulty of understanding waves in a 3D world from a 2D observational grid. Nevertheless, the convolutional modeling may help in identifying the observed events, as illustrated here by comparing observed and modeled events in Figure 9.3c and Figure 9.3d. The modeled raypaths are especially helpful in identifying the wave modes and ray coverage.

In another example of convolutional modeling, Figure 9.4 illustrates a study of reflection events from the steep flank of a salt-diapirs model in the North Sea (Jones, 2008). Here two types of salt flank reflections under examination are that bounced from the inner side of the salt body and that double-bounced at flat and dipping horizons. The double-bounced reflections are called **prismatic reflections** because of the shape of their ray trajectory. The convolutional modeling helps us recognize complicated events by providing both raypaths in model space (Figure 9.4b) and moveout patterns in the CMP (Figure 9.4c) or in other types of data displays.

This example shows an advantage of ray modeling owing to the ease of modeling one event at a time. In contrast, it is a challenging task to study overlapping events in most waveform modeling results. For instance, one can appreciate the difficulty of differentiating various events in waveform modeling results like those shown in Figure 9.4d without knowing their paths in the model space. In this case, the two CMPs in Figure 9.4c and 9.4d demonstrate the value of verifying the ray-traced events with waveform-modeled events. Box 9.1 shows the use of conventional modeling to make synthetic seismograms based on well logs.

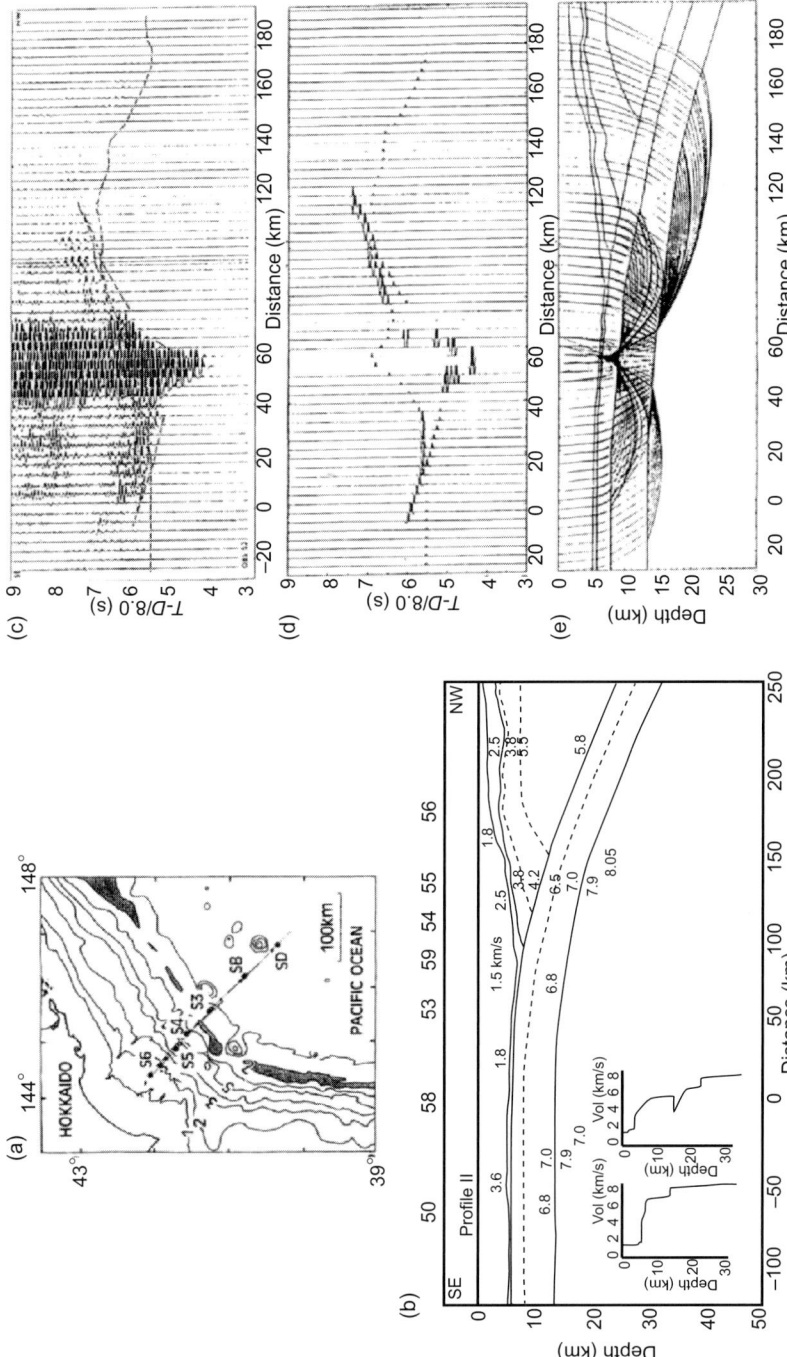

Figure 9.3 A ray modeling study of the subducted oceanic lithosphere east of Japan (Iwasaki *et al.*, 1989). (a) Map of the seismic profile. (b) 2D velocity model. (c) A common receiver gather recorded at OBS site S3. (d) Synthetic seismograms. (e) Raypaths.

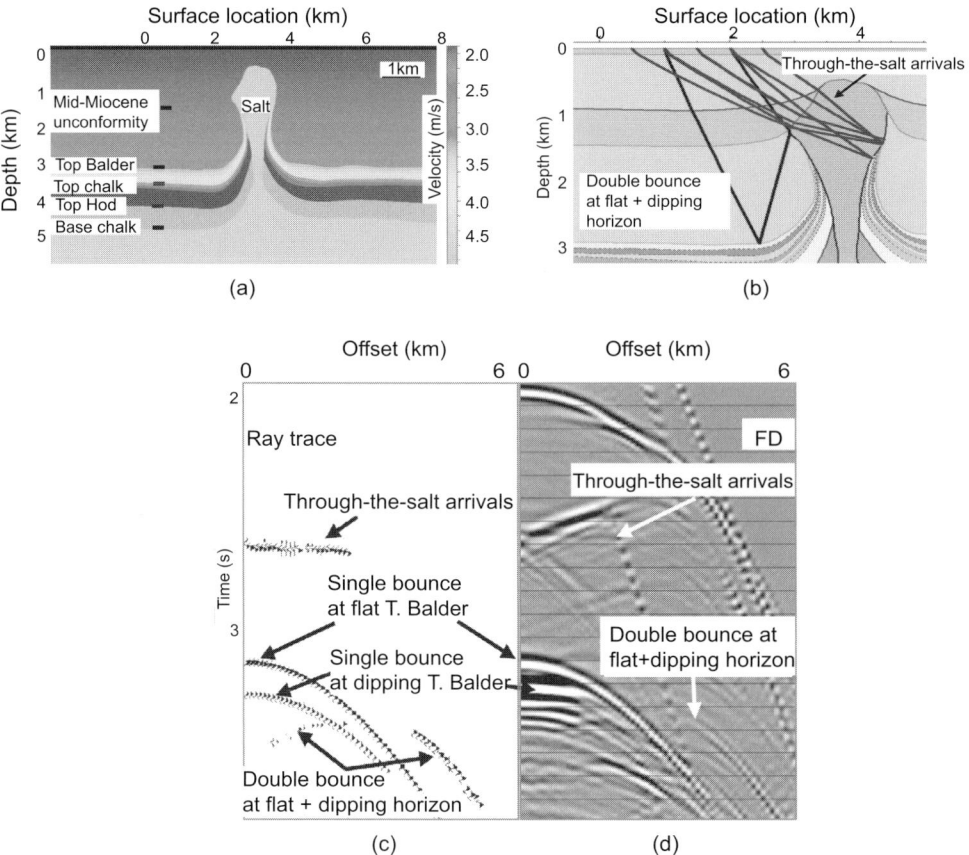

Figure 9.4 A ray modeling study of reflections from the steep salt flank (Jones, 2008). (a) Interval-velocity model. (b) Reflection raypaths through the salt and double bounce at flat and dipping horizons. (c) Ray-traced events on a CMP gather at surface location 1.5 km. (d) Finite-difference (FD) modeled events on the same CMP gather.

9.1.3   Seismic waveform modeling

9.1.3.1   Seismograms and snapshots

Seismic waveform modeling is useful to study subsurface structures and/or seismic sources. A basic understanding of seismic waveform modeling is necessary for everyone involved in seismic data processing because at the core of most seismic imaging and inversion methods there is a forward modeling method based on waveform or ray theories. A graph of seismic waves recorded at a fixed location as a function of time is called a **seismogram**. A graph of seismic waves recorded at a fixed time moment over a model is called a **snapshot**. How we model wave propagation in computers represents how we approximate the seismic waves in the real world.

While the records from a seismic sensor are seismograms, it is usually more difficult to understand events on seismograms than on snapshots because the latter show the wave

Box 9.1 Modeling of multiple reflections in well logs using convolutional modeling

A practical use of the convolutional modeling in exploration seismology is the making of synthetic seismograms using well logs. This is a necessary step for relating reflection events seen on well logs to those seen on seismic profile imageries. An issue of concern is the role of multiple reflections. Although there are various ways to suppress multiple reflections in processing surface seismic data, the resultant images may still contain influences from multiple reflections, especially interbed multiples. Hence there is a need to study the relative amplitude of multiple reflections versus primary reflections in synthetic seismograms based on well log data.

Box 9.1 Figure 1 shows such a study conducted by Schoenberger and Levin (1974). The top two traces are density log and velocity log from one well, which are used to derive the third trace, showing the reflection coefficient. The remaining four traces are synthetic seismograms of the input trace, direct trace, multiple trace, and total trace. Here the input trace includes only primary reflections without transmission losses at interfaces, and the direct trace includes transmission losses and intrabed reflections. Most of the difference between these two traces is due to transmission losses. The multiple trace includes only multiple reflections, and the total trace is a stack of the direct trace and the multiple trace. We can see in this case that the primary reflections have much higher amplitude than the multiples, though the latter is gaining power with increasing recording time.

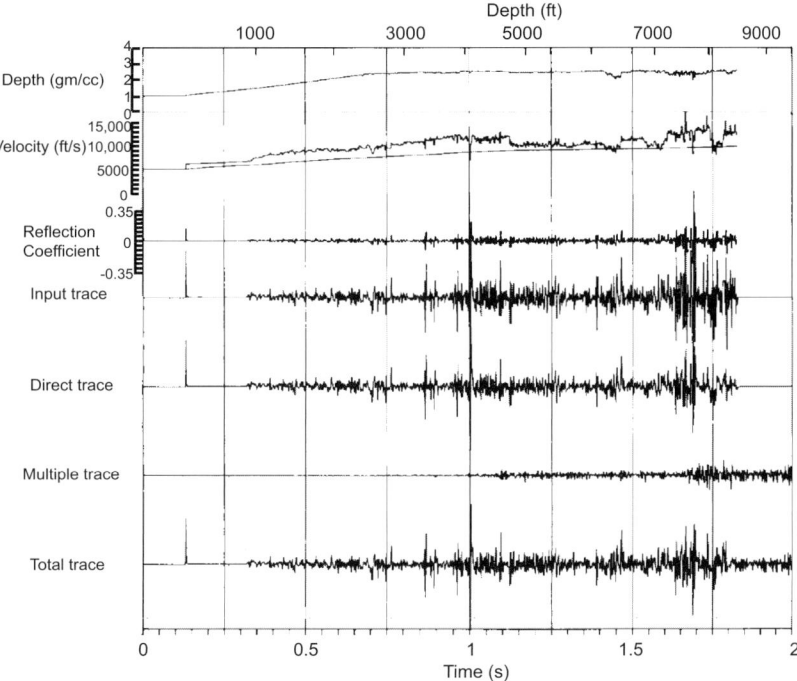

Box 9.1 Figure 1 Density log, velocity log, reflectivity log, and four synthetic seismograms based on convolutional modeling (Schoenberger & Levin, 1974).

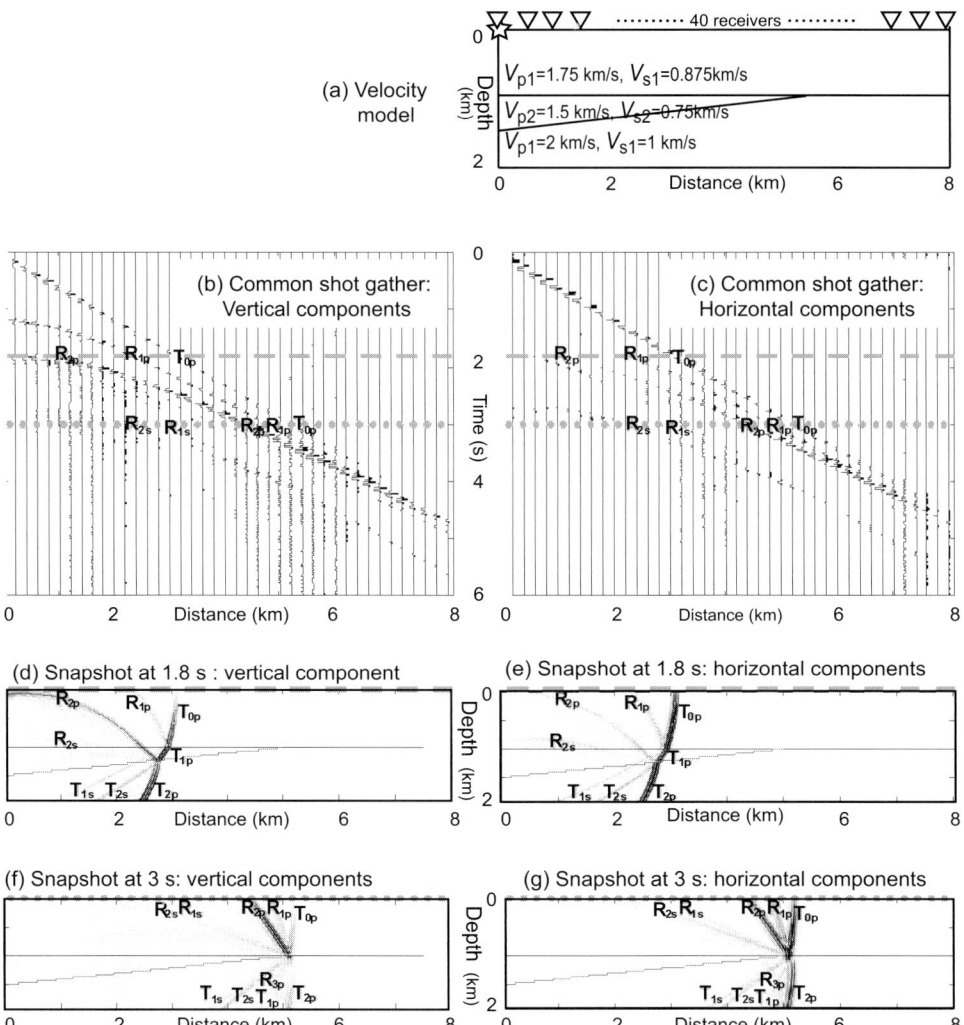

Figure 9.5 (a) A low-velocity wedge model, with a shot (star) and 40 receivers (triangles) along the surface. (b), (c), Modeled common shot gathers of vertical and horizontal components, respectively. (d)–(g), Modeled snapshots at 1.8 s and 3 s after the shot onset time, respectively. Symbols T and R denote the transmitted and reflected waves, respectively, and their subscripts denote order of the interface emitting P or S waves.

propagation history similar to what we see in real life. As an example, Figure 9.5 shows elastic waveform modeling for a simple low-velocity wedge model using a pseudo-spectral method (e.g., Kosloff & Baysal, 1982). In the simple model shown in panel (a), the P-wave velocities are twice as high as the corresponding S-wave velocities. From the modeling, the vertical and horizontal components of a common shot gather of seismograms are shown in panels (b) and (c), respectively; snapshots at 1.8 and 3.0 seconds after the onset of the shot are shown in panels (d) to (g). On each snapshot symbols T and P denote the transmitted and reflected waves, respectively, and their subscripts denote the order of the model interface where the P or S waves are transmitted or reflected. For instance, T_{0p} is the original P-wave transmitted from the shot that is located on the 0th interface, the surface of the model.

Looking at the vertical or horizontal components of the common shot gather, we should be able to identify the direct wave (T_{0p}) and P-wave reflections from the two interfaces, R_{1p} and R_{2p}. On the other hand, it is not easy to identify additional wave types on the seismograms. In contrast, it is much easier to recognize major transmission and reflection waves on the snapshots. It is even better to view a time sequence of snapshots as a movie of the wave propagation history in a model. In the snapshots in panels (d) to (g) of Figure 9.5, since the source is located in the upper-left corner of the model, all reflection waves dip toward the upper-left direction from their reflectors, and all transmitted waves dip toward down-left or upper-right directions from their transmission interfaces. Most primary transmission and reflection waves are denoted on these snapshot panels.

We may visualize a 3D volume to record the modeling result of this 2D model, in which the x-axis denotes distance, the z-axis denotes depth, and the y-axis denotes the recording time. Then each time slice is a snapshot, and each depth slice is a collection of seismograms at the depth. It is then easy to realize that on each snapshot the waves appearing near the model surface should be recorded on the seismograms at the time moment of each snapshot. To illustrate this notion, the gray dashed and dotted lines in the seismograms of panels (b) and (c) denote 1.8 and 3 seconds, respectively, the time moments for the snapshots in panels (d) to (g). There is indeed good agreement between events T_{0p}, R_{1p} and R_{2p} on the gray dashed lines in panels (b) and (c) and the corresponding events along the top surface of snapshots in panels (d) and (e). Equally good agreement is seen between events T_{0p}, R_{1p}, R_{2p}, R_{1s}, and R_{2s} on the gray dotted lines in panels (b) and (c) with these events along the top surface of snapshots in panels (f) and (g). It may not be easy for many people to recognize the reflected S-waves R_{1s} and R_{2s} on the common shot gather seismograms in panels (b) and (c), but the snapshots may be helpful.

Field seismic records are shot gathers or other types of seismograms rather than snapshots. Geophysicists must be able to come up with simple velocity models based on geologic and geophysical information, and then use waveform modeling to assess the validity of the models. Once a reasonably good velocity model is established, seismic migration methods can be carried out to map the reflectors. As discussed in Chapter 7, the first step in seismic migration is downward continuation, which is a process to produce snapshots at depth from the surface-recorded seismograms.

9.1.3.2 Common seismic waveform modeling methods

In practice the propagation of seismic waves is modeled using various versions of wave equations. While analytical solutions of the differential equations may exist for some simple models, nearly all applied geophysicists use numerical methods to carry out seismic waveform modeling. The most common ones are the **finite-difference (FD)**, **finite-element (FE)**, and **pseudo-spectral (PS)** methods. These methods can be designed for both acoustic and elastic applications in models of different dimensions.

By approximating differentiation operations with difference operations, the FD method is the most straightforward and therefore the most popular method of waveform modeling. For example, if q is the variable of a single-variable function, $f(q)$, the first derivative

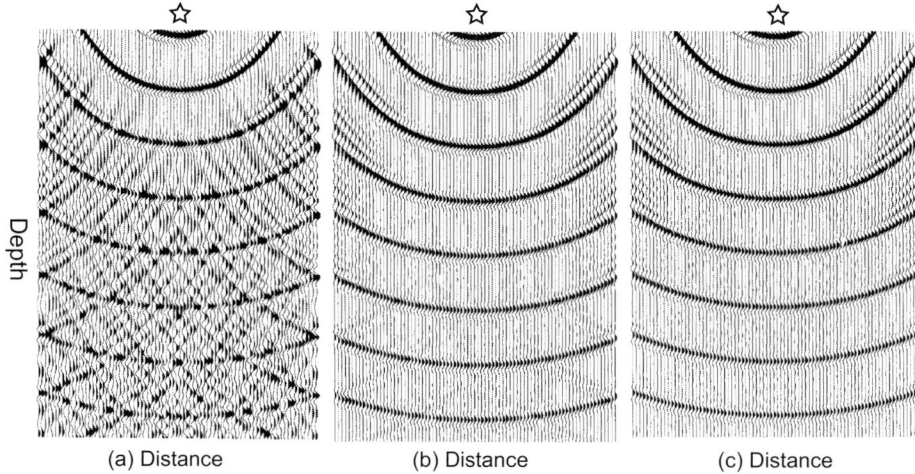

(a) Distance   (b) Distance   (c) Distance

Figure 9.6 Snapshots of three cross-sections showing some absorbing boundary conditions in finite-difference modeling of monochromatic waves from a point source denoted by the stars (Clayton & Engquist, 1980). The left and right model edges are: (a) perfect reflectors; (b) absorbing reflections of high angles of incidence; and (c) using an improved absorbing boundary condition.

of the function with respect to q can be approximated by a forward difference operator

$$D^+ f(q) = [f(q + \Delta q) - f(q)] / \Delta q \qquad (9\text{--}1a)$$

or by a backward difference operator

$$D^- f(q) = [f(q) - f(q - \Delta q)] / \Delta q \qquad (9\text{--}1b)$$

The second derivative can be approximated by a second-order difference operator

$$D^+ D - f(q) = D^- D^+ f(q) = [f(q + \Delta q) - 2f(q) + f(q - \Delta q)] / \Delta q^2 \quad (9\text{--}2)$$

Along the q-axis the odd-ordered difference operators are off-centered by one-half of the sample interval Δq, and the even-ordered difference operators are centered properly.

For developers and users of FD methods, considerable attention is devoted toward three issues: bookkeeping, absorbing boundary conditions, and numerical dispersion. The bookkeeping involves setup and management of various model parameters to facilitate the FD computation. To convert a velocity model in the form of Figure 9.1e into the finite-difference form of Figure 9.1f, for example, we need to map the layered model values into a finite-difference grid and to keep track of the gridded values.

The unwanted reflections from the outer boundaries of the model are annoying artifacts. As shown in panel (a) of Figure 9.6, the intended circular wavefronts from a point source are masked by the unwanted reflections from the left and right edges of the model. Panels (b) and (c) of this figure show the results of applying two types of analytical absorbing boundary conditions. There are also practical ways to damp the unwanted boundary reflections, such as padding arrays of small holes along model edges.

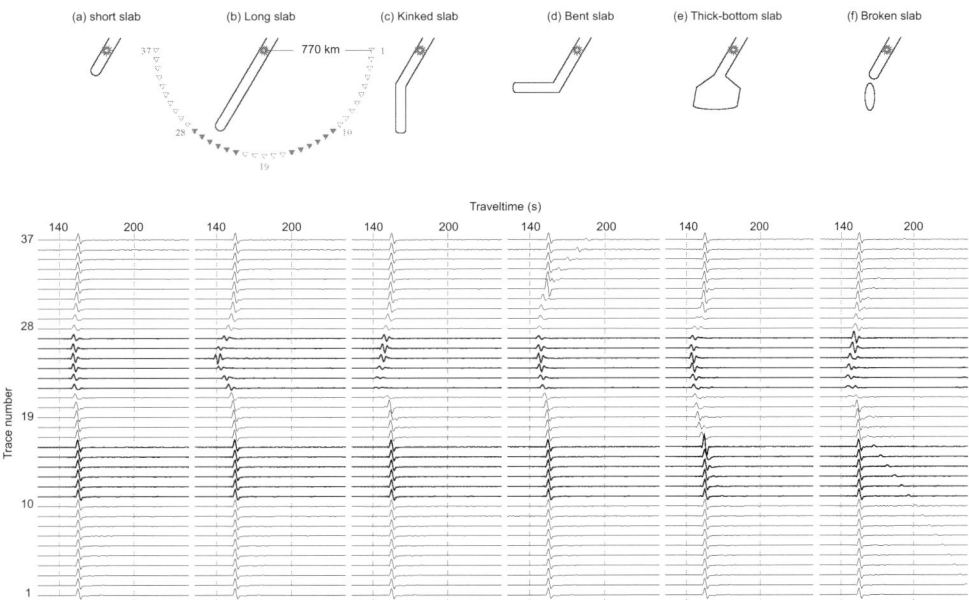

Figure 9.7 Synthetic SH-wave seismograms for six 2D models of subducted lithospheric slabs of different shapes. The receiver positions, as shown in (b), are the same for all six models. The filled triangles in (b) and heavy seismograms denote the take-off angles of mantle P and S waves that may be seen at seismologic stations. (After Zhou and Chen, 1995.)

Numerical dispersion is a computational artifact due to inadequate discretization of the model in simulations like FD waveform modeling. For instance, when insufficiently fine temporal or spatial sampling rates are used, the simulated wavefields will behave in more anomalous ways as the propagating time increases. On the other hand, finer sampling will require larger computational resources of memory and computing time. In practice we may choose to use the coarsest sampling rate that produces a tolerable level of artifacts.

In addition to those due to model boundaries and numerical dispersion, many other artifacts can be produced from seismic waveform modeling. For instance, a careful examination of panel (f) of Figure 9.5 will reveal some ringing waves below and to the right of reflections R_{2s} and R_{1s}. You can find similar ringing waves in other snapshot panels of this figure. These waves are artifacts with respect to the intended signals, the transmitted and reflected waves that are labeled in this figure. They are caused by the step function approximation of the lower dipping interface of the pinchout middle layer of the model. The ringing waves will be reduced if we use a much finer grid step function to approximate the dipping interface. On the other hand, if that interface is intended to be a coarse grid step function, then the corresponding ringing waves will be signal rather than artifacts.

The finite-element or FE method of seismic waveform modeling solves for differential equations of the propagating waves over some spatial elements designed according to the targeted structures. The method is elegant but requires more theoretical devotion. A variant of the FE method is the boundary element (BE) method, which can save much memory by reducing the dimension of the problem by one. Figure 9.7 shows a study of the impacts on SH-wave waveforms due to different morphologic shapes of the subducted lithospheric

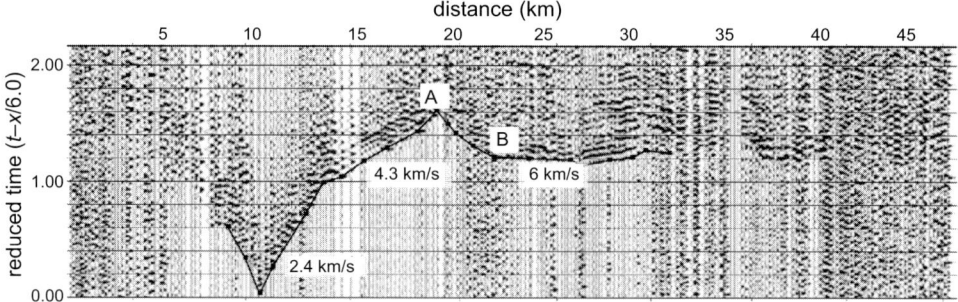

Figure 9.8 A shot gather from a seismic refraction study across the San Gabriel fault (Yan *et al.*, 2005). Traveltimes are reduced by 6 km/s. The first arrival picks are denoted by the black line with apparent velocities at 2.4, 4.3, and 6 km/s for three segments. The AB segment has a negative slope indicating an apparent velocity too large for the crust.

slabs (Zhou & Chen, 1995). One may take such a modeling approach to decipher the possible fates of deeply subducted lithospheric slabs. A limitation of the BE method is its difficulty in handling models with gradient variation of properties.

In recent years the desire for more efficient waveform modeling methods has attracted much attention to the pseudo-spectral or PS method. In contrast to the FD methods which approximate the derivatives by difference equations using local values, a PS method uses global information in the spectral space. The PS method solves differential equations using a combination of time-space and spectral domains. It effectively conducts finite-difference operations in the spectral space. Figure 9.5 is a waveform modeling based on the PS method.

In Figure 9.5 the multiple reflections are not seen clearly, because they have smaller amplitude with respect to the primary transmission and reflection waves in a relatively short time span of wave propagation. As was illustrated in Box 9.1 Figure 1, as time increases from the onset moment of the shot, the amplitude of multiple reflections increases gradually, while the amplitude of primary waves decreases gradually. Observations like these have motivated the development of various ways to ignore multiply scattered waves, such as the Born approximation. In scattering theory and quantum mechanics, the Born approximation applies the perturbation method to scattering by an extended body. It takes the incident field in place of the total field as the driving field at each point in the medium. The Born approximation is accurate if in the medium the scattered field is small in comparison with the incident field.

9.1.3.3   Fitting refracted waveforms for crustal velocity structure

The value of waveform modeling is illustrated here using an example of seismic refraction profiling that is often applied in crustal seismology and near-surface geophysics. This example is based on seismic refraction data recorded over the north central Transverse ranges and adjacent Mohave Desert in California (Yan *et al.*, 2005). Figure 9.8 shows a field shot gather in which the traveltime are reduced by 6 km/s. Though the data are very noisy, we can pick the first arrival waves, and measure their apparent velocities as discussed

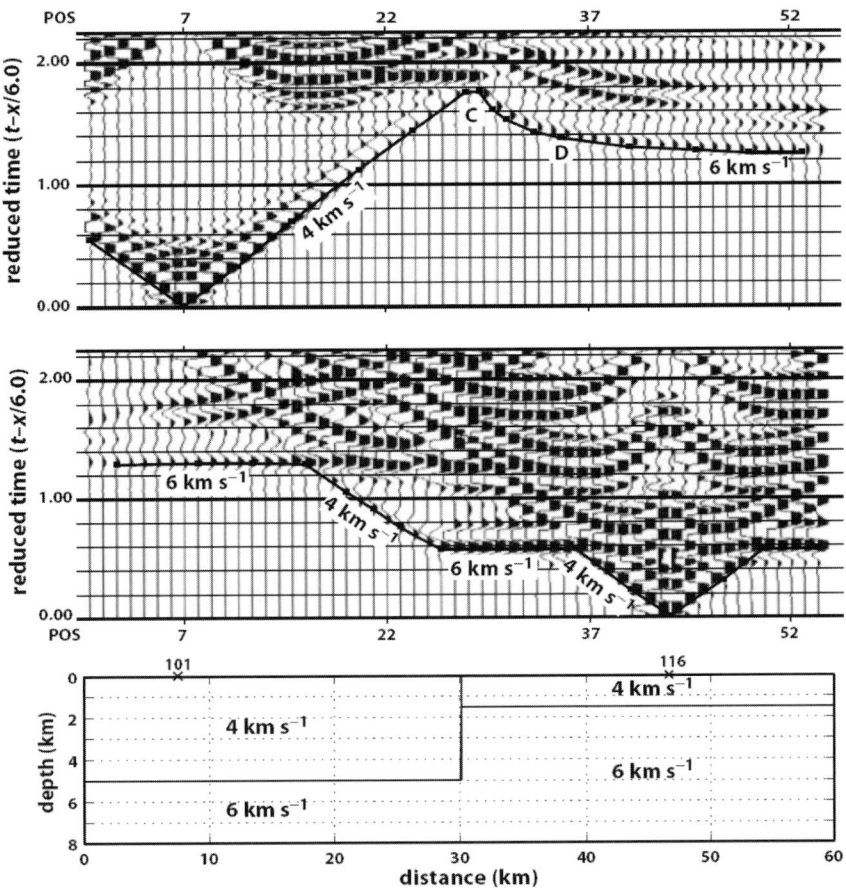

Figure 9.9 Two finite-difference synthetic shot gathers at shot positions 101 and 116 that are marked by ×s on the surface of a 2D velocity model shown in the lower panel (Yan *et al.*, 2005). In the synthetic shot gather of the upper panel, the CD segment of the first arrivals resembles the negative slope seen in the field shot gather shown in the previous figure.

in Section 8.1.4.3. A strange feature in the data is the negative slope of the AB segment which indicates an apparent velocity too large for the crust.

Yan *et al.* (2005) took a waveform modeling approach to compute synthetic shot gathers for a series of 2D crustal velocity models using a finite-difference method, and to improve the models by monitoring the fitness between the field data and model-based waveform predictions of the first arrival waves. For instance, Figure 9.9 shows two synthetic shot gathers calculated for one of the 2D velocity models. In the synthetic shot gather of the upper panel in Figure 9.9, the CD segment of the first arrivals resembles the negative slope of the AB segment of the field data in Figure 9.8.

Figure 9.9 shows a clear correspondence between the lateral positions of the fault in the 2D velocity model and the corresponding changes in the moveout trend of the first arrivals in the two synthetic shot gathers. Such correspondence represents the constraining power of the waveform modeling approach. By requiring a good fit between the first arrival trends

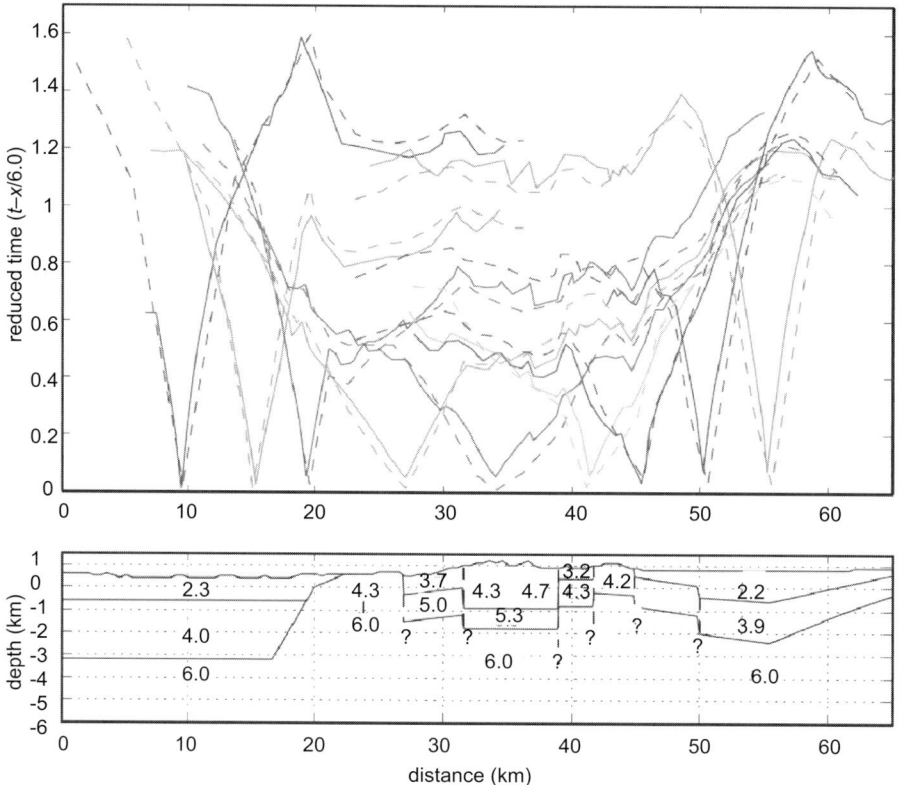

Figure 9.10 (Upper) Comparison of the first arrivals of the field data (solid curves) with synthetic first arrivals (dashed curves) for several shots. (Lower) The 2D velocity model for the synthetic modeling. The fit between the field data and predicted curves is reasonable, especially at far offset.

of the field data and the model predictions of a number of shot gathers, as shown in Figure 9.10, Yan *et al.* (2005) derived a reasonable crustal velocity model for the region, as shown in the lower panel of Figure 9.10.

9.1.4   Seismic physical modeling

Seismic physical modeling involves recording controlled shots in real-world physical models. The main advantage of physical modeling is that it is a true physical experiment with real elements. Physical modeling is useful in cases when the theory, a prerequisite of numerical modeling, is incomplete or unreliable. Since all physical modeling experiments are conducted in the real world rather than in computers, the results include real-world issues such as errors in the model building and acquisition processes as well as phenomena such as seismic wave conversion, scattering, and attenuation. In fact, we may regard all field data acquisitions as physical modeling experiments. In contrast, it is always a concern for numerical modeling whether all of the waves and noises are computed properly. Data acquisition in physical modeling is usually much cheaper than in numerical modeling. The

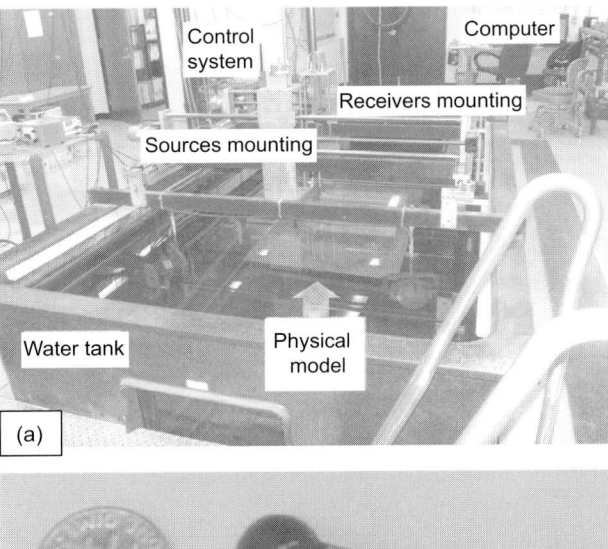

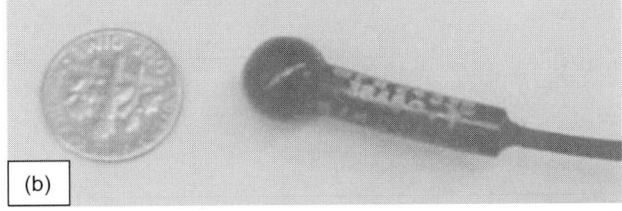

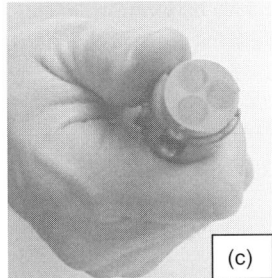

Figure 9.11 Photographs of: (a) the seismic physical modeling lab at University of Houston. (b) A spherical transducer for acoustic experiments. (c) A three-component transducer for elastic experiments.

drawback of physical modeling is that it is much more difficult to make physical models than numerical models. When the model gets complicated, it is not easy to find materials of physical properties matching the study requirements.

Physical models can consist of statistical properties, such as porosity of sandstone, but most physical models are deterministic, using synthetic materials such as rubber and Plexiglas to meet the planned specifications in geometry, velocity, and density values. In seismic physical modeling studies in the laboratory, the spatial sizes of the real structures are scaled down by several thousand times. The spatial down-scaling is compensated by frequency up-scaling using the same scaling factor. If the central frequency of a field dataset is 40 Hz, and we use a scaling factor of 10 000, then the central frequency of the physical experiment will be 400 kHz in the ultrasonic frequency range.

Figure 9.11a is a photograph of the acoustic physical modeling device at the Allied Geophysical Laboratories in University of Houston. The device has produced many useful datasets. During the modeling experiment, the physical model is submersed in a water tank, and the sources and receivers are fixed on the two separated mounting devices whose positions and motions are controlled by a control system through a computer. Under such laboratory conditions most of the sources and receivers are piezoelectric transducers that are able to convert electrical voltage changes into acoustic vibrations or vice versa. Figure 9.11b shows a spherical transducer to emit P waves for acoustic experiments. Figure 9.11c shows a three-component transducer for elastic physical experiments.

Figure 9.12 A CT-scan of a physical model with an interface between white rubber and Plexiglas. Traces of fine brass power were added along the flat parts of the interface, as indicated by two arrows, to enhance its image. The circle highlights a gap in the interface without brass powder.

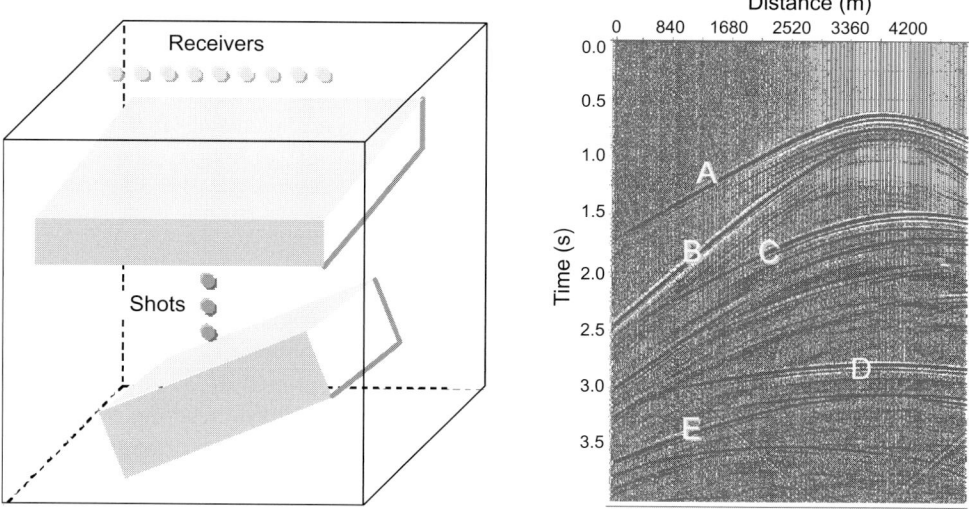

Figure 9.13 A physical modeling example. (Left) Model setup for a VSP experiment. (Right) A common receiver gather. Five events are labeled as A, B, C, D, and E.

While all physical models are produced based on designed parameters, it is often necessary to determine the model geometry and interval velocities after building the model. Figure 9.12 shows a CT scan of a physical model. In this case, two different kinds of Plexiglas were used to make the model, but the two materials have the same appearance on the CT scan. Fortunately, we have placed traces of brass powder along the interface between the two Plexiglas materials, as indicated by the light-colored lines marked with two arrows in Figure 9.12. The circle highlights a ramp in the interface, representing a fault.

In Figure 9.13 the model setup of a VSP survey and the result of a common receiver gather from physical modeling are shown. Following the setup shown in the left panel, the physical model was created by placing two blocks of Plexiglas into a water tank, and using two spherical transducers. A deep transducer placed between the two Plexiglas blocks functioned as shot, and another shallow transducer placed above the two Plexiglas blocks functioned as receiver. The recorded common receiver gather as shown in the right panel resembles some characteristics of real data, such as the ringing of the wavelet due to narrow bandwidth and complexity of the seismic events, though the model is relatively simple. Some of the events are revealed by ray tracing modeling. Event A is the first arrival from the shot to all receivers through the shallow flat block and water layers. Events B and C are reflections from the top and bottom of the deep dipping block layer. Event D is reflection from the bottom of the water tank. Most interestingly, Event E is a prismatic reflection that is bounced twice in the deep dipping block, once from the bottom and another from the lower left facet of the block.

Exercise 9.1

1. Under what conditions may data fitting be superior to model inversion? What are the conditions for the reverse case?

2. How common is it in the Earth that the density is much more slowly varying than the velocity, like that shown in the first two logs in Box 9.1 Figure 1? Given a slowly varying function and a rapidly varying function, which one is easier to invert for and why?

3. In Figure 9.8, if the velocity follows a layer-cake model, what is the value of the velocity corresponding to the AB segment of the first arrivals?

9.2   Data fitting via regression

In this section some basic information is introduced about fitting measured data to a simple function representing the trend of the underlying physics; the process is generally called a **regression**. We will first see some common ways to measure the misfits in the data fitting process. Then, the process of regress is illustrated using a polynomial function. Because we may better manage the misfits in the data fitting by treating them as random variables, we will examine a couple of commonly occurring probabilistic treatments of misfits. The statistical values of the data-fitting misfits can be assessed using a theory called maximum likelihood estimate.

9.2.1   Misfit measures

Data fitting via regression is a statistical approach to minimize the misfit between data and functions representing either general trends or model predictions. Geophysicists seem to spend most of their time fitting predictions of models to data, like those shown in Figures 9.2 and 9.10. The **misfit** measures quantify the fitness between the data and model predictions. One way to improve predictions from the model is to identify some trends in the data, and then explain the causes of the trends by properties of the model. For example, a seismologist may gather arrival times of various seismic waves recorded at stations at different distances and azimuths from the sources; the trends of the traveltimes of seismic waves as functions of source-to-receiver offset and azimuth may be used to constrain the underlying velocity models.

Given a discrete dataset $\{d_i\}$, where the index i is from 1 to N, we may estimate the trend of the data by fitting a function curve to the data. Here fitting means minimizing a predefined error function that describes the difference between the data value $\{d_i\}$ and the corresponding values of the function, $\{p_i\}$. The values of $\{p_i\}$ often represent predictions based on a model. For instance, the predictions may be traveltimes of reflection waves in a velocity model. In an N-dimensional Cartesian space, $\{d_i\}$ and $\{p_i\}$ denote the locations of two points. The misfit between these two points can be measured by the **Euclidean distance**, or the **L$_2$ norm** between the two points:

$$L_2 \ \text{norm} \equiv \left[\sum_{i=1}^{N} (d_i - p_i) \right]^{1/2} \tag{9-3}$$

The term "norm" means a measure of length as defined for a specific mathematical domain. We may generalize the above equation so that, for a positive real value k (not necessarily an integer), the **L$_k$ norm** is defined as

$$L_k \ \text{norm} \equiv \left[\sum_{i=1}^{N} (d_i - p_i)^k \right]^{1/k} \tag{9-4}$$

The concept of such an L_k norm is quite general, and the only requirement is that k be positive. We often refer to $r_i = (d_i - p_i)$ as a **residue**, indicating that it is one of the remains that have not yet been fitted to the model predictions.

The most commonly used norm is the L_2 norm. If the population of all residues $\{r_i\}$ follows a Gaussian distribution, then we can prove that the L_2 norm is the most suitable way to measure the misfit level of the residues. Because the inverse formulation for the L_2 norm exists analytically, it is easy to use the L_2 norm in many applications to minimize misfits. In general, smaller L-norms, such as L_1 or $L_{0.3}$, give nearly equal weight to all residues regardless of their magnitudes, $|r_i|$. In contrast, successively higher L-norms give higher weight to residues of larger magnitudes. In the extreme case, $L_\infty = k \, |r_{max}|$, where $|r_{max}|$ is the maximum magnitude of all residues, and k is the number of the residues of the maximum magnitude. Those residues that have large magnitude and lie away from their peers are called **outliers**. While some studies have tried to suppress them, outliers are often important indicators that some fundamental aspects of the real world have not been accommodated by the models.

9.2.2   Regression

Regression is the process of fitting a function to a discrete dataset. For example, the discrete dataset may be denoted by $\{d_i = d(x_i)\}$, and the prediction function may be a polynomial of x with coefficients $\{c_i\}$:

$$p(x) = c_0 + c_1 x + c_2 x^2 + \cdots \tag{9-5}$$

Here the goal of regression is to determine the coefficients $\{c_i\}$ of the prediction function $p(x)$ that will minimize the residues. It is up to the user to decide how many terms of the polynomial to use for a given problem. If we use a polynomial of order 1, for instance, the function is a straight line with two coefficients, c_0 and c_1; this is a linear regression. If we use a polynomial of order 2, which describes a parabola with three coefficients, then we have a parabolic regression. Once we have decided the order of the polynomial, the number of coefficients to be resolved is fixed. This number of the coefficients is also called the **degree of freedom** of the regression. In general, the degree of freedom for an inversion is the number of unknown model parameters to be determined.

9.2.2.1 Regression as an inverse problem

To solve an inverse problem, the first step is to specify the relationship between model variables and the prediction function by parameterizing the model. In the regression case here, this first step includes a decision on the type and order of polynomials to use. Because most geophysical data are acquired and processed in discrete forms, the model is formulated in a discrete form. Here, we may formulate a $(M-1)$th order polynomial regression with N data points:

$$\begin{bmatrix} d_1 \\ d_2 \\ \vdots \\ d_N \end{bmatrix} = \begin{bmatrix} 1 & X_1 & \cdots & X_1^{M-1} \\ 1 & X_2 & \cdots & X_2^{M-1} \\ \vdots & \vdots & \vdots & \vdots \\ 1 & X_N & \cdots & X_N^{M-1} \end{bmatrix} \begin{bmatrix} c_1 \\ c_2 \\ \vdots \\ c_M \end{bmatrix} \tag{9–6}$$

The above linear system may be written in matrix form

$$\mathbf{d} = \mathbf{Am} \tag{9–7}$$

As we have discussed previously, matrix $\mathbf{A}$ consists of **Fréchet differential kernels** relating the variations in the data to the variations in the model. The element in the ith row and jth column of $\mathbf{A}$ is $a_{ij} = \frac{\partial d_i}{\partial c_j}$. Matrix $\mathbf{A}$ is also referred to as the **kernel matrix** or forward modeling operator.

The next step in an inversion is to determine the parameters of the model vector $\mathbf{m}$. If the inverse of the matrix $\mathbf{A}$ exists, we simply have

$$\mathbf{m} = \mathbf{A}^{-1}\mathbf{d} \tag{9–8}$$

Unfortunately, this formula almost never works in geophysics because the exact inverse $\mathbf{A}^{-1}$ generally does not exist, owing to the fact that $\mathbf{A}$ is usually rank defective. The causes of the situation include:

1. Inaccuracies and inconsistencies in the data vector $\mathbf{b}$;
2. Degrees of freedom of the model $\mathbf{x}$ being poorly constrained by the given data.

Consequently, geophysicists almost always deal with "**probabilistic inversions**" rather than "**exact inversions**". The probabilistic inverse theory assumes imprecise and incomplete data, and proceeds with some assumptions on the statistic behaviors of the data and models. For instance, we may be able to assume that the misfits are mostly due to the noise in the data and follow a Gaussian probability distribution. Then, the least squares solution to be discussed in the next section is the proper answer to the inverse problem; and the solution is

$$\mathbf{m} = (\mathbf{A}^T\mathbf{A})^{-1}\mathbf{A}^T\mathbf{d} \tag{9-9}$$

Let us try this solution for the linear regression case

$$\begin{bmatrix} d_1 \\ d_2 \\ \vdots \\ d_N \end{bmatrix} = \begin{bmatrix} 1 & X_1 \\ 1 & X_2 \\ \vdots & \vdots \\ 1 & X_N \end{bmatrix} \begin{bmatrix} C_1 \\ C_2 \end{bmatrix} \tag{9-10}$$

Forming the matrix products

$$\mathbf{A}^T\mathbf{A} = \begin{bmatrix} N & \sum x_i \\ \sum x_i & \sum x_i{}^2 \end{bmatrix} \tag{9-11}$$

$$\mathbf{A}^T\mathbf{d} = \begin{bmatrix} \sum d_i \\ \sum d_i x_i \end{bmatrix} \tag{9-12}$$

Hence the least squares solution for this case is

$$\mathbf{m} = \begin{bmatrix} c_1 \\ c_2 \end{bmatrix} = \begin{bmatrix} N & \sum X_i \\ \sum X_i & \sum X_i{}^2 \end{bmatrix}^{-1} \begin{bmatrix} \sum d_i \\ \sum d_i x_i \end{bmatrix} \tag{9-13}$$

This means

$$c_1 = \bar{d} - c_2 \bar{x} \tag{9-14a}$$

$$c_2 = l_{dx}/l_{xx} \tag{9-14b}$$

For the above equations, the mean of the data vector is

$$\bar{d} = \frac{1}{N} \sum_{i=1}^{N} d_i \tag{9-15a}$$

The mean of the model vector is

$$\bar{x} = \frac{1}{N} \sum_{i=1}^{N} x_i \tag{9-15b}$$

The auto-covariance coefficient is

$$l_{xx} = \sum_{i=1}^{N} (x_i - \bar{x})^2 \tag{9-15c}$$

And the cross-covariance coefficient is

$$l_{dx} = \sum_{i=1}^{N} (d - \bar{d})(x - \bar{x}) \tag{9-15d}$$

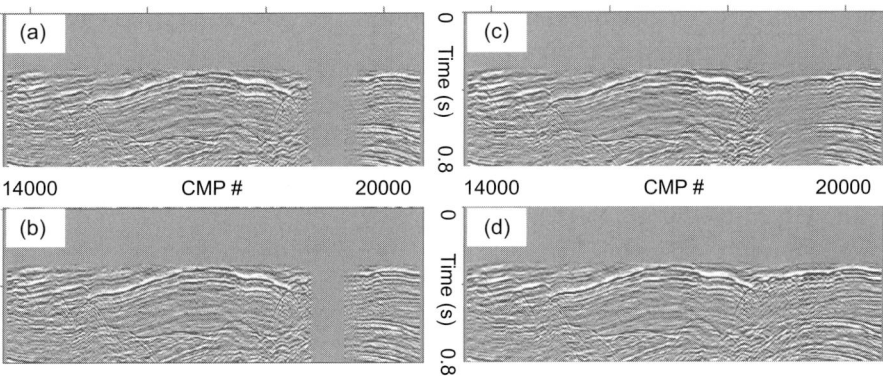

Figure 9.14 Filling a gap on a stack section from the North Sea (Biondi, 2004). (a) and (b) are based on a nearest neighborhood interpolation without and with weights, respectively. (c) and (d) are based on a linear interpolation without and with weights, respectively.

9.2.2.2   Interpolation

As discussed in Section 3.5, **interpolation** is a widely applied method in geophysical data processing, and it is partially based on regression. One application of interpolation is to fill the acquisition gaps in seismic imaging. Figure 9.14 shows examples of filling acquisition gaps (Biondi, 2004) based on nearest neighborhood and linear interpolation schemes.

To see the process of interpolation, take a function $f(x)$ as an example. We may have measured its values at N sites $\{x_n\}$ with measured values $\{f(x_n)\}$. Now we want to interpolate, or predict the function value at a general position x away from the measured sites. This can be done in several steps. First, we choose a set of **basis functions**, like the powers of x in the polynomial expansion of (9–5). We prefer orthogonal basis functions like that shown in Section 3.5. Then the function is expanded into the basis functions $\{\psi_k(x)\}$:

$$f(x) = \sum_{k \in K} c_k \psi_k(x) \tag{9–16}$$

where c_k are the coefficients. Second, we determine the coefficients using the measured values through a regression or inversion process. Finally, we can predict the function values using (9–16) and the determined coefficients.

Interpolation with convolutional bases is an interesting case when the basis function is in form of a function of $(x - k)$:

$$\psi_k(x) = \beta(x - k) \tag{9–17}$$

Hence the original function is a convolution between the coefficient and the basis function:

$$f(x) = \sum_{k \in K} c_k \beta(x - k) \tag{9–18}$$

Now when $x = n$, we have

$$f(n) = \sum_{k \in K} c_k \beta(n - k) \tag{9–19}$$

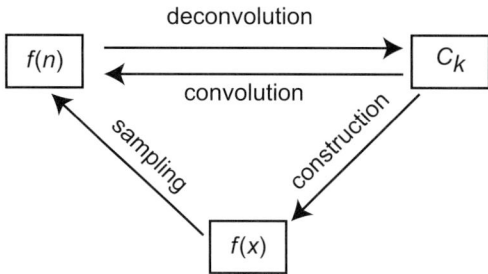

Figure 9.15 A schematic diagram illustrating relationship among different variables for interpolation with convolutional bases (Fomel, 2001).

which is a discrete convolution. We can invert the above equation for the coefficients c_k using the measured values of $f(n)$. This interpolation process is depicted in Figure 9.15 by Fomel (2001).

9.2.3   Probabilistic distributions of misfits

When fitting data of large samples, we often treat the misfits as random variables so that their behavior can be analyzed using statistics. Among many probabilistic distributions seen in geophysical studies, the Gaussian or normal distribution is the most common. We will discuss the Gaussian distribution and the double exponential distribution here. The latter is a good characterization of the residues of many geophysical inverse applications.

9.2.3.1   Gaussian distribution

For each random variable, its chance of occurring is quantified by its **probability density function (PDF)** over its value range. Tossing a dice, for instance, may result in one of the integers from 1 to 6 shown on the top-facing side of the dice. Here the value range consists of six integers from 1 to 6, and the PDF function is 1/6 for each of the six integers. The PDF of such a random variable follows a **uniform distribution** with a boxcar shape.

Now suppose we have a number of random variables $\{\alpha_i\}$, each with a boxcar-shaped PDF. The PDF of combining any two such variables is the convolution of two individual boxcar-shaped PDFs

$$P(\alpha_1 + \alpha_2) = P(\alpha_1) * P(\alpha_2) \tag{9–20}$$

and its shape is a symmetric triangle. When we add more random variables of uniform distributions, the combined PDF becomes

$$P(\alpha_1 + \alpha_2 + \alpha_3 + \cdots)$$
$$= P(\alpha_1) * P(\alpha_2) * P(\alpha_3) * \cdots => \text{Gaussian distribution} \tag{9–21}$$

The above notion is a basic form of the **central limit theorem** that "linear operations on random variables tend to produce a Gaussian distribution". This theorem reveals the

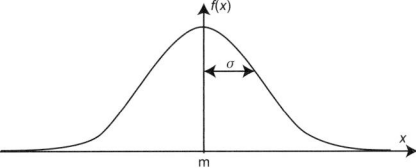

Figure 9.16 Sketch of the PDF of a Gaussian distribution.

fundamental reason that the Gaussian distribution is widely applicable. The Gaussian distribution function of a variable x is

$$f(x) = \frac{1}{\sqrt{2\pi}\sigma} \exp[-(x-m)^2/\sigma^2] \tag{9-22}$$

where m is the mean and σ^2 is the variance. It is graphically shown in Figure 9.16.

For two random variables x and y following a Gaussian distribution, if they are independent or uncorrelated with each other, their combined PDF is a Gaussian distribution in 2D:

$$f(x,y) = \frac{1}{2\pi\sigma_x\sigma_y} \exp[-(x-\bar{x})^2/(2\sigma_x^2) - (y-\bar{y})^2(2\sigma_y^2)] \tag{9-23}$$

If x and y are correlated, what happens to the PDF? Note that when x and y are uncorrelated the exponent of the PDF is

$$-(x-\bar{x})^2/(2\sigma_{x^2}) + (y-\bar{y})^2/(2\sigma_{y^2}) = -\frac{1}{2}\gamma^T \begin{pmatrix} 1/\sigma_x^2 & 0 \\ 0 & 1/\sigma_y^2 \end{pmatrix} \gamma \tag{9-24}$$

where $\gamma = (x-\bar{x}, y-\bar{y})^T$. Now the two random variables are correlated, we may expect that the new exponent is

$$-\frac{1}{2}\gamma^T \mathbf{C}^{-1}\gamma \tag{9-25}$$

where $\mathbf{C}$ is the covariance matrix between the two random vectors. Then the new PDF function $f(x,y)$ is

$$f(x,y) = \frac{1}{2\pi\sqrt{|\mathbf{C}|}} \exp\left[-\frac{1}{2}\gamma^T \mathbf{C}^{-1}\gamma\right] \tag{9-26}$$

The covariance matrix $\mathbf{C}$ is a positive definite matrix. We can decompose it using eigenvalue–eigenvector decomposition (see Section 9.4.1):

$$\mathbf{C} = \mathbf{U}^T \Lambda \mathbf{U} \tag{9-27}$$

where $\mathbf{U}$ is an orthogonal matrix (i.e., $\mathbf{U}^T\mathbf{U} = \mathbf{U}\mathbf{U}^T = \mathbf{I}$) and

$$\Lambda = \begin{pmatrix} \lambda_1 & 0 \\ 0 & \lambda_2 \end{pmatrix} \tag{9-28}$$

with $\lambda_1, \lambda_2 > 0$. Now, from

$$\gamma^T \mathbf{C}^{-1}\gamma = \gamma^T \mathbf{U}^T \Lambda^{-1} \mathbf{U}\gamma = (\mathbf{U}\gamma)^T \Lambda^{-1}(\mathbf{U}\gamma)$$

we have

$$f(x, y) = \frac{1}{2\pi\sqrt{|\mathbf{C}|}} \exp[-(\mathbf{U}\gamma)^{\mathrm{T}}\mathbf{\Lambda}^{-1}(\mathbf{U}\gamma)/2] \tag{9-29}$$

The new variables in $(\mathbf{U}\gamma)$ are independent, or uncorrelated, with each other. The Gaussian distribution is useful for methods such as the least squares, which minimizes the L_2 norm.

9.2.3.2 Exponential distribution

For many processing methods in geophysics we need to use the L_1 norm. The corresponding PDF is closely related to the exponential distributions. For the exponential distribution of one random variable x, its PDF is given by

$$f(x) = \frac{1}{2\sigma} \exp\left[-|x - \bar{x}|/\sigma\right] \tag{9-30}$$

This function is sketched in Figure 9.17.

For n random variables following the exponential distribution, if these variables are independent from each other, then their combined PDF is

$$f(x_i) = \frac{1}{2^n \sigma^n} \exp\left[-\sum_{i=1}^{n} |x_i - \bar{x}_i|/\sigma\right] \tag{9-31}$$

When we compare the PDFs of Gaussian versus exponential distributions, we see that the two distributions differ mostly when the variable x is within a short distance from the mean. The exponential distribution is like overlapping the Gaussian distribution with extra data around the mean value.

One well known case of the exponential distribution is the distribution of the errors from picking the arrival times of phase arrivals in global seismologic networks. Figure 8.9 compared the observed arrival times compiled by the International Seismologic Centre with predictions from the JB model. When we plot the misfits between such observed and predicted traveltimes, we find that the misfit distribution is similar to that shown in Figure 9.17, with x denoting the traveltime misfit, and $f(x)$ denoting the number of picks for each misfit value. What are the causes for such behavior of the misfit distribution?

One explanation is that, because the global traveltimes of seismic phases can be explained very well by a 1D depth-varying global model like the JB model or PREM model, some operators may pick the phase arrivals with the guidance of traveltime tables based on a 1D global model. It is understandable that we may need such guidance to identify a wide time window for a careful hand picking. However, it would be terrible to use such theoretical

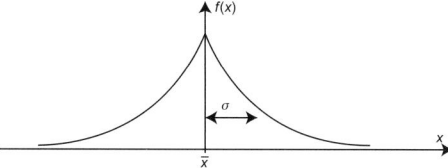

Figure 9.17 Sketch of the PDF of an exponential distribution.

guidance to choose the actual wave train to pick the arrivals. This issue will be addressed again in one of the questions in Exercise 9.3.

9.2.4   Maximum likelihood estimate

In practice, different PDFs will result in different forms of common statistic parameters such as mean and variance. The maximum likelihood estimate is a useful way to find statistical values based on the given PDF; hence the key issue is to assess the PDF. Once the PDF is found, we can find the formula for each statistical parameter by minimizing the variation of the PDF with respect to the parameter.

Suppose we have N samples from a Gaussian distribution with a mean m and a variance σ^2. The probability of these N observed values $\{x_i\}$ $(i = 1, 2, \ldots, N)$ is

$$P = \frac{1}{(2\pi)^{N/2} \sigma^N} \exp\left[-\frac{1}{2\sigma^2} \sum_{i=1}^{n} |x_i - m|^2 \right] \qquad (9\text{--}32)$$

We can estimate the mean and the variance of the sampled dataset by maximizing the above probability function. Define a **likelihood function** $L = \ln P$:

$$L = -\frac{N}{2} \ln(2\pi) - N \ln \sigma - \frac{1}{2\sigma^2} \sum_{i=1}^{n} |x_i - m|^2 \qquad (9\text{--}33)$$

Now letting $\frac{\partial L}{\partial m} = 0$, we have

$$\frac{\partial L}{\partial m} = \frac{1}{\sigma^2} \sum_{i=1}^{n} |x_i - m| = 0 \qquad (9\text{--}34a)$$

or

$$m = \frac{1}{N} \sum_{i=1}^{n} x_i \qquad (9\text{--}34b)$$

This is the conventional definition of the average or mean. In addition, from $\frac{\partial L}{\partial \sigma} = 0$, we have

$$\frac{\partial L}{\partial \sigma} = -N/\sigma + \frac{1}{\sigma^3} \sum_{i=1}^{n} |x_i - m|^2 = 0 \qquad (9\text{--}35a)$$

Therefore,

$$\sigma^2 = \frac{1}{N} \sum_{i=1}^{n} |x_i - m|^2 \qquad (9\text{--}35b)$$

This is the conventional form of variance. Hence, the conventional forms of mean and variance assume that the PDF follows a Gaussian distribution.

Now let us examine the maximum likelihood estimates for an exponential distribution. Suppose we have n observations sampled from an exponential distribution, and we want to estimate the mean and the variance. This time based on (9–31) we find the likelihood function L as

$$L = \ln f(x_i) = -n \ln 2 - n \ln \sigma - \sum_{i=1}^{n} |x_i - m| / \sigma]$$ (9–36)

Note that here we have used m as the mean for all variables. Setting $\frac{\partial L}{\partial m} = 0$, we have

$$\frac{\partial L}{\partial m} = \frac{2}{\sigma} \sum_{i=1}^{n} \text{sign}(x_i - m) = 0$$ (9–37)

where $\text{sign}(x)$ is the sign function. The above formula means that

$$\text{if } n \text{ is odd, } n = (2p + 1), m = x_{p+1}$$ (9–38a)

$$\text{and if } n \text{ is even, } n = (2p)m \text{ is any value between } x_p \text{ and } x_{p+1}.$$ (9–38b)

Next, differentiating the likelihood function with respect to the variance

$$\frac{\partial L}{\partial \sigma} = -n/\sigma + \frac{1}{\sigma^2} \sum_{i=1}^{n} |x_i - m| = 0$$ (9–39)

hence

$$\sigma = \frac{1}{n} \sum_{i=1}^{n} |x_i - m|$$

Then the variance is

$$\sigma = \frac{1}{n^2} \left(\sum_{i=1}^{n} |x_i - m| \right)^2$$ (9–40)

Exercise 9.2

1. Why is the definition of length measure or norm an important issue for inverse theory? How does the issue connect with real signal and noise properties?

2. Discuss the benefits of expressing a function as the combination of basis functions like that shown in (9–16). What are the preferred properties of the basis functions?

3. An experiment yielded the following measures of a function $v(t)$:

t:	0	2	3	3	6	8	8	8	10
$v(t)$:	5	2	3	1	2	5	3	4	6

(a) Graph the data;

(b) Write the forward system of a linear fit, and its L_2 inverse system;

(c) If we want to force the linear trend to go through a point (t_0, v_0), what would be the forward system?

(d) Following (c) and for $t_0 = 0.0$ and $v_0 = 3.0$, find the L_2 inverse system;

(e) Solve for the line in (d). Which data point is the largest outlier?

9.3   Least squares inversion

Since deriving images of the subsurface is a common goal of seismic data processing, seismic inversion is synonymous with the process of inverse imaging. The inverse approach has the advantage of objectively determining the model properties based on the given data and model parameterization. A common approach of inversion, as described in Section 8.4.1.2, is to linearize the relationship between the variation of model variables and variation of data variables, and to invert for model variables. For a number of geophysical applications such as inverse filtering and tomographic velocity analysis, seismic inversion is a preferred method owing to its objectiveness in obtaining the solutions.

9.3.1   The least squares (LS) solution

The least squares (LS) method is a traditional linear inversion approach that works by minimizing the L_2 norm of the predefined misfit or error function. This implies that we have assumed that the misfit errors follow a Gaussian distribution. Previously in Section 6.1.2, we have seen applications of the LS method in deconvolution. The LS method is a widely used inverse and deserves special attention.

9.3.1.1   Derivation of the least squares solution

Let us write (9–7) using $\mathbf{x}$ as the model vector

$$\mathbf{d} = \mathbf{Ax} \qquad (9\text{--}7')$$

From what we learned in the previous section, the square power of the L_2 norm of misfits (9–3) for the above system is

$$E = (\mathbf{d} - \mathbf{Ax})^\mathsf{T}(\mathbf{d} - \mathbf{Ax}) = \sum_i \left(d_i - \sum_j a_{ij}x_j \right)^2 \qquad (9\text{--}41)$$

where E stands for the misfit error function or objective function, which is a scalar that we want to minimize. We can express E in different forms as

$$\begin{aligned}
E &= (\mathbf{d} - \mathbf{Ax})^\mathsf{T}(\mathbf{d} - \mathbf{Ax}) \\
&= \sum_i \left(d_i - \sum_j a_{ij}x_j \right)\left(d_i - \sum_k a_{ik}x_k \right) \qquad (9\text{--}42) \\
&= \sum_i d_i^2 - 2\sum_i d_i \sum_j a_{ij}x_j + \sum_j \sum_k x_j x_k \sum_i a_{ij}a_{ik}
\end{aligned}$$

Since the coefficients $\{a_{ij}\}$ and $\{d_i\}$ are given, the LS misfit error E is a function of the unknown model vector $\{x_j\}$. Hence, our objective is to find a set of $\mathbf{x}$ that will minimize E. Mathematically, the minimization can be achieved by setting the differentiation of E with respect to the model $\mathbf{x}$ to zero: $\frac{\partial E}{\partial x_m} \to 0$, where $m = 1, 2, \ldots, M$, and M is the dimension of $\mathbf{x}$. Notice here $\frac{\partial x_j}{\partial x_m} = \delta_{jm}$, an assumption that the model parameters are independent of each other. Thus

$$\frac{\partial E}{\partial x_m} = -2 \sum_i d_i \sum_j a_{ij} \delta_{jm} + \sum_j \sum_k (\delta_{km} x_j + \delta_{jm} x_k) \sum_i a_{ij} a_{ik}$$

$$= -2 \sum_i d_i a_{im} + 2 \sum_k x_k \sum_i a_{im} a_{ik} = 0 \qquad (9\text{--}43)$$

The above equation can be written in vector form as

$$2\mathbf{A}^{\mathrm{T}}\mathbf{A}\mathbf{x} - 2\mathbf{A}^{\mathrm{T}}\mathbf{d} = 0 \qquad (9\text{--}44)$$

Another way to derive the above equation is to differentiate the matrix form of (9–41) with respect to the model variable vector

$$\frac{\partial}{\partial \mathbf{x}} E = \frac{\partial}{\partial \mathbf{x}}[(\mathbf{d} - \mathbf{A}\mathbf{x})^{\mathrm{T}}(\mathbf{d} - \mathbf{A}\mathbf{x})]$$

$$= \frac{\partial}{\partial \mathbf{x}}[(\mathbf{d}^{\mathrm{T}} - \mathbf{x}^{\mathrm{T}}\mathbf{A}^{\mathrm{T}})^{\mathrm{T}}(\mathbf{d} - \mathbf{A}\mathbf{x})]$$

$$= \frac{\partial}{\partial \mathbf{x}}(\mathbf{d}^{\mathrm{T}}\mathbf{d} - \mathbf{d}^{\mathrm{T}}\mathbf{A}\mathbf{x} - \mathbf{x}^{\mathrm{T}}\mathbf{A}^{\mathrm{T}}\mathbf{d} + \mathbf{x}^{\mathrm{T}}\mathbf{A}^{\mathrm{T}}\mathbf{A}\mathbf{x})$$

Since E is a scalar, all terms in the above are scalars, hence $\mathbf{d}^{\mathrm{T}}\mathbf{A}\mathbf{x} = \mathbf{x}^{\mathrm{T}}\mathbf{A}^{\mathrm{T}}\mathbf{d}$. Thus

$$\frac{\partial}{\partial \mathbf{x}} E = \frac{\partial}{\partial \mathbf{x}}(\mathbf{d}^{\mathrm{T}}\mathbf{d} - 2\mathbf{x}^{\mathrm{T}}\mathbf{A}^{\mathrm{T}}\mathbf{d} - \mathbf{x}^{\mathrm{T}}\mathbf{A}^{\mathrm{T}}\mathbf{A}\mathbf{x})$$

$$= -2\left(\frac{\partial}{\partial \mathbf{x}}\mathbf{x}^{\mathrm{T}}\right)\mathbf{A}^{\mathrm{T}}\mathbf{d} + \left(\frac{\partial}{\partial \mathbf{x}}\mathbf{x}^{\mathrm{T}}\right)\mathbf{A}^{\mathrm{T}}\mathbf{A}\mathbf{x} + \mathbf{x}^{\mathrm{T}}\mathbf{A}^{\mathrm{T}}\mathbf{A}\left(\frac{\partial}{\partial \mathbf{x}}\mathbf{x}\right)$$

By the same scalar argument, the last two terms of the above equation equal, that is

$$\frac{\partial}{\partial \mathbf{x}} E = -2\left(\frac{\partial}{\partial \mathbf{x}}\mathbf{x}^{\mathrm{T}}\right)\mathbf{A}^{\mathrm{T}}\mathbf{d} + 2\left(\frac{\partial}{\partial \mathbf{x}}\mathbf{x}^{\mathrm{T}}\right)\mathbf{A}^{\mathrm{T}}\mathbf{A}\mathbf{x} = 0$$

This verifies (9–44), which means

$$\mathbf{A}^{\mathrm{T}}\mathbf{A}\mathbf{x} = \mathbf{A}^{\mathrm{T}}\mathbf{d} \qquad (9\text{--}45)$$

Even though this equation is the same as if we left-multiply $\mathbf{A}^{\mathrm{T}}$ to both sides of (9–7′), we have derived the above relation through a process of minimizing the L_2 norm. Therefore

$$\mathbf{x}_{\mathrm{LS}} = (\mathbf{A}^{\mathrm{T}}\mathbf{A})^{-1}\mathbf{A}^{\mathrm{T}}\mathbf{d} \qquad (9\text{--}46)$$

is the LS inversion, or LS inverse solution.

Here $(\mathbf{A}^{\mathrm{T}}\mathbf{A})$ is a symmetric matrix of dimension of the model space. In most cases the size of the model space M is much smaller than the size of the data space N. For example, in seismic tomography N is usually tens or hundreds times M. Therefore, the form of the

LS solution in these cases uses a much smaller computer memory than that for the straight inverse $\mathbf{x} = \mathbf{A}^{-1}\mathbf{d}$, making many inverse applications feasible.

9.3.1.2 Determinacy of the LS solution

The LS solution (9–46) regards misfits as noise, and assumes that the noise distribution follows a Gaussian distribution. It is one choice among the solutions for inverting the linear system (9–7) when there is no exact inverse for $\mathbf{A}$. Of course, if $\mathbf{A}$ has a unique inverse, then $(\mathbf{A}^T\mathbf{A})^{-1}\mathbf{A}^T = \mathbf{A}^{-1}$. When we use LS inversion, it implies that the exact inverse does not exist, hence a unique inverse solution is not guaranteed. This gives rise to the topic of the determinacy of inverse solutions.

The determinacy of inverse solutions depends on data coverage, data quality, and model parameterization. Common practical situations include choosing among different model parameterizations for a given dataset, or when the model parameterization has been decided and we need to design a proper data acquisition scheme. Analysis of the determinacy of the system helps us to reach a reasonable decision in these situations.

Similar to the terms used for discrete Fourier transforms in Section 3.2.2, when there are not enough data measures to constrain all model parameters uniquely, the inverse problem is **under-determined**. This situation happens when the kernel matrix in Equation (9–7) has more rows than columns. In contrast, when the number of data constraints is greater than the model parameter, the model parameters cannot satisfy all the data constraints, and then the system is **over-determined**. An **even-determined** system is a special case when the number of data constraints is equal to the number of model parameters. The **mixed-determined** systems belong to a subset of the under-determined class when parts of the model parameters are well constrained. The rest of the under-determined systems, where the model parameters are totally unconstrained, are purely under-determined.

The LS solution is unique only for over-determined or even-determined systems. However, for most applications in geophysics this does not apply. In many practical cases, common intuition is insufficient to sense the determinacy of the system, and under-determinacy will result in various artifacts like those discussed in Section 8.5.1, although the fitting error could be zero. Detection of the inversion artifacts is not a simple matter in most cases. We need to examine the data resolution matrix and model resolution matrix in order to verify the artifacts.

There are commonly two ways to conduct inversions of under-determined of mixed-determined systems. We may re-partition the model space so that we can conduct the inversion in a subspace where the system is over-determined. Alternatively, we may use *a priori* information to constrain the inversion. We will examine those situations in more detail later. Box 9.2 illustrates the application of LS method in seismic migration.

9.3.2 Damped and weighted LS inversions

In practice the forward linear system (9–7) is often ill-conditioned owing to poor data coverage; then it will be difficult to conduct the inversion for meaningful solutions. A practical aid for the inversion is applying constraints based on *a priori* information. Here, damped and weighted inversions are introduced as the two traditional ways of applying

Box 9.2 The least squares (LS) migration and an approximate solution

The LS migration is a direct application of the LS inversion to seismic migration, and its main benefit is obtaining adequate amplitudes of migrated images, which is important to seismic attribute analyses. Following the discussion in Section 7.1.1, if we regard forward modeling as a mapping from the model vector $\mathbf{m}$ to data vector $\mathbf{d}$ via equation (9–7), then migration is the adjoint of the modeling operator $\mathbf{A}$ (Claerbout, 1985)

$$\mathbf{m}_{\text{mig}} = \mathbf{A}^{\text{T}}\mathbf{d} \tag{7–3'}$$

where $\mathbf{m}_{\text{mig}}$ is the migrated model. In contrast, the LS migrated model $\mathbf{m}_{\text{LS}}$ is just the LS solution (9–46), $\mathbf{m}_{\text{LS}} = (\mathbf{A}^{\text{T}}\mathbf{A})^{-1}\mathbf{A}^{\text{T}}\mathbf{d}$. This means

$$\mathbf{m}_{\text{LS}} = \mathbf{B}_0\mathbf{m}_{\text{mig}} \tag{9–47}$$

where $\mathbf{B}_0 = (\mathbf{A}^{\text{T}}\mathbf{A})^{-1}$. However, LS migration is an expensive proposition because the dimensions of the data and model vectors are very large.

Guitton (2004) proposed a low-cost approximation to LS migration via adaptive filtering. The idea is that the right-hand side of (9–47) may be viewed as a convolution. His procedure is listed in the following, and it is illustrated in Box 9.2 Figure 1 using a dataset based on the Marmousi model.

1. Migrate the original data to yield $\mathbf{m}_{\text{mig}}$, as shown in panel (a).
2. Forward modeling from the migrated model to create a new dataset $\mathbf{d}_1$ as in

$$\mathbf{d}_1 = \mathbf{A}\mathbf{m}_{\text{mig}} \tag{9–48}$$

3. Conduct a second migration using $\mathbf{d}_1$

$$\mathbf{m}_2 = \mathbf{A}^{\text{T}}\mathbf{d}_1 = \mathbf{A}^{\text{T}}\mathbf{A}\mathbf{m}_{\text{mig}} \tag{9–49}$$

which is shown in panel (b). The above equation means

$$\mathbf{m}_{\text{mig}} = \mathbf{B}_0\mathbf{m}_2 \tag{9–50}$$

The similarity between equations (9–47) and (9–50) has motivated this method, and the key is to solve for $\mathbf{B}_0$.

4. Form a matching filter system with respect to (9–50)

$$\mathbf{m}_{\text{mig}} = \mathbf{M}_2\mathbf{b} \tag{9–51}$$

where $\mathbf{M}_2$ is the matrix form of the non-stationary convolution with $\mathbf{m}_2$; and solve for the adaptive filters $\mathbf{b}$ using a minimum-norm LS solution of (9–51). Panel (c) shows the solution of the adaptive filters from the two migrated models in panels (a) and (b).

5. Approximating $\mathbf{B}_0$ with a bank of the adaptive filters $\mathbf{b}$, convolve this $\mathbf{B}_0$ with $\mathbf{m}_{\text{mig}}$ to obtain the approximated LS solution according to (9–47). Panel (d) is such a solution.

In the figure, the four migration results from Guitton (2004) in panels (a), (b), (d) and (e) differ mostly in amplitude rather than geometry; this underscores the notion that LS migration is an effort to calibrate the amplitude distortion by the migration operator. If the

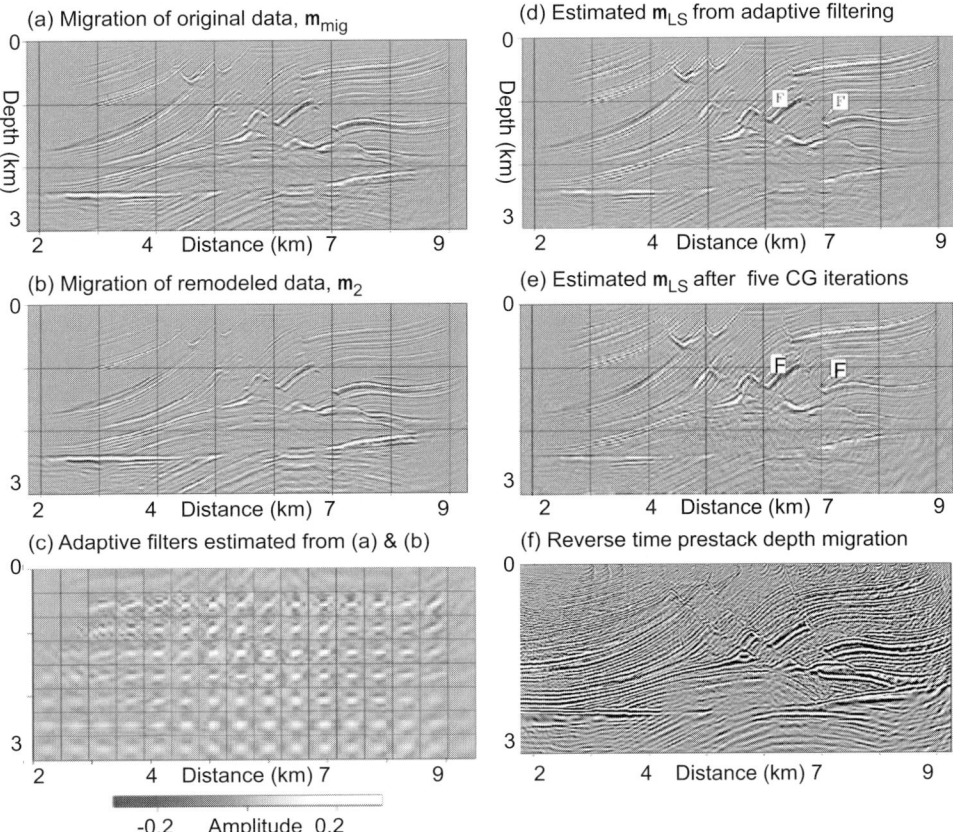

(a) Migration of original data, m_{mig}

(d) Estimated m_{LS} from adaptive filtering

(b) Migration of remodeled data, m_2

(e) Estimated m_{LS} after five CG iterations

(c) Adaptive filters estimated from (a) & (b)

(f) Reverse time prestack depth migration

-0.2   Amplitude 0.2

Box 9.2 Figure 1 Illustration of an approximated LS migration via adaptive filters (Guitton, 2004). See captions of the panels and text for explanation. In (d) and (e) letters F denote fault planes. In (e) CG stands for conjugate gradient inversion. Panel (f) is the result of a reverse time pre-stack depth migration (Youn & Zhou, 2001).

migration operator has a biased effect on amplitude, this effect will be amplified by the second migration in panel (b) relative to the first migration in panel (a). Their difference is the base for the adaptive filters in panel (c) to estimate the adjustment for this biased effect, and the resulting migration in panel (d) should have suppressed the effect. Here the estimated LS migration from the adaptive filtering in (d) is compared with a LS migration in (e) from five iterations of conjugate gradient (CG) inversion.

The quality of a LS migration depends on the quality of the migration operator. If the operator is limited, then the LS migration will be limited. In the current case Guitton (2004) used a pre-stack split-step double-square root migration method with one reference velocity; this will not be able to handle sharp lateral velocity variations in the Marmousi model. Panel (f) shows the result from a reverse time pre-stack depth migration (Youn & Zhou, 2001) that is suitable for strong lateral velocity variations. Although there are some artifacts near the shot locations near the surface due to the use of just 10% of the data (24 out of 240 shot gathers), this last migration result has very high quality in resolving the reflector geometry and amplitude of the Marmousi model.

a priori information in LS inversions. The topic of constrained inversion will be elaborated further in Section 9.5.3.

The previous derivation of the LS solution in Section 9.3.1.1 aims to minimize the L_2 norm of misfits

$$\min\{(\mathbf{d} - \mathbf{Ax})^T(\mathbf{d} - \mathbf{Ax})\} \tag{9–52}$$

The problem is that, when the connection between the data space and model space is weak, the above minimization in the data space does not translate into pushing for a more correct solution in the model space. Of course, the connection between the two spaces is the kernel matrix $\mathbf{A}$. However, this matrix is usually unknown because it is typically a function of the solution model. A common practice is to use the reference model to generate a kernel matrix as an approximation, and then carry out the inversion iteratively. However, when the initial reference model is far from the correct model, the iterative inversion process may move towards an incorrect solution.

Hence it is important to find a good initial reference model. In addition, we shall constrain the difference between the solution model and the reference model. Since this difference is just the model vector $\mathbf{x}$ that we invert for, we can impose the constraint by minimizing the following

$$\min\{(\mathbf{d} - \mathbf{Ax})^T(\mathbf{d} - \mathbf{Ax}) + \varepsilon^2\mathbf{x}^T\mathbf{x}\} \tag{9–53}$$

where the second term is the L_2 norm of the model vector and ε^2 is the damping factor which governs the relative weights of the two terms to be minimized. For instance, if $\varepsilon = 0$, we are back to the simple LS inversion in (9–52).

The idea of the above damped LS inversion was introduced by Levenburg (1944). Similar to the derivation of LS inversion in Section 9.3.1.1, we take the derivatives of (9–53) with respect to the model vector and then set it to zero. This leads to the following damped LS solution

$$\mathbf{x}_{\text{DLS}} = (\mathbf{A}^T\mathbf{A} + \varepsilon^2\mathbf{I})^{-1}\mathbf{A}^T\mathbf{d} \tag{9–54}$$

where $\mathbf{I}$ is the identity matrix. Because such an inversion aims at minimizing the norm of model perturbations, it is also called the minimum-norm LS solution or **minimum-norm inverse**. The effect of damping allows it to suppress high-amplitude oscillations in the solution model.

Often we have some ideas about the quality of the data and the relative influence on the model from different pieces of the data. Thus we may want to impose variable weights on data in the inverse formulation. We may weigh the ith data point by w_i and introduce a diagonal matrix $\mathbf{W}$ with w_i as its ith element. Then we can carry out the minimization of a new error function

$$\begin{aligned}
E &= (\mathbf{d} - \mathbf{Ax})^T\mathbf{W}(\mathbf{d} - \mathbf{Ax}) \\
&= \sum_i w_i\left(d_i - \sum_j a_{ij}x_j\right)\left(d_i - \sum_k a_{ik}x_k\right) \\
&= \sum_i w_i d_i^2 - 2\sum_i d_i \sum_j w_i a_{ij}x_j + \sum_j \sum_k x_j x_k \sum_i w_i a_{ij}a_{ik}
\end{aligned} \tag{9–55}$$

Following the same minimization procedure, we arrive at the relation

$$\mathbf{A}^T \mathbf{W} \mathbf{A} \mathbf{x} = \mathbf{A}^T \mathbf{W} \mathbf{d} \tag{9–56}$$

Hence, the weighted LS solution is

$$\mathbf{x}_{WLS} = (\mathbf{A}^T \mathbf{W} \mathbf{A})^{-1} \mathbf{A}^T \mathbf{W} \mathbf{d}. \tag{9–57}$$

We can easily see that the damped and weighted LS solution is

$$\mathbf{x}_{DWLS} = (\mathbf{A}^T \mathbf{W} \mathbf{A} + \varepsilon^2 \mathbf{I})^{-1} \mathbf{A}^T \mathbf{W} \mathbf{d} \tag{9–58}$$

In practice, different versions of damped and weighted LS inversions might be used.

9.3.3 Resolution matrices

A question for any inverse solution is its resolution and error, or how good it is. Several practical ways to evaluate the resolution and error of seismic imaging have been discussed previously in Sections 4.4.2 and 4.4.3. Here we examine these issues again from the view point of inversion. The error of processed solutions can be evaluated by re-sampling as discussed in Section 4.4.3 and checking the consistency among solutions from different data samples.

Constructing the resolution matrices is one way to estimate the soundness of the inverse solutions. The first type is the **data resolution matrix**, which measures how sensitive a model can affect the prediction; this is not used much in practice. The second is the **model resolution matrix**, which measures how well the model can be resolved by the given data. The model resolution matrix is often used and referred to as the "resolution matrix" in the literature.

Rewriting (9–7) here

$$\mathbf{d} = \mathbf{A} \mathbf{m} \tag{9–7}$$

If we are able to find a practical inverse solution, we may denote it $\tilde{\mathbf{m}}$, and express it as

$$\tilde{\mathbf{m}} = \tilde{\mathbf{A}}^{-1} \mathbf{d} \tag{9–59}$$

Then the prediction from this solution model is

$$\mathbf{d} = \mathbf{A} \tilde{\mathbf{m}} = (\mathbf{A} \tilde{\mathbf{A}}^{-1}) \mathbf{d} \tag{9–60}$$

Hence the data resolution matrix is

$$\mathbf{N} = \mathbf{A} \tilde{\mathbf{A}}^{-1} \tag{9–61}$$

which indicates how well the predictions from the solution model match the data.

The dimension of the data resolution matrix is the square of the data space length. The rows of the data resolution matrix describe how well neighboring data can be independently predicted, while the diagonal elements indicate how much weight a datum has in its own prediction. Clearly, the best data resolution matrix is a unit matrix $\mathbf{I}$.

Similarly, we have

$$\tilde{\mathbf{m}} = \tilde{\mathbf{A}}^{-1} \mathbf{d} = (\tilde{\mathbf{A}}^{-1} \mathbf{A}) \mathbf{m} \tag{9–62}$$

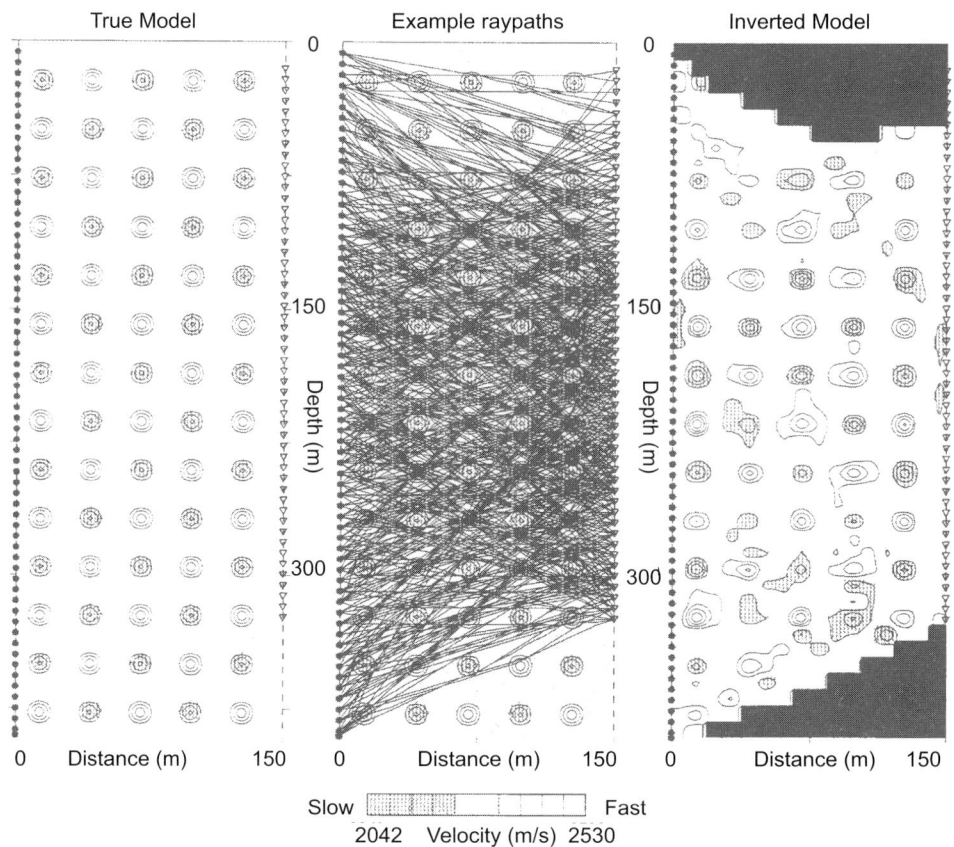

Figure 9.18 A checkerboard resolution test for a cross-well tomographic study of a shallow unconsolidated oil reservoir in southern Texas (Zhou *et al.*, 1993). Comparison of the true velocity impulses (left panel) with the inverted results (right panel) indicates the resolution level. Areas of no ray coverage in the right panel are in black.

Thus the model resolution matrix is

$$\mathbf{R} = \tilde{\mathbf{A}}^{-1}\mathbf{A} \tag{9–63}$$

The dimension of the model resolution matrix is the square of the model space length. Equations (9–62) and (9–63) indicate that $\mathbf{R}$ is a mapping matrix from the true model to the inverted model. Although we do not know the true model, the resolution matrix gives us an estimate of how good the inverted model is. The best resolution is achieved when $\mathbf{R} = \mathbf{I}$. We often describe a good resolution matrix as being "spiky" in any part of the model space. The above equation indicates that the quality of the resolution is related to the kernel matrix which depends on data coverage.

The resolution matrices are important means to QC the inverse solutions. In practice, however, many kernel matrices are too large for constructing the resolution matrices in a meaningful way. Practically, the resolution of a given data coverage may be estimated through an impulse resolution test at each model location. As shown in Section 4.2.2, a

checkerboard resolution test with multiple impulses is often conducted to increase the computation efficiency. A checkerboard resolution test consists of the following steps:

1. Making a synthetic "checkerboard" model with regularly distributed impulses of velocity perturbation with respect to a background velocity model that is usually laterally homogeneous;
2. Generating traveltimes of the checkerboard model using real source and receiver locations;
3. Inverting the traveltimes to reconstruct the checkerboard model, and the solution displays the resolution of the data kernel at each location with impulses of velocity perturbation.

Figure 9.18 shows a checkerboard resolution test from a cross-well tomography study of a shallow unconsolidated oil reservoir in southern Texas (Zhou *et al.*, 1993). Following the above test procedure, the synthetic true model in the left panel consists of a series of fast and slow velocity anomalies on a background 1D velocity model. The middle panel shows an example of ray coverage. Using a set of synthetically generated traveltimes with added noise, a tomographic inversion resulted in the model shown in the right panel. This result amplifies the raypath smear artifacts, which are especially insidious because of the tendency of researchers to interpret linear anomalies in seismic images as true features. A simple rule of thumb is that all linear tomographic anomalies are questionable if the anomalies show a geometric pattern similar to that of the raypaths.

9.3.4 Least squares inversion in hypocenter determination

9.3.4.1 LS inversion for hypocenter position and origin time

An earthquake hypocenter is the focus where the elastic energy radiation is initiated during the quake. A hypocenter can be quantified as a model hypocentral vector

$$\mathbf{m} = (T_0, h, \theta, \phi) \tag{9-64}$$

where the elements are the origin time, the depth, the colatitude, and the longitude, respectively. Commonly the hypocenter of each earthquake is determined based on the traveltimes of P-wave or S-wave arrivals through a LS inversion. Determining hypocenters is important not only for solid Earth geophysics, but also for exploration geophysics where we want to determine the hypocentral vectors of microseismics caused by injections, such as in the unconventional shale gas plays today.

Using the time–distance relationship and assuming a simple velocity model, we can obtain a rough estimate of the hypocenter quickly. This initial guess $\mathbf{m}_0$ can be regarded as an initial reference model. Next we can search for an improved solution by minimizing the residues between the observed traveltimes and modeled traveltimes, which can be expressed as a traveltime residual vector

$$\delta\mathbf{T} = \mathbf{A}\,\delta\mathbf{m} \tag{9-65}$$

where $\mathbf{A} = \{a_{i\alpha}\}$ is a matrix of partial derivatives, $i = 1, 2, \ldots, N$ as the indices of traveltime residues, and $\alpha = 1, 2, 3, 4$ as the indices of the four hypocentral parameters.

The LS solution of the above equation minimizes

$$E = \sum_{i=1}^{N} \left(\delta T_i - \sum_{\alpha=1}^{4} a_{i\alpha} \delta m_\alpha \right)^2 \qquad (9\text{--}66)$$

The minimization is accomplished by differentiating E with respect to the model variables and setting the result equal to zero. Following the argument for (9–54), we may use a damped LS solution:

$$\delta \mathbf{m} = (\mathbf{A}^T \mathbf{A} + \varepsilon^2 \mathbf{I})^{-1} \mathbf{A}^T \delta \mathbf{T} \qquad (9\text{--}54')$$

Because the relationship (9–65) is valid only when $\delta \mathbf{m}$ is small, we may take an iterative inversion approach and constrain the magnitude of $\delta \mathbf{m}$ during each step. During the kth iteration, the new position $\mathbf{m}^{(k)} = \mathbf{m}^{(k+1)} + \delta \mathbf{m}$ will serve as the new reference model. Iteration may be repeated until $\delta \mathbf{m}$ is sufficiently small and we need to verify that the process is converging.

9.3.4.2   Equal differential time (EDT) surfaces

In applications of traveltime inversion for hypocenters, we have to account for velocity model variations, picking errors, and data coverage. In particular, the effects on traveltimes from the focal depth and the origin time may cancel each other in common cases when most rays take off steeply from the source. This produces a depth and origin time tradeoff in hypocentral determination. One remedy was suggested by Zhou (1994) to construct a so-called equal differential time (EDT) surface in 3D model space.

An EDT surface is the collection of all spatial points satisfying the time difference between a pair of arrivals from a hypocenter. Suppose there are two observed phase arrivals T_j and T_k from a hypocenter at position $\mathbf{r}$, then the hypocenter must satisfy the following condition:

$$T_j - T_k = \tau_{rj} - \tau_{rk} \qquad (9\text{--}67)$$

where τ_{rj} and τ_{rk} are computed traveltimes from hypocenter $\mathbf{r}$ in a proper velocity model. The above equation is the expression of the corresponding EDT surface. An EDT surface can be constructed using any two phase arrivals such as the P-wave arrivals at two stations, or P-wave and S-wave arrivals at one station.

Figure 9.19 shows 3D sliced views of some EDT surfaces in a 3D velocity model for southern Californian crust. The model area is shown in the map view of Figure 4.21a. The map view in Figure 9.19 is a horizontal slice at sea level. The positions of the two vertical slices are indicated by the two dashed lines in the map view. The P-wave from this $M_L = 1.11$ event was recorded by four stations, shown as small triangles with numbers. The variations in the EDT surfaces are due to the effects of the 3D velocity model on the traveltimes.

9.3.4.3   Master station method of hypocenter determination

Zhou (1994) devised a master station method of hypocenter determination in two steps. First, an initial hypocenter solution is found by searching all joint positions of the EDT

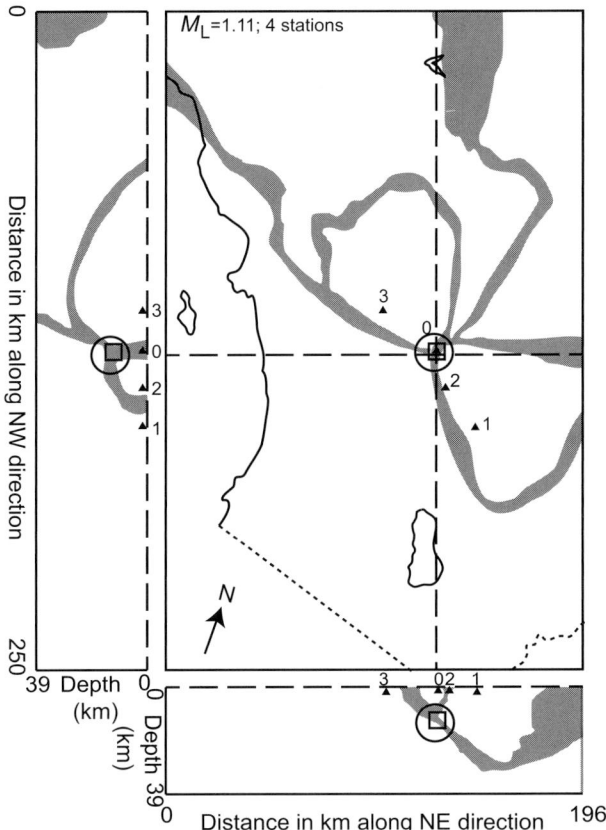

Figure 9.19 3D sliced views of equal differential time (EDT) surfaces for determining a hypocenter in southern California (Zhou, 1994). The area was shown in Figure 4.19a. An EDT surface is constructed based on a pair of phase arrivals from an earthquake. The hypocenter shall be positioned at the joint position of all EDP surfaces. Here the EDT surfaces plotted are based on P-wave arrivals at four stations denoted by triangles. The circle is centered on the bulletin hypocenter. The cross and square, respectively, are centered on the initial and final master station solutions.

surfaces available. Because modeled traveltimes from all stations can be pre-computed and stored as reference files, the search can be done very efficiently. Second, the final hypocenter solution is determined by a joint minimization of the variance of traveltime residuals and the origin time error. In Figure 9.19, the circle is centered on the hypocenter of a $M_L = 1.11$ event as shown in the bulletin compiled by the southern California Earthquake Center. The cross and square, respectively, are centered on the initial and final solutions of the master station method.

Figure 9.20 compares the solutions for a $M_L = 1.77$ event and confidence contours of the inversion. The two pairs of 3D slices in this figure show the confidence ellipsoids of the traveltime residual variance and origin time error of the master station method. Each of the slices is centered on the final solution of the master station method. The circle is centered at the bulletin hypocenter, and the cross is centered at the initial solution from searching

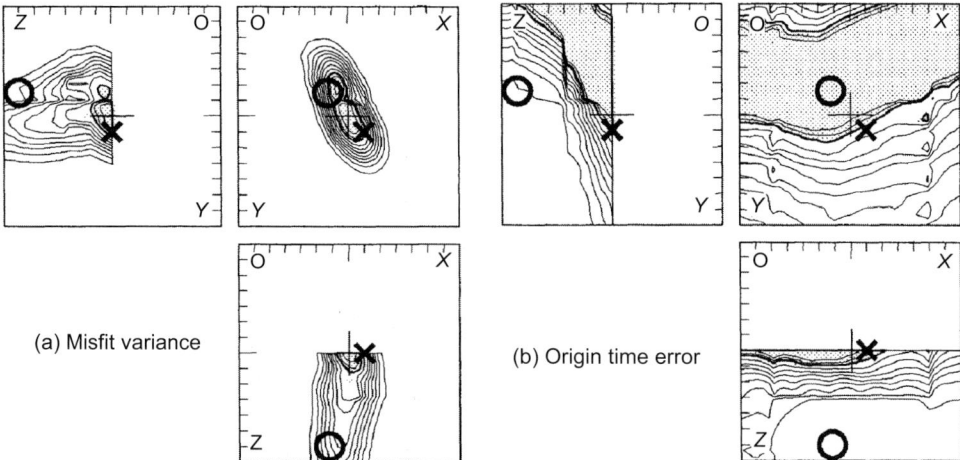

(a) Misfit variance

(b) Origin time error

Figure 9.20 3D slices of confidence ellipsoid for inverting a hypocenter in southern California: (a) for the variance of traveltime residuals; (b) for the origin time error. Each slice is centered on the inverted hypocenter. X, Y, and Z denote east–west, north–south, and depth directions, respectively. The tick marks are at 1 km intervals. The circle and cross denote the positions of the bulletin and initial hypocenters, respectively. Shaded areas have confidence higher than 90%, and the two contours in the shaded region are at 95% and 99%. Contours in the white areas have an equal interval of 10%.

the EDT surfaces. Clearly the final solution is at the position of the highest confidence in terms of traveltime residue variance and origin time error.

We can see that the EDT surface expressed by (9–67) is a difference between two traveltime residuals. Interestingly, the idea of the EDT surfaces was later taken into the double-difference hypocenter inversion method (Waldhauser & Ellsworth, 2000), which has become the most popular way to determine hypocenters today.

Exercise 9.3

1. If you have a least squares algorithm already, how will you use this algorithm to solve for the L_1 solution? In other words, how will you minimize the L_1 error using this least squares algorithm?

2. When Sir Harold Jeffreys analyzed seismic traveltime residuals, he noticed that the distribution of the residuals was not quite Gaussian. He expressed it as

$$f(t) = \frac{1 - \varepsilon}{\sqrt{2\pi}\sigma} \exp\left[-\frac{(t - \bar{t})^2}{2\sigma^2}\right] + \varepsilon g(t),$$

where $g(t)$ is a smooth function representing a gradual decay in the traveltime residuals. Now for a given dataset of observation t_i ($i = 1, 2, \ldots, n$), apply the maximum likelihood method to show that the mean $\bar{t} = \frac{w_i t_i}{w_i}$. and $\sigma^2 = \frac{w_i (t_i - \bar{t})^2}{w_i}$, where $1/w_i = 1 + \mu \exp[(t - \bar{t})^2/(2\sigma^2)]$ and μ is the ratio of the tail amplitude to the peak value of $f(t)$. In practice we use iterations to solve for $\bar{t}$ and σ^2 because they appear in the expression for w_i.

3. Why it is difficult to derive the focal depth of earthquakes? Can we use first arrivals at three stations to determine four focal parameters (x, y, z, T_0)?

9.4   Matrix decompositions and generalized inversion

The previous section has given plenty of illustrations of the importance of the kernel matrix in the linear forward system (9–7). The quality of the inverse solution depends on how well we can invert this matrix. In this section, several matrix decomposition techniques are discussed. These techniques are useful not only for inverse problems, but also for applications like principal decomposition that are used widely in seismic attribute studies. Our goal is to understand the principles behind each method.

The matrix was introduced as an algebraic operator around the middle of the nineteenth century by Cayley (Lanczos, 1961). Beyond being just a simplification of a set of algebraic equations, matrix notation inspired interest in the structure of the equation set and its effect on general solutions rather than a narrow interest in any particular solution values. When a matrix is pre-multiplied to a vector, the resultant vector can be regarded as a linear transformed version of the original vector. Therefore a multiplication of matrices is a linear transformation. Any matrix can generally be decomposed into transformations of some other matrices.

9.4.1   Eigenvalue–eigenvector decomposition (EED) of a square matrix

For any square matrix $\mathbf{B}$ of dimension N, we call the equation

$$\mathbf{B}\,\mathbf{x} = \lambda\mathbf{x} \tag{9–68}$$

the eigenvalue problem associated with $\mathbf{B}$. The prefix "eigen" means characteristic or representing the essential properties of $\mathbf{B}$. The non-zero scalars $\lambda_1, \lambda_2, \ldots, \lambda_N$ are called the **eigenvalues** of $\mathbf{B}$, each λ_1 satisfying the **characteristic equation**

$$|\mathbf{B} - \lambda_i\mathbf{I}| = 0 \tag{9–69}$$

Any vector $\mathbf{x}_i$ satisfying equation (9–68) is an **eigenvector** or a principal axis of matrix $\mathbf{B}$. Each eigenvector $\mathbf{x}_i$ has a corresponding eigenvalue λ_i. Putting all eigenvalues into a diagonal matrix $\mathbf{L}$ and setting all the eigenvectors as the columns of a matrix $\mathbf{U}$, we have the **eigenvalue–eigenvector decomposition (EED)** of square matrix $\mathbf{B}$:

$$\mathbf{B}\,\mathbf{U} = \mathbf{U}\,\Lambda \tag{9–70}$$

If a square matrix of dimension N can be decomposed into N independent eigenvectors, the matrix is called a **full-rank** matrix.

We focus on a symmetric matrix here because many matrix inverse problems can be converted into inverse of symmetric matrices. For instance, the LS inversion contains an inverse of $(\mathbf{A}^\mathsf{T}\mathbf{A})$ which is symmetric. Let us consider the following proposition about the eigenvalue–eigenvector problem of a real symmetric matrix $\mathbf{S}$. The eigenvectors of $\mathbf{S}$ form an orthogonal matrix $\mathbf{U}$, and the eigenvalues of $\mathbf{S}$ form a diagonal matrix $\mathbf{\Lambda}$. Then

$$\mathbf{S}\mathbf{U} = \mathbf{U}\mathbf{\Lambda} \tag{9-71a}$$

$$\mathbf{U}^T\mathbf{U} = \mathbf{U}\mathbf{U}^T = \mathbf{I} \tag{9-71b}$$

The above proposition can be proved using the so-called Schur's canonical decomposition, which is beyond the scope of this book.

For a square matrix of small dimensions, the practical ways of conducting EED has two steps:

1. Solve characteristic equation (9–69) for all eigenvalues of the matrix;
2. For each non-zero eigenvalue, solve (9–68) for the corresponding eigenvector.

9.4.2 Singular value decomposition (SVD) of a general matrix

The decomposition of a general rectangular matrix requires the use of singular value decomposition, or SVD (Lanczos, 1961). The SVD decomposes any rectangular matrix $\mathbf{A}$ of m rows and n columns into a multiplication of three matrices of useful properties:

$$\mathbf{A}_{m \times n} = \mathbf{U}_{m \times m} \mathbf{\Lambda}_{m \times n} \mathbf{V}^T_{n \times n} \tag{9-72}$$

In the above equation $\mathbf{\Lambda}_{m \times n}$ is a rectangular diagonal matrix containing p **singular values** (or principal values), $\{s_i\}$, of the matrix $\mathbf{A}$:

$$\mathbf{\Lambda}_{m \times n} = \text{diag}(s_1, s_1, \ldots, s_p, 0, \ldots, 0)_{m \times n} = \begin{bmatrix} s_1 & & & & & & \\ & s_1 & & & & & \\ & & \ddots & & & \text{all} \quad 0 & \\ & & & s_p & & & \\ & & & & 0 & & \\ & \text{all} \quad 0 & & & & \ddots & \\ & & & & & & 0 \end{bmatrix}_{m \times n} \tag{9-73}$$

and $\mathbf{U}$ and $\mathbf{V}$ are two square **unitary matrices**. The number of non-zero singular values, p, is called the **trace** or **rank** of the matrix.

A unitary matrix means its inverse is simply its transpose conjugate. In other words,

$$\mathbf{U}^{-1} = (\mathbf{U}^T)^* \tag{9-74a}$$

$$\mathbf{V}^{-1} = (\mathbf{V}^T)^* \tag{9-74b}$$

If $\mathbf{U}$ and $\mathbf{V}$ are real matrices, we have

$$\mathbf{U}^{-1} = (\mathbf{U}^T) \tag{9-75a}$$

$$\mathbf{V}^{-1} = (\mathbf{V}^T) \tag{9-75b}$$

Figure 9.21 shows SVD in a traveltime inversion of a highly simplified crosswell seismic experiment with five sources in the left well and five receivers in the right well. The forward system is (9–65), in which the kernel matrix $\mathbf{A}$ is decomposed using SVD. The data vector

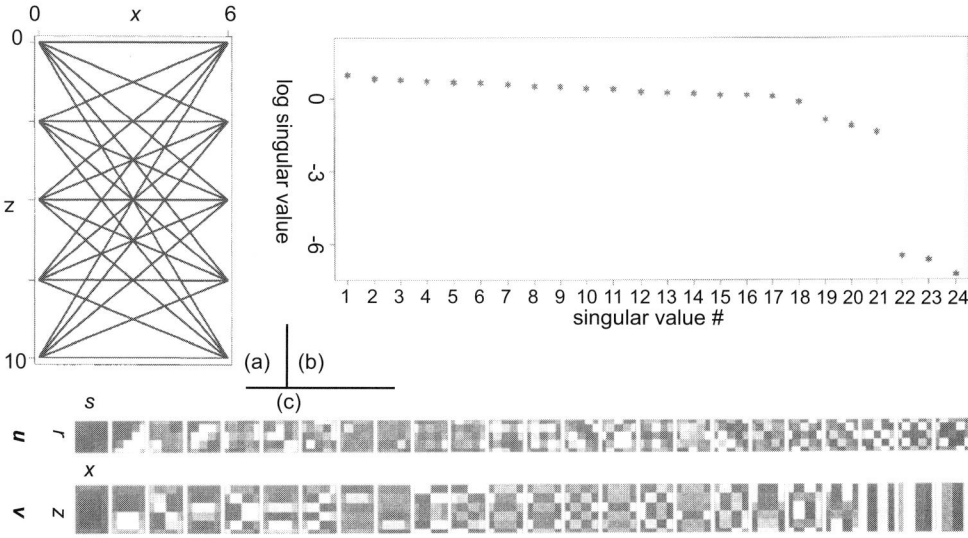

Figure 9.21 Illustration of SVD in a simple crosswell setup (Michelena, 1993). (a) Straight raypaths from five sources in the left well to five receivers in the right well. The model area consists of 6×4 equal-sized cells. (b) Twenty-four singular values of SVD of the kernel matrix for this setup. (c) Eigenvector matrices **U** and **V** of the SVD. The size of singular values decreases as the amount of lateral variation in the model space increases.

has 25 elements corresponding to the 25 first arrivals from the five sources to five receivers. The model vector has 24 variables as the slowness perturbations of the 24 model cells. So the 25 by 24 kernel matrix has 24 singular values as shown in Figure 9.21b. Figure 9.21c shows the 25 eigenvectors in **U** and 24 eigenvectors in **V**.

The SVD manifests the intrinsic properties of the matrix **A**. For instance, the trace of **A** is simply the number of non-zero elements in $\boldsymbol{\Lambda}$; and the inverse of **A** is

$$\mathbf{A}^{-1} = (\mathbf{U} \cdot \boldsymbol{\Lambda} \cdot \mathbf{V}^{\mathrm{T}})^{-1} = (\mathbf{V}^{\mathrm{T}})^{-1}\boldsymbol{\Lambda}^{-1}\mathbf{U}^{-1} = \mathbf{V}\boldsymbol{\Lambda}^{-1}\mathbf{U}^{-1} \tag{9-76}$$

The inverse of the diagonal matrix $\boldsymbol{\Lambda}$ in SVD is done by inversing its diagonal elements, or replacing each singular value s_i by $1/s_i$. This will cause overflow if s_i is zero or very small. A generalized inverse of diagonal matrix has been suggested to leave the zero diagonal value unchanged, and only the non-zero singular values are inversed:

$$\left(\Lambda_g^{-1}\right)_{n \times m} = \mathrm{diag}(1/s_1, 1/s_2, \ldots, 1/s_p, 0, \ldots, 0)_{n \times m} \tag{9-77}$$

For an arbitrary matrix $\mathbf{A}_{m \times n}$, its SVD can be obtained from the EED result; this offers a practical way to obtain SVD for matrices of small dimensions. Suppose the rank of matrix **A** is p. Let us introduce a $(m+n) \times (m+n)$ matrix **S**:

$$\mathbf{S} = \begin{bmatrix} \mathbf{0} & \mathbf{A} \\ \mathbf{A}^* & \mathbf{0} \end{bmatrix} \tag{9-78}$$

When **A** is real, **S** is clearly symmetric. We can prove that the singular values of **A** are also the eigenvalues of **S**.

For the ith eigenvector $\mathbf{w}_i$ and ith eigenvalue λ_i of $\mathbf{S}$, we have

$$\mathbf{Sw}_i = \lambda_i \mathbf{w}_i \tag{9–79}$$

If we rewrite the conjugate vector $\mathbf{w}_i^* = (\mathbf{u}_i^* \ v_i^*)$ where $\mathbf{u}_i$ is a $n \times 1$ vector and v_i is a $m \times 1$ vector, we have

$$\begin{bmatrix} \mathbf{0} & \mathbf{A} \\ \mathbf{A}^* & \mathbf{0} \end{bmatrix} \begin{bmatrix} \mathbf{u}_i \\ \mathbf{v}_i \end{bmatrix} = \lambda_i \begin{bmatrix} \mathbf{u}_i \\ \mathbf{v}_i \end{bmatrix} \tag{9–80}$$

Therefore

$$\mathbf{A}\mathbf{v}_i = \lambda_i \mathbf{u}_i \tag{9–81a}$$

$$\mathbf{A}^*\mathbf{u}_i = \lambda_i \mathbf{v}_i \tag{9–81b}$$

It is easy to see that if λ_i and $(\mathbf{u}_i^* \ \mathbf{v}_i^*)$ are one eigenvalue–eigenvector pair of $\mathbf{S}$, then $(-\lambda_i)$ and $(-\mathbf{u}_i^* \ \mathbf{v}_i^*)$ will be another pair. The orthogonality of all the eigenvectors means that, when $i \neq k$, for the first eigenvalue–eigenvector pair of $\mathbf{S}$

$$\mathbf{w}_i^*\mathbf{w}_k = \mathbf{u}_i^*\mathbf{u}_k + \mathbf{v}_i^*\mathbf{v}_k = 0 \tag{9–82a}$$

and for the second eigenvalue–eigenvector pair of $\mathbf{S}$

$$-\mathbf{u}_i^*\mathbf{u}_k + \mathbf{v}_i^*\mathbf{v}_k = 0 \tag{9–82b}$$

From the above two equations we have

$$\mathbf{u}_i^*\mathbf{u}_k = 0 \quad \text{and} \quad v_i^* v_k = 0 \tag{9–83}$$

which means that the $\mathbf{u}_i$s and v_is independently form $(n \times n)$ and $(m \times m)$ orthogonal sub-matrices in $\mathbf{S}$.

Based on (9–81) we obtain

$$\mathbf{A}^*\mathbf{A}\,\mathbf{v}_i = \lambda_i \mathbf{A}^* \mathbf{u}_i = \lambda_i^2 \mathbf{v}_i \tag{9–84a}$$

$$\mathbf{A}\mathbf{A}^* \mathbf{u}_i = \lambda_i \mathbf{A}\,\mathbf{v}_i = \lambda_i^2 u_i \tag{9–84b}$$

Because both $\mathbf{A}^*\mathbf{A}$ and $\mathbf{A}\,\mathbf{A}^*$ are symmetric (hence squared) matrices, we have an $m \times m$ eigensystem

$$\mathbf{A}^*\mathbf{A}\,\mathbf{V} = \mathbf{V}\mathrm{diag}\left(\lambda_1^2, \ldots, \lambda_p^2, 0, \ldots, 0\right), \tag{9–85a}$$

and an $n \times n$ eigensystem

$$\mathbf{A}\mathbf{A}^*\mathbf{U} = \mathbf{U}\,\mathrm{diag}\left(\lambda_1^2, \ldots, \lambda_p^2, 0, \ldots, 0\right) \tag{9–85b}$$

where $\mathbf{V} = (v_1 \ v_2 \ldots v_m)$ and $\mathbf{U} = (\mathbf{u}_1 \ \mathbf{u}_2 \ldots \mathbf{u}_n)$. Since the rank of $\mathbf{A}$ is p, the number of zero eigenvalues for $\mathbf{A}^*\mathbf{A}$ and $\mathbf{A}\mathbf{A}^*$, respectively, are $(m-p)$ and $(n-p)$.

In summary, when λ_i is the singular value of $\mathbf{A}$, λ_i^2 will be an eigenvalue of $\mathbf{A}^*\mathbf{A}$ or $\mathbf{A}\mathbf{A}^*$. Therefore, the singular values of a matrix $\mathbf{A}$ are the non-negative square roots of the eigenvalues of $\mathbf{A}^*\mathbf{A}$ or $\mathbf{A}\mathbf{A}^*$, whichever has fewer rows and columns.

Example

Decomposing an extremely simple 3×2 matrix $\mathbf{A} = \begin{bmatrix} 0 & 0 \\ 1 & 0 \\ 0 & 0 \end{bmatrix}$.

Since $\mathbf{A}^{\mathrm{T}} \mathbf{A} = \begin{bmatrix} 1 & 0 \\ 0 & 0 \end{bmatrix}$, from (9–84a) the eigenequation is

$$\mathbf{A}^{\mathrm{T}} \mathbf{A} \mathbf{v}_i = \begin{bmatrix} 1 & 0 \\ 0 & 0 \end{bmatrix} \mathbf{v}_i = \lambda_1^2 \mathbf{v}_i$$

which can be solved to obtain two eigenvalue–eigenvector pairs:

$$\lambda_1 = 1 \qquad \mathbf{v}_1 = (1 \quad 0)^{\mathrm{T}}$$

and

$$\lambda_2 = 0 \qquad \mathbf{v}_2 = (0 \quad 1)^{\mathrm{T}}$$

This means $\mathbf{V} = (\mathbf{v}_1 \ \mathbf{v}_2) = \begin{bmatrix} 1 & 0 \\ 0 & 1 \end{bmatrix}$.

Similarly, since $\mathbf{A} \mathbf{A}^{\mathrm{T}} = \begin{bmatrix} 0 & 0 & 0 \\ 0 & 1 & 0 \\ 0 & 0 & 0 \end{bmatrix}$, from (9–84b) the eigenequation is

$$\mathbf{A} \mathbf{A}^{\mathrm{T}} \mathbf{u}_i = \begin{bmatrix} 0 & 0 & 0 \\ 0 & 1 & 0 \\ 0 & 0 & 0 \end{bmatrix} \mathbf{u}_i = \lambda_i^2 \mathbf{u}_i$$

which can be solved to obtain three eigenvalue–eigenvector pairs:

$$\lambda_1 = 1, \ \mathbf{u}_1 = (0 \ 1 \ 0)^{\mathrm{T}}$$
$$\lambda_2 = 0, \ \mathbf{u}_2 = (1 \ 0 \ 0)^{\mathrm{T}}$$

and

$$\lambda_3 = 0, \ \mathbf{u}_3 = (0 \ 0 \ 1)^{\mathrm{T}}$$

This means

$$\mathbf{U} = (\mathbf{u}_1 \mathbf{u}_2 \mathbf{u}_3) = \begin{bmatrix} 0 & 1 & 0 \\ 1 & 0 & 0 \\ 0 & 0 & 1 \end{bmatrix}$$

Putting things together, the SVD is

$$\mathbf{A} = \mathbf{U} \boldsymbol{\Lambda} \mathbf{V}^{\mathrm{T}} = \begin{bmatrix} 0 & 1 & 0 \\ 1 & 0 & 0 \\ 0 & 0 & 1 \end{bmatrix} \begin{bmatrix} 1 & 0 \\ 0 & 0 \\ 0 & 0 \end{bmatrix} \begin{bmatrix} 1 & 0 \\ 0 & 1 \end{bmatrix}$$

9.4.3   Generalized inversion

We have considered modeling and inverse as two reverse mapping processes between data space and model space through the kernel matrix $\mathbf{A}$ in (9–7) and its inverse. For a given inverse problem with data, data kernel, and target model vector, there could be a part of the data space that is unrelated to any part of the model space. Conversely, there could also be a part of the model space that is unrelated to any part of the available data. Those subspaces are called **null spaces**, because the corresponding information cannot be mapped between the data space and model space.

We cannot obtain the exact inverse of most forward problems $\mathbf{Gm} = \mathbf{d}$, because the inverse matrix $\mathbf{G}^{-1}$ does not exist. Then these problems become non-deterministic because the inverse solution will be approximated and non-unique. The goal of inverse for such non-deterministic problems is to find a "generalized" inverse, $\mathbf{G}_g^{-1}$, which maintains the minimum misfit under some given criteria. Using the LS inversion of $\mathbf{Ax} = \mathbf{d}$ as an example, we know from (9–46) that

$$x_{LS} = (\mathbf{A}^T\mathbf{A})^{-1}\mathbf{A}^T\mathbf{d} \tag{9–46'}$$

We can take $\mathbf{G} = \mathbf{A}^T\mathbf{A}$, and $\mathbf{b} = \mathbf{A}^T\mathbf{d}$, then the generalized inverse will be

$$\mathbf{x}_g = \mathbf{G}_g^{-1}\mathbf{b} \tag{9–86}$$

The term **generalized inversion** means, regardless of whether an exact inverse exists or not, the one inverse form which can be adopted in general, and sometimes in ad hoc fashion. The most common generalized inverse is the one drawn from the SVD:

$$\mathbf{G} = \mathbf{U}\mathbf{\Lambda}\mathbf{V}^T \tag{9–87}$$

Following the description of SVD in the previous section, if matrix $\mathbf{G}$ has a rank p, there will be p non-zero diagonal elements in the singular value matrix $\mathbf{\Lambda}$:

$$\mathbf{\Lambda} = \mathrm{diag}(s_1, s_2, \ldots, s_p, 0, \ldots, 0) \tag{9–88}$$

We devise a generalized inverse of $\mathbf{\Lambda}$ as

$$\mathbf{\Lambda}_g^{-1} = \mathrm{diag}(1/s_1, 1/s_2, \ldots, 1/s_p, 0, \ldots, 0) \tag{9–89}$$

Note that only the non-zero elements are reversed, and zero elements are kept as zeros. Then, the generalized inverse solution of $\mathbf{G}$ is

$$\mathbf{G}_g^{-1} = \mathbf{V}_p\mathbf{\Lambda}_g^{-1}\mathbf{U}_p^T \tag{9–90}$$

where $\mathbf{V}p$ and $\mathbf{U}p$ are the first p rows of matrices $\mathbf{V}$ and $\mathbf{U}$, respectively. One can easily extend the above idea to exclude very small singular values.

Based on what we have learnt in the last section on SVD, the two unitary matrices consist of

$$\mathbf{U} = [\mathbf{U}_p, \mathbf{U}_0] \tag{9–91a}$$

$$\mathbf{V} = [\mathbf{V}_p, \mathbf{V}_0] \tag{9–91b}$$

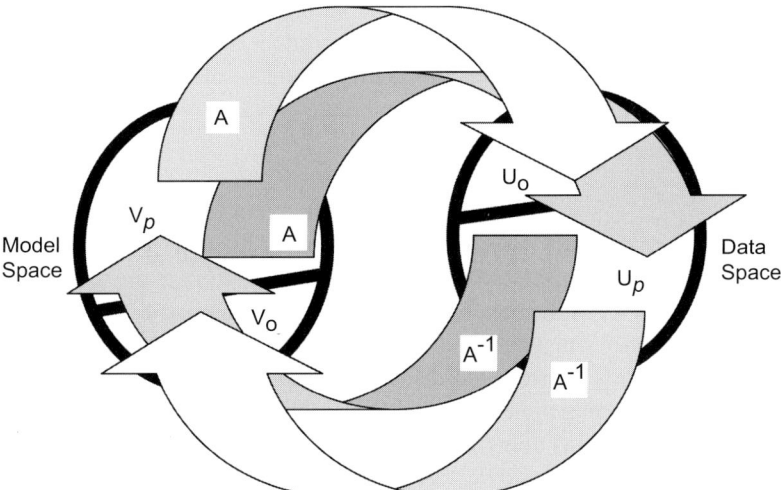

Figure 9.22 A schematic illustration of the relationship between model space and data space via singular value decomposition. Information corresponding to the null spaces $\mathbf{U}_0$ and $\mathbf{V}_0$ cannot be mapped into the opposite space.

where $\mathbf{U}_0$ and $\mathbf{V}_0$ are collections of eigenvectors corresponding to zero singular values. This means

$$\mathbf{G} = \mathbf{U}\mathbf{\Lambda}\mathbf{V}^{\mathrm{T}} = [U_p, U_0] \begin{bmatrix} \Lambda_p & 0 \\ 0 & 0 \end{bmatrix} \begin{bmatrix} V_p^{\mathrm{T}} \\ V_0^{\mathrm{T}} \end{bmatrix} \tag{9–92}$$

The null space matrices $\mathbf{U}0$ and $\mathbf{V}_0$ are subspaces of the data space and model space, respectively. The relationship between the model space and data space via the kernel matrix A and its SVD is sketched in Figure 9.22. From model space to data space, the forward modeling takes only information corresponding to $\mathbf{V}_p$; the mapping of information from those parts of the model space corresponding to $\mathbf{V}_0$ is blocked by the zero singular values. Thus the existence of $\mathbf{V}_0$ is the cause of **model non-uniqueness**. Conversely, from data space to model space, the inverse mapping is valid only for those data corresponding to $\mathbf{U}_p$; the mapping of information from those data corresponding to $\mathbf{U}_0$ is blocked by the zero singular values. Hence the existence of $\mathbf{U}_0$ is the cause of **data inconsistency**.

9.4.4 Conjugate gradient (CG) method

The conjugate gradient (CG) method, or conjugate direction (CD) method in general sense, is an iterative method to solve the inverse of a linear system like (9–7). In many applications, the matrix to be inverted is large and sparse. A direct inverse like that using SVD becomes impossible because the storage of the whole matrix may be of the order of over 10^{10} elements in the computers' randomly accessible memory. Iterative inversions such as the CG method are good remedies in this situation because they require access to only one row of the matrix at a time, hence the name of "row-action" methods.

The first detailed account of the CG method was by Hestenes and Stiefel (1952). The iterations of this method terminate at most in M steps, where M is the number of model parameters. We want to examine the principle of the CG method in analogy with the LS inversion. Let us use $(\mathbf{x}, \mathbf{y})$ to represent a dot product of two vectors $\mathbf{x}$ and $\mathbf{y}$. As shown in Section 9.3, the LS inversion aims to minimize the L_2-norm of misfits

$$
\begin{aligned}
E &= (\mathbf{Ax} - \mathbf{d})^*(\mathbf{Ax} - \mathbf{d}) \\
&= (\mathbf{x}^*\mathbf{A}^* - \mathbf{d}^*)(\mathbf{Ax} - \mathbf{d}) \\
&= \mathbf{x}^*\mathbf{A}^*\mathbf{Ax} - \mathbf{x}^*\mathbf{A}^*\mathbf{d} - \mathbf{d}^*\mathbf{Ax} + \mathbf{d}^*\mathbf{d} \\
&= (\mathbf{x}^*, \mathbf{A}^*\mathbf{Ax}) - 2(\mathbf{x}^*, \mathbf{A}^*\mathbf{d}) + (\mathbf{d}, \mathbf{d}^*)
\end{aligned}
\tag{9-93}
$$

where the asterisk signs denote complex transpose of the matrix.

In the model space where the model parameters become axes of the coordinate, the above expression represents a series of ellipsoids in data space with the fitting errors as the radii. Of course, all ellipsoids must surround the prediction of the final model, $\mathbf{xg}$, which will have the minimum radius $\min(E)$.

For a linear operator $\mathbf{G}$, two vectors $\mathbf{p}$ and $\mathbf{q}$ are said to have **conjugate directions** if

$$
(\mathbf{p}, \mathbf{Gq}) = 0
\tag{9-94}
$$

If we take $\mathbf{G}$ as the forward modeling or mapping matrix from model space to data space, then $\mathbf{p}$ is in the data space, $\mathbf{q}$ is in the model space. Hence the vectors are in conjugate directions when $\mathbf{p}$ is perpendicular to the mapped direction of $\mathbf{q}$ in the data space.

If the kernel matrix $\mathbf{G}$ is a square matrix, the data space and model space have the same dimension. Without a loss of generality, we can have a square matrix by taking the LS result where $\mathbf{A}^*\mathbf{A}$ is the new matrix $\mathbf{G}$ and $\mathbf{A}^*\mathbf{d}$ is the new $\mathbf{d}$. Now suppose that $\mathbf{p}$ in data space is at the direction tangent to one solution ellipsoid for $\mathbf{x}_i$, then the conjugate direction $\mathbf{q}$ (the direction in model space for the next solution $\mathbf{x}_{i+1}$) for the operator $\mathbf{G}$ will be in a direction toward the center of this ellipsoid. This forms the base for the CG method.

Starting with an initial estimate model $\mathbf{x}_0$, the CG method is a way to determine successively new estimates $\mathbf{x}_0, \mathbf{x}_1, \mathbf{x}_2, \ldots$, each constructed such that $\mathbf{x}_i$ is closer to the true model $\mathbf{x}$ than $\mathbf{x}_{i-1}$. At each step the residual vector $\mathbf{r}_i = \mathbf{d} - \mathbf{G}\mathbf{x}_i$ is computed. Normally the residual vector can be used as a measure of the "goodness" of the estimate $\mathbf{x}_i$. However, this measure is not reliable because it is possible to construct cases in which the squared residual $|\mathbf{r}_i|^2$ generally increases at each step while the length of the error vector $|\mathbf{x} - \mathbf{x}_i|$ decreases monotonically (Hestenes & Stiefel, 1952).

For the symmetric and positive definite matrix $\mathbf{G} = \mathbf{A}^*\mathbf{A}$, the following are the CG formulas to construct the estimates:

Initial step: Select an estimate $\mathbf{x}_0$ and compute the residual $\mathbf{r}_0$ and the direction $\mathbf{p}_0$ by

$$
\mathbf{p}_0 = \mathbf{r}_0 = \mathbf{d} - \mathbf{G}\mathbf{x}_0
\tag{9-95}
$$

General iterations: From the previous estimate $\mathbf{x}_i$, the residual $\mathbf{r}_i$, and the direction $\mathbf{p}_i$, compute new estimate $\mathbf{x}_{i+1}$, residual $\mathbf{r}_{i+1}$, and direction $\mathbf{p}_{i+1}$ by

$$
\text{(in data space)} \quad \mathbf{r}_{i+1} = \mathbf{r}_i - \frac{(\mathbf{p}_i, \mathbf{r}_i)}{(\mathbf{p}_i, \mathbf{Gp}_i)}\mathbf{Gp}_i
\tag{9-96a}
$$

$$(\text{in model space}) \quad \mathbf{x}_{i+1} = \mathbf{x}_i + \frac{(\mathbf{p}_i, \mathbf{r}_i)}{(\mathbf{p}_i, \mathbf{Gp}_i)} \mathbf{p}_i \qquad (9\text{--}96\text{b})$$

$$(\text{in model space}) \quad \mathbf{p}_{i+1} = \mathbf{p}_i - \frac{(\mathbf{r}_{i+1}, \mathbf{Gp}_i)}{(\mathbf{p}_i, \mathbf{Gp}_i)} \mathbf{p}_i \qquad (9\text{--}96\text{c})$$

The residues $\mathbf{r}_0, \mathbf{r}_1, \ldots$ are mutually orthogonal

$$(\mathbf{r}_i, \mathbf{r}_j) = 0 \quad (i \neq j) \qquad (9\text{--}97\text{a})$$

and the direction vectors $\mathbf{p}_0, \mathbf{p}_1, \ldots$ are mutually conjugate

$$(\mathbf{p}_i, \mathbf{Gp}_j) = 0 \quad (i \neq j) \qquad (9\text{--}97\text{b})$$

To get the final model estimate and the inverse matrix, once we obtained the set of M mutually conjugate vectors $\mathbf{p}_0, \ldots, \mathbf{p}_{M-1}$ the model solution vector is

$$\mathbf{x} = \sum_{i=0}^{M-1} \frac{(\mathbf{p}_i, \mathbf{d})}{(\mathbf{p}_i, \mathbf{Gp}_i)} \mathbf{p}_i \qquad (9\text{--}98)$$

If we denote by p_{ij} the jth component of $\mathbf{p}_i$, then

$$g'_{jk} = \sum_{i=0}^{M-1} \frac{(p_{ij}, p_{ik})}{(\mathbf{p}_i, \mathbf{Gp}_i)} \qquad (9\text{--}99)$$

is the element in the jth row and kth column of the inverse $\mathbf{G}^{-1}$. On the other hand, the general formulas (9–98) and (9–99) are not used in actual machine computation because they need to store all the vectors $\mathbf{p}_i$ and they are much more influenced by rounding-off errors. A step-by-step routine given by formulas (9–95) and (9–96) is actually used.

Exercise 9.4

1. Show that if λ is an eigenvalue of the problem $\mathbf{Ax} = \lambda \mathbf{x}$, it is also an eigenvalue of the "adjoint" problem $\mathbf{A}^{\mathrm{T}} \mathbf{y} = \lambda \mathbf{y}$.

2. For a 2×2 matrix $\mathbf{A} = \begin{pmatrix} 10 & 2 \\ -10 & 2 \end{pmatrix}$

   (a) Take an eigenvalue–eigenvector decomposition, i.e., find $\mathbf{X}$ and $\mathbf{\Lambda}$ as in $\mathbf{A} = \mathbf{X} \mathbf{\Lambda} \mathbf{X}^{-1}$;

   (b) Take a singular value decomposition, e.g., find $\mathbf{U}$ and $\mathbf{\Lambda}_s$ and $\mathbf{V}$ as in $\mathbf{A} = \mathbf{U} \mathbf{\Lambda}_s \mathbf{V}^{\mathrm{T}}$;

   (c) Find the solution $\mathbf{m}$ in $\mathbf{d} = \mathbf{Am}$, where $\mathbf{d}^{\mathrm{T}} = (1, 2)$.

3. For a linear equation set $\begin{cases} x_1 + x_3 = 1 \\ x_2 = 2 \\ -x_2 = 1 \end{cases}$

   (a) Form the least squares inverse;

   (b) Compute the model parameters using the generalized inverse;

   (c) Evaluate the resolution for the model in connection with the original equation set.

9.5   Practical solutions in geophysical inversion

Geophysical inversion is still a research topic in many aspects, so practitioners are always interested in practical solutions. The solutions discussed in this section are selected from several classic deterministic inversion approaches that are used widely today.

9.5.1   The Backus–Gilbert method

We have repeated that an inverse solution minimizes some measures of misfit or error functions. Alternatively, we can minimize the spread of the resolution and model covariance. In the late 1960s, George Backus and Freeman Gilbert (1967, 1968, and 1970) demonstrated that the set of Earth models that yield the physically observed values of any independent set of gross Earth data is either empty or of infinite dimensions, meaning a high degree of non-uniqueness in geophysical inverse problems. Consequently, they devised a new approach to the inverse problem through optimizing the resolution matrices. Instead of minimizing the difference between data and predictions from models, an inverse can be derived by minimizing the spread function, the difference between the resolution matrix and a function of desired shape. This **Backus–Gilbert method** is useful for solving under-determined inversions, such as using surface wave data to map 3D velocity structures on a global scale.

In Section 9.3 we have defined the data resolution matrix $\mathbf{N} = \mathbf{A}\tilde{\mathbf{A}}^{-1}$ and the model resolution matrix $\mathbf{R} = \tilde{\mathbf{A}}^{-1}\mathbf{A}$, where $\mathbf{A}$ is the forward kernel matrix and $\tilde{\mathbf{A}}^{-1}$ is the practical inverse matrix. If we wish the resolution matrix to be in the shape of a Dirichlet delta, the spread function for data resolution matrix is

$$\mathrm{spread}(\mathbf{N}) = \|\mathbf{N} - \mathbf{I}\|^2 = \sum_{i=1}^{N}\sum_{j=1}^{N}(N_{ij} - I_{ij})^2 \tag{9–100a}$$

and that for the model resolution matrix is

$$\mathrm{spread}(\mathbf{R}) = \|\mathbf{R} - \mathbf{I}\|^2 = \sum_{i=1}^{M}\sum_{j=1}^{M}(R_{ij} - I_{ij})^2 \tag{9–100b}$$

where $\|\cdot\|$ denotes the L$_2$ norm; N_{ij}, R_{ij}, and I_{ij} are the elements on the ith row and jth column of the corresponding matrices $\mathbf{N}$, $\mathbf{R}$, and $\mathbf{I}$. Expressions in (9–100) are called the **Dirichlet spread functions**.

When we minimize only the spread function of the data resolution matrix (9–100a), we end up at exactly the same formula as the LS generalized inverse. In general, we may minimize the following linear combination of three terms covering spreads in data space, model space, and deviation from the reference model

$$\alpha_1 \, \mathrm{spread}\,(\mathbf{N}) + \alpha_2 \, \mathrm{spread}\,(\mathbf{R}) + \alpha_3 \, \mathrm{size}\,(<\mathbf{m}\mathbf{m}^{\mathsf{T}}>) \tag{9–101}$$

where α_1, α_2, and α_3 are weighting coefficients for each term, and the third term is the covariance size of the solution model, which can be defined as the sum of the square power of all diagonal elements of $<\mathbf{m}\,\mathbf{m}^T$. A LS minimization of the above linear combination results in an equation for a general inverse problem:

$$\alpha_1[\mathbf{A}^T\mathbf{A}]\mathbf{A}^{-1} + \mathbf{A}^{-1}\{\alpha_2\mathbf{A}\,\mathbf{A}^T + \alpha_3 < \mathbf{d}\,\mathbf{d}^T >\} = [\alpha_1 + \alpha_2]\,\mathbf{A}^T \quad (9\text{--}102)$$

Most inverse solutions are specific cases of inverting (9–102) for given weights on each of the three components of (9–101). For example, if we have a complete weight on the data resolution ($\alpha_1 = 1$), no weight on the model resolution ($\alpha_2 = 0$), and a partial weight on the covariance size ($\alpha_3 = \varepsilon^2$ and $<\mathbf{d}\,\mathbf{d}^T >= \mathbf{I}$), we get

$$\left[\mathbf{A}^T\mathbf{A}\right]\mathbf{A}^{-1} + \mathbf{A}^{-1}\varepsilon^2\mathbf{I} = \mathbf{A}^T$$

or $\qquad\qquad\qquad\qquad\qquad\qquad\qquad\qquad\qquad\qquad\qquad\quad$ (9–103)

$$\mathbf{A}^{-1} = \left[\mathbf{A}^T\mathbf{A} + \varepsilon^2\mathbf{I}\right]^{-1}\mathbf{A}^T$$

which is precisely the damped LS inverse matrix in (9–54).

For discrete cases, the Backus–Gilbert method is the inverse solution from minimizing the following Backus–Gilbert spread function for model resolution:

$$\text{spread}(\mathbf{R}) = \sum_{i=1}^{M}\sum_{j=1}^{M} w(i,\,j)R_{ij}^2 \quad (9\text{--}104)$$

where the weighting factor $w(i,\,j)$ is non-negative and symmetric in i and j. Specifically, $w(i,\,i) = 0$. The Dirichlet spread function (9–100b) is just a special case when $w(i,\,j) = 1$. Other examples of the spread function include a "blurred" delta shape. The objective in devising the variable weighting in this new spread function is to suppress **side-lobes** in the resolution matrix, or high-amplitude areas away from the diagonal in the matrix.

Typically the model resolution matrix $\mathbf{R}$ is self-normalized, meaning

$$\sum_{j=1}^{M} R_{ij} = 1 \quad (9\text{--}105)$$

Needless to say, the new resolution matrix formed by minimizing the new spread function should also satisfy the above equation. Consequently, the derivation of the Backus–Gilbert generalized inverse is to minimize (9–104) subject to the condition of (9–105), using the method of Lagrange multipliers (to be discussed in Section 9.5.3.2).

The kth Lagrange function corresponding to the kth model parameter is Φ_κ:

$$\Phi_\kappa = \sum_{j=1}^{M} w(k,\,j)R_{kj}^2 - 2\lambda \sum_{j=1}^{M} R_{kj} \quad (9\text{--}106)$$

where -2λ is the kth Lagrange multiplier.

Using $R_{kj} = \sum_{i=1}^{N} (A_g^{-1})_{ki} A_{ij}$ where the subscript denotes that A_g^{-1} is a generalized inverse matrix, we have

$$\Phi_k = \sum_{j=1}^{M} w(k,j) \left[\sum_{i=1}^{N} (A_g^{-1})_{ki} A_{ij} \right] \left[\sum_{l=1}^{N} (A_g^{-1})_{kl} A_{lj} \right] - 2\lambda \sum_{j=1}^{M} \left[\sum_{l=1}^{N} (A_g^{-1})_{kl} A_{lj} \right]$$

$$= \sum_{i=1}^{N} (A_g^{-1})_{ki} \sum_{l=1}^{N} (A_g^{-1})_{kl} \left[\sum_{j=1}^{M} w(k,j) A_{ij} A_{lj} \right] - 2\lambda \sum_{l=1}^{N} (A_g^{-1})_{kl} \left[\sum_{j=1}^{M} A_{lj} \right]$$

$$= \sum_{i=1}^{N} (A_g^{-1})_{ki} \left[\sum_{l=1}^{N} (A_g^{-1})_{kl} [S_{ij}]^{(k)} \right] - 2\lambda \sum_{l=1}^{N} (A_g^{-1})_{kl} u_l \tag{9-107}$$

where $[S_{ij}]^{(k)} = \sum_{l=1}^{M} w(k,1) A_{il} A_{lj}$, which is a weighted (by the kth weighting factor) sum of the ith row of $\mathbf{A}$ multiplying the jth column of $\mathbf{A}$; and $u_1 = \sum_{j=1}^{M} A_{lj}$, which is the sum of the lth row of $\mathbf{A}$. The expression (9-107) can be written in matrix form

$$\Phi_k = \mathbf{v}_k^{\mathrm{T}} \mathbf{S}^{(k)} \mathbf{v}_k - 2\lambda \mathbf{v}_k^{\mathrm{T}} \mathbf{u} \tag{9-108}$$

where $\mathbf{v}_k^{\mathrm{T}}$ is the kth row of A_g^{-1}, $\mathbf{S}^{(k)}$ is a matrix with elements $[S_{ij}]^{(k)}$, and $\mathbf{u}$ is a vector of elements u_1.

Differentiating the expression (9-108) with respect to the pth element of $\mathbf{v}_k$, $\mathbf{v}_{kp}$, and then setting the result equal to zero, we yield

$$2 \left(\frac{\partial}{\partial v_{kp}} \mathbf{v}_k^{\mathrm{T}} \right) \mathbf{S}^{(k)} \mathbf{v}_k - 2\lambda \left(\frac{\partial}{\partial v_{kp}} \mathbf{v}_k^{\mathrm{T}} \right) \mathbf{u} = 0 \tag{9-109}$$

Because there is only one non-zero element in vector $(\frac{\partial}{\partial v_{kp}} \mathbf{v}_k^{\mathrm{T}})$, the above leads to

$$\mathbf{S}^{(k)} \mathbf{v}_k - \lambda \mathbf{u} = 0 \tag{9-110}$$

The above equation system needs to be solved in conjunction with the condition (9-105). These two together are equivalent to a $(N+1)(N+1)$ matrix equation

$$\begin{bmatrix} \mathbf{S}^{(k)} & \mathbf{u} \\ \mathbf{u}^{\mathrm{T}} & 0 \end{bmatrix} \begin{bmatrix} \mathbf{v}_k \\ -\lambda \end{bmatrix} = \begin{bmatrix} 0 \\ 1 \end{bmatrix} \tag{9-111}$$

This equation set will be solved for the inverse expression contained in $\mathbf{v}_k$. Notice that the matrix on the left contains only contributions from $\mathbf{A}$ and the weighting factors. The final solution is in the form of

$$(A_g^{-1}) = \left(\sum_{i=1}^{N} [S_{il}]_k^{-1} u_i \right) \Big/ \left(\sum_{i=1}^{N} \sum_{j=1}^{N} u_i u_j [S_{il}]_k^{-1} \right) \tag{9-112}$$

Figure 9.23 shows an example of gravity inversion using the Backus–Gilbert method (Green, 1975). The non-uniqueness in such inverse problems makes it important to use prior information and to emphasize long-wavelength model components. In this case, the initial model was constructed using the information that the batholith extends for 21 km along the profile and its density contrast is -0.15 g/cm^3.

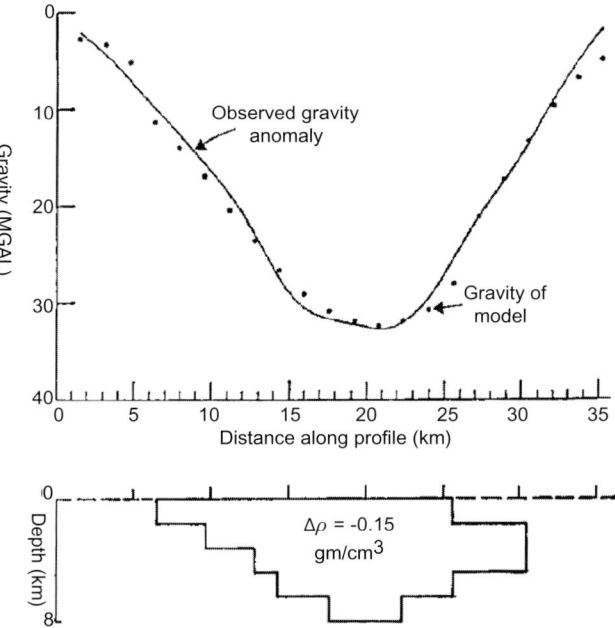

Figure 9.23 A Backus–Gilbert inversion of the gravity profile across the Guichon Creek Batholith, British Columbia (Green, 1975). (Top) Observed gravity anomaly (solid curve) in comparison with model predictions (dots). (Bottom) 2D model of density anomaly from the inversion. The initial model used *a priori* information that batholith extends for 21 km along the profile and its density contrast is –0.15 gm/cm^3.

9.5.2   Iterative inversion

In practice an inversion is often applied iteratively, for several reasons. First, for inversion systems with a large number of model variables, it is impractical to carry out their matrix inversion. Usually, the inverse solutions are derived through iterations of "row-action" methods like conjugate gradient or back-projection which have been applied extensively in medical CT scans.

Second, a major challenge for seismic inversions is the fact that wave and raypaths vary with velocity variations. This is referred to as the *coupling between raypaths and velocities*. In fact, this is the fundamental difficulty making seismic tomography more difficult than its counterpart in the medical field. The coupling makes the inverse non-linear and dependent on the initial reference model. A common solution is to use a good reference model based on *a priori* information. Hence, the problem is translated into using traveltime residues to invert slowness perturbations with respect to the reference model. When the slowness perturbation is small, or when the reference model is very good, this approach often results in reasonable solutions. Furthermore, the whole inverse process can be reiterated so that more realistic raypaths can be achieved based on slowness solutions from previous inversion iterations.

In the 1970s, three types of iterative reconstruction algorithms were introduced in medical CT scanning, namely back projection (BP), algebraic reconstruction technique (ART), and simultaneous iterative reconstruction technique (SIRT). In the 1980s, the LSQR algorithm was invented. These methods will be introduced using a discrete form of traveltime problem:

$$\Delta \mathbf{t}_{N \times 1} = \mathbf{L}_{N \times M} \Delta \mathbf{s}_{M \times 1} \tag{9–113}$$

where the subscripts denote the dimensionality, $\Delta \mathbf{t}_{N \times 1}$ is traveltime residual vector of length N, $\Delta \mathbf{s}_{M \times 1}$ is slowness perturbation vector of length M, and the data kernel $\mathbf{L}$ contains the portions of ray lengths traversing through the model cells.

9.5.2.1   Back-projection

Back-projection (BP) is an iterative method similar to the well-known Gauss–Seidel and Jacobi methods of linear algebra. It begins by writing each element of the data kernel in (9–113) as

$$L_{ij} = h_i F_{ij} \tag{9–114}$$

where h_i is the length of the ith ray and F_{ij} is the fractional value of the ith ray in the jth model cell. Using $\Delta \mathbf{t}'$ whose element is $\Delta t'_i = \Delta t_i / h_i$, the linear system (9–113) is converted to

$$\Delta \mathbf{t}' = \mathbf{F} \Delta \mathbf{s} \tag{9–115}$$

The LS system of the above is

$$(\mathbf{F}^{\mathrm{T}} \mathbf{F}) \Delta \mathbf{s} = \mathbf{F}^{\mathrm{T}} \Delta \mathbf{t}' \tag{9–116}$$

By an expansion of the matrix $(\mathbf{F}^{\mathrm{T}} \mathbf{F}) = \mathbf{I} - (\mathbf{I} - \mathbf{F}^{\mathrm{T}} \mathbf{F})$, we have the estimated solution

$$\Delta \mathbf{s}^{\mathrm{est}} = \mathbf{F}^{\mathrm{T}} \Delta \mathbf{t}' + (\mathbf{I} - \mathbf{F}^{\mathrm{T}} \mathbf{F}) \Delta \mathbf{s}^{\mathrm{est}} \tag{9–117}$$

The back-projection applies the above equation recursively by expressing the kth model estimate $\Delta \mathbf{s}^{(k)}$ in terms of the $(k–1)$th model estimate $\Delta \mathbf{s}^{(k-1)}$:

$$\Delta \mathbf{s}^{(k)} = \mathbf{F}^{\mathrm{T}} \Delta \mathbf{t}' + (\mathbf{I} - \mathbf{F}^{\mathrm{T}} \mathbf{F}) \Delta \mathbf{s}^{(k-1)} \tag{9–118}$$

If the above iteration is started with $\Delta \mathbf{s}^{(0)} = \mathbf{0}$, then the jth element of the first solution is

$$\Delta s_j^{(1)} = \sum_{i=1}^{N} \frac{F_{ij} d_i}{h_i} \tag{9–119}$$

The above expression earns the name of back-projection. If there is only one ray and one model cell, the amount of the slowness perturbation is the traveltime residue divided by the raypath length. If there are several model cells, then the traveltime residue is back-projected equally among them; so those cells with the shorter raypath lengths are assigned with larger slowness perturbations. If there are several rays, a given model cell's total slowness perturbation is just the sum of the estimates for the individual rays. This last step is quite unphysical because it causes the estimates of the model parameters to grow with the number of rays. Remarkably, this problem introduces only long-wavelength errors into the image, so that a high-pass-filtered version of the image can often be useful.

    ## ART and SIRT

The **algebraic reconstruction technique** (**ART**) was introduced by Gordon *et al.* (1970), and later was modified to the **simultaneous iterative reconstruction technique** (**SIRT**) by Gilbert (1972). There is only a small difference between the two: ART updates the new model values immediately after each row-action whereas SIRT updates the new model values after each iteration when all row-actions are completed. As a result, ART depends on the order of the equations while SIRT does not. The early application of SIRT to geophysical problems was on cross-borehole seismic tomography (Dines & Lytle, 1979). A modified SIRT inverse (Comer and Clayton, unpublished manuscript, 1984) is introduced in the following.

In (9–118) replacing $\Delta \mathbf{t}'$ by $\Delta \mathbf{t}$ and $\mathbf{F}^{\mathrm{T}}$ by $\mathbf{L}^{\mathrm{B}}$, the kth iteration SIRT solution is

$$\Delta \mathbf{s}^{(k)} = \mathbf{L}^{\mathrm{B}} \Delta \mathbf{t} + \left| \mathbf{I} - \mathbf{L}^{\mathrm{B}} \mathbf{L} \right| \Delta \mathbf{s}^{(k-1)} \tag{9–120}$$

In the above recursion the new matrix is

$$\mathbf{L}^{\mathrm{B}} = \mathbf{S} \mathbf{L}^{\mathrm{T}} \mathbf{D} \tag{9–121a}$$

$$\mathbf{S} = \mathrm{diag} \left| \frac{1}{\mu + L_j^M} \right|_{N \times N} \tag{9–121b}$$

where μ is a damping parameter, $L_j^M = \sum_{i=1}^{N} l_{ij}$ is the sum of all path lengths through the jth model cell; and

$$\mathbf{D} = \mathrm{diag} \left| \frac{1}{L_i^N} \right|_{M \times M} \tag{9–121c}$$

where $L_i^N = \sum_{j=1}^{M} l_{ij}$ is the total path length of the ith ray.

We may again take the initial model as $\Delta \mathbf{s}^{(0)} = 0$. In each iteration step the traveltime residuals of all rays that pass through a model cell are accumulated and averaged with respect to their path lengths to update the slowness perturbation of that cell. If a cell is intersected by fewer than a certain number of rays, it may be counted as an uncovered cell by setting its slowness perturbation to zero. A generalized scheme of SIRT inverse may be multiplying in front of equation (9–120) by a normalization factor, and $\mathbf{L}^{\mathrm{B}}$ may be defined as the complex conjugate transpose of $\mathbf{L}$.

To adopt a method to accelerate the inversion's convergence (Olson, 1987), a modification on the matrix $\mathbf{S}$ is necessary. Comer and Clayton (unpublished) have shown that

$$\Delta \mathbf{s}^{(k)} = \mathbf{L}^{\mathrm{B}} \sum_{l=0}^{k-1} \left| \mathbf{I} - \mathbf{L}^{\mathrm{B}} \mathbf{L} \right|^l \Delta \mathbf{t} \tag{9–122}$$

Now for a matrix $\mathbf{A}$

$$\sum_{l=0}^{k-1} \left| \mathbf{I} - \mathbf{A} \right|^l \mathbf{A} = \sum_{j=0}^{k-1} C_k^{j+1} \left| -\mathbf{A} \right|^j \mathbf{A} = \mathbf{I} - \left| \mathbf{I} - \mathbf{A} \right|^k \tag{9–123}$$

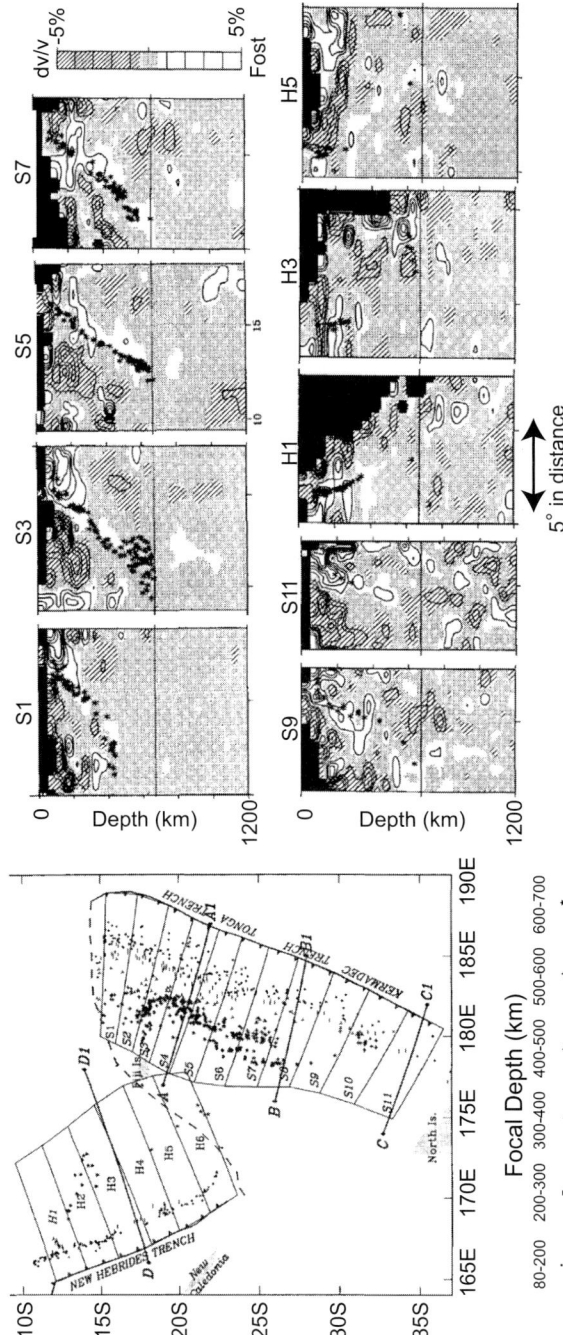

Figure 9.24 A map view and nine cross-sections of a mantle P-wave tomographic study in the Tonga, Kermadec, and New Hebrides region using SIRT method (Zhou, 1990). The earthquake foci are denoted as small stars on the map and cross-sections.

for a general matrix $\mathbf{A}$ we have

$$\Delta \mathbf{s}^{(n)} = \mathbf{L}^{B} \sum_{m=0}^{n-1} \left| \mathbf{I} - \mathbf{L}^{B} \mathbf{L} \right|^{m} \mathbf{L} \Delta \mathbf{s}$$

$$= [\mathbf{I} - (\mathbf{I} - \mathbf{L}^{B} \mathbf{L})^{n}] \Delta \mathbf{s} \qquad (9\text{-}124)$$

Hence, $\mathbf{I} - (\mathbf{I} - \mathbf{L}^{B} \mathbf{L})n$ is the conventional resolution matrix, the mapping matrix between the inverted solution and the real solution.

During the 1980s and 1990s SIRT was among the most popular inversion methods to map mantle velocity anomalies associated with subduction zones and mantle plumes. As an example, Figure 9.24 shows the result of applying a SIRT inversion to map 3D mantle P-wave velocity structure in a large region containing the Tonga, Kermadec, and New Hebrides arcs in the southwestern Pacific Ocean using traveltime data from earthquakes. The deep subduction zones as indicated by the deep earthquake foci are often associated with high velocity anomalies as shown in this figure. A prominent slow-velocity anomaly often appears above the subduction zone at shallow mantle depths. This wedge-shaped slow anomaly below the back-arc volcanoes is interpreted as the signature of the back-arc mantle flow in association with the subduction of the lithospheric slab.

9.5.2.3 LSQR algorithm

The LSQR algorithm developed by Paige and Saunders (1982) is an efficient numerical scheme to solve the minimum-norm LS inversion. To invert linear systems like (9–113) for the solution vector, it is a conjugate gradient inversion that is somewhat similar to the LS inversion using SVD. The SVD solution is constructed in a p-dimensional subspace of the model space, spanned by the eigenvectors of $\mathbf{G} = \mathbf{L}^{T} \mathbf{L}$ belonging to p non-zero singular values. Since the variance of model parameters is inversely proportional to the magnitude of the smallest singular values, often only large singular values are allowed in inversions. This means that the inverse solutions are heavily smoothed.

The LSQR algorithm first tri-diagonalizes $\mathbf{G}$ into $\mathbf{T}p$ with a simple scheme that works as follows. First normalize the new data vector $\mathbf{L}^{T} \Delta \mathbf{t}$ with the norm $\beta_1 = |\mathbf{L}^{T} \Delta \mathbf{t}|$, and take it as the first column $\mathbf{v}^{(1)}$ of the transformation orthogonal matrix $\mathbf{V}$. Thus $\mathbf{v}^{(1)}$ is the first 'basis' vector of a subspace to be constructed in the model space. The next basis vectors are now essentially determined by repeated multiplications with $\mathbf{G}$ and subsequent orthogonalization and normalization. To find the next vector we first construct $\mathbf{w}^{(1)} = \mathbf{G}\mathbf{v}^{(1)}$ $- \alpha_1 \mathbf{v}^{(1)}$ and choose $\alpha_1 = \mathbf{v}^{(1)T} \mathbf{G} \mathbf{v}^{(1)}$ so that $\mathbf{w}^{(1)T} \mathbf{v}^{(1)} = 0$, and then set $\mathbf{v}^{(2)} = \mathbf{w}^{(1)} / |\mathbf{w}^{(1)}|$. To construct $\mathbf{v}^{(3)}$ we must orthogonalize it to the first two vectors using

$$\mathbf{w}^{(2)} = \mathbf{G}\mathbf{v}^{(2)} - \alpha_1 \mathbf{v}^{(2)} - \beta_2 \mathbf{v}^{(1)} \qquad (9\text{-}125a)$$

$$\mathbf{v}^{(3)} = \mathbf{w}^{(2)} / \left| \mathbf{w}^{(2)} \right| \qquad (9\text{-}125b)$$

where $\alpha_2 = \mathbf{v}^{(2)T} \mathbf{G} \mathbf{v}^{(2)}$ and $\beta_2 = | \mathbf{w}^{(1)}|$. One can show that a three-term recursion of this kind suffices to orthogonalize the whole set of vectors $\mathbf{v}^{(1)}, \ldots, \mathbf{v}^{(p)}$, constructed using

$$\gamma_{j+1} \mathbf{v}^{(j+1)} = \mathbf{G}\mathbf{v}^{(j)} - \alpha_j \mathbf{v}^{(2)} - \beta_j \mathbf{v}^{(j-1)} \qquad (9\text{-}126)$$

Multiplying (9–126) by $\mathbf{v}^{(j+1)\mathrm{T}}$ easily shows that the normalizing factor $\gamma_{j+1} = \beta_{j+1}$. Reorganizing (9–126) and assembling $\mathbf{v}^{(1)}, \ldots, \mathbf{v}^{(p)}$ as the columns of a matrix $\mathbf{V}_p$ we obtain

$$\mathbf{GV}_p = \mathbf{V}_{p+1}\mathbf{T}_p \tag{9–127}$$

where $\mathbf{T}_\mathrm{p}$ is a tri-diagonal $(p+1) \times p$ matrix with upper subdiagonal $(\beta_2, \ldots, \beta_{p+1})$, diagonal $(\alpha_1, \ldots, \alpha_p)$, and lower subdiagonal $(\beta_2, \ldots, \beta_p)$.

In analogy with the SVD inversion, the LSQR solution is found by expanding the pth approximation $\mathbf{x}^{(p)}$ to $\mathbf{x}$ in terms of the basis vectors $\mathbf{v}^{(1)}, \ldots, \mathbf{v}^{(p)}$

$$\mathbf{x}^{(p)} = \mathbf{V}_p \mathbf{y}_p \tag{9–128}$$

so that the LS system $\mathbf{A}^\mathrm{T}\mathbf{A}\Delta\mathbf{m} = \mathbf{A}^\mathrm{T}\Delta\mathbf{t} = \beta_1\mathbf{v}^{(1)}$ is reduced to a system $\mathbf{A}^\mathrm{T}\mathbf{A}\mathbf{V}_p\,\mathbf{y}_p = \mathbf{V}_{p+1}\mathbf{T}_p\,\mathbf{y}_p = \beta_1\mathbf{v}^{(1)}$. After pre-multiplication with $\mathbf{V}_{p+1}^\mathrm{T}$, we get

$$\mathbf{T}_p\,\mathbf{y}_p = \beta_1\bar{e}_1 \tag{9–129}$$

which is a tri-diagonal system of $(p+1)$ equations and p unknowns, where $\bar{e}_1 = \mathbf{V}_{p+1}^\mathrm{T}\mathbf{v}^{(1)}$. The system (9–129) can be solved using the simple or damped LS with very little extra computational effort. In the LSQR algorithm, Paige and Saunders (1982) avoid explicit use of $\mathbf{G} = \mathbf{A}^\mathrm{T}\mathbf{A}$ and further reduce $\mathbf{T}p$ to a bi-diagonal matrix.

Figure 9.25 shows horizontal slices of P- and S-wave velocity variations in the depth range of 0–3 km from the sea level. These are slices of 3D crustal velocity models in southern California from tomographic inversions employing the LSQR algorithm (Zhou, 1996). The model region as shown in panel (a) has an area of 300×480 km^2. Each model consists of $30 \times 48 \times 13$ model cells, with a constant cell size of $10 \times 10 \times 3$ km^3. The 3D

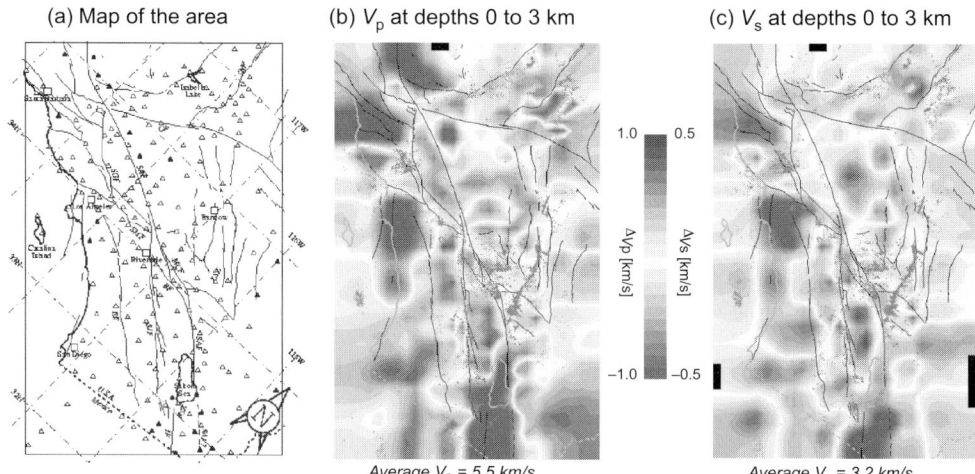

Figure 9.25 (a) Map view of major faults (black curves) and seismologic stations (triangles) in southern California. (b), (c) P- and S-wave velocities in the depth range 0–3 km from inversions using a LSQR algorithm (Zhou, 1996). Areas of insufficient data coverage are blacked out. The correlation coefficient is 66% between the lateral variations of the P- and S-velocities of this layer. Pink dots are earthquake foci. For color versions see plate section.

velocity models are inverted from over 1 000 000 P-wave first-arrival times and over 130 000 S-wave first arrivals from over 37 000 local earthquakes and quarry blasts from 1981 to 1995. 3D ray tracing is employed to determine hypocenters and 3D velocity variations. Considering the large undulation of the Moho in the region, first-arrival times of turning rays are used to gradually update the 3D velocities and raypaths. The near-surface layers of the P and S velocity models show good correlation with the surface geology. Major sedimentary basins exhibit slow velocities and mountainous areas of crystalline rocks show fast velocities.

9.5.3 Constrained inversion

Previously, Section 6.1.2.3 showed the constrained LS method in determining the prediction error operator. Section 8.5.5 further demonstrated the benefits of applying constraints in inversion problems. The damped and weighted LS inversions in Section 9.3.2 are also good examples of constrained inversion. Here this topic is elaborated in two aspects, using *a priori* information and Lagrangian multipliers.

9.5.3.1 *A priori* information

Information known prior to a geophysical inversion is one of the most influential factors for the success of the inversion. One reason is that most inverse problems are imposed for determining the most appropriate model values in pre-defined model setups; hence existing information can be useful in designing model setups and value ranges prior to inversion. Another reason is the widely existing non-uniqueness in geophysical inverse problems due to limited data coverage; thus using prior information increases the chance of choosing the geologically most plausible model among solutions that fit the data equally well. A reasonable constraint based on *a priori* information may help stabilizing the inversion. For instance, first-arrival tomography has limited ability to handle low-velocity anomalies (see Box 9.3 later). Then we may constrain the inverse solutions by assuming that the velocity only increases with depth in all parts of the model.

It is convenient to use prior information because it is widely available, ranging from known physical relations between data and model variables, to reasonable ranges of data and model values, and to smoothness of patterns in model solutions. Information about things such as a non-zero mean of the solution or smoothness (flatness or roughness) of solution patterns, can be included directly as weighting factors into the LS inverse. The new scalar function to be minimized is

$$\Phi(m) = E + \varepsilon^2 L \tag{9--130}$$

where the inner product $E = \mathbf{e}^T \mathbf{e}$ is the original error function, ε^2 as a weighting factor here is the damping factor in (9–54), and L is the Euclidean length of model variables that we want to constrain. L can be modified from its form in the **minimum-norm inverse** (9–54) $L = \mathbf{m}^T \mathbf{m}$, to

$$L = (\mathbf{m} - <\mathbf{m}>)^T \mathbf{W}_m (\mathbf{m} - <\mathbf{m}>) \tag{9--131}$$

where $<\mathbf{m}>$ is the *a priori* expected value of the model parameters, and $\mathbf{W}_m$ is a weighting factor that enters into the calculation for the length of the solution vector.

For example, the flatness of a discrete model of M variables can be expressed by a vector $\mathbf{f}$ of length $(M-1)$:

$$\mathbf{f} = \mathbf{D}(\mathbf{m}- <\mathbf{m}>) \tag{9–132}$$

where $\mathbf{D}$ is a bi-diagonal matrix of dimension $(M-1) \times M$ with a constant diagonal element -1 and a constant superdiagonal element 1. The kth element of $\mathbf{f}$ is $f_k = (m_{k+1}- <m_{k+1}>) - (m_k- <m_k>)$ for k $= 0, 1, \ldots, M$–1. Then, to maximize the flatness of the model solution we can minimize the additional solution norm:

$$\mathbf{L} = \mathbf{f}^\mathsf{T}\mathbf{f} = (\mathbf{m}- <\mathbf{m}>)^\mathsf{T} \mathbf{D}^\mathsf{T}\mathbf{D}(\mathbf{m}- <\mathbf{m}>). \tag{9–133}$$

Comparing (9–133) with (9–131), the weight matrix $\mathbf{W}_\mathrm{m}$ for this case is tri-diagonal with a constant value -1 for both super- and subdiagonal elements, and 2 for all diagonal elements except the first and last one which have a value of 1.

In another example, the roughness of a discrete model can be denoted by the second derivative which, in a discrete sense, is the matrix $\mathbf{D}$ in (9–132) in the form of a tri-diagonal matrix with subdiagonal, diagonal, and superdiagonal elements of values 1, -2, and 1, respectively. Then the corresponding weighting matrix $\mathbf{W}_\mathrm{m}$ is a penta-diagonal matrix with a constant value of -4 for the first super- and first subdiagonal elements, a value of 1 for both the second super- and second subdiagonal elements, and 6 for all diagonal elements except the first and last, with a value of 5. Therefore, by suitably choosing the *a priori* model vector $<\mathbf{m}>$ and the weighting matrix $\mathbf{W}_\mathrm{m}$, we can quantify a wide variety of measures of simplicity for the model solution.

One may, of course, place a diagonal weighting matrix $\mathbf{W}_\mathrm{e}$ on the data vector, according to knowledge of the error in data. Thus, the general minimizing function can be expressed as

$$\Phi(\mathbf{m}) = \mathbf{e}^\mathsf{T}\mathbf{W}_\mathrm{e}\mathbf{e} + \varepsilon^2(\mathbf{m}- <\mathbf{m}>)^\mathsf{T}\mathbf{W}_\mathrm{m}(\mathbf{m} - <\mathbf{m}>) \tag{9–134}$$

There are three common types of inverse solutions corresponding to minimizing (9–134) with different values of the damping factor ε^2. If the forward equation $\mathbf{G}\,\mathbf{m} = \mathbf{d}$ is completely over-determined, we do not need to constrain the model norm, so $\varepsilon^2 = 0$ in (9–134). The inverse solution is

$$\tilde{\mathbf{m}} = (\mathbf{G}^\mathsf{T} \mathbf{W}_\mathrm{e} \mathbf{G})^{-1} \mathbf{G}^\mathsf{T} \mathbf{W}_\mathrm{e} \mathbf{d} \tag{9–135}$$

Second, if the forward problem is completely under-determined, we only have to constrain the model norm so $\varepsilon^2 = \infty$ in (9–134). In other words, we can ignore the first right-hand term of (9–134). The inverse solution in this case is

$$\tilde{\mathbf{m}} = <\mathbf{m}> +\mathbf{W}_\mathrm{m}\mathbf{G}^\mathsf{T} (\mathbf{G}\,\mathbf{W}_\mathrm{m}\,\mathbf{G}^\mathsf{T})^{-1} (\mathbf{d} - \mathbf{G}<\mathbf{m}>) \tag{9–136}$$

Finally, if the forward problem is mixed-determined, we only have to constrain the model with a general (9–134) (i.e., ε^2 is a finite non-zero value). The inverse solution in this case is

$$\tilde{\mathbf{m}} = <\mathbf{m}> +\mathbf{G}^\mathsf{T} \mathbf{W}_\mathrm{e} \mathbf{G} + \varepsilon^2\mathbf{W}_\mathrm{m})^{-1}\mathbf{G}^\mathsf{T} \mathbf{W}_\mathrm{e} (\mathbf{d} - \mathbf{G}<\mathbf{m}>) \tag{9–137}$$

which is equivalent to

$$\tilde{\mathbf{m}} = <\mathbf{m}> +\mathbf{W}_\mathrm{m}^{-1}\mathbf{G}^\mathsf{T} (\mathbf{G}\,\mathbf{W}_\mathrm{m}^{-1}\mathbf{G}^\mathsf{T} + \varepsilon^2 \mathbf{W}_\mathrm{e}^{-1})^{-1} (\mathbf{d} - \mathbf{G}<\mathbf{m}>) \tag{9–138}$$

In all cases, one must take care to ascertain whether the inverses actually exist. Depending on the choice of the weighting matrices, sufficient *a priori* information may or may not have been added to the problem to suppress the under-determinacy.

Another commonly seen form of *a priori* information is that of linear equality constraints in the form of p new equations $\mathbf{F\,m} = \mathbf{h}$. This means that some linear functions of the model parameters equal a constant. One way to think of this is as if we try to force a portion of the equations with exact predictions. The inverse problem now becomes:

$$\text{Solve } \tilde{\mathbf{m}} \text{ from } G - m = d$$

by minimizing

$$|\mathbf{d} - \mathbf{G}\,\tilde{\mathbf{m}}|^2 \tag{9–139}$$

with conditions $\mathbf{F\,m} = \mathbf{h}$.

One approach to solve the above problem (Lawson & Hanson, 1974) is to include the new p equations $\mathbf{F\,m} = \mathbf{h}$ as additional rows to $\mathbf{G\,m} = \mathbf{d}$ and then adjust the new weighting matrix $\mathbf{W}_e$ of dimension $(M + p) \times (M + p)$ so that the new rows are given much more weight than the original rows. The prediction errors of the new equations are therefore zero at the expense of increasing the prediction error of the old equations. Another approach is through the use of Lagrange multipliers: this is a classic method to deal with conditioned extrema problems as discussed in the next section.

9.5.3.2   Constraining by Lagrangian multipliers

The method of the Lagrangian multipliers is illustrated in many books and papers. A simple two-variable example is shown below:

$$\text{Minimizing } E(x, y), \tag{9–140a}$$

$$\text{with condition } \Phi(x, y) = 0. \tag{9–140b}$$

One way to deal with this problem is to solve the conditional equation for an expression of y as a function of x (or vice versa), and then substitute the expression $y(x)$ into $E(x, y)$. So a function with a single variable $E[x, y(x)]$ can then be minimized by setting $dE/dx = 0$.

The method of Lagrange multipliers deals with the constraints in their implicit forms, and hence has advantages when the condition equation is hard to solve. At the values of x and y where the function E is minimized, E is stationary with respect to x and y, i.e., small changes in x and y lead to no change in the value of E:

$$dE = (\partial E/\partial x)dx + (\partial E/\partial y)dy = 0 \tag{9–141}$$

The condition equation relates the perturbations dx and dy:

$$d\Phi = (\partial \Phi/\partial x)dx + (\partial \Phi/\partial y)dy = 0 \tag{9–142}$$

Notice that (9–141) equals zero owing to the original minimization, while (9–142) equals zero because the original condition equation equals zero.

The Lagrange multipliers method uses a weighted sum of the above two equations

$$dE + \lambda\,d\Phi = (\partial E/\partial x + \lambda\,d\Phi/\partial x)dx + (\partial E/\partial y + \lambda\,\partial \Phi/\partial y)dy = 0 \tag{9–143}$$

Box 9.3 Challenge of low-velocity layers to first-arrival tomography: the Yilmaz model

The presence of low-velocity layers (LVLs) is a challenge to first-arrival tomography because the raypaths stay above the LVLs. This point can be generalized to a notion that it is challenging for seismic tomography, especially traveltime tomography, to resolve low-velocity anomalies because raypaths tend to stay away from them. Following a study by Liu *et al.* (2010), here we examine the impact of the LVLs on first-arrival tomographic velocity model building of the near surface using a synthetic 2D near-surface velocity model from Yilmaz (2001) as shown in Box 9.3 Figure 1. In panel (a), dotted curves in white indicate four reversed-velocity interfaces (RVIs) across which the overlying velocity is higher than the underlying velocity. In panel (b) there are no raypaths traversing along the RVIs, so in the corresponding areas A, B and C the tomographic solutions may be corrupted.

Two inversion methods have been applied to the first arrival data generated from the model in Box 9.3 Figure 1b, a commercial grid tomography and a deformable layer tomography (DLT) discussed in Section 9.5.4. Box 9.3 Figure 2 compares three solutions and their differences with respect to the true model in Figure 1. The first two solutions were derived by applying the two methods to data without added noise, and the third solution was from applying the DLT to data with added noise. Interestingly, the two LVLs below areas A and B in Figure 1a are well resolved by the DLT method, probably because their lateral extents are smaller than the source-to-receiver offset. In contrast, these LVLs cannot be resolved at all by the grid tomography as used in the commercial software.

To see the impact of velocity errors on migration results, depth migrations of the two deep reflectors in Figure 1a were conducted using three velocity models created by merging the three solutions of the near-surface velocities shown in the left column of Figure 2 with the deep part of the true velocity model in Figure 1a. The depth migration results are shown

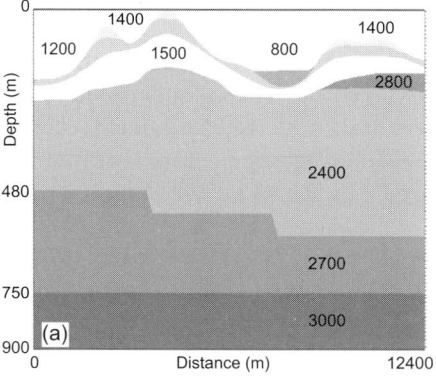
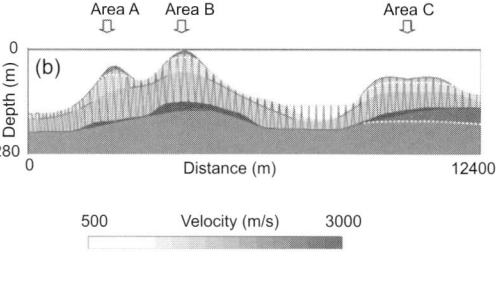

Box 9.3 Figure 1 (a) The Yilmaz 2D velocity model with four reverse-velocity interfaces (RVIs) indicated by white dotted curves. The numbers indicate velocities in m/s. The plot is vertically exaggerated by 12 times. (b) The near-surface portion of the Yilmaz model, serving as the true model for the first-arrival tomography tests. The curves are first-arrival raypaths from some sources to receivers along the surface. Areas A, B, and C are expected to have more velocity errors due to the presence of RVIs.

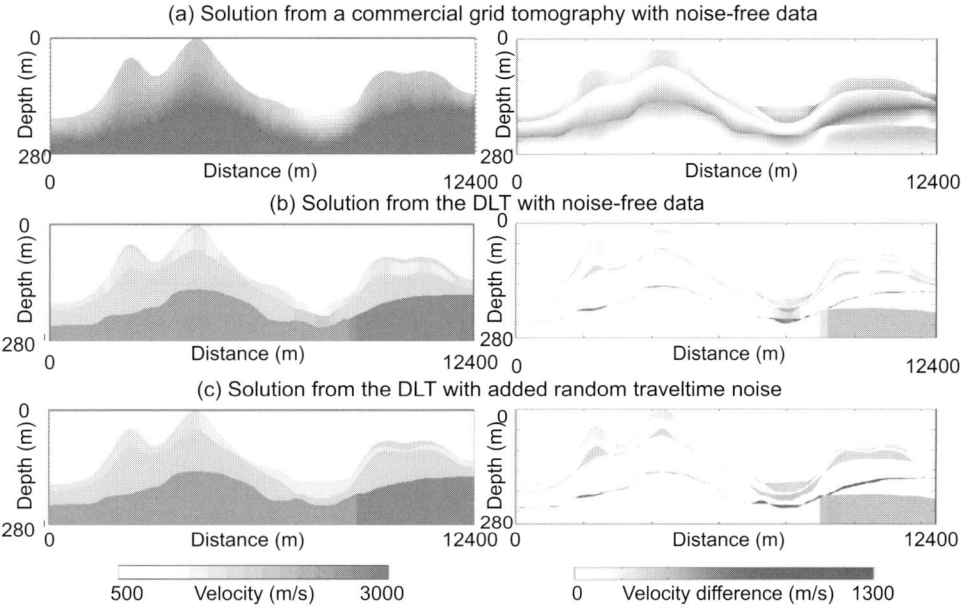

Box 9.3 Figure 2 Three tomographic solutions in the left column and their difference with respect to the true model in the right column. (a) Solution from a commercial grid tomography with noise-free data. (b) Solution from the DLT with noise-free data. (c) Solution from the DLT with added random traveltime noise.

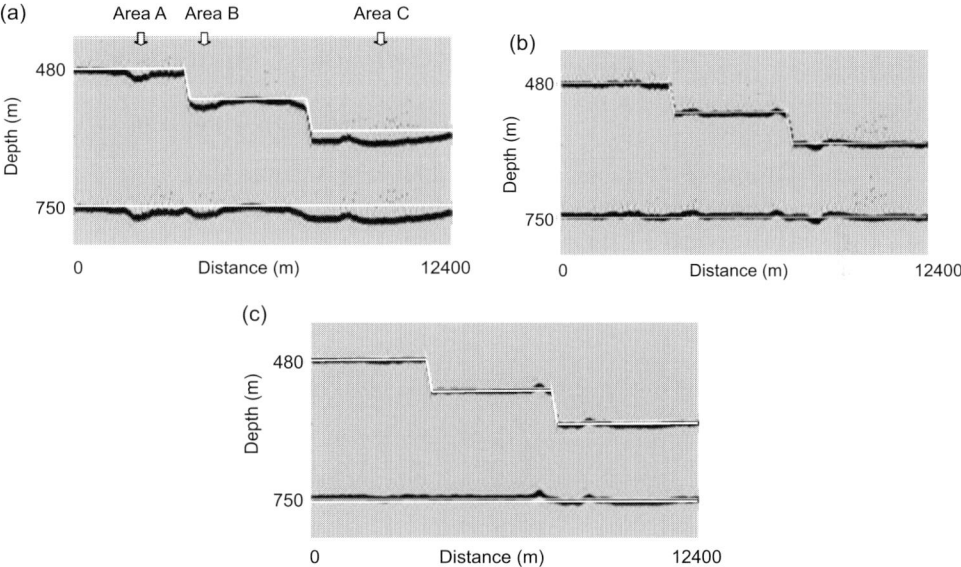

Box 9.3 Figure 3 Migrated sections of two deep reflectors using models containing the near-surface velocity solutions shown in Box 9.3 Figure 1 based on: (a) commercial tomography on noise-free data; (b) DLT on noise-free data; (c) DLT on noise-added data. White lines denote the true positions of the reflectors. Areas A, B, and C in (a) show that the events are over-migrated because the near-surface velocities are too fast.

in Box 9.3 Figure 3, in comparison with the correct positions of the two deep reflectors denoted by white lines. Box 9.3 Figure 3a shows the migrated result using the near-surface velocity model from grid-based tomography, in which, below areas A, B, and C, the events are over-migrated owing to faster velocity at those areas in near-surface. Especially below area C, where the RVI is wider, there is a significant impact of near-surface velocity error on imaging the deep reflectors. Comparing with the results from a commercial grid-based tomography, the DLT method delivers much better near-surface velocity solutions and less error in the images of deep reflectors.

where the weighting factor λ is called the Lagrange multiplier. Since both perturbations dx and dy are variables, the two expressions in the two parentheses of the above equation must all be zero. We now have three simultaneous equations:

$$\partial E/\partial x + \lambda\, \partial\Phi/\partial x = 0$$
$$\partial E/\partial y + \lambda\, \partial\Phi/\partial y = 0$$
$$\Phi(x, y) = 0 \tag{9-144}$$

for three unknowns x, y, and λ.

Therefore, the method converts the conditional minimization problem into an unconditional minimization problem with one more unknown and one more equation. The method can be easily extended to problems with more than two variables, such as (9–139), in the last section, which can be rewritten as

$$\text{Minimizing } E(m) = (\mathbf{d} - \mathbf{G}\,\mathbf{m})^{\mathrm{T}}(\mathbf{d} - \mathbf{G}\,\mathbf{m}) \tag{9-145a}$$

$$\text{with conditions } \Phi(\mathbf{m}) = \mathbf{h} - \mathbf{Fm} = \mathbf{0}. \tag{9-145b}$$

There are M minimizing equations with M unknowns mi ($i = 0, 1, \ldots, M$), and q conditional equations. By this method, there are $M + q$ simultaneous equations for M unknowns and q Lagrange multipliers:

$$\partial E/\partial m_i + \sum_{j=1}^{q} \lambda_j\, \partial\Phi_j/\partial m_i = 0 \quad \text{and } \Phi_j(m) = 0 \tag{9-146}$$

The above $M + q$ simultaneous equations can be put in matrix form as

$$\begin{pmatrix} \mathbf{G}^{\mathrm{T}}\mathbf{G} & \mathbf{F}^{\mathrm{T}} \\ \mathbf{F} & \mathbf{0} \end{pmatrix} \begin{pmatrix} \mathbf{m} \\ \lambda \end{pmatrix} - \begin{pmatrix} \mathbf{G}^{\mathrm{T}}\mathbf{d} \\ h \end{pmatrix} = \mathbf{0} \tag{9-147}$$

9.5.4   Joint inversion: an example for hypocenters and focal mechanisms

A joint inversion involves inverting two or more different types of data when there is sufficient commonality linking the solutions together. A simultaneous joint inversion uses different types of data simultaneously to determine the solution models, such as building a salt velocity model using seismic and gravity data simultaneously. A progressive joint inversion uses each type of data at each step in an iterative fashion, like an iterative process of hypocentral determination using the current velocity model and velocity inversion using the current hypocenter positions. Here joint inversion is illustrated using a simultaneous determination of earthquake hypocenters and focal mechanisms.

Currently earthquake hypocenters and focal mechanisms are determined separately. As shown in Section 9.3.4, hypocentral determination is a process of constraining the hypocenter and source origin time based on arrival times of seismic waves at stations. The determination of focal mechanisms is done traditionally by fitting first-motion polarity data or fitting waveforms. A challenge for hypocentral determination using traveltimes is the trade-off between focal depth and origin time, because most seismic rays take off downwards from the source. In contrast, the first-motion polarity data on a focal sphere are dependent on hypocentral position and velocity model but independent of the origin time. Hence, a joint inversion of both hypocenter and focal mechanism may help to decouple the tradeoff between focal depth and origin time, and the quality of focal mechanism solutions may be improved by using more accurate hypocenters.

9.5.4.1 Impacts of velocity model and hypocenter position on focal mechanisms

First-motion polarities picked from waveforms are independent of traveltime readings. However, different velocity models will result in different ray azimuth and takeoff angles on the focal sphere and therefore different focal polarity patterns. Because most focal mechanism studies are based on depth-dependent 1D velocity models, it is important to analyze the dependency of focal polarity pattern on velocity models.

Figure 9.26 shows the focal polarity patterns and mechanism solutions of three earthquakes in two velocity models in southern California. The events in the upper row used a 3D velocity model, and the events in the lower row used a 1D model that is the average velocities of the 3D model at each epicenter. The left, middle, and right columns, respectively, are for an aftershock of the 1987 Whittier earthquake, an aftershock of the 1992 Landers earthquake, and an aftershock of the 1994 Northridge earthquake. The smaller polarity symbols in the figure correspond to those stations that might be involved with polarity reversals. The polarity patterns in 1D velocity models appear as circular girdles on the focal sphere, indicating that many takeoff angles are insensitive to the epicentral distance. In contrast, takeoff angles calculated from the 3D velocity model (upper row in this figure) generally do not exhibit patterns of circular girdles. For each plot the stress axes are for the focal mechanism determined from the polarity pattern.

Clearly, using different velocity models can cause significant angular rotations of the stress axes, and determination of focal mechanisms using a layered 1D velocity model may be hampered by the fact that many calculated takeoff angles are insensitive to the change in epicentral distance. The insensitivity is due to the fact that most polarity data follow refracted raypaths whose takeoff angles often become constant in a 1D velocity model. In contrast, the lateral velocity variations in 3D velocity models tend to vary the azimuth and takeoff angles even at the same epicentral distance.

Similar to the influence from changing the velocity models, when a hypocenter moves its position, it may significantly alter the focal polarity pattern and hence the mechanism solution, especially when 3D velocity models are used. Figure 9.27 demonstrates this point by slightly moving the hypocenter of a magnitude 3.3 earthquake that occurred on August 6, 1996, on the Mission Creek fault near the San Gorgonio pass. The same 3D velocity model used in Figure 9.26 is used. The focal mechanism in the center of Figure 9.27 uses

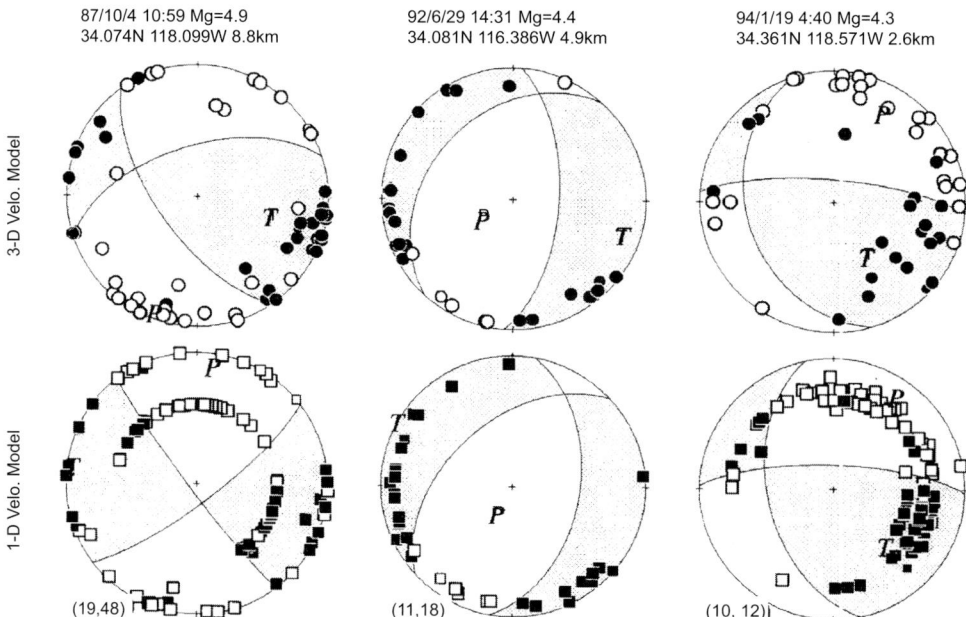

Figure 9.26 Impact of changing velocity models on the focal polarity pattern of three events shown in three columns. Plots in the upper row are based on a 3D velocity model, and those in the second row are based on 1D layer averages of the 3D velocity model. Solid and open symbols, respectively, denote compressions and dilatations. P and T denote stress axes. The two values in parentheses in the lower row are the rotation angles of the P- and T-axes in degrees relative to those of the upper row.

the catalog hypocenter, and the other four mechanisms have their hypocenters moved by 1 km in latitude, longitude, and depth from the catalog hypocenter. Though the absolute hypocentral movement is only 1.73 km, considerable alterations on the focal polarity pattern and mechanism are observed. The P- and T-axes rotate by more than $10°$ in two out of the four cases. The relationship between the focal polarity pattern and hypocentral position appears to be highly non-linear.

This synthetic experiment demonstrates the dependency of focal polarity pattern on velocity model and hypocenter position. The sensitivity of the focal polarity pattern to the small changes in hypocentral position is the main reason for conducting a joint determination of hypocenters and focal mechanisms determined from first motion data. For instance, the perturbed hypocenter in the lower-left case of Figure 9.27 gives a much better fit ($r = 90\%$) to the polarity data than the catalog hypocenter ($r = 83\%$). If this perturbed hypocentral position also delivers a better fit to the traveltime data, we should move the hypocentral solution to this new position.

9.5.4.2   Joint determination of hypocenters and focal mechanisms

The objective now is to search for the hypocenter that will minimize the variance of traveltime residues and also maximize the fitness between the focal polarity data and the

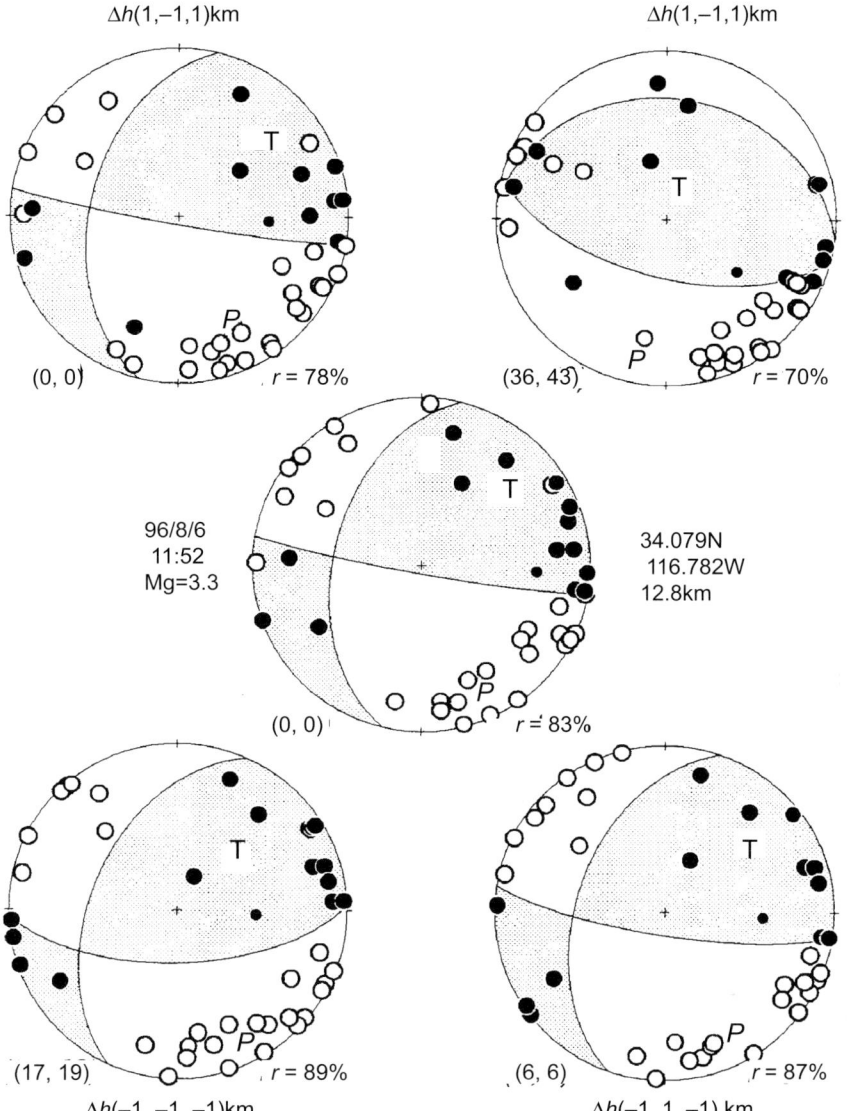

Figure 9.27 Effect of slightly moving hypocenter on focal polarity pattern and focal mechanisms. The center plot is at the catalog hypocenter. The date, magnitude (Mg), and hypocentral coordinates are shown on the left and right sides. The other four plots are at hypocenters that are moved by the amount shown above or below in latitude, longitude, and depth directions. The two values in the lower-left parentheses for each sphere are the degrees of rotation of the P- and T-axes from those of the catalog hypocenter in the center. The r value to the lower right of each sphere is the correlation between each focal mechanism and the polarity data.

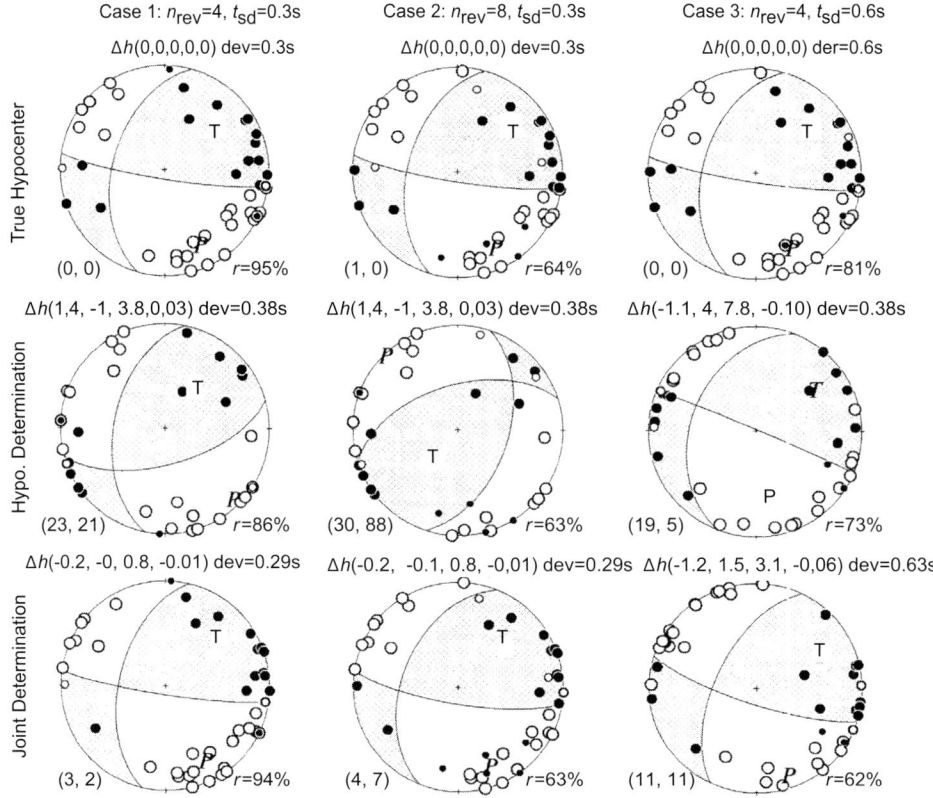

Figure 9.28 Focal mechanism and hypocenter solutions from synthetic tests of a magnitude 3.3 event in three cases (columns) of different noise levels. The number n_{rev} is the number of reversed polarity readings, and t_{sd} is standard deviation of Gaussian traveltime noise added. The top row shows the polarity data and focal mechanism solutions at the true hypocenter. The middle row shows solutions obtained at hypocenters determined using traveltime data alone. The bottom row shows solutions determined jointly using traveltime and polarity data. The hypocentral misfit vector, Δh, contains misfits along (x, y, z) directions in km, and origin time error in sec. *dev* is standard deviation of traveltime residues. Two numbers in the parenthesis are the rotation of the *P*- and *T*-axes from the true axes in degrees. The *r* value is the correlation between the focal mechanism and polarity data.

focal mechanism solution. Considering the non-linear relationship between the data and model variables, we conduct a joint determination by a forward search as used in the master station method discussed in Section 9.3.4.3. In a forward search, it is easy to combine two optimization objectives into a joint determination. The master station method has two steps. In the first step all potential hypocenters are identified by searching for all potential positions that will minimize the traveltime residues. In the second step, a local search is conducted around each potential hypocenter following the misfit gradient in the reference 3D velocity model. Now a maximization of the focal polarity fitting is added to the second step. A new hypocenter is acceptable not only when it reduces the misfit of traveltime residues and increases the first-motion polarity fitting.

The joint determination is illustrated in Figure 9.28 using synthetic simulations taking the catalog hypocenter and the central focal mechanism shown in Figure 9.27 as the true model solutions. Forty-six first-motion polarity readings and 62 P-wave traveltimes were generated using a 3D velocity model. Noise was added by reversing some (n_{rev}) randomly selected polarity readings: four reversed readings for the first and third cases (the left and right columns), and eight reversed readings for the second case (the middle column). Gaussian random noise of zero mean was added to the model traveltimes; the standard deviation of the noise is 0.3 s for the first two cases, and 0.6 s for the third case. The magnitude of the added traveltime noise is greater than the estimated real noise level, because the documented picking error is 0.02–0.05 s, and the standard deviation of over one million first-break residues in the 3D velocity model is below 0.2 s.

The top row of Figure 9.28 shows the polarity data and focal mechanisms obtained at the true model hypocenter, the middle row shows focal mechanisms obtained at hypocenters that were determined using traveltime data alone, and the bottom row shows focal mechanisms and hypocenters that were jointly determined using traveltime and polarity data. There are two types of misfit measures. The first measure concerns the model fitness, including the hypocentral misfit vector Δh and the rotation degrees from the true strain axes. Such model misfits can be gauged directly in a synthetic simulation, but cannot be measured in real applications. The second measure concerns the data fitting level, including the standard deviation of traveltime residues for hypocentral determination and the correlation between the polarity data and the focal mechanism. We can always measure the fitting level in the data space in real applications. However, we should keep in mind that the data fitting level is not always proportional to the model fitness, owing to factors such as non-linearity between the data and model, limitations of the theory, and presence of noise. For instance, the joint determination in the first two cases obtained an erroneous hypocenter with a 0.9 s traveltime deviation which is smaller than the 0.3 s deviation at the true hypocenter. Also in the third case the joint determination gives a focal mechanism that differs from the true answer but better fits the focal polarity data.

The simulations in Figure 9.28 show that the error in the hypocenter location is related to the error in the focal mechanism. As a result, the joint determination performs better than the separate determinations. Even with the high level of added noise in the traveltimes and polarity data, the joint determination converges to focal mechanism solutions that are close to the true answer. The joint determination also reduces the hypocentral misfit vector, including errors in the origin time and focal depth. Considering the relative impact of noise in real traveltime and polarity data, the joint inversion seems to be less affected by the traveltime noise because of its small magnitude and relatively small influence on the focal mechanism solutions. However, the joint inversion could be impaired if many first-motion readings have polarity error.

Exercise 9.5

1. The Backus–Gilbert method has been applied to derive 3D phase or group velocity structures based on surface wave data. Describe how this is done by searching out and reading publications on this topic.

2. The damping factor ε^2 in the minimum-norm LS solution (9–54) appears again in Section 9.5.3.1 on *a priori* information. Explain why the damping factor is able to preferentially suppress high-amplitude oscillations in the solution model. You are encouraged to explore this effect by creating a numerical simulation.

3. A joint inversion can be regarded as a constrained inversion, and vice versa. However, not all joint inversions achieve better results than separated inversions. Make a list of criteria for choosing a joint inversion over separated inversions.

9.6   Summary

- Data fitting and model inversion are two complementary approaches to relate the data space to the model space in order to determine a model that best represents the reality. Data fitting is based on forward modeling to search for models that fit well with the observed data and satisfy our scientific intuition. Model inversion uses our scientific intuition to set up rules about how the models should behave, and then determines the model variations that fit best with the available data.

- Forward modeling methods set the foundation for all model inversion methods. Seismic forward modeling methods include ray modeling, waveform modeling, and physical modeling. Understanding of basic seismic modeling methods such as ray tracing and finite-difference waveform modeling is necessary for all expert geophysicists.

- One way to carry out data fitting is regression, the process of fitting measured data to a simple function representing the trends of the underlying physics. We may quantify misfits in data or model spaces either using norm measures such as L_1-norm, or using statistical quantities such as standard deviation.

- The least squares method as a classic linear inversion is widely applicable in geophysics and other disciplines. It assumes that the misfits follow a Gaussian distribution, and often uses damping and weighting in practice.

- For many geophysical applications, inversion is preferred because of its objectiveness in determining the values of model properties based on the given model parameterization and data. For small inverse problems, a generalized inversion is applicable based on singular value decomposition of the kernel matrix. Large inverse problems require "row-action" solvers such as the LSQR method.

- The main challenges to geophysical inversion are the non-uniqueness of the solutions due to poor data coverage, the non-linear relationship between data and model, and dependency of that relationship on the solution. There are a number of practical ways to constrain the solutions and the resolution, such as the Backus–Gilbert method and various constrained inversions.

- A joint inversion may outperform separated inversion when the null spaces can be reduced by complementary effects from the joint use of multiple data types.

- Practical data fitting and model inversion often require *a priori* information to build reasonable model parameterization and reference models, followed by iterative applications of modeling or inversion, and careful QC of the solutions.

FURTHER READING

Kosloff, D. and E. Baysal, 1982, Forward modeling by a Fourier method, *Geophysics*, 47, 1402–1422.

Menke, W., 1989, *Geophysical Data Analysis: Discrete Inverse Theory*, Academic Press.

Moser, T. J., 1991, Shortest path calculation of seismic rays, *Geophysics*, 56, 59–67.

Ulrych, T. J., Sacchi, M. D. and Woodbury, A., 2001, A Bayes tour of inversion: A tutorial, *Geophysics* 66, 55–69.

Zhou, H., 1994, Rapid 3-D hypocentral determination using a master station method, *J. Geophys. Res.*, 99, 15439–15455.

10 Special topics in seismic processing

Chapter contents

10.1 Some processing issues in seismic data acquisition

10.2 Suppression of multiple reflections

10.3 Processing for seismic velocity anisotropy

10.4 Multi-component seismic data processing

10.5 Processing for seismic attributes

10.6 Summary

Further reading

From the previous chapters the reader should have become acquainted with many of the basic skills of seismic data analysis. Any practice of seismic data processing utilizes some of these skills to solve particular problems, and uses special tools to address more focused issues. Everyone in this field will encounter special issues in his/her career; hence knowing the common features of some special topics is very useful. In this chapter several special processing topics are reviewed to show the use of the basic data processing skills that we have learned, and to expose the reader to some widely seen processing topics. Each of these topics deals with issues associated with a particular problem or property. The first section introduces the issues involved in four aspects of seismic data acquisition: monitoring of source signals including fracking-induced micro-seismicity; monitoring background noises; seismic illumination analysis; and preservation of low-frequency signals. The second section is on suppression of multiple reflections, which is of service to many conventional seismic imaging methods that use only primary reflections. After defining common types of multiples, three classes of multiple suppression methods are introduced. The first is based on the differential moveout between primaries and multiples; the second exploits the periodicity of the multiples; and the third reduces all surface-related multiple energy via pre-stack inversion. The next section reviews the basics in seismic anisotropy, a property of the medium that causes a variation of the speed of seismic waves as a function of the traversing angle. Information on seismic anisotropy helps in improving the fidelity of seismic imagery in fault imaging, and in detecting the dominant orientations of fractures. The fourth section briefly covers multi-component seismic data processing, with an analysis of its pros and cons and with illustrations in wavefield separation, converted wave processing, and VSP data processing. The final section introduces the processing aspect of seismic attributes, including a variety of localized attributes, geometric attributes, and texture attributes, plus related processing in seismic-to-well tie and impedance inversion. To become an expert in the practice of these and other topics in seismic data processing, the reader must learn the fundamentals of seismic wave and ray theory, common issues in seismic data acquisition, processing and interpretation, and spend some time in processing and utilizing field seismic data.

10.1   Some processing issues in seismic data acquisition

In Chapter 1 of this book we observed some processing issues in seismic data acquisition. This section exposes the reader to more processing-related issues in seismic acquisition that dictate the quality of seismic data and imagery. These issues are important in designing seismic surveys.

10.1.1   Monitoring of source signals

Seismic data consist of contributions from four chief factors: seismic sources, impacts of media properties on propagating waves, receiver characteristics, and background noises of various origins. While the purposes of most seismic studies are to understand aspects of either media properties or seismic sources, we have to assess the impacts and characters of all four factors. Considering that most seismic studies aim to map media properties while seismic receivers can be characterized under controlled conditions, this and the next subsections review some basic considerations in monitoring source signals and background noises.

10.1.1.1   Basics of source monitoring

Seismic sources include natural events such as earthquakes due to fault breaks, landslides, and fluid movements, and man-made events such as explosions, fluid injections, and other controlled sources like small dynamite arrays, weight drops, Vibroseis hammers, airguns, water-guns, and sparkers. Seismic studies are called **passive** if using natural sources, and **active** if using man-made sources. For the majority of seismic studies to map subsurface structures as well as fluid and rock properties, we need to assess and suppress inhomogeneities in the seismic data due to different sources, receivers, and noises. For such studies, the ideal signature of seismic sources will be consistent and homogeneous in spatial, temporal, and frequency dimensions. At a given frequency range and time window, a spatially consistent and homogenous source is a sphere of the same amplitude in different azimuth and take-off angles. Such a source signal is called **omnidirectional**, meaning it does not vary over different spatial angles.

Unfortunately, it is nearly impossible to make a perfectly omnidirectional source in the real world. Hence we have to monitor source signals in all seismic experiments. We want to measure the source signal in order to distinguish its signature in seismic data from the effects of media properties. The radiation patterns of an airgun array, as shown in Figure 1.14, for instance, indicate that an omnidirectional source signature can be approximated only within a range of about 0 to 70° in the take-off angle at frequencies much lower than 60 Hz. At higher frequencies, such as the 90 Hz frequency panel in this figure, the signature of an individual airgun emerges, and the notches or low-amplitude stripes in the total source signature will show up in the seismic records. We can expect more artifacts in seismic imagery if we leave these source inhomogeneities in the data.

Further reasons for monitoring source signals arise from the requirement for temporal consistency. This means that the source signature needs to be repeatable during a seismic survey, and that the variations in the source signature must be within a required range. In those seismic surveys using a multitude of controlled sources, we have to make sure that different sources have the "same" (or sufficiently similar) source signals during the survey.

When a controlled seismic source such as an airgun or dynamite is fired, seismic waves are excited in the Earth with significant energy over a frequency range. Suppose the central frequency of the source waves is 30 Hz and the average media velocity around the source is 1500 m/s, then the wavelength of the central frequency is 50 m. Since the physical size of most controlled seismic sources is much smaller than 50 m, the **equivalent source dimension** is usually much larger than that of the controlled seismic sources. Hence, Earth materials around the controlled seismic source, or the **near-field** media, are part of this equivalent seismic source. One of the reasons that offshore seismic data are generally of higher quality than onshore seismic data is that the near-field medium has much more uniform properties in offshore than onshore cases.

Consequently, at lower frequencies the source signals tend to be more omnidirectional, as shown in Figure 1.14 for the offshore cases. However, the influence of the near-field media increases at lower frequencies owing to the increase in the equivalent source dimension. This notion is particularly relevant to onshore seismic surveys because of the inhomogeneity of rocks. The level of the inhomogeneity is usually at a maximum near the surface, or near major lithological boundaries such as sediments versus salt and igneous rocks. In onshore seismic surveys, we prefer source sites consisting of rocks of high integrity, and it is highly desirable to minimize the difference in the rock properties and conditions between different source sites in a survey.

Data processing for monitoring seismic sources requires effective ways to quantify the source signatures over the spatial, temporal, and frequency ranges of the survey. The main objective is to capture the source functions and analyze their variations. Monitoring seismic source signals requires the deployment of a near-field array of receivers in order to record the source signal over as wide a spatial angle as possible. At the data processing stage, a practically useful way is to study a number of common shot gathers over the survey

1700 ms   2100 ms

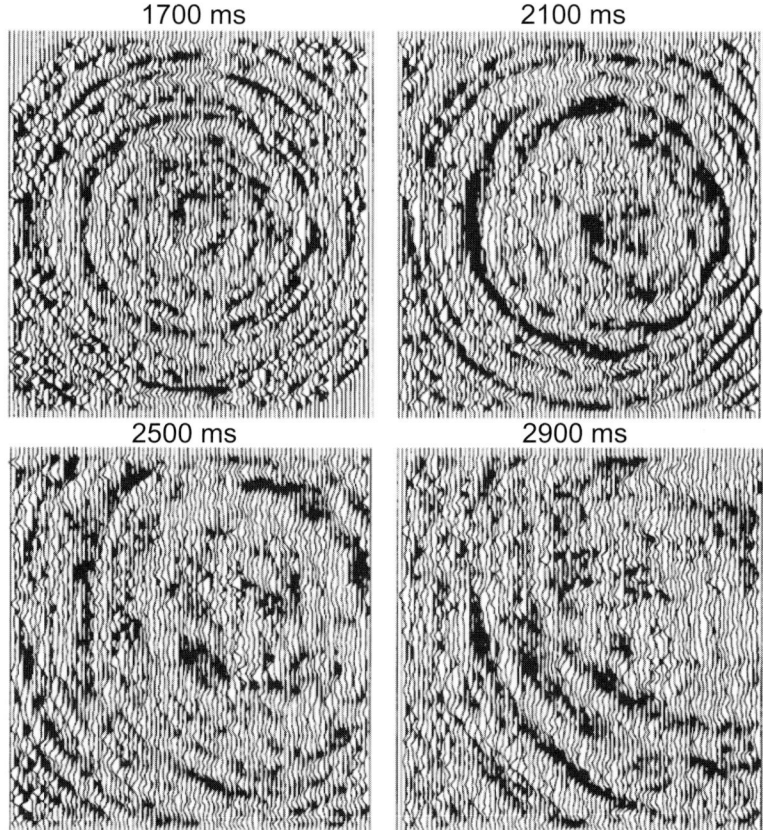

2500 ms   2900 ms

Figure 10.1 Four time slices through a square cross-spread of a common shot gather (Vermeer, 1998). The number above each panel is the two-way time of the time slice.

area, such as the time slices of a common shot gather shown in Figure 10.1. Although the situation here is complicated by factors such as velocity variation and interactions between different wave modes in the subsurface structure, the circular wavefronts can be easily seen in the shallow time slices. At places of reasonably homogeneous media, such time slices at near-field distance offer good assessments of the quality of the source signal, particularly about biases in its spatial radiation pattern.

10.1.1.2 Monitoring microseismicity induced by fracking

One hot topic in recent years is microseismicity induced by **hydraulic fracturing and stimulations**, or **fracking** and **frac** for short, owing to fluid injections in tight petroleum reservoirs as well as to naturally occurring fault and fracture activations. This is of particular importance for reservoir characterization in brittle rocks such as carbonates and has become very popular recently in shales and tight sands of unconventional oil and gas reservoirs. Like the earthquakes induced by impoundments of water reservoirs, the main triggers of the induced microseismicity are changes in stress regime and rigidity due to fluid movements. Therefore, the positions, occurrence times, and mechanisms of the induced seismicity become potential indicators of reservoir conditions in response to the fluid injection programs.

The positions and the origin times of microseismicity are routinely determined using the arrival times of P and S waves, and some common techniques were introduced in Section 9.3.4. The accuracy of such methods has two requirements: a suitable spatial distribution of stations with respect to the event locations, and a robust way to pick the P- and S-wave arrival times. While many microseismicity monitoring programs using sensors placed in wellbores offer high SNR, surface sensors provide much better spatial coverage but much lower SNR than the wellbore data. Picking body-wave arrival times from micro-earthquakes is a labor-intensive job owing to the low signal amplitude and the likelihood of multiple events occurring in the same temporal and spatial windows. A feasible way is to verify the hypocentral solutions by comparing observed waveforms with forward modeling results.

The **focal mechanisms** of large earthquakes (typically magnitude 4.0 or larger) are routinely determined via centroid moment tensor inversions of waveform data. Most of the mechanisms are of shear-faulting or double-couple (DC) nature. For micro-earthquakes we may determine focal mechanisms in a similar fashion if we have good-quality waveform data and sufficient station coverage. As illustrated in Section 9.5.4, we may be able to determine the DC mechanisms when we have first-motion polarity data of sufficient spatial coverage over the focal sphere. In cases when a cluster of microseismic events are likely to have similar mechanisms, DC mechanism solutions may be obtained based on composite first-motion data of events in the same group. An example of such focal mechanism solutions is shown in Figure 10.2.

Figure 10.3 demonstrates the results of a focal mechanism study using a star-like surface geophone array over the horizontal section of a treatment well that is about 1.8 km below the surface (Duncan & Eisner, 2010; Eisner et al., 2010). The DC solutions of induced microseismic events near the horizontal section of the treatment well are shown in a map view and a vertical-section view. The solutions show two main sets of fault planes of different mechanisms: a steeply dipping set that demonstrates normal or reverse dip-slip motion and a less steeply dipping set that seems to fail only with reverse motion. Because normal and reverse motion are unlikely in the same tectonic setting, the authors postulate that the events associated with dip-slip mechanisms are most likely caused by hydraulic-fracture loading, whereas the reverse faulting along the less steeply dipping planes is more likely the result of reactivation of pre-existing faults. This work shows that focal mechanisms may be used to differentiate microseismic events due to new fracturing from other events induced on pre-existing fractures and faults.

While over 90% of large earthquakes have shear-faulting or double-couple (DC) mechanisms, many other earthquakes, especially small events, have non-DC mechanisms. For example, the event shown in Figure 9.2 due to the tragic collapse of a coal mine has a closing-crack mechanism. Besides collapsing events, other examples of non-DC mechanisms include non-planar faulting, tensile failure under high fluid pressure, explosion, volcanic eruption, and landslides. Seismologists typically use moment tensor to represent focal mechanisms, and use moment tensor inversion to determine focal mechanisms based on seismic waveform data. Agencies such as the US Geological Survey routinely publish moment tensor solutions of earthquakes of magnitude greater than 4.0.

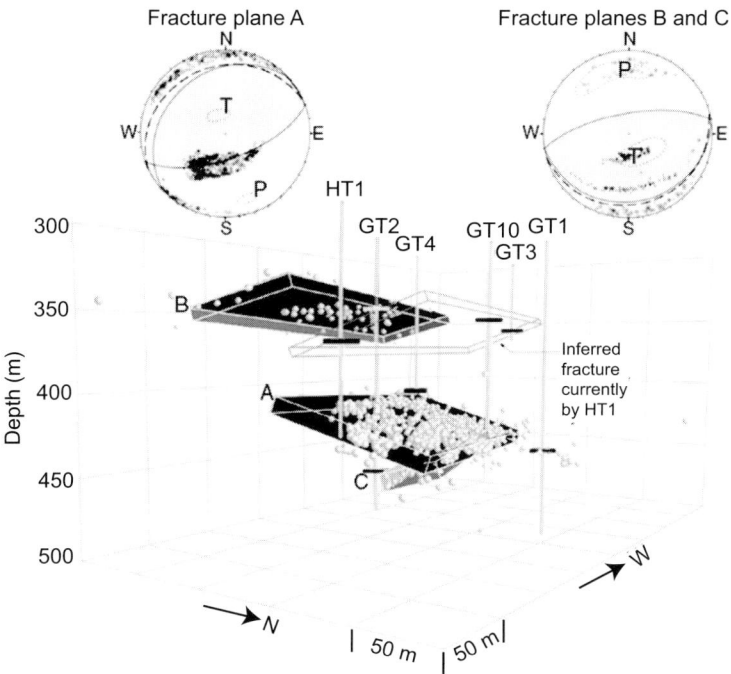

Figure 10.2 Perspective view of fracture planes defined by microseismic event locations in Clinton County, Kentucky (Rutledge *et al.*, 1998). The two fault plane solutions displayed at the top indicate that the seismically active fractures correspond to reverse faulting. The mechanisms were solved as composites by grouping the events from common planes. The dashed curves on the focal hemispheres show the orientations of the planes determined from the respective source locations.

10.1.2   Monitoring of background noises

Even in the absence of known seismic sources, seismic sensors are able to record incoming energies known as the **background noises** or **ambient noises**. These are Earth waves of unidentified origin. While we can expect that movements of any part of the Earth produce seismic waves, a chief component of the background noise is **cultural noise** due to human activities, such as traffic and machinery. Studies (e.g., McNamara & Buland, 2004) indicate that cultural noise propagates mainly as surface waves of high frequencies (more than several hertz) that attenuate within several kilometers in distance and depth. Usually cultural noise is significantly reduced in boreholes, deep caves, and tunnels. A good indication of cultural noise is its strong diurnal variations and its frequency dependence on the source of the disturbance. In addition to earthquakes, other natural sources of background noise include ocean waves and wind noises.

Since background noises are always present, they can be quantified as functions of spatial location and frequency. We can even extract signals from the ambient noise, as shown in Box 10.1. Figure 10.4 shows the USGS **New Low Noise Model**, or **NLNM**, which defines the lowest observed vertical seismic noise levels throughout the seismic

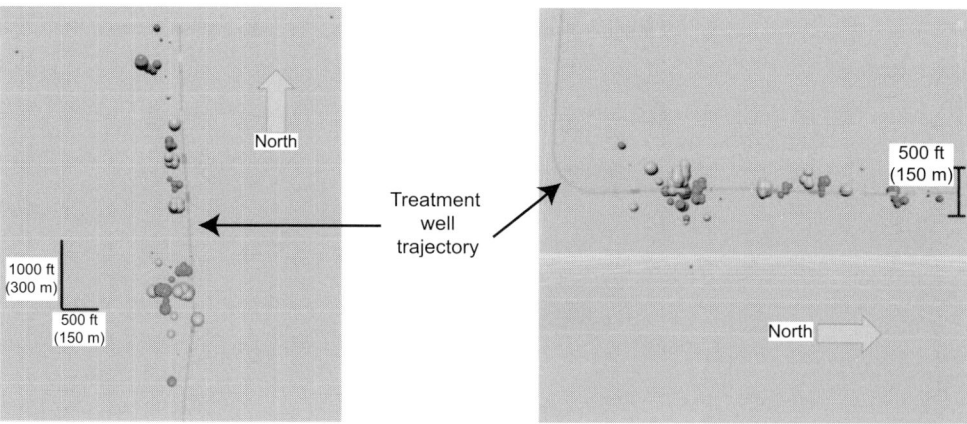

Figure 10.3 (Left) Map view and (right) vertical-section view of microseismic events in a focal mechanism study. The induced events are predominantly located west of the treatment well. Light spheres are the shallow-dipping, reverse-faulting events, showing significant vertical growth, and likely representing the reactivation of pre-existing faults. Dark spheres are dip-slip events that are more confined in depth to be near the treatment well, and are probably newly created by the fracking process (Duncan & Eisner, 2010).

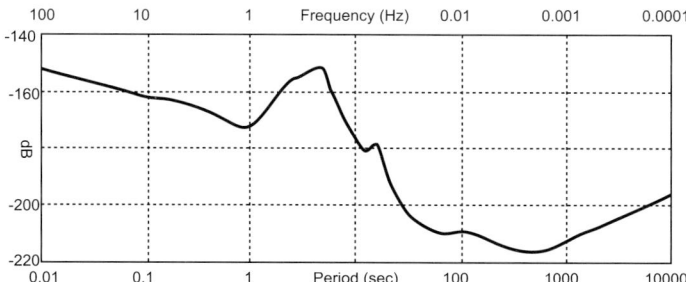

Figure 10.4 The USGS New Low Noise Model (NLNM), expressed as the effective amplitudes of ground acceleration in a constant relative bandwidth of one-sixth decade.

frequency band (Peterson, 1993). The NLNM is very valuable for assessing the quality of seismic stations/sensors and the detectability of small signals. This figure is one of several possible representations of the NLNM. The amplitudes in this figure may also be interpreted as average peak amplitudes in a bandwidth of one-third octave. For instance, the minimum vertical ground noise between the periods of 10 and 20 s is at −180 dB.

By definition of the low noise model, nearly all sites have noise levels above the NLNM, such as at the two sites shown in Figure 10.5. The noise level in a borehole such as at Station ANMO is clearly much lower than that at the surface as at Station HLID, especially at high frequencies. At high frequencies, a noise level no more than 20 dB above the NLNM may be considered as very good in most areas. The marine noise level between 2 and 20 s has large seasonal variations and may be 50 dB above the NLNM during times of winter storms. At longer periods, the vertical ground noise is often within 10 or 20 dB of the NLNM even at noisy stations. The horizontal component of the noise is usually worse than

Box 10.1 Ambient-noise seismology

Although the ambient noise recorded at each seismologic station behaves more or less randomly, cross-correlation of ambient-noise records between a pair of seismologic stations will result in a waveform that resembles the Green function between the receivers. The reason is that the only coherent signal in the cross-correlation function between the ambient-noise records of the two stations is the Green function, which is the impulse response between the two stations. A similar idea is explored in Exercise 5.4.3.

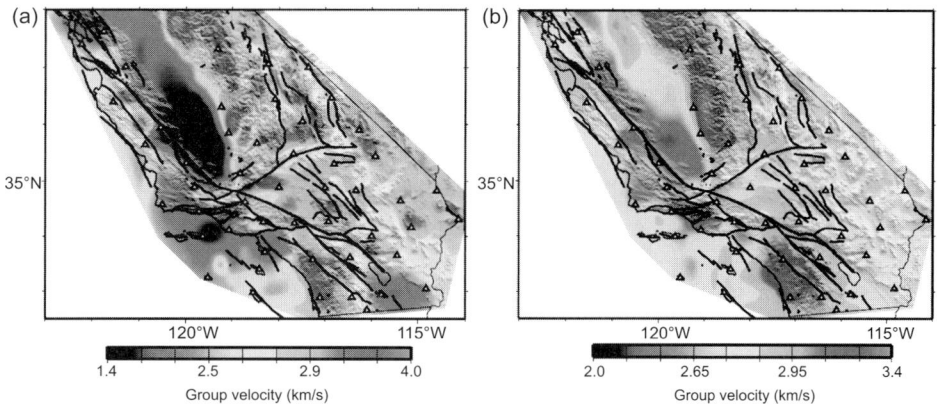

Box 10.1 Figure 1 Maps of Rayleigh-wave group velocities at (a) 7.5 s period and (b) 15 s period in southern California, based on cross-correlating 30 days of ambient noise between USArray stations denoted by triangles (Shapiro *et al.*, 2005). Black curves are major faults. For color versions see plate section.

Box 10.1 Figure 1 shows two maps of Rayleigh-wave group velocities at periods of 7.5 and 15 seconds, respectively, based on cross-correlating 30 days of ambient seismic noise recorded at some USArray stations in southern California. The patterns of these velocity maps are geologically plausible, such as the low-velocity sedimentary basins and high-velocity igneous mountains, similar to the patterns of the bodywave tomography maps of near-surface depths in Figure 9.25 using earthquake sources. The map on the left has shorter period, hence shallower sampling depths, than the map on the right. The examples here verify the feasibility of using ambient noise records to conduct seismic mapping of the subsurface structures, a useful solution for situations when few seismic sources are available. However, the records need to be long enough to ensure their randomness.

the vertical component owing to tilt-gravity coupling. The horizontal noise level can be considered good when it is within 20 dB above the vertical noise level at the same station.

A major advance in seismic data acquisition during the last century was the use of arrayed sensors to suppress noise and enhance coherent signal at each survey node. Consequently a common practice of industry onshore seismic surveys employs a group of about 10

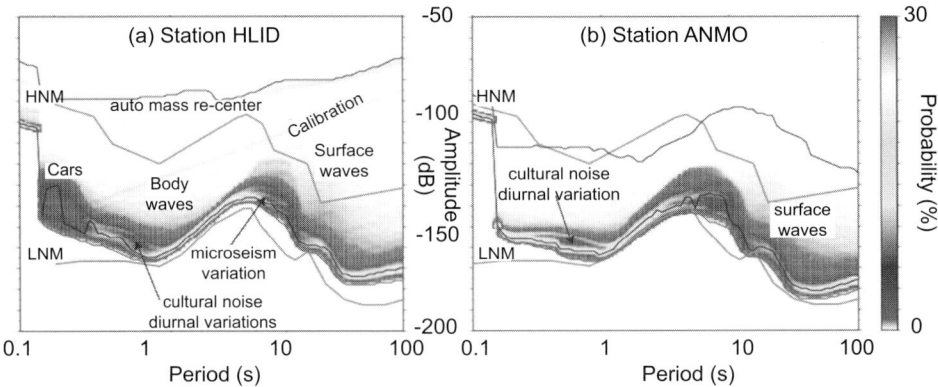

Figure 10.5 Background noise levels at two sites (McNamara & Buland, 2004). (A) Station HLID where the automobile traffic along a dirt road 20 m from the station creates a 20–30 dB increase in power around 0.3 s in period. (B) Station ANMO in a borehole with very low noise level. HNM and LNM are the high and low noise models, respectively, by Peterson (1993).

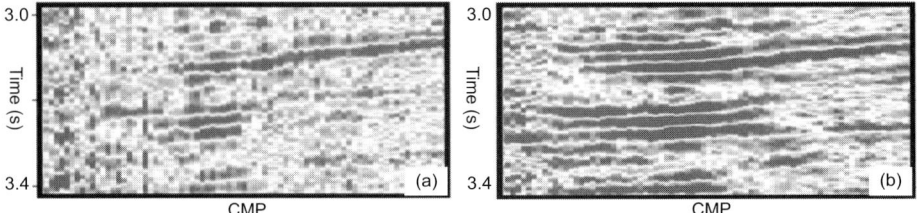

Figure 10.6 Comparison between stacks based on: (a) a simulated 20-m array; and (b) single-sensors with 5 m interval (Strobbia et al., 2009). Improvement in (b) is due to its better accounting for the large and rapidly varying statics in the survey area, using a finer spatial sample interval than that in (a).

geophones at each station location, and the stacked trace of the group is the record of the station. In recent years the demand for better resolution and the much-improved quality of modern sensors has generated a debate between the acquisitions using single sensors versus arrayed sensors. With a fixed number of sensors, single-sensor surveys offer denser spacing between the receivers than the arrayed surveys, but the latter usually provide more effective noise suppression. In practice, a chief criterion is which arrangement better serves the business and scientific objectives.

While seismic surveys with arrayed sensors work well in perhaps most cases, single-sensor surveys can be superior when the quality of individual sensors can be maintained. The single-sensor surveys are especially attractive when denser receiver interval is a key objective. Figure 10.6 compares the stack sections from two types of surveys in a permafrost area (Strobbia et al., 2009). In this case, the strong and short-wavelength static variations due to the permafrost variations deteriorated the 20 m array solution shown in Figure 10.6a. In contrast, the single-sensor solution with 5 m intervals shown in Figure 10.6b shows good improvements. In practice we must justify the extra cost in time and money for denser arrays against the improvements they provide in imaging quality.

10.1.3   Seismic illumination analysis

10.1.3.1   Factors behind seismic illumination

The illumination analysis of seismic data is a quantification of the spatial coverage of the traversing seismic waves through the subsurface target. It is a valuable tool for assessing the resolution and biases of the imaging solutions for the given data. There are five chief factors influencing the illumination: (1) data wave type and SNR; (2) acquisition geometry; (3) overburden structure; (4) complexity of target; and (5) analysis method. For the first factor, the propagation of different wave types may follow different physical principles that govern their wavepaths and therefore the illumination and SNR. Seismic reflection waves, which constitute the majority of seismic data in petroleum exploration, require the presence of coherent reflectors which are equivalent to alignments of seismic impedance contrasts. Nearly all industry reflection seismic imaging studies use subcritical reflections and shots and receivers along the surface, meaning that the reflection angles are generally within 30–40° from the vertical orientation.

The second factor, the acquisition geometry, defines the distributions of and spacing between shots and receivers. Here a key parameter is the range of shot-to-receiver offset, which dictates the range of reflection angles for reflection waves, or depth range of wavepaths or raypaths for turning waves. Another key parameter is the azimuthal range of shot-to-receiver directions for 3D surveys. For 2D seismic surveys, we prefer to align the shot-to-receiver direction along the regional dip direction of the overburden and target, assuming that they have a consistent regional dip direction. When either the topography, or the overburden, or the target is of three-dimensional nature, a consistent regional dip direction will not exist. In this situation, we have to use a 3D seismic survey rather than a 2D survey. In 3D seismic surveys, we prefer the range of the azimuthal angle of the shot-to-receiver direction to be as wide as possible. The shot spacing and receiver spacing dictate whether we can have enough multiplicity of data coverage to perform adequate statics analysis, velocity analysis, and imaging.

The third factor, the overburden structure, refers to the impact of the heterogeneity of the overburden on the illumination. Because seismic waves tend to spend a minimum amount of their traveltime in places of locally slower velocities, wavefronts bend backward over slow velocities and forward over fast velocities. The backward and forward bendings of wavefronts produce convergence and divergence in wave amplitudes, respectively. If there is a flat reflector of uniform reflectivity below an overburden of high- and low-velocity anomalies, the imaged result of the flat reflector may have **time pull-ups** of low-amplitude reflections beneath high-velocity anomalies in the overburden, and **time sags** of high-amplitude reflections beneath low-velocity anomalies in the overburden. The kinetics of the situation is similar to the case of fault shadow as portrayed in Figures 7.29 and 7.30. In addition, the reflectivity in the overburden will partition the traversing wave energy and limit the angular range of reflection waves passing through the overburden.

Figure 10.7 shows an example of evaluating the effect of overburden on seismic illumination by Rickett (2003). Panel (a) is a synthetic velocity model, which is part of the so-called "Canadian foothills over-thrusting onto the North Sea" model. The model contains a flat reflector of uniform amplitude at 3.9 km depth, with an overburden structure of significant complexity. This "true" velocity model is used to derive the depth migration results shown

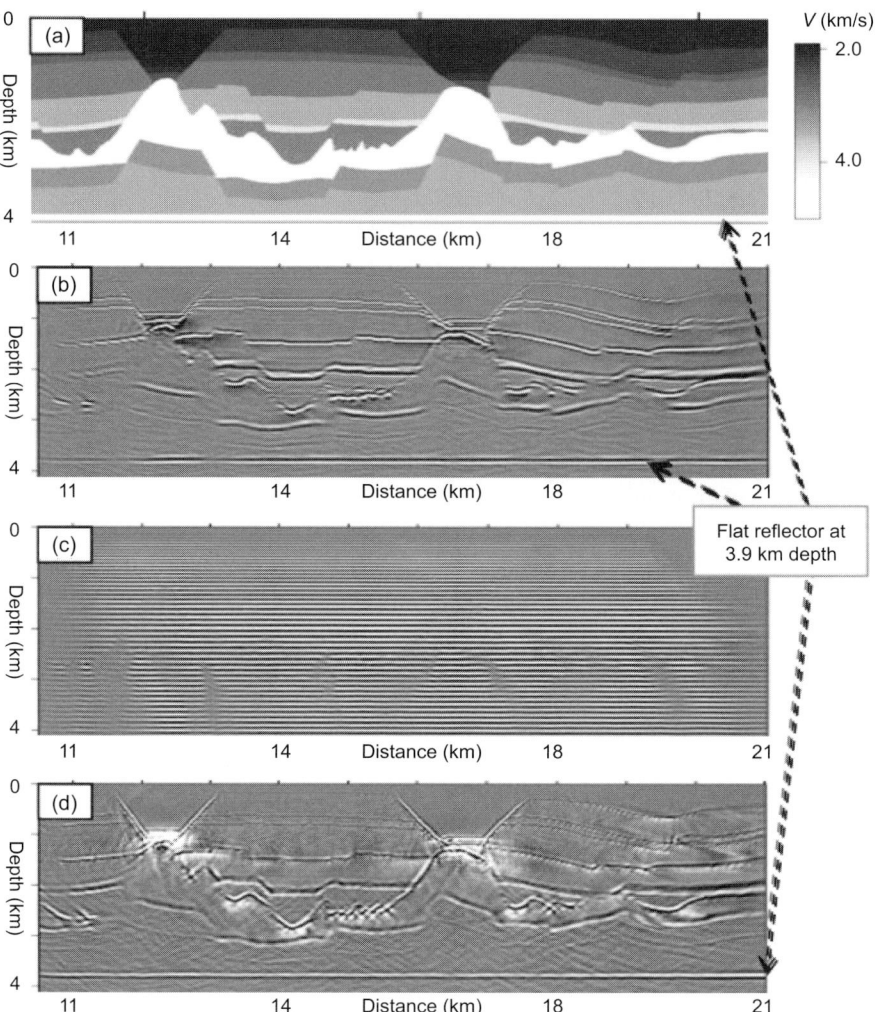

Figure 10.7 (a) Velocity model. (b) Shot-profile migration. (c) Result of modeling and migration with a flat-reflector model. (d) Least squares migration. (From Rickett, 2003.)

in the remaining panels of this figure. Panel (b) is the result of a shot-profile migration using numerically modeled waveforms as the data. We can see the variations in the amplitude and geometry of the flat reflector at 3.9 km depth, indicating the influence of the overburden. Panel (c) is the migration result using forward modeled data based on a reflectivity model consisting of many flat and uniform reflectors. This panel shows an increase in distortions of the geometry and amplitude of the flat reflectors with increasing depth. Finally, panel (d) is the solution after 10 iterations of a least squares (LS) migration using the same data as for (b). As discussed in Box 9.2, LS migration is a direct application of the LS inversion to seismic migration. From the point of view of seismic illumination, the amplitude and geometry of the flat reflector at 3.9 km depth are much improved in panel (d) with respect

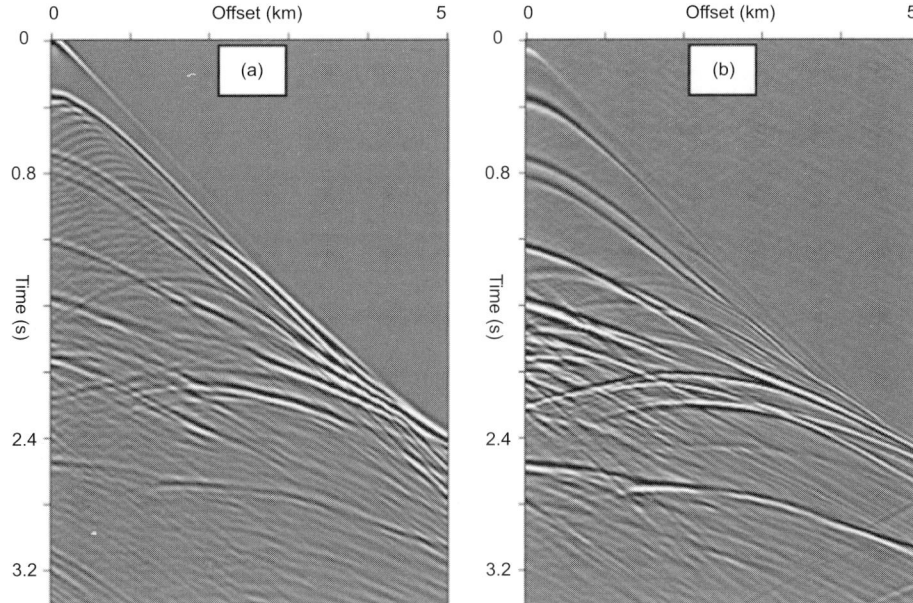

Figure 10.8 Two synthetic shot gathers at a surface shot location of 18.2 km in the velocity model shown in Box 10.1 Figure 1a (Rickett, 2003) using: (a) full 3D two-way finite-difference modeling, (b) 2D linear one-way modeling.

to (b). Interestingly, the migrated result with better accounting for the illumination gives a clearer picture of both the signal (reflectors) and various imaging artifacts. A problem of the inverted result in panel (d) is that the phase angle of the reflectors has been altered to nearly being reversed.

The fourth factor, the complexity of target, refers to the geometrical complexity of reflectors when the data are reflection waves. Since the majority of seismic surveys use shots and receivers deployed along the surface of the Earth, the complexity of the reflectors means the dipping angle, number of reflectors, and variations in the geometry and reflectivity of the reflectors. Due to the unknown nature of the complexity of reflectors in the real world, simplified reflectors such as flat and linear reflectors are often used to assess the general trend. As an example, flat reflectors of uniform reflectivity are used in Figure 10.7 so that we can easily evaluate the impact of different factors on the illumination.

The last factor, the analysis method, often plays a vital role in the quality of the seismic illumination analysis. Figure 10.8 compares two synthetic shot gathers (Rickett, 2003) using two different 3D finite-difference modeling algorithms. Panel (a) is from full 3D two-way finite-difference modeling, and (b) is from 2D linear one-way modeling. Although the timing of most events is similar, they differ significantly in amplitude and phase angle. Panel (a) shows more ringing patterns which may be accurate reflections due to digitizing dipping reflectors as step functions. Panel (b) contains more artifacts in its upper-right portion before the first breaks. We should keep in mind that none of the numerical modeling methods can include all the wave propagation effects contained in the field data. Hence an expert in seismic illumination also needs to be an expert in seismic wave propagation.

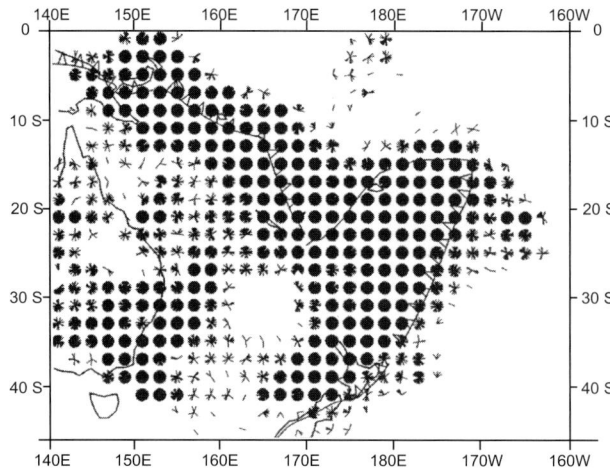

Figure 10.9 Azimuthal distribution of ray hit counts at 500–550 km depth range in a tomographic study of the Tonga, Kermadec, and New Hebrides region (Zhou, 1990). For each model cell, all traversing rays are denoted as short lines from the center of the cell.

10.1.3.2   Quantification of seismic illumination

Quantification of seismic illumination typically involves forward modeling using actual survey geometry in a realistic velocity model. The assessment of the illumination may be done kinematically using ray theory, or dynamically using the wave equation. Owing to practical constraints such as the project time, computational resources available, and lack of accurate velocity models, simplifications in the computation methods and models are often taken. The most aggressive simplification uses ray theory in a homogeneous or a layer-cake velocity model, which leads to the use of source-to-receiver midpoints to assess the hit counts over the target reflectors. Following this approach we count the folds of reflections as the numbers of midpoints in CMP bins.

Figure 10.9 shows an example of using ray theory to map hit counts in a 3D traveltime tomography study of the subducted lithospheric slabs beneath the Tonga, Kermadec, and New Hebrides region. The model volume consists of a 3D mesh of cubic cells. After 3D ray tracing, the ray segments in each cell can be collected to show the number of rays and their distributions over the azimuthal and dip angles. In this figure, the ray hit counts of a layer of the cells at a depth range of 500–550 km are shown. All ray segments in each cell are shown as short lines from the center of the cell along the directions of the raypaths. We have seen in Chapter 8 that the quality of traveltime tomography depends on the ray hit counts and crisscrossing level of the rays. Maps of ray hit counts like the one shown here give a good indication of whether the targets can be mapped with sufficient resolution.

To evaluate the resolution of waveform images such as the solutions of seismic migration, we shall take a wave-equation approach in the illumination analysis. Xie *et al.* (2006) suggested a local plane-wave analysis of the target illumination using an illumination matrix based on localized, directional energy fluxes for both source and receiver wavefields. Figure 10.10 is a sketch showing the structure of a 2D local illumination matrix. Each element in the matrix corresponds to an independent scattering observation of the

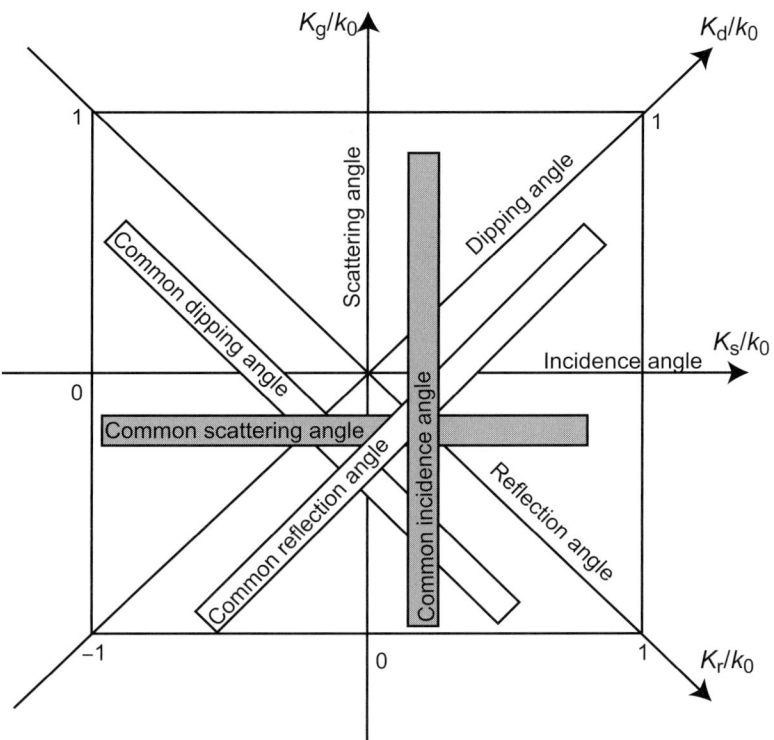

Figure 10.10 Structure diagram of a 2D local illumination matrix. The horizontal and vertical coordinates are the horizontal components K_s and K_g of incidence and scattering wavenumbers, or the incidence and scattering angles. The main and auxiliary diagonal directions are horizontal-reflection wavenumber K_r and dipping wavenumber K_d, respectively. Different angle gathers are shown as strips with different orientations.

target. Such an approach can handle forward multiple-scattering phenomena such as focusing/defocusing, diffraction, and interference effects in complex velocity models.

Figure 10.11 shows local illumination matrices at four locations in the 2D SEG/EAGE salt model using normalized values of the energy fluxes. The end-on spread of the data acquisition (e.g., Figure 1.2b) results in a shift of the illumination toward the upper-left direction in all locations, and a greater shift at the shallowest location (a). The model is illuminated by wider effective apertures at shallow depth, and the illumination spans a relatively narrow aperture at deeper depth, particularly in the subsalt location (c). Owing to the shadowing effect of the salt overburden, the illumination level in the subsalt region like location (c) is weak and apparently missing for some dipping and reflection directions.

10.1.4   Preservation of low-frequency signals

We have seen in Chapter 4 that the resolution and fidelity of seismic data depend largely on the data bandwidth, which is the range of the data frequencies that are of sufficiently high SNR. Nearly all seismic processing and imaging projects require data to have broad

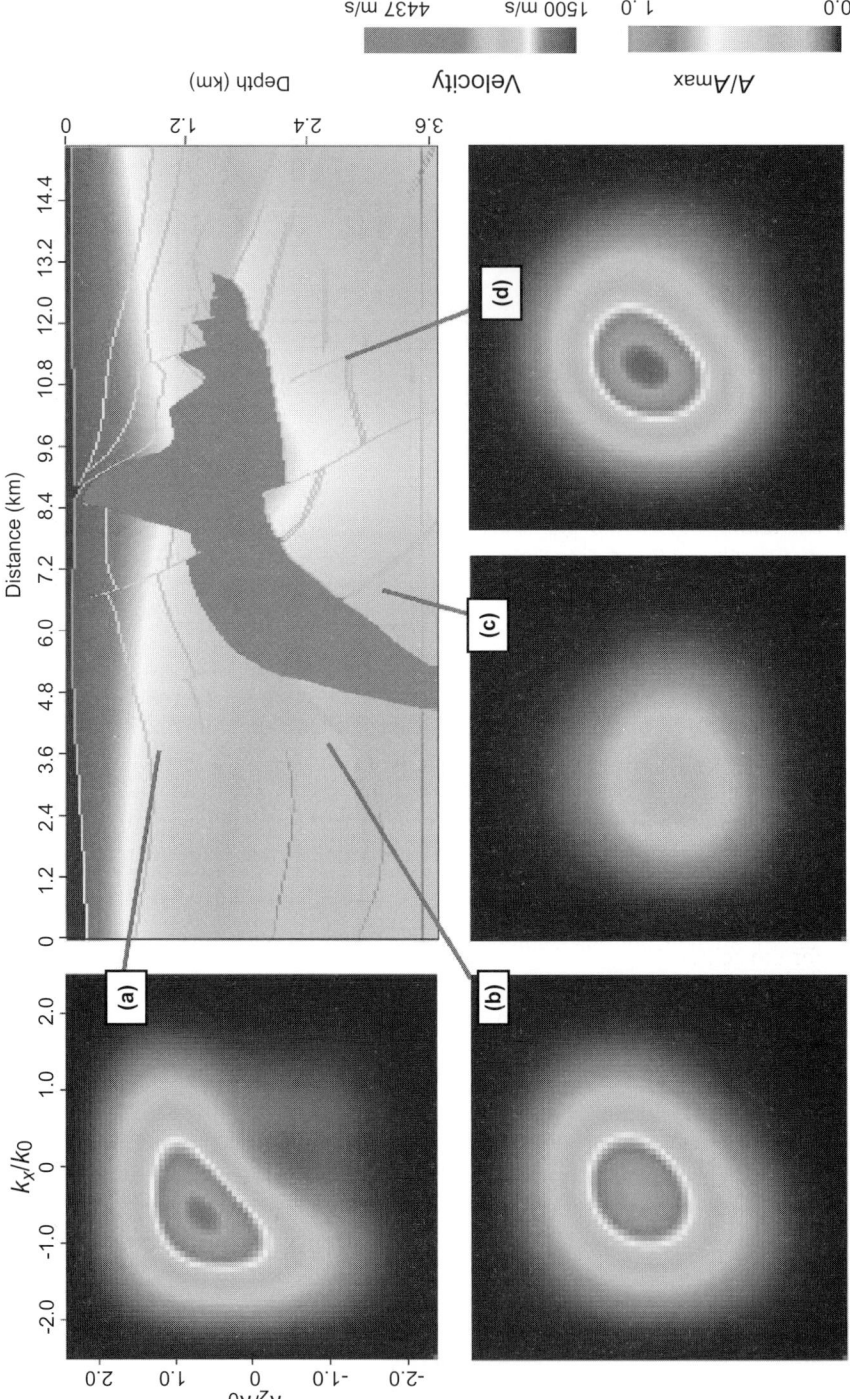

Figure 10.11 Local illumination matrices at four locations in the 2D SEG/EAGE salt model shown in the upper right panel (Xie *et al.*, 2006). For color version see plate section.

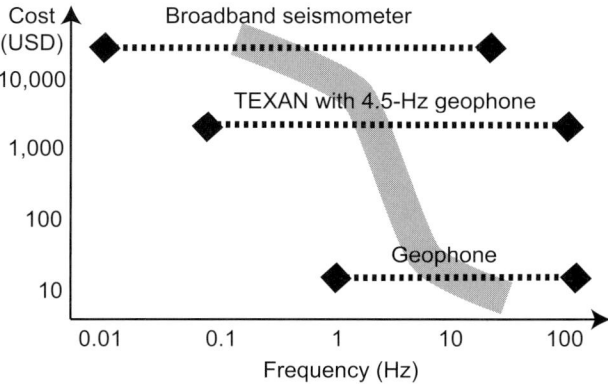

Figure 10.12 Current cost ranges versus marked frequencies of three types of portable seismometers. The lower the frequency the higher price of the seismometers.

enough bandwidth, such as two or higher octaves. Though in theory the bandwidth can be maintained by expanding either the higher or lower frequencies of the data, the way that rocks and fluids attentuate the propagation of seismic waves means the high-frequency signals quickly lose their SNR. Hence, to preserve the resolution and fidelity of seismic data, it is more practical to preserve the low-frequency signals.

In the early days of exploration seismology, a rule of thumb is that the SNR will decrease to below the acceptable level after 100 cycles of wave oscillations in young or weathered sedimentary rocks that are often called "soft rocks". For instance, if we want to map an oil reservoir that is buried below soft rocks of 2 km in thickness, the two-way traveling distance of the reflection waves is over 4 km, so the shortest data wavelength is 4 km / 100 = 40 m, if the SNR is unacceptable beyond 100 cycles. If the average velocity is 2 km/s, then reflection data of frequencies higher than 20 Hz will not have sufficient SNR in this case. Although modern acquisition instruments and signal-enhancement technology have increased the range of wave propagation with sufficient SNR beyond 100 cycles, there is a physical limit for the achievable SNR at high frequencies.

Hence for highly attenuating media such as most soft sedimentary strata, the only practical means to maintain the bandwidth of seismic data is to acquire and retrieve the low-frequency signals whose SNR decreases much more slowly than that of the high frequencies. After some of the largest earthquakes, for example, long-period Rayleigh waves can maintain good SNR for many months after each occurrence. However, the cost of seismometers increases exponentially with their ability to record lower frequencies. Recording lower-frequency seismic waves requires instruments with larger dimensions and/or higher sensitivity; either way will increase the cost of the instruments. Consequently, the **marked frequency** of a seismometer is typically its **low-corner frequency** below which the SNR starts to decrease. The spectral response of a digital seismograph is typically flat from its marked frequency to the **high-corner frequency**. The high-frequency limit of a digital sensor is set by its highest sampling rate.

A practical limit for acquiring low-frequency seismic signals is the cost of the instruments. Figure 10.12 shows the relationship between the current costs and marked frequencies of three types of portable seismometers. Clearly, the cost of seismic sensors rises

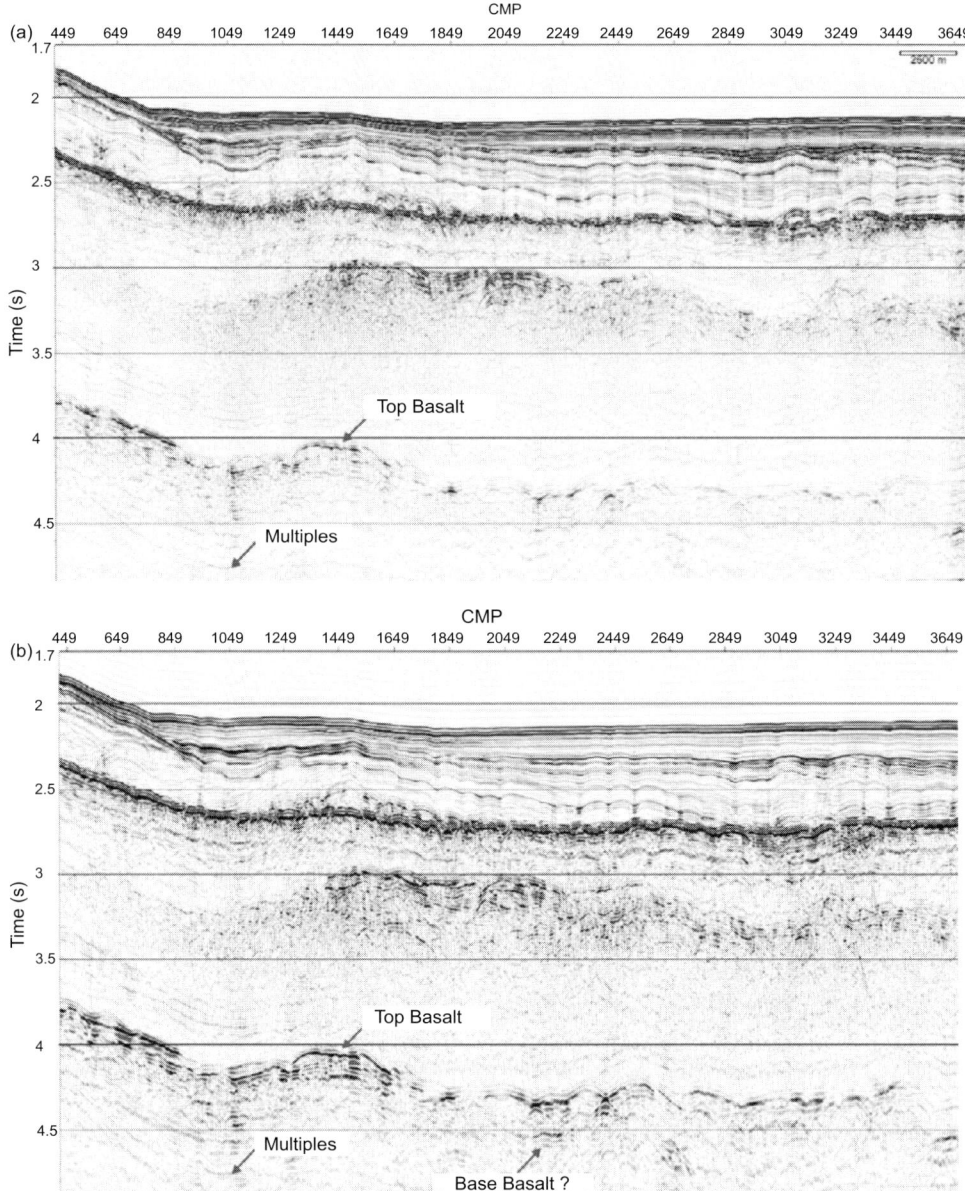

Figure 10.13 Comparison between seismic imageries in an area with basalt layers. (a) Based on conventional data and (b) based on low-frequency data (Ziolkowski *et al.*, 2003).

exponentially with the increase in their low-frequency sensitivity. The broadband portable seismometers, such as those manufactured by Guralp Systems in Britain, and Nanometrics Seismological Instruments in Canada, routinely acquire data of good SNR over the frequency range of 0.01 to 50 Hz. The drawback, however, is their high cost. Currently each such broadband seismometer costs more than $10 000, which is impractical for seismic surveys for petroleum reservoirs that require tens of thousands of sensors. Consequently, only the low-cost short-period seismic sensors such as geophones and hydrophones are used in exploration seismology. **Short-period seismometers** are those whose marked frequencies are higher than 1 Hz. The TEXAN shown in Figure 10.12 is a seismic recorder manufactured by Refraction Technology. Such an instrument is often used in crustal seismology where each TEXAN is hooked with a low-frequency geophone.

Retrieving low-frequency signals from the short-period seismic data is of critical value for many areas of exploration seismology and crustal seismology. As an example, Figure 10.13 compares two images in an area with basalt layers whose high interval velocities and irregular surfaces make it one of the most challenging places for seismic imaging. Because most lava basalts are in thin layers, the low-frequency seismic waves will penetrate through them much more effectively than the high-frequency waves. In this figure the image from the low-frequency data shows substantial improvement over that based on the conventional data.

Exercise 10.1

1. Explain or refute a notion that in early days of reflection seismology people took –70 dB as the noise level when there were no other references available.

2. Conduct a literature search to summarize the principles of ambient-noise seismology discussed in Box 10.1. What are the pros and cons?

3. There are possible errors in the phase angle of the reflectors shown in panels (b) and (d) of Figure 10.7. How will you evaluate errors in the phase angle? What are your estimates of phase angle errors for these two panels?

10.2 Suppression of multiple reflections

10.2.1 Definition of multiples

Multiple reflections or **multiples** are seismic waves that have been bounced more than once from subsurface reflectors or scatters, in contrast to the primary reflection waves, or **primaries**, which have been bounced only once. Suppression of multiples is motivated by the fact that most traditional reflection seismic methods use only the primary reflections to map the temporal and spatial distributions of subsurface reflectors. It is a practical decision to use only primaries because they are the easiest to recognize and process for subsurface imagery, and they typically have much higher amplitudes and SNR than the multiples. However, this decision requires the suppression of all non-primary reflection waves, including first breaks, surface waves, converted waves, and of course multiples. Of all waves other than the primaries, multiples stand out as the most difficult to suppress because they usually overlap with and are often indistinguishable from the primaries. Multiples left in the data will generate artifacts that mask the images supposedly based on primaries, making interpretation difficult.

All seismic reflection waves are bounced from reflectors that are interfaces characterized by a discontinuity in **seismic impedance**, the product of seismic velocity and density. Seismic impedance becomes acoustic when only the P-wave is involved, but elastic when both P- and S-waves are concerned. For sedimentary rocks, reflectors often correspond to unconformities owing to consistent changes in the alignment of rock fabrics across the hiatuses. A reflector of primary reflections is also a reflector of multiple reflections. Although multiples have lower amplitude and typically longer traveltimes than the corresponding primaries (e.g., see Box 9.1 Figure 1), they are colored noise that may be mistaken for primaries. The multiples are particularly strong at coherent and strong reflection interfaces such as the top and bottom surfaces of water bodies, salt bodies, basalt layers, and carbonate platforms, as well as major unconformities and lithological boundaries. Because the top and bottom of a sea water body are coherent and strong reflectors, multiples in marine seismic data are classified into **surface multiples** and **internal multiples**. The former have at least one downward reflection initiated at the water surface which is among the most reflective interfaces in nature, and the latter have all of their downward reflections occurring at the water bottom or below.

The naming of multiples considers their raypaths, as shown in the examples in Figure 10.14 from Dragoset and Jeričević (1998). The **order of the multiples** refers to the number of downgoing reflections (or refractions), particularly when the reflection occurs at the water surface in marine data owing to the strong amplitude of such multiples. For example, panels (a), (c), (d), (g) in this figure show first-order multiples, and the remaining panels show second-order multiples. **Water-bottom** multiples are those reflected either upwards or downwards from the bottom of the water body. **Refracted multiples** have at least one segment of their raypaths traveling along one of the interfaces. The surface multiples, those reflected downwards from the water top or ground surfaces, may be reflected upwards from water bottom or other interfaces. The strength of the multiples from each reflector depends mainly on two factors: the reflectivity of the reflector giving rise to the amplitude of all reflections, and the lateral continuity of the reflector giving rise to the coherency of both the primary and multiple reflections. Hence, the most easily recognizable multiples are reflected from laterally coherent and generally smooth interfaces such as water bottom and smooth salt–sediment boundaries.

We will be able to remove multiples if we can identify them, or at least recognize the difference between the primary and multiple reflections. We usually cannot do much for those multiples that follow the same behaviors as the primaries. One consistent pattern of field seismic data is the differences in the characteristics and patterns of different seismic events such as primaries and multiples. Hence, the word "demultiple" usually does not mean the complete removal of all multiples, particularly for field data. Practically, words like demultiple and multiple elimination mean the same as attenuating or suppressing the multiples.

10.2.2   Demultiple via transforms

10.2.2.1   Demultiple via moveout differences

A traditional approach to separate multiples from primaries exploits the difference between their moveout patterns. This approach uses methods such as NMO stacking and f–k filtering. In the situation of using NMO stacking, we want to choose a stacking velocity corresponding to the primary reflections, which will become flattened after the NMO. Because velocity

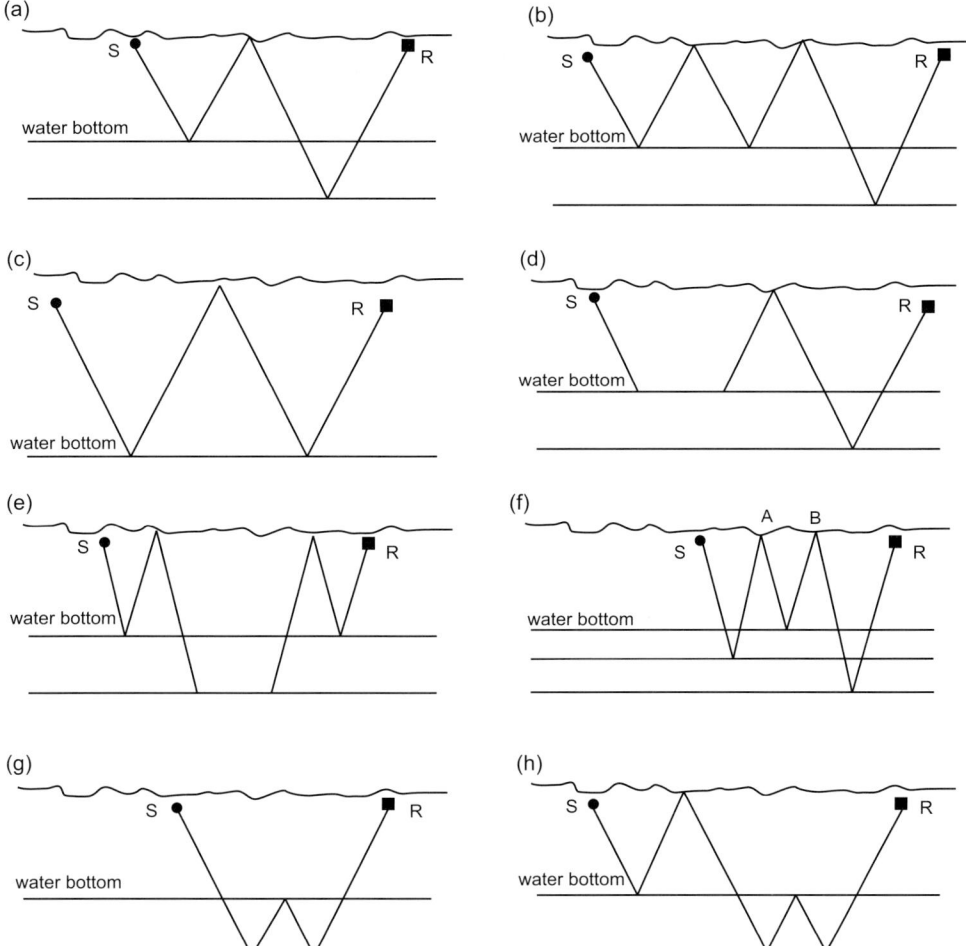

Figure 10.14 Raypaths of common multiples (Dragoset & Jeričević, 1998). (a) First-order surface multiple. (b) Second-order surface multiple. (c) Water-bottom multiple. (d) Refracted first-order surface multiple. (e) Refracted second-order surface multiple. (f) Second-order surface multiple. (g) Internal multiple. (h) Surface multiple that includes an internal multiple.

usually increases with depth, multiples tend to have lower NMO stacking velocity than the primaries, because the latter traverse much deeper (e.g., Figure 2.8). Hence, the multiple reflection events will not be flattened after the NMO that flattens the primary events. Thus a stacking after the NMO will constructively enhance the primaries and destructively suppress the multiples. The advantages of demultiple using NMO stacking are its simplicity and efficiency; the disadvantages are that the process produces only post-stack data and it does not work for multiples with moveout velocities similar to that of the primaries, such as in the near-offset traces.

To use f–k filtering to suppress multiples, a CMP gather is first NMO corrected, and then 2D Fourier transformed into the f–k domain. The primary and multiple energies ideally occupy different portions in the f–k space due to their different moveout patterns. Hence, some form of muting of the portions containing the multiples can be applied, followed by inversion 2D Fourier transformation to yield a new CMP gather in the t–x domain with multiples suppressed. We need to apply taper(s) between the pass and reject zones in the f–k space to reduce the artifacts due to Gibb's ringing. One may extend the f–k filtering approach to suppress multiples in other pre-stack and post-stack domains, as long as the multiples and primaries fall in different areas after the f–k transform.

10.2.2.2   Demultiple via Radon transform

The idea of separating the primaries from the multiples using a specific transform is pursued by many workers, and one of the improved methods is via Radon transform. The idea was introduced by Thorson and Claerbout (1985) and Hampson (1986), using parabolic Radon transform. As outlined in Section 2.4, the Radon transform is a quasi-reversible transformation that sums data events exhibiting a linear, parabolic, or hyperbolic trajectory in one domain to a single event in the transformed domain. Consequently, filters can be designed in the Radon domain to remove much of the energy associated with primaries, leaving behind the multiples. After inverse transformation of the filtered data back to the input domain, the multiple reflections remain and can be subtracted from the original data.

Figure 10.15 compares the effectiveness of multiple suppressions by the f–k approach and by the parabolic Radon transform (Alvarez & Larner, 2004). In this case, the f–k approach is not very effective, particularly for the near-offset traces. In contrast, the parabolic Radon transform approach works more effectively, especially for shallow multiples when the multiple and primary have enough differential moveout; it is less effective for multiples at later zero-offset times.

Figure 10.16 shows an example of Radon demultiple by Foster and Mosher (1992). In practice, after the Radon transform, the primary energy region in the τ–p domain is suppressed first so the inverse Radon transform actually recovers the multiples, like that shown in Figure 10.16c. Subtracting the multiples-only events from the original data in the t–x domain (Figure 10.16a) finally produces the primaries-only solution (Figure 10.16b). This procedure is used to minimize an artificial appearance in the solution of the primary reflections. Since the Radon transform is quasi-reversible, the inversely Radon transformed image has an artificial appearance. Hence the subtraction of the multiples-only inverse Radon transformed events from the original data may retain more characteristics of the original data.

The use of the parabolic Radon transform to remove multiples from seismic data has difficulties with the peg-leg multiples from the water bottom and the top of salt in deep water settings (Ver West, 2002). This is a result of the underlying assumption of the Radon transform that the model can be represented by parabolas of various curvatures centered at zero offset. It has been shown that this assumption results in a significant near-offset remnant of the peg-leg multiple in the data after Radon demultiple. One effective way to attack this multiple using Radon-like techniques is to expand the underlying model to include parabolas that are shifted away from zero offset as well as those centered on it. This necessitates the expansion of the Radon transform domain from 2D to 3D, and a careful choice of the parameterization of the shift component.

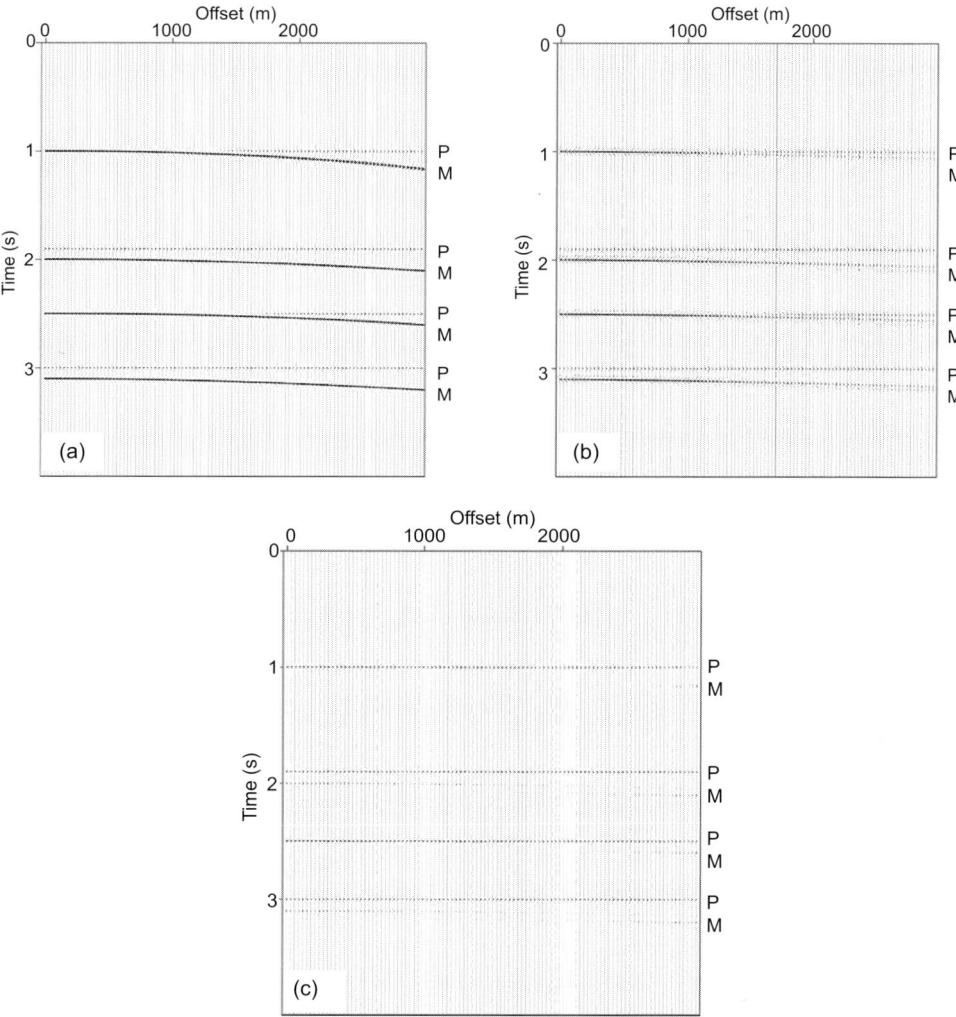

Figure 10.15 Synthetic tests of two demultiple methods (Alvarez & Larner, 2004). (a) Synthetic NMO-corrected CMP gather. The peak amplitude of the multiples (M) is four times that of the primaries (P), and both are invariant with offset. (b) The *f–k* multiple suppression leaves considerable residual multiple energy, particularly on the near-offset traces. (c) The parabolic Radon transform approach works well when the multiple and primary have enough differential moveout, such as for the earliest multiple. The approach is less effective for multiples at later zero-offset times.

10.2.3 Demultiple via predictive deconvolution

10.2.3.1 Predictive deconvolution and the demultiple flow

The deconvolution approach explores the periodicity of the multiples in order to suppress them. A good example is shown in Box 6.1 Figure 1 where the records of an **ocean bottom**

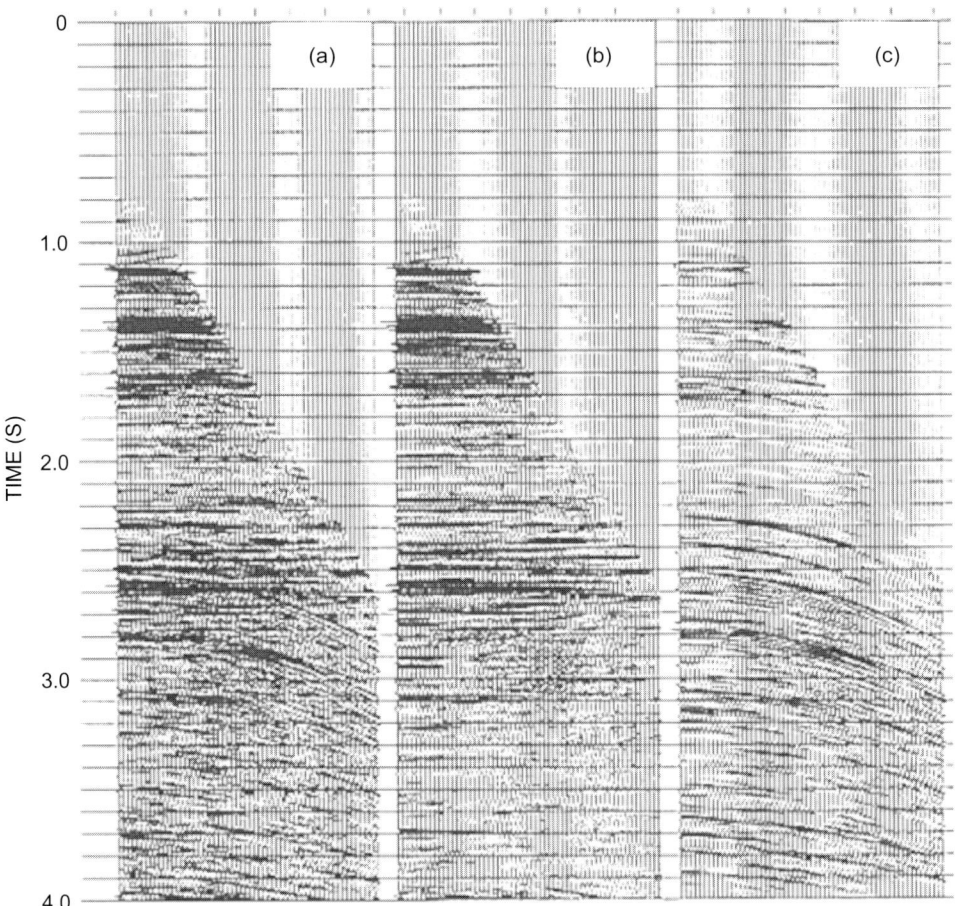

Figure 10.16 A field data example of multiple suppression using Radon transform (Foster & Mosher, 1992). (a) A NMO-corrected gather. (b) The gather after multiples subtraction. (c) Estimated multiples.

seismometer (OBS) show extremely strong water bottom multiples. In places like offshore California and Florida, the water bottom consists of crystalline rocks or carbonates whose density and velocity are significantly higher than that of water. Over ten orders of multiple reflections have been reported in such places, as shown in the input panel of Box 6.1 Figure 1. There are also many vertical stripes on this input panel, suggesting that the multiples have a trace-by-trace behavior; hence a trace-by-trace processing may be suitable. After removing the periodic component of the records, the output panel in Box 6.1 Figure 1 appears to be much more plausible geologically.

The predictive deconvolution approach is effective for suppressing multiples for zero-offset data and for non-zero-offset data acquired in water depths less than 100 m. It usually takes a trace-by-trace operation in the processing that implicitly assumes a 1D layer-cake Earth model. However, differences clearly exist in the source signature, receiver response,

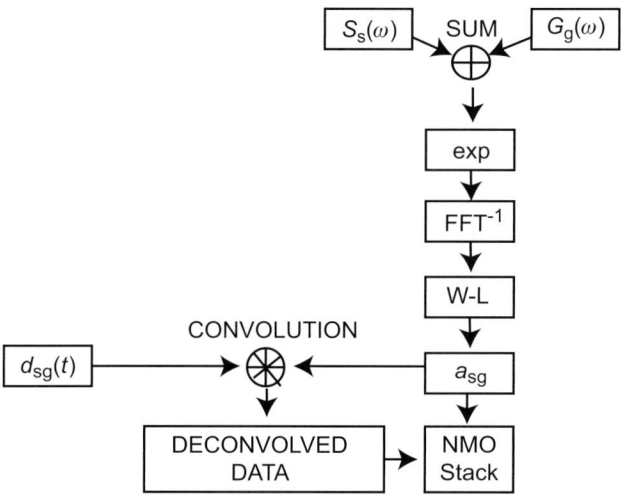

Figure 10.17 A predictive deconvolution procedure (Morley & Claerbout, 1983) for processing the input seismic trace d_{sg} recorded at sth shot and gth geophone location. For each trace d_{sg} the model spectrum is formed by adding the log-amplitude spectra of the shot and geophone, $S_s(\omega)$ and $G_g(\omega)$, and exponentiating the sum. The inverse Fourier transform yields an autocorrelation. A Weiner–Levinson (W–L) predictive deconvolution then obtains a causal inverse filter a_{sg}, which convolves with the input trace d_{sg} to produce the multiple-suppressed output trace.

and changes in the media properties for different locations and offsets. Hence, efforts have been made to account for these variations.

As an example, a processing flow devised by Morley and Claerbout (1983) is shown in Figure 10.17. These authors account for the variations in the reflectivity and water depths at source and receiver locations by taking each seismic trace as a convolution of an average frequency response with specific shot, geophone, midpoint, and offset responses. The log-amplitude spectra of each shot and geophone, $S_s(\omega)$ and $G_g(\omega)$, are solved by linear least-squares using all traces in "shot–receiver" space. Then the multiple reverberation response for each shot–receiver pair is identified with the product of the shot and geophone responses, and the corresponding causal inverse filter is solved using a standard Weiner–Levinson (W–L) predictive deconvolution algorithm. The multiple-suppressed output trace is the result of convolving the inverse filter with the input trace.

An important issue in the use of predictive deconvolution is the choice of prediction distance and the filter length, or time gate. In the above example, the time gate was chosen to be slightly less than the minimum seafloor time across the section, in order to leave the phase of the bubble pulse unchanged. The filter length was chosen to be just long enough to include the maximum seafloor time across the section. Note in the right panel of Figure 10.18 that the primary near 3 s stands out very clearly after suppressing the multiples. A weaker primary some 200 ms below the strongest primary is also visible. In the previous example shown in Box 6.1 Figure 1, a time gate nearly 1600 ms long was used.

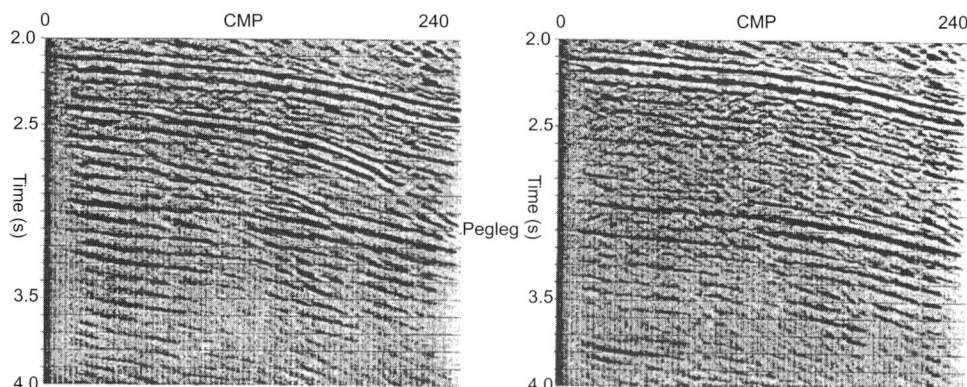

Figure 10.18 The CDP stacks of data from the Flemish Cap area of the Labrador Sea before (left) and after (right) the processing sequence of Figure 10.4 (Morley and Claerbout, 1983). The multiples around 3 s are mostly eliminated.

10.2.3.2 Offset effect on demultiple via predictive deconvolution

The efficacy of demultiple via predictive deconvolution degrades rapidly with offset owing to the non-stationarity of the primary-to-multiple traveltime separation, which becomes comparable with the main data period at far offset. To overcome this problem, we can design some types of data transformation that will maintain the periodicity in time between the primary and multiple reflections. A well-known example is discussed in Yilmaz (1987) using the tau–p transform.

Another approach, devised by Schoenberger and Houston (1998), is to apply a stationarity transform that will make the separation time between the primary and multiple invariant with offset, thus enabling the predictive deconvolution to large offset. Figure 10.19a shows their synthetic CMP gather with shallow, mid-depth, and deep primary reflections and first-order peg-leg multiples following simple hyperbolic traveltime trajectories. The predictive deconvolution as shown in Figure 10.19b is effective only for small offset. The stationarity transform has allowed successful suppression of the multiples at far offset by predictive deconvolution (Figure 10.19c).

Figure 10.20 shows the application of the deconvolution methods to field data in terms of far-offset stacks. Panel (a) is a zoom-in window of the stack without demultiple, on which the primary (P) and multiple (M) are denoted. Panel (b) is the stack after applying the conventional predictive deconvolution. Though the amplitude of the multiple is attenuated, much of the multiple remains in place. Panel (c) is the stack after the joint use of stationarity transform and predictive deconvolution, which resulted in the highest level of multiple suppression in this case.

To minimize the impact of random noise on the predictive deconvolution, Hornbostel (1999) proposed a noise-optimized objective (NOO) function. The filter is optimized through the application of the filter. The NOO operators have the property of maximizing the amplitude of the multiples or minimizing that of the primary or random noise in the data. Examples of linear operators with such a property include stacking, band-pass filtering, dip filtering, muting, and scaling. Stacking is useful to minimize the predictable energy on a stacked trace, and the pre-stack filters are less affected by random noise. NOO stacking

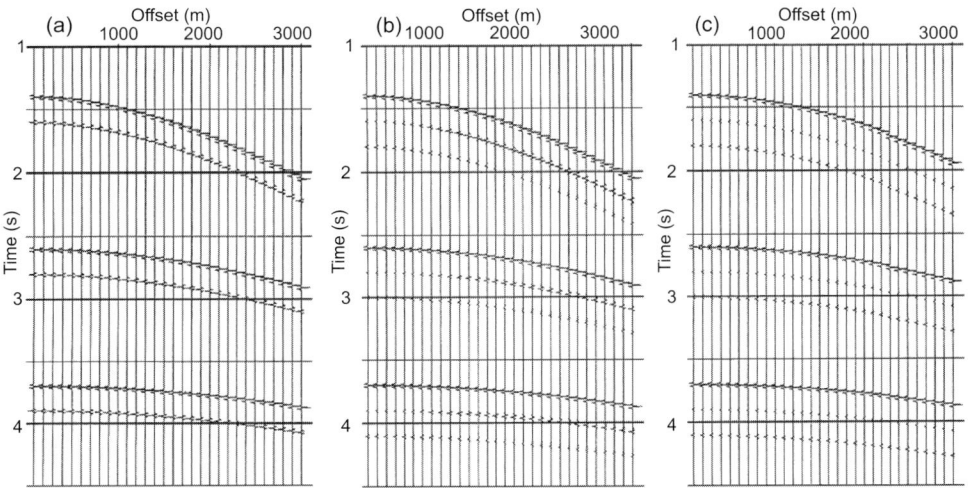

Figure 10.19 (a) A synthetic CMP gather containing three primary reflections and first-order peg-leg multiples that follow simple hyperbolic traveltime trajectories. (b) Data after predictive deconvolution. (c) Data after stationarity-transform, predictive deconvolution, and reverse stationarity-transform (Schoenberger & Houston, 1998).

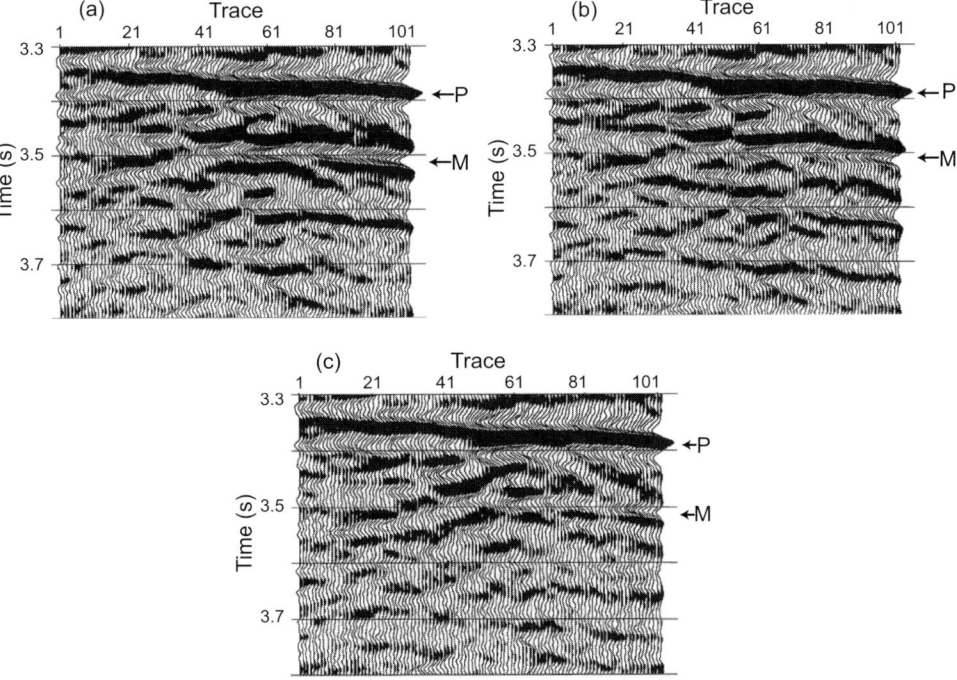

Figure 10.20 Comparison of the far-offset stacks (Schoenberger & Houston, 1998). (a) Stack without demultiple. (b) Predictive deconvolution before stack. (c) Stationarity transformation, predictive deconvolution, inverse stationarity transformation, NMO, and stack.

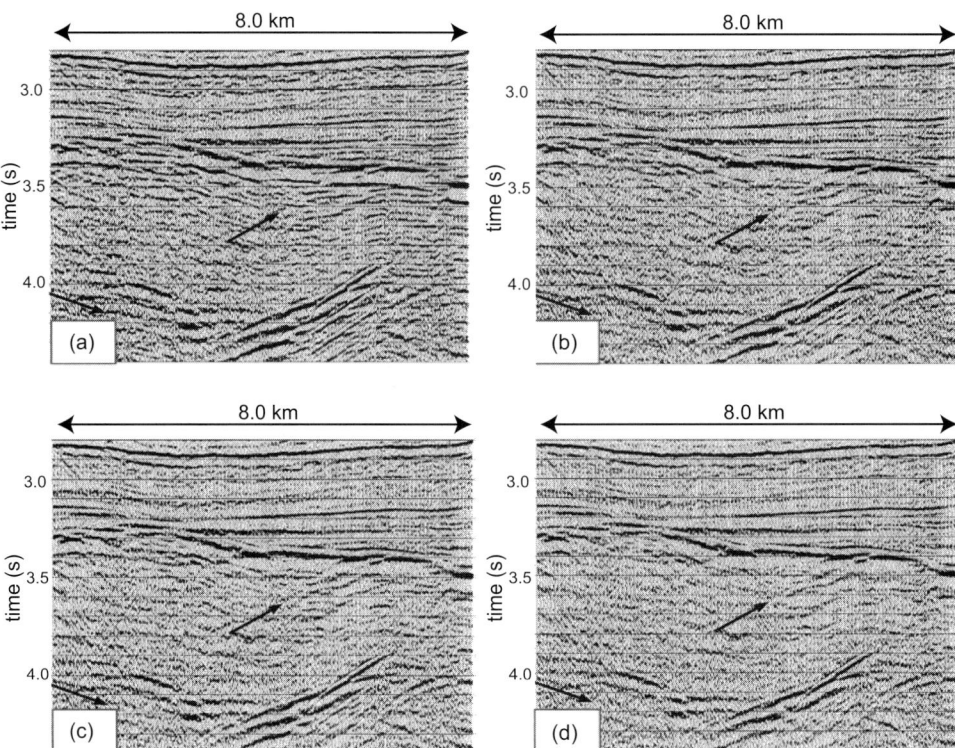

Figure 10.21 Comparison of stacks of a portion of North Sea field data (Hornbostel, 1999). (a) Without demultiple. (b) With standard gap predictive deconvolution. (c) With time-varying gap predictive deconvolution. (d) With time-varying gap stack-minimization NOO filter.

differs from the traditional ones because the filter is designed for pre-stack data. It differs from the standard pre-stack prediction filter because it minimizes the predictable energy on the stacked traces. Figure 10.21 provides a field data comparison between standard gap predictive deconvolution, time-varying gap predictive deconvolution, and time-varying gap stack-minimization NOO filter.

10.2.4   Surface-related multiple elimination (SRME)

There is a group of so-called **surface-related multiple elimination (SRME)** methods aiming to remove all surface-related multiple energy via pre-stack inversion. Many traditional multiple suppression methods rely on velocity discrimination, and therefore are less effective in cases of small moveout differences between primaries and multiples. This challenge motivated the SRME method to take advantage of the idea that each surface-related multiple is a superposition of several primary reflections from some surfaces and interfaces. When the above idea holds true as in most cases, the SRME takes part of recorded seismic data to predict other parts via inversion, and the predicted portions can be regarded as multiples and removed. The approach is particularly advantageous in the sense that no subsurface information and velocity model are required at all.

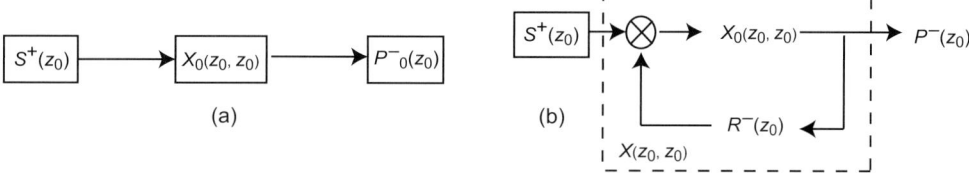

Figure 10.22 Two block-diagram models of the upgoing wavefield $\mathbf{P}^-(z_0)$ as convolution of the source wavefields $\mathbf{S}^+(z_0)$ with two subsurface responses: (a) $\mathbf{X}_0(z_0, z_0)$ only; and (b) $\mathbf{X}(z_0, z_0)$, a combination of $\mathbf{X}_0(z_0, z_0)$ and surface-related multiples $\mathbf{R}^-(z_0)$. (After Verschuur *et al.*, 1992.)

Implementation of the SRME considers the multiple elimination process as a seismic inversion by minimizing the energy in the data after the elimination. The input data of the method are upgoing reflected waves related to downgoing source waves. It first calculates a field of predicted multiples, and later subtracts the predicted multiples from the input data. In this prediction-and-subtraction process, the subtraction step supposedly compensates for the prediction errors. The SRME processing is effective as long as all relevant data are recorded within the aperture and offset ranges of the seismic survey. This means a well sampled dataset is required.

10.2.4.1 The process of SRME

The SRME method is introduced here following the description by Verschuur *et al.* (1992). The SRME and related multiple removal methods rely on a simple but powerful concept: every surface-predictable multiple consists of segments that, from a surface perspective, are primary events such as direct waves and primary reflections. If ignoring surface-related multiples, the upgoing pressure wavefield $\mathbf{P}^-_0(z_0)$ at the surface z_0 can be expressed as a convolution

$$\mathbf{P}^-_0(z_0) = \mathbf{X}_0(z_0, z_0)\mathbf{S}^+(z_0) \qquad (10-1)$$

where $\mathbf{S}^+(z_0)$ is the downgoing source wavefields from the surface, and $\mathbf{X}_0(z_0, z_0)$ is the response matrix of the subsurface without surface-related multiples. Note that $\mathbf{X}_0(z_0, z_0)$ contains all primary reflections and internal multiples of the subsurface. As we have seen in Chapter 5, the convolution can be expressed in a block diagram like that in Figure 10.22a.

In the presence of a free surface, any upgoing wave arriving at the surface will be reflected into a downgoing wave. Taking $\mathbf{R}^-(z_0)$ as the reflectivity matrix of the free surface, equation (10–1) should be modified to

$$\mathbf{P}^-(z_0) = \mathbf{X}_0(z_0, z_0)[\mathbf{S}^+(z_0) + \mathbf{R}^-(z_0)\mathbf{P}^-(z_0)] \qquad (10-2)$$

Moving the two upgoing wavefield terms to the left side, we have

$$\mathbf{P}^-(z_0) = [\mathbf{I} - \mathbf{X}_0(z_0, z_0)\mathbf{R}^-(z_0)]^{-1}\mathbf{X}_0(z_0, z_0)\mathbf{S}_0(z_0) \qquad (10-3)$$

Alternatively, we can define a new response matrix of the subsurface $\mathbf{X}(z_0, z_0)$ that includes the surface-related multiples. Then the upgoing wavefield is a new convolution

$$\mathbf{P}^-(z_0) = \mathbf{X}(z_0, z_0)\mathbf{S}^+(z_0) \qquad (10-4)$$

Comparing (10–3) with (10–4), we find

$$X(z_0, z_0) = X_0(z_0, z_0)/[I - X_0(z_0, z_0)R^-(z_0)] \tag{10–5}$$

Recalling the recursive filters that we studied in Chapter 5, the new response matrix has the form of a recursive filter. In the block diagram shown in Figure 10.22b, the surface reflectivity matrix constitutes the recursive feedback loop. Such a recursive feedback loop describes the continuous process of generating multiple reflections by the free surface.

The inverse matrix in (10–3) can be expanded into a series

$$[I - X_0(z_0, z_0)R^-(z_0)]^{-1} = I + [X_0(z_0, z_0)R^-(z_0)] + [X_0(z_0, z_0)R^-(z_0)]^2$$
$$+ [X_0(z_0, z_0)R^-(z_0)]^3 + \cdots \tag{10–6}$$

Hence,

$$P^-(z_0) = P_0^-(z_0) + [X_0(z_0, z_0)R^-(z_0)]X_0(z_0, z_0)S^+(z_0)$$
$$+ [X_0(z_0, z_0)R^-(z_0)]^2 X_0(z_0, z_0)S^+(z_0) + \cdots \tag{10–7}$$

Comparing with (10–1) reveals that the extra terms in (10–7) generate all surface-related multiples.

The objective of SRME is to turn the input data $P^-(z_0)$ into $P^-_0(z_0)$ which is free from surface-related multiples. This can be achieved by inverting (10–2) for $X_0(z_0, z_0)$

$$X_0(z_0, z_0) = P^-(z_0)[S^+(z_0) + R^-(z_0)P^-(z_0)]^{-1} \tag{10–8}$$

or, using (10–4)

$$X_0(z_0, z_0) = X(z_0, z_0)[I + R^-(z_0)X(z_0, z_0)]^{-1} \tag{10–9}$$

Similar to the expansion of (10–6), we can expand the above inverse matrix as

$$X_0(z_0, z_0) = X(z_0, z_0)\{I - R^-(z_0)X(z_0, z_0)$$
$$+ [R^-(z_0)X(z_0, z_0)]^2 - [R^-(z_0)X(z_0, z_0)]^3 + \cdots\} \tag{10–10}$$

In the presence of strong multiple reflections such as water reverberations at post-critical angles, straightforward inversion as described by (10–9) is unstable. Hence, the practice of SRME takes only a limited number of terms in (10–10) to stabilize the inversion. It is clear that the above two equations do not use any model of the subsurface. Only $X(z_0, z_0)$, the seismic data after deconvolution for the source wavefield, and the free surface reflectivity matrix $R^-(z_0)$ are used. The data are taken as a multiple prediction operator, containing all necessary information about the subsurface in order to predict the multiples. This notion requires data be the true unit-valued impulse response of the medium, a requirement that is too strict in practice where the source wavefield is not available. This motivated an adaptive procedure suggested by Verschuur et al. (1992) to let the predicted multiples adaptively match in amplitude and phase with the multiples presented in the data. By minimizing the energy in the data after the multiple suppression procedure, an estimate of the source signature is obtained as well.

10.2.4.2   Examples of SRME

A field data example of SRME is shown in Figures 10.23 and 10.24, comparing the stack sections and shot records before and after the multiple suppression process. In practice, a

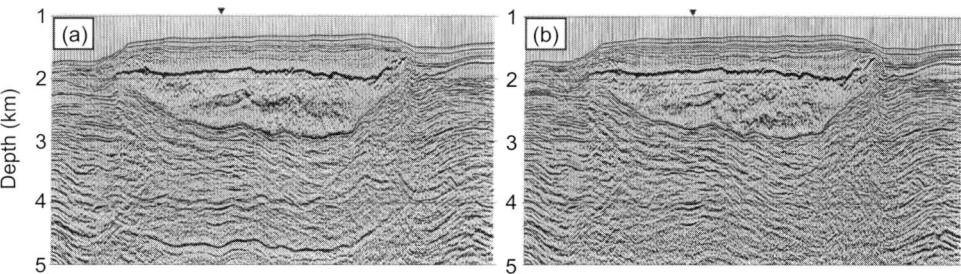

Figure 10.23 Comparison of pre-stack depth migrations of 25-km-long 2D data from the Gulf of Mexico (Dragoset & Jeričević, 1998). (a) No treatment of strong surface multiples due to the water bottom and a salt layer lying between depths of 1.8 and 2.8 km. (b) After SRME. The small triangles mark the location of the shot records shown in the next figure.

number of practical issues have to be solved for the implementation of SRME. For instance, the input for the procedure was supposed to be the upgoing pressure wavefield $\mathbf{P}^{-}(z_0)$ at the free surface, but this is not the actual measured data. In the case of marine data, the total pressure is measured below the free surface, while for land data the vertical component of the total particle velocity is measured at the free surface. Hence, before starting the SRME, a "deghosting" process of decomposition should be applied to arrive at upgoing reflected wavefields. Of course, real-world situations never match our theories exactly. As an example, some remnant multiples are present in Figure 10.24b, suggesting that field data imperfections such as 3D effects may reduce the performance of the method.

Figure 10.25 shows another comparison between time stacks of a 2D data from the North Sea before and after SRME. As a data-driven procedure, SRME does not make any assumptions about the subsurface. It suppresses both surface-related multiples as well as the source signature. By taking the output of the SRME method as the input data again, we can re-apply SRME to suppress internal multiples by extrapolating the "surface" and iteratively applying the multiple suppression procedure. However, the SRME method requires a complete internal physical consistency between primary and multiple events. Its 3D implementation requires dense sampling of the wavefield on the surface, meaning in theory the deployment of a shot and a receiver at every surface location. Typical marine acquisition geometries deliver much sparser surface coverage, which results in severe shot- and receiver-domain aliasing in the crossline direction.

10.2.5   Imaging with multiples

We may improve the quality of velocity model and seismic imaging by using multiples. Clearly multiple reflections contain a wealth of information that can used to improve the resolution of subsurface images. In fact, this idea has been proven effective in Section 7.4.3, where the multiples were used together with the primary reflections to generate the imageries of the subsurface via the reverse time migration method. Two-way wave equation migration methods consider the whole seismic wavefield as the input, without distinguishing the primaries from the multiples. The challenge to this approach is its high sensitivity to the accuracy of the velocity models. On the other hand, we should be able to take advantage of this sensitivity to create a new way of velocity model building using

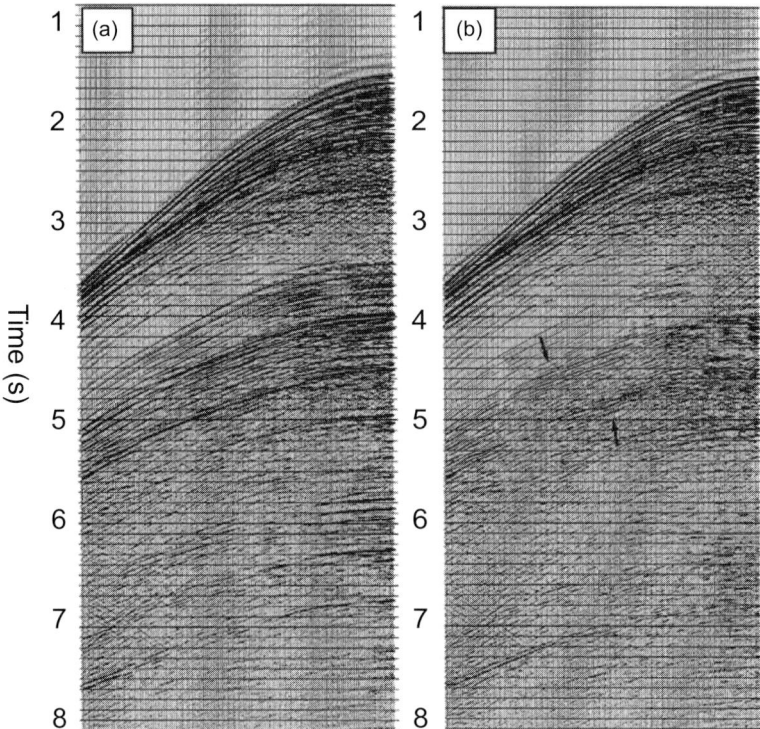

Figure 10.24 A common shot gather from the data shown in previous figure (Dragoset & Jeričević, 1998). (a) At zero-offset time the events starting at 3.5 s are water-bottom multiples, at 4.0 s is a multiple that reflects upward at the top of salt and at the water bottom, at 4.5 s is a multiple that reflects upward twice from the top of salt, and at 5.75, 6.25, and 6.75 s are second-order surface multiples. (b) The demultiple method has removed many of the surface multiples, but some remnants are present (arrows), suggesting that field data imperfections may reduce the performance of the method.

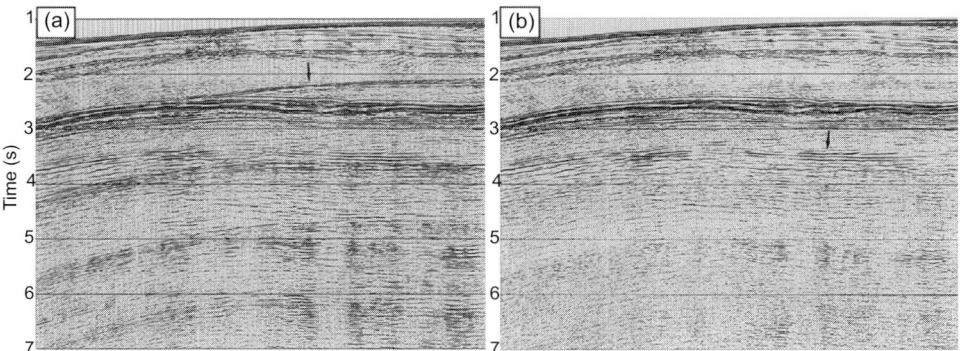

Figure 10.25 Comparison of time stacks of a 30-km-long 2D data from the North Sea (Dragoset & Jeričević, 1998). (a) No treatment of the multiples. Four orders of surface multiples are visible, beginning with the first water-bottom multiple indicated by an arrow. (b) After SRME. The primary events around 3.5 s (arrow) are unaffected by the multiple suppression.

multiple reflections. Particularly for moveout-based velocity model building such as those based on NMO and CIG, multiple reflections offer extra spatial coverage and constraining power.

A different way of imaging multiple reflections was suggested by Berkhout and Verschuur (2006) to transform multiples in primary reflections. They refer to the approach of removal of surface-related multiples by weighted convolution, described in the previous subsection, as SRME1. If regarding SRME1 as a forward prediction, then a backward prediction process can be realized as a weighted cross-correlation (WCC) between primaries and multiples. This backward prediction process results in an alternative for multiples removal called SRME2, as well as a mechanism to transform multiples into primaries. Then imaging of multiple reflections is achieved in three steps. First, separate the primaries into multiples using a combination of the SRME1 and SRME2. Second, transform the multiples into primaries using WCC. Finally, image the transformed multiples using traditional one-way imaging methods.

10.2.5 Exercise 10.2

1. Make a table to summarize major attributes of three groups of multiple suppression methods in Sections 10.2.2 to 10.2.4. The attribute list should include assumptions, principles, main method(s), applicable situations, and limitations.

2. Come up with three different ways to quantify the effectiveness of the two multiple suppression methods shown in Figure 10.20, panels (b) and (c).

3. Devise a way to construct a common image gather (CIG) using multiple reflections. What type of migration operator should be used? How will such a CIG differ from a CIG based on primaries?

10.3   Processing for seismic velocity anisotropy

10.3.1   Basics of seismic velocity anisotropy

Anisotropy is the variation of a material property with the observational angle. According to Anderson (1989) the concept of anisotropy was documented originally in studying the acoustic speed and heat conduction of mineral crystals. In seismic data analysis it is velocity anisotropy that is of concern, as discussed in Chapters 1 and 8. Processing for seismic velocity anisotropy involves detecting its presence, compensating for its effects in the data, and building velocity models to account for velocity anisotropy.

Following "the principle of parsimony", we always want to use the simplest set of parameters and models in the practice of seismic data analysis in order to achieve the most plausible solutions. In exploration geophysics, seismic velocity anisotropy has been regarded as one of the "necessary devils" that could be important but difficult to handle. Hence, although it is common in nature and has been studied for many years, seismic velocity anisotropy has not been included in routine seismic data processing flows. On the

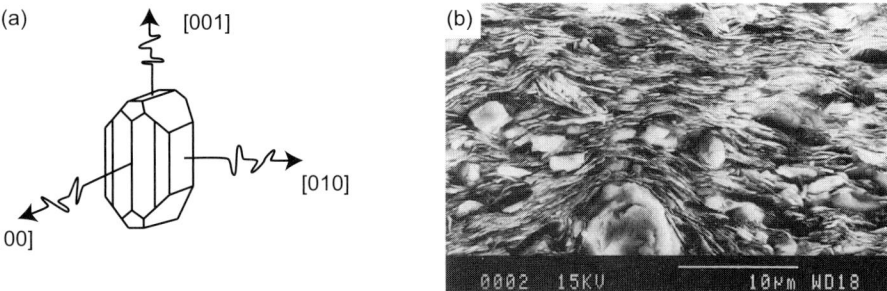

Figure 10.26 (a) An olivine crystal showing intrinsic velocity anisotropy. (b) A scanning electron microphotography of shale (Hornby *et al.*, 1994) showing extrinsic velocity anisotropy at the microscale. (c) A photograph of layered sediments in San Lorenzo Canyon, New Mexico, suggesting extrinsic velocity anisotropy at the macroscale.

other hand, azimuthally varying velocity anisotropy is routinely analyzed in solid Earth geophysics to infer the flow directions of crustal and mantle rocks over geologic time. Such analyses are typically based on the orientation and traveltime difference of **shear wave splitting**, the occurrence of two shear waves through an anisotropic medium, a fast traveling shear wave oscillating along the main orientation of rock beddings or fractures, and a slow traveling shear wave oscillating perpendicular to the main orientation of rock beddings or fractures.

10.3.1.1   Intrinsic and extrinsic anisotropy

In recent years the topic of seismic velocity anisotropy has been gaining attention in exploration geophysics due to the need to account for it in several situations. For instance, where the velocity anisotropy is significant, failure to account for it in velocity model building may result in serious error in the positions of the hydrocarbon reservoirs. In searching for the presence and orientation of fractures in brittle rocks, geophysicists often rely on clues about anisotropic behavior of the data. Strictly speaking, seismic velocity anisotropy is divided into that which is intrinsic or extrinsic in nature. As shown in Figure 10.26a, **intrinsic anisotropy** is due to the mineralogical structure, and it exists at all length scales. In contrast, extrinsic anisotropy is created by the alignment of rock grains at the microscale (Figure 10.26b) or rock bedding at the macroscale (Figure 10.26c). Extrinsic anisotropy varies with length scale.

In the practice of exploration seismology today, it is typically unclear how much of the observed velocity anisotropy is due to intrinsic anisotropy of rocks versus the extrinsic anisotropy of aligned rock strata. However, considering the fact that the wavelength of seismic data is usually much greater than the grain size of minerals in the upper crust, it is safe to say that we mostly deal with extrinsic velocity anisotropy in exploration geophysics and crustal seismology. The presence of extrinsic velocity anisotropy requires the presence and alignment of anisotropic crystals, and/or alignment of rock layers of different velocities. In this section, the readers are exposed to some basic issues on the processing of seismic data in the presence of seismic velocity anisotropy.

A first concern here is about uniqueness in dealing with indications of velocity anisotropy. When we encounter an observed indication or evidence of velocity anisotropy, how sure are we that it is really due to velocity anisotropy? What other factors could have caused the observation? How can we distinguish between the effects of different causes? The answers requires us to characterize and quantify the cause–effect relationship for each factor and its effects. Currently the exploration geophysics community as a whole has not been able to answer these questions in most cases. This means that anisotropy in seismic velocity will remain a research topic for many years.

One of the pioneer papers on seismic anisotropy was published by Nur and Simmons (1969), who studied stress-induced anisotropy in laboratory experiments. They showed clear evidence of velocity anisotropy when anisotropic stresses were applied to a granite rock sample. Their paper explains the physical process behind stress-induced anisotropy. As exemplified by this paper, the first approach to understanding velocity anisotropy is to conduct an experimental study.

10.3.1.2 Transverse isotropy and effective medium theories

In 1981, two important papers on the subject of seismic velocity anisotropy were published, one based on observation and the other on theory. First, Helbig (1981) categorized anisotropic models according to crystal structures. This work developed the geometric theory of seismic anisotropy as the foundation for layer-induced **transverse isotropy**, or **TI**. As exemplified by this paper, a second approach to study velocity anisotropy is to compile and analyze observed data. Second, Hudson (1981) evaluated the transverse isotropy model with effective stiffness parameters. This is one of the original theoretical treatments of velocity anisotropy and attenuation of elastic waves in cracked media. As exemplified by Hudson's paper, a third approach to velocity anisotropy is to conduct a theoretical study or numerical modeling.

From years of studying velocity anisotropy, several concepts or models were proposed and proven to be important. As a classic example, Hsu and Schoenberg (1993) introduced the linear slip fracture model to deal with complex fracture structure. Using this conceptual model, media with different fractures can be described by a single matrix. Another important concept is a whole set of **effective medium theories**. Such **medium-average** or **equivalent medium** concepts are among the most common ways for geophysicists to simplify the real phenomena. An example is the use of stacking velocity to represent the average velocity

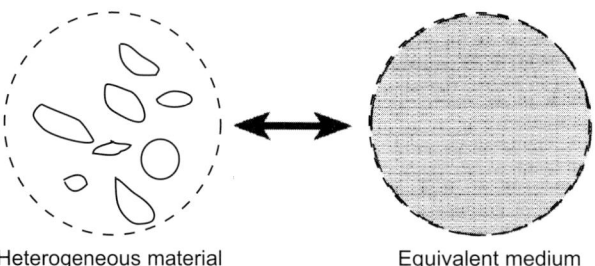

Heterogeneous material   Equivalent medium

Figure 10.27 An illustration of effective medium theories: finding the elastic properties of a homogenous solid having the equivalent elastic properties of a heterogeneous solid with a complex microstructure (from Hornby *et al.*, 1994).

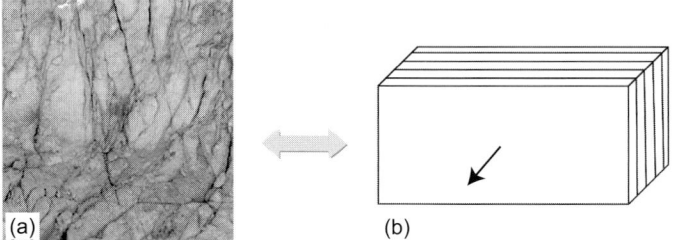

Figure 10.28 (a) Rocks with nearly vertical fractures. (b) A model of vertically aligned fractures as a medium of horizontal transverse isotropy, or HTI. Such a model may explain the main characteristics of seismic responses to the real rock formation shown in (a).

above a time horizon in semblance velocity analysis. Figures 10.27 and 10.28 give two examples of effective medium theories. In Figure 10.27 the heterogeneous properties of shales are modeled by a homogeneous equivalent medium model.

Figure 10.28 shows a rock formation with nearly vertical fractures, and a model with vertically aligned fractures of regular spacing. If the simplified model of **horizontal transverse isotropy** (**HTI**) can predict much of the seismic response of the fractured rocks in the real world, we can use such a simplified model to assess real situations. Hornby *et al.* (1994) demonstrated the differential effective medium theory and the self-consistent theory. Their paper is a good reference on shale anisotropy and how to model it. It gives detailed mathematical formulations to calculate shale anisotropy, and shows elastic wave propagation in anisotropic media.

Another classical paper on velocity anisotropy was by Rathore *et al.* (1995) who studied P- and S-wave anisotropy of synthetic sandstone with controlled crack geometry. Their paper demonstrates elastic wave propagation in a poro-elastic medium based on laboratory measurements and theoretical modeling. It shows how fluid flow in pore space with pore pressure in equilibrium affects seismic anisotropy. The poro-elasticity concept is extremely important for the petroleum industry and is an exciting and challenging research topic.

In practice, most workers focus on **transverse isotropy** (**TI**) which is the simplest type of velocity anisotropy with a single symmetry axis. According to the orientation of the

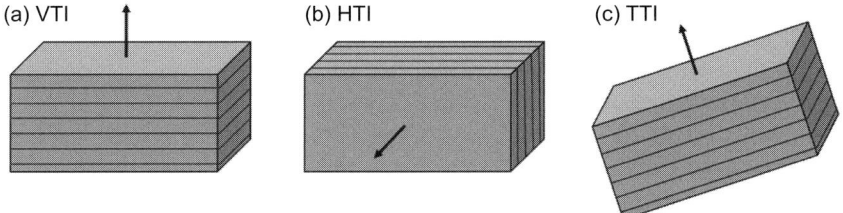

(a) VTI   (b) HTI   (c) TTI

Figure 10.29 Three common transverse isotropy models: (a) vertical transverse isotropy (VTI); (b) horizontal transverse isotropy (HTI); and (c) tilt transverse isotropy (TTI).

symmetry axis, three types of TI models are shown in Figure 10.29: **vertical transverse isotropy (VTI)**, **horizontal transverse isotropy (HTI)**, and **tilt transverse isotropy (TTI)**. The VTI is representative of the most common geophysical model, the layer-cake model. A more realistic situation with dipping beds is represented by the TTI model. The HTI shown in Figure 10.28 is representative of vertically fractured rock formations.

10.3.2   Thomsen's description of weak anisotropic media

Among the rich literature on seismic anisotropy, the paper by Thomsen (1986) has played a pivotal role in disseminating the general aspects of velocity anisotropy to the exploration geophysics community. Rather than bring new ideas to researchers on seismic velocity anisotropy, Thomsen delivered a useful and elegant expression of the anisotropic behavior using the so-called Thomsen parameters, epsilon ε, delta δ, and gamma γ. These parameters are widely used in the geophysics community today. Thomsen's paper focuses on TI media with a weak level of isotropy, meaning the values of the Thomsen parameters are within about 20% in their magnitudes.

Following Thomsen's paper, the following is a mathematic description of the VTI model. The generalized Hook's law relates the stress σ_{ij} to the strain ε_{kl} of a medium:

$$\sigma_{ij} = \sum_{k=1}^{3} \sum_{l=1}^{3} C_{ijkl} \varepsilon_{kl} \qquad (10\text{--}11)$$

where $i, j = 1, 2, 3$, and C_{ijkl} is the $3 \times 3 \times 3 \times 3$ elastic modulus tensor, which completely characterizes the elasticity of the medium. Using symmetry of the stress and strain tensors, a more compact form due to Voigt is

$$
\begin{array}{ccccccc}
ij \ \text{ or } \ kl = & 11 & 22 & 33 & 32 = 23 & 31 = 13 & 12 = 21 \\
\Downarrow & \Downarrow & \Downarrow \ \ \Downarrow \ \ \Downarrow & \Downarrow & \Downarrow & \Downarrow \\
\alpha & \beta & 1 \ \ 2 \ \ 3 & 4 & 5 & 6
\end{array} \qquad (10\text{--}12)
$$

Hence, the modulus tensor can be expressed by the 6×6 elastic modulus matrix $C_{\alpha\beta}$. Notice that in the compact form, the first three indices, 1, 2, and 3, denote the three axial components of the stress and strain tensors, and indices 4, 5, and 6 denote the three deviatoric components of the stress and strain tensors. Bear in mind that an axial component describes extensional or compressional change of the stress or strain along one of the three coordinate directions, while a deviatoric component describes rotational change of the stress or strain with respect to one of the three coordinate directions.

With the above notions, each element of the elastic modulus $C_{\alpha\beta}$ simply relates a stress component to its corresponding strain component. For instance, C_{11} relates the axial stress and axial strain in the first coordinate direction; C_{33} relates the axial stress and axial strain in the third coordinate direction; C_{44} relates the deviatoric stress and deviatoric strain with respect to the first coordinate direction; and C_{66} relates the deviatoric stress and deviatoric strain with respect to the third coordinate direction.

Using the modulus matrix, we can now describe VTI, the most common type of velocity anisotropy (Figure 10.29a). A VTI medium can be described by five elastic parameters:

$$V_{P0} \equiv \sqrt{C_{33}/\rho} \qquad (10\text{–}13\text{a})$$

$$V_{SH0} \equiv \sqrt{C_{44}/\rho} \qquad (10\text{–}13\text{b})$$

$$\varepsilon \equiv \frac{C_{11} - C_{33}}{2C_{33}} \qquad (10\text{–}13\text{c})$$

$$\gamma \equiv \frac{C_{11} - C_{44}}{2C_{44}} \qquad (10\text{–}13\text{d})$$

$$\delta^* \equiv \frac{2(C_{13} + C_{44})^2 - (C_{33} - C_{44})(C_{11} + C_{33} - 2C_{44})}{2C_{33}^2} \qquad (10\text{–}13\text{e})$$

where V_{P0} and V_{SH0} are the P-wave and SH-wave velocities along the symmetry axis, and the remaining three equations describe the three dimensionless Thomsen anisotropy parameters.

Based on the means of the corresponding elastic modulus, ε and γ can be viewed as the elliptic coefficients of the P-wave and SH-wave velocities. In other words, they describe the fractional difference between velocities along the vertical direction (slow or along the symmetry axis) and horizontal direction (fast or perpendicular to the symmetry axis). The two equations associated with the SH-wave are dropped in many applications that assume the existence of only P-waves, leaving only three parameters V_{P0}, ε, and δ^*. It is important to note that the last two dimensionless parameters are derived without the weak anisotropy assumption.

The final Thomsen parameter, δ^*, is much more complicated because it involves cross-terms in the elastic modulus. Simplification of this and other parameters using weak anisotropy, or small magnitudes (under $\sim$20%) of the three anisotropy parameters, is one of the objectives of Thomsen's 1986 paper. Such a simplification leads to the following expression for the second Thomsen anisotropy parameter:

$$\delta \equiv \frac{(C_{13} + C_{44})^2 - (C_{33} - C_{44})^2}{2C_{33}(C_{33} - C_{44})} \qquad (10\text{–}14)$$

In addition, it leads to the following phase velocities as functions of the phase angle θ, which is the angle between the symmetry axis of the VTI medium and normal direction to the wavefront:

$$V_P(\theta) = V_{P0}(1 + \delta \sin^2 \theta \cos^2 \theta + \varepsilon \sin^4 \theta) \qquad (10\text{–}15\text{a})$$

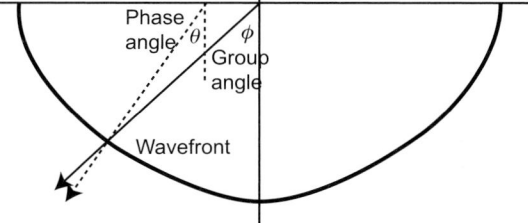

Figure 10.30 For a seismic wave traveling in an anisotropic medium, the group (ray) vector is from the origin of the wavefront, while the phase vector (dashed line) is perpendicular to the wavefront.

$$V_{SV}(\theta) = V_{SH0}\left[1 + \frac{V_{P0}^2}{V_{SH0}^2}(\varepsilon - \delta)\sin^2\theta\cos^2\theta\right] \tag{10–15b}$$

$$V_{SH}(\theta) = V_{SH0}(1 + \gamma\sin^2\theta) \tag{10–15c}$$

As shown in Figure 10.30, the phase angle generally differs from the group angle of seismic waves originating from the source at the center. Notice that, based on Equation (10–15a), when $\delta = \varepsilon$, the P-wave velocity is elliptical to the second order with respect to the phase angle θ; when $\delta = 0$, the P-wave velocity is elliptical to the fourth order with respect to the phase angle θ.

Thomsen's paper serves another important purpose: it is an example of a good scientific paper. This paper was written with a balance of background information, motivational reasoning, and a new view on velocity anisotropy in terms of the Thomsen parameters. Though the paper is quite succinct, it did a great job of covering the theoretical background and the main assumptions associated with weak anisotropic media. Above all, his assumptions and theoretical development are based on observational data analyzed in the paper.

10.3.3   Time-domain estimates of velocity anisotropy

The main objective of seismic data processing in anisotropic media is to estimate anisotropic velocities from surface reflection data. For data processing in the time-domain, work on velocity anisotropy has traditionally been associated with NMO velocity analysis as well as different ways of time migration imaging. One of the original reasons to introduce velocity anisotropy is the observation of "hockey-stick" patterns in NMO analysis and in the CIG migration velocity analysis in Chapter 8. Much of the past effort has gone into expressing stacking velocities using anisotropic parameters. Assuming straight-ray reflections from a flat reflector below a VTI overburden medium, Thomsen (1986) gives the NMO velocity of the reflected P-wave as

$$V_{NMO}^P(0) = V_{P0}\sqrt{1 + 2\delta} \tag{10–16}$$

and the NMO velocity of the reflected SH-wave as

$$V_{NMO}^{SH}(0) = V_{SH0}\sqrt{1 + 2\gamma} \tag{10–17}$$

where V_{P0} and V_{SH0} are vertical P-wave and S-wave velocities, respectively.

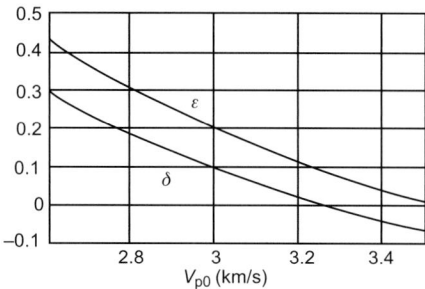 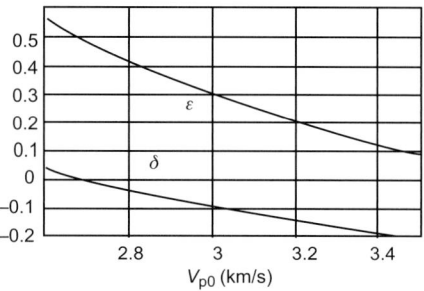

Figure 10.31 Estimated Thomsen parameters ε and δ from synthetic tests in VTI media (Alkhalifah & Tsvankin, 1995). (a) Inverted values as function of V_{P0} from NMO velocities for a horizontal reflector with a ray parameter of 0.23 s/km at 50° dip. The model parameters are $V_{P0} = 3$ km/s, $\varepsilon = 0.2$, and $\delta = 0.1$. (b) Inverted values as function of V_{P0} for the model with $V_{P0} = 3$ km/s, $\varepsilon = 0.3$, and $\delta = -0.1$.

Equation (10–16) means that the influence of the VTI anisotropy on the P-wave NMO is due entirely to the parameter δ. Assuming the validity of straight raypaths in a layer-cake model with VTI anisotropy, we can extract the vertical P-wave velocity V_{P0} from a check shot survey, and then determine the dimensionless parameter δ from the P-wave NMO velocity $V_{NM0}^{P}(0)$. Using the derived V_{P0} and δ parameters, we can then proceed to determine ε from pre-stack data with different incidence angles.

Similarly, if we have multi-component data that allow us to conduct NMO velocity analysis for a pure SH wave, we can then follow the same procedure using Equation (10–17). In other words, we can first extract V_{SH0} from a check shot survey, and then use the SH-wave NMO velocity $V_{NMO}^{SH}(0)$ and the equation to determine γ for different reflection times.

A more rigorous study on NMO velocity analysis in the presence of VTI anisotropy but dipping reflectors was conducted by Alkhalifah and Tsvankin (1995). Their sensitivity study showed that the three parameters in P-wave velocity anisotropy, V_{P0}, ε, and δ, cannot be determined uniquely from the surface reflection data alone. As an alternative solution, they introduced the anellipticity parameter η in order to relate the horizontal velocity to the NMO velocity:

$$V_{h} = V_{NM0}(0)\sqrt{1 + 2\eta} \tag{10–18}$$

The similarity between the expressions in equations (10–16), (10–17), and (10–18) was one of the motivations to introduce the new parameter. This anellipticity parameter was introduced as a normalized difference between the two Thomsen parameters δ and ε in media with weakly limited anisotropy:

$$\eta = \frac{\varepsilon - \delta}{1 + 2\delta} \tag{10–19}$$

If $\eta = 0$, or $\delta = \varepsilon$, the P-wave velocity is elliptical to the second order with respect to the phase angle, then the horizontal velocity is the same as the NMO velocity, and the moveout curves become hyperbolic. This is why we call η the anellipticity parameter. In the isotropic case, we have $\delta = \varepsilon = 0$. Figure 10.31 shows two graphs of two Thomsen parameters as a

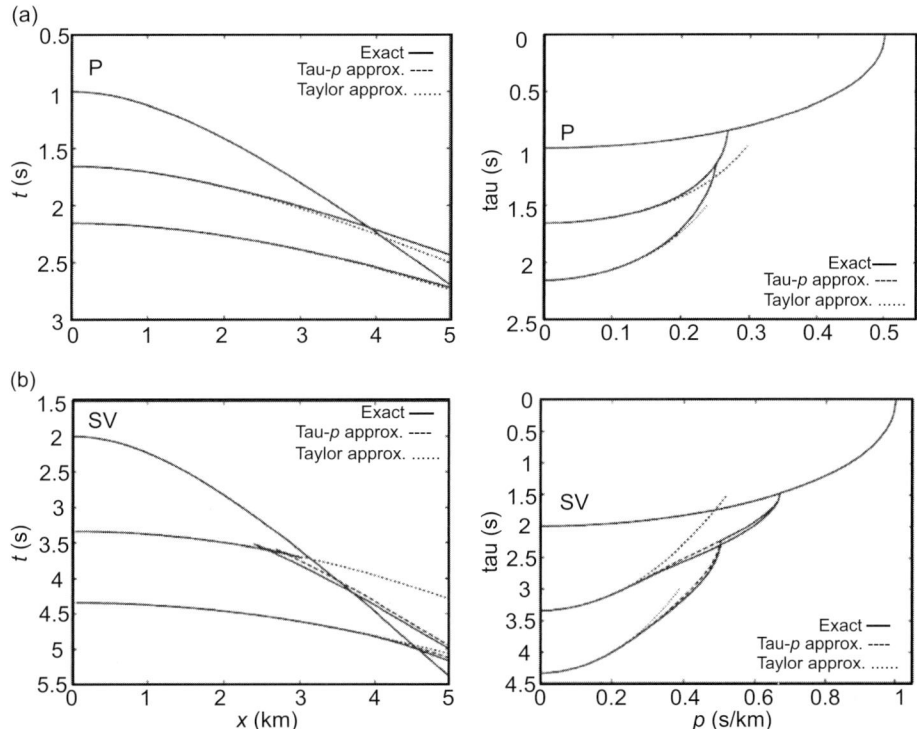

Figure 10.32 Moveout curves (left) and $\tau-p$ curves (right) of a three-layer model for (a) P-waves and (b) SV-waves (van der Baan & Kendall, 2002). Only the middle layer is anisotropic. Solid line: exact curves; long dashes: $\tau-p$ method using reduced-parameter expression; short dashes: Taylor series approximation.

function of the normal velocity. In both graphs the curves of the inverted values pass the model values, but the non-uniqueness of the inversion is clearly seen. Note in both cases the curves of δ and ε are somewhat parallel with each other, meaning a slow change in the value of η.

Taking advantage of the $\tau-p$ transform to separate events of different slowness values (see Section 2.4.2), many workers extract anisotropic parameters in the $\tau-p$ domain. As an example, Figure 10.32 shows the moveout curves on the traveltime-offset (left) plots and the associated $\tau-p$ (right) plots for a three-layer model having an anisotropic middle layer of shale sandwiched by two isotropic layers (van der Baan & Kendall, 2002). These authors developed a $\tau-p$ transform method to compute P and SV reflection moveout curves in stratified, laterally homogeneous, anisotropic media. Hyperbolic curves in the traveltime-offset domain map onto ellipses in the $\tau-p$ domain by summing the contributions of the individual layers. The impact of the anisotropy is clearly indicated by the deviation from the two lower ellipses in the $\tau-p$ domain.

To combat the non-uniqueness in inverting for interval anisotropic parameters, a layer stripping approach is often taken to estimate the effective or interval anisotropic parameters. For the modeled data shown in Figure 10.32, a layer stripping of the SV-wave reflections

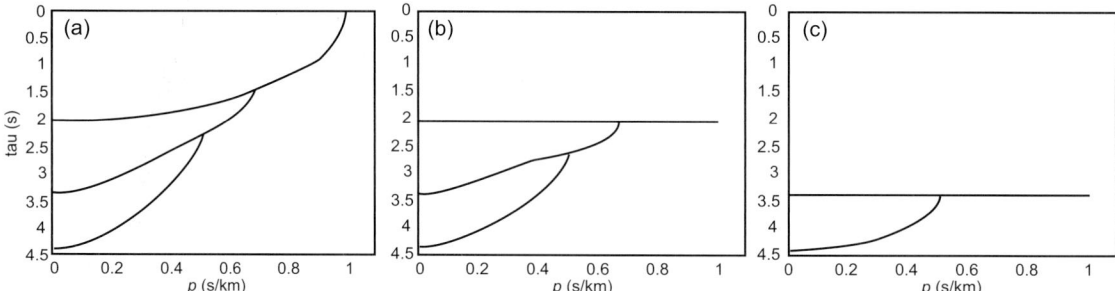

Figure 10.33 Layer stripping in the τ–p domain (after van der Baan & Kendall, 2002). (a) τ–p curves of all three SV-waves, as in the lower-right panel of the previous figure. (b) First layer removed. (c) Top two layers removed. The strong deviation of the moveout of the middle reflector in (a) and (b) is caused by the anisotropic middle layer, whereas the other two layers are isotropic.

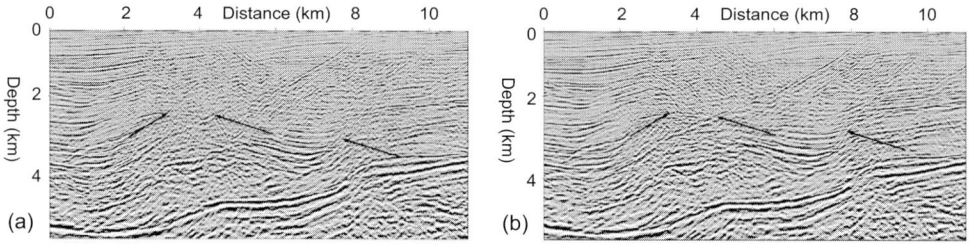

Figure 10.34 A 2D line from West Africa after (a) isotropic and (b) VTI time imaging (Tsvankin *et al.*, 2010). The processing sequence includes NMO and DMO corrections and post-stack phase-shift time migration. Both time sections were converted to depth. The arrows point to the main improvements made by the anisotropic processing.

in the τ–p domain is shown in Figure 10.33 (van der Baan & Kendall, 2002). From the top layer downwards in the τ–p domain, the traveltime difference between every two adjacent reflection curves is calculated and then removed. In panels (b) and (c) of Figure 10.33, each of the stripped reflectors is flattened to the corresponding zero-offset time. At each step in the stripping process, the moveout curve of the reflector right below the flattened one is used to estimate the interval parameters of the layer.

A recent review by Tsvankin *et al.* (2010) pointed out that anisotropic migration with the estimated Thomsen parameters typically improves the fidelity of seismic imagery, with better focusing and positioning of reflectors of different dips, including steep interfaces such as salt flanks. Figure 10.34 from their review was used by Alkhalifah *et al.* (1996) to illustrate the improvement achieved by anisotropic time processing of data from offshore West Africa, where thick TI shale formations cause series imaging problems. As highlighted by the right arrow in this figure, the fault plane at midpoint 7.5 km and depth 3 km is nearly absent on the isotropic section but imaged well by the VTI time imaging. The fault plane highlighted by the middle arrow is much more coherent on the VTI section than on the isotropic section. Accurate imaging of faults beneath the shales has played a major role in prospect identification in the area.

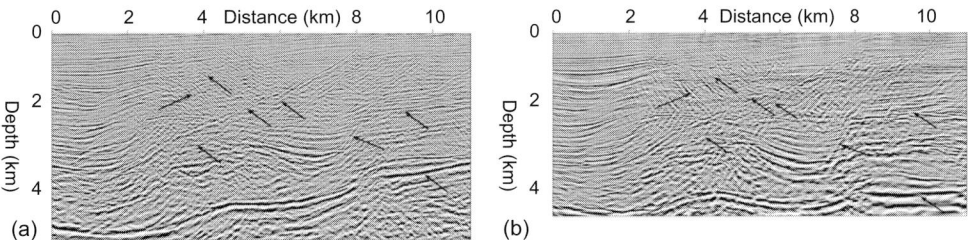

Figure 10.35 Comparison between imaging solutions: (a) the same as in Figure 10.34b from VTI time processing and (b) VTI migration velocity analysis and pre-stack depth migration. The arrows point to the main differences (Sarkar & Tsvankin, 2006).

10.3.4 Depth-domain estimates of velocity anisotropy

10.3.4.1 Impact of velocity anisotropy on depth imaging

Seismic velocity anisotropy is a property of the medium; hence we will ultimately express it in depth models. This is a challenging task because, as shown in the previous section, the problem is highly non-unique and far from being solved even in the time domain. Nevertheless, it is important to study the problem in the depth domain for several reasons. First, the depth domain is much more "physical" than the time domain; hence it will increase the chance of understanding the physical meaning of velocity anisotropy. Second, even when we achieve a perfect fit in the time domain, the data space, there is no guarantee that the solution parameters are unique in the model space. Tackling the problem in the depth domain may give us a much better chance to apply extra constraints to narrow down the solutions in a geologically plausible manner.

In the presence of velocity anisotropy, proper accounting for it may result in more significant improvement in depth imaging than in time imaging; this is because depth imaging is much more sensitive to velocity variation than time imaging. This notion is especially true in the case of lateral velocity variation, which cannot be handled by time-domain techniques. Using the same 2D field data from West Africa shown in Figure 10.34, Sarkar and Tsvankin (2006) made a comparison between VTI time imaging and depth imaging shown in Figure 10.35. In the depth processing, migration velocity analysis was carried out using a factorized VTI block model. Comparing the two images shown in Figure 10.35b reveals a significant improvement in the geologic plausibility of the imaged structures from the depth imaging with respect to the time imaging. The depth imaging result has more continuous fault planes, and a reduction in the cross-over events associated with kinked or joining points in the structure.

The factorized VTI block model for the depth imaging in the previous figure is shown in Figure 10.36. Each of the model blocks has constant δ and ε, and the velocity V_{P0} is a linear function of the spatial coordinates. Sarkar and Tsvankin (2006) suggested that factorized VTI is the simplest model to include both anisotropy and heterogeneity, and such a model requires minimal *a priori* information to constrain the relevant parameters. These authors argue that, in the absence of pronounced velocity jumps across layer boundaries, knowledge of the vertical velocity at the top of a piecewise-factorized VTI medium is sufficient to

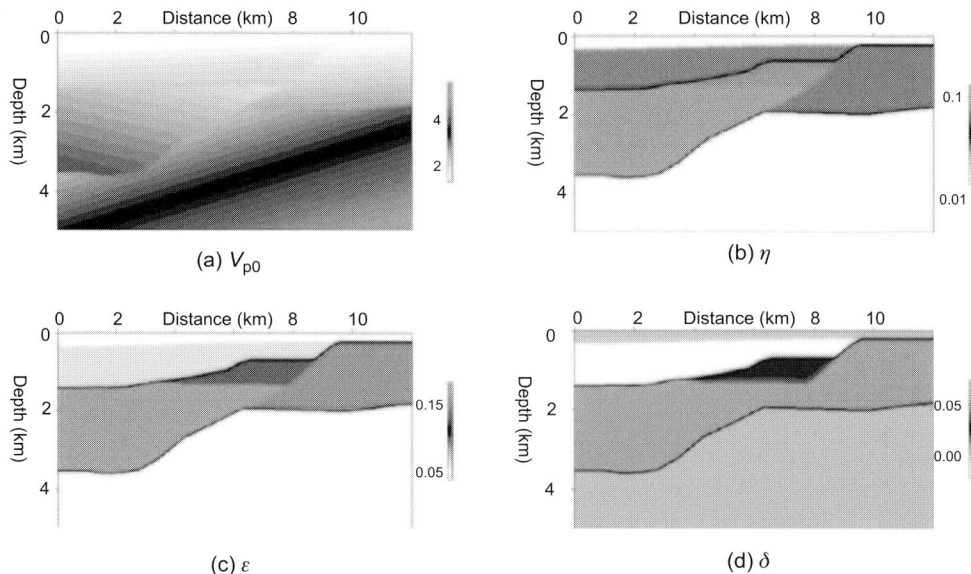

Figure 10.36 Estimated parameters (a) V_{P0}, (b) η, (c) ε, and (d) δ used to generate the depth migrated section in Figure 10.35b. (After Sarkar & Tsvankin, 2006.)

estimate the parameters δ, ε, V_{P0} and the velocity gradient throughout the cross-section using only P-wave data.

Because most variables in anisotropic media are dependent on each other, we need to develop ways to untangle such dependency when estimating the anisotropic parameters. Box 10.2 shows an order for determining layered anisotropic parameters in TTI models based on analyzing the sensitivity of traveltime data with respect to anisotropic parameters.

10.3.4.2   Assessing velocity anisotropy in depth domain

The challenges to depth-domain estimates of velocity anisotropy force us to be very cautious in estimating the anisotropic model parameters. In such processing works, we need to keep the model as simple as possible, and check the quality of the model from as many different angles as possible. In conducting tomographic inversion for anisotropic depth models, it is popular to adopt the layer-stripping approach to use layered models and update the model parameters in an interactive and constrained manner.

Let us see an example of residual velocity analysis for layered VTI models (Koren *et al.*, 2008). This software updates interval anisotropic parameters by applying local tomography to residual values picked from common image gathers (CIGs) generated by anisotropic curved-ray Kirchhoff time migration. The model consists of locally 1D, spatially varying VTI with the interval vertical V_{P0} and the two Thomsen anisotropy parameters δ and ε. The interval velocity δ is updated from short-offset reflection events, and ε is updated from available long-offset data. The model parameters are updated from the top down vertically and layer by layer, one parameter at a time.

Box 10.2 Sensitivity of traveltime to anisotropic parameters

Because the information for velocity anisotropy comes chiefly from traveltimes of seismic waves, it is useful to compare the relative influence of different anisotropic parameters on the traveltime of a traversing wave. We can quantify this sensitivity using the derivative of the traveltime with respect to each model parameter. For a simple model of a homogeneous tilted transverse isotropy (TTI), Box 10.2 Figure 1 shows the sensitivity of P-wave traveltime for four model parameters, the normal slowness $SW_{P0} = 1/V_{P0}$, ε, δ, and $\sin \phi$, where ϕ is the tilt angle of the symmetry angle of the TI from vertical.

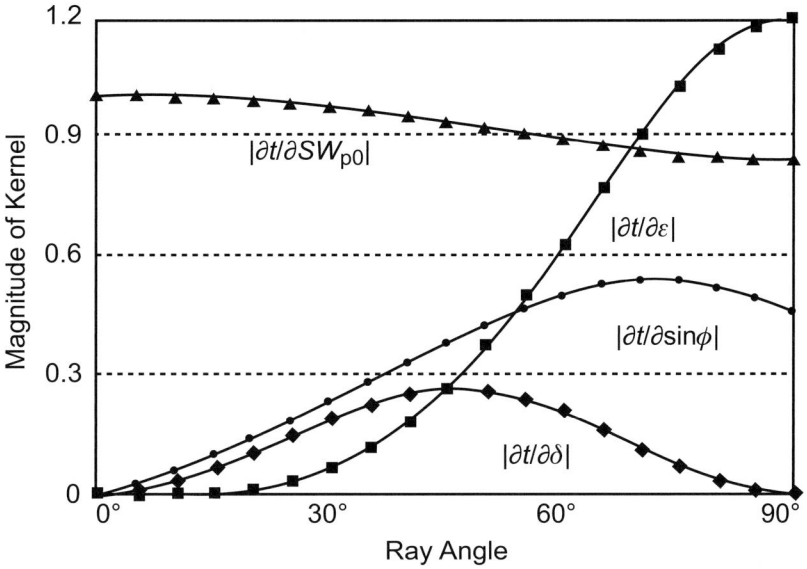

Box 10.2 Figure 1 Magnitudes of traveltime derivatives with respect to four TTI parameters as functions of ray angle (Jiang & Zhou, 2011). SW_{P0} is the normal P-wave slowness; ϕ is the tilt angle of the TI symmetry axis from vertical. In this case $\varepsilon = 15\%$, $\delta = 10\%$, and $\phi = 45°$.

The traveltime derivatives are kernels for tomographic inversion, because they relate the data space and model space. The relative magnitude of these kernels measures the level of connection between the data and the corresponding model variables. In this case, although the derivatives are computed using a specific set of model values ($\varepsilon = 15\%$, $\delta = 10\%$, and $\phi = 45°$), the relative trends of the derivatives do not change much for different model values. The figure shows that the kernel for slowness has a high level throughout the ray angle range, meaning slowness of velocity can be determined from traveltimes of rays from all angles. The amplitude of the kernels for ε and $\sin \phi$ raises toward higher ray angle, meaning it can be better constrained using rays perpendicular to the TI symmetry axis. The kernel for δ reaches its peak value at 45° in ray angle and decreases to zero at 0° and 90°, meaning it can only be determined using rays of incidence angle around 45° from the TI symmetry axis.

The amplitude level of each kernel and its variation with ray angle depict the traveltime sensitivity of the model parameter. The above figure indicates that, from high to low sensitivity, the parameters are in the order: SW_{P0} (or V_{P0}), ε, ϕ, and δ. After analyzing the relationships and influences of errors between the TTI model parameters in synthetic inversion tests in VSP and cross-well setups, Jiang and Zhou (2011) suggested the order of ease of determining layered anisotropic parameters was: V_{P0}, layer geometry, ε, δ, and ϕ (Box 10.2 Figure 2). Following this order to extract the anisotropic parameters may minimize the influence of the errors. For each model setup in practice, we should conduct tests to quantify a priority order like this, to guide the anisotropic process flow.

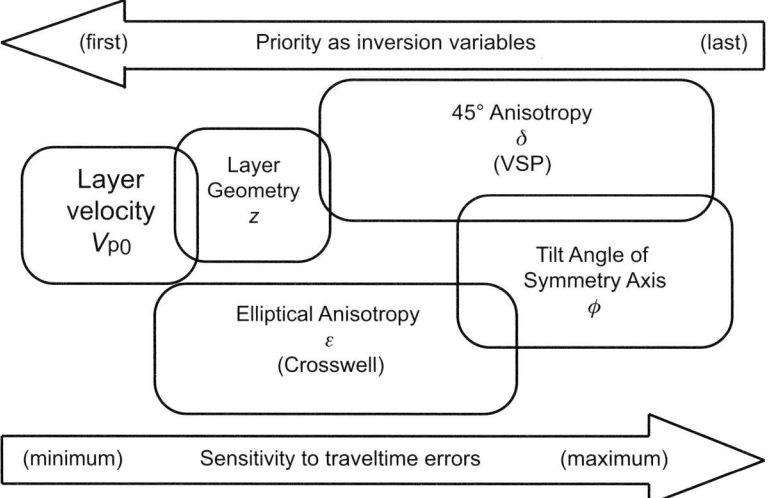

Box 10.2 Figure 2 From analyzing the influences of errors between anisotropic parameters in synthetic VSP and cross-well tests, the order of ease in determining anisotropic parameters is: V_{P0}, layer geometry, ε, δ, and ϕ (Jiang & Zhou, 2011).

Figure 10.37 displays the interactive panels of this software for updating the value of ε which varies linearly in each model layer. Panel (a) shows the migrated section in which the reflection rays from a horizon under examination are highlighted. Panel (b) shows the vertical variation of ε at the examination location (#10400), as well as the vertical profiles of V_{P0} and δ at the same location. Panel (c) shows the offset CIG at the location. A user may edit the interval values interactively, and the panel displays will be updated correspondingly. Panel (d) shows the semblance across the horizon as a function of the lateral location and the value of ε, on which the background, residual, and updated values of ε are plotted. Panel (e) shows the semblance variation percentage (horizontal axis) versus residual ε (vertical axis) at the model point under examination, enabling interactive picking of the residual

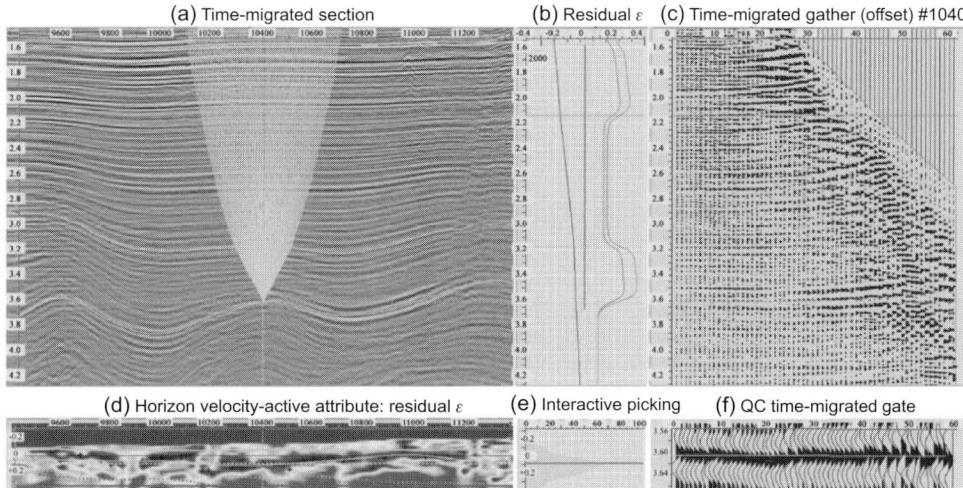

Figure 10.37 Interactive panels of residual analysis for ε in a software package (Koren *et al.*, 2008). Here the value of ε varies linearly in each model layer. The multiple-panel setting can be used to build the model from the top and layer by layer. (a) Migrated stack showing reflection rays at the CIG under analysis. (b) Model layer parameters that are editable by the user. (c) The CIG under analysis. (d) Semblance of velocity versus residual ε. (e) Interactive picking of residual ε. (f) A gate of time-migrated traces.

ε. Panel (f) shows a zoom window of the time-migrated CIG around the horizon under examination, allowing quality control of the detail in the moveout pattern.

Using the interactive panels, a user can pick residual-anisotropy parameters corresponding to the residual-moveout curves that best fit the migrated reflection events. This software treats the residual moveout at a given model point as two contributions, one from the overburden residual parameters and the other from the actual picked residual parameters. Such local tomography allows a layer-stripping and interactive estimation of the long-wavelength trends of the anisotropy parameters, and the reliable portions of the results can be used as the background model for a global tomography to update all model parameters with all data available. Figure 10.38 compares time-migrated sections at two locations before and after updating ε (Koren *et al.*, 2008). The inclusion of the anisotropy improves the continuity and amplitude of reflectors on these sections. The imaging of faults is also improved, as shown in the highlighted areas on the lower panels.

Exercise 10.3

1. Name several geologic settings that are suitable for HTI and VTI models, respectively. What are the impacts of HTI and VTI media on surface seismic reflection data? Comment on the effectiveness of using wellbore measurements to estimate HTI and VTI media.

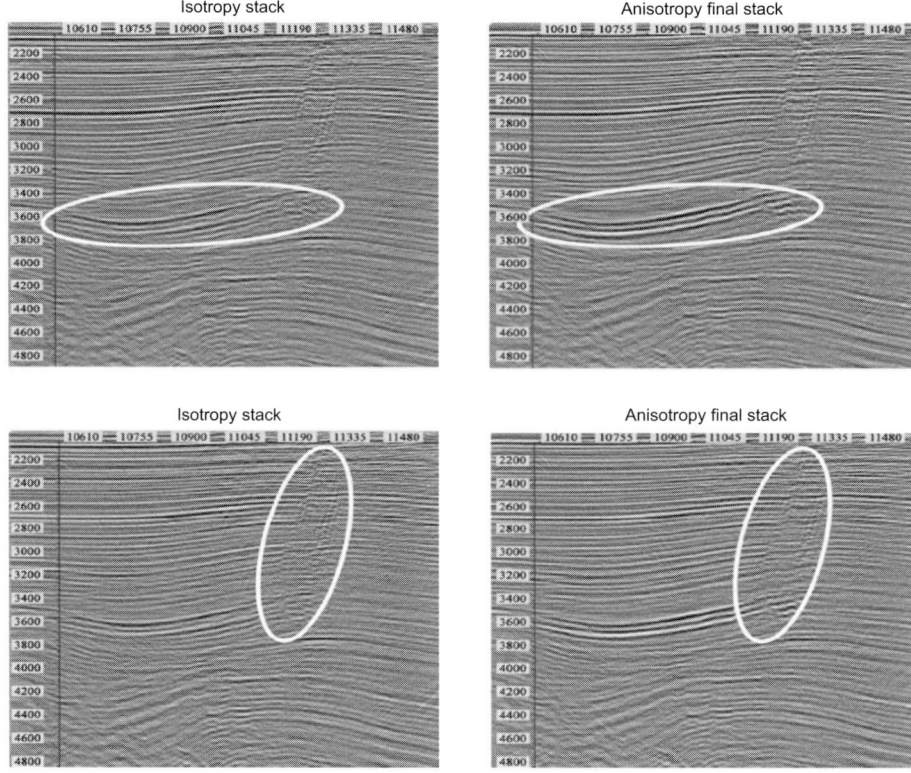

Figure 10.38 Time-migrated sections at two locations (upper and lower panels) before (left) and after (right) updating ε (Koren *et al.*, 2008). The horizontal axis shows the CRP numbers. Vertical time is in milliseconds. The ellipses highlight areas of significant change.

2. Use a spreadsheet to compile a list of diagnostics for distinguishing velocity anisotropy from velocity heterogeneity using seismic and other geophysical and geologic information.

3. Prove via theoretical derivation or numerical calculation that a raypath in a homogeneous TI medium follows a straight line.

10.4 Multi-component seismic data processing

10.4.1 Benefits and challenges of multi-component seismic data

In all types of media, seismic waves propagate with particle motions in all directions. Thus, from the birth of the science of seismology, nearly all seismographs have recorded three-component seismograms of both scalar fields such as pressure waves and vector fields such as particle-motion waves. For practical and economic reasons, however, most exploration seismic surveys record only single-component data, examples being the water pressure field recorded by hydrophones offshore and the vertical component of the velocity of particle motion recorded by geophones onshore. Such single-component data may simplify data acquisition and processing, but often severely limit the usage of data. Multi-component seismology aims to acquire, process, and interpret data acquired using multi-component sensors and/or multi-component sources. This section discusses the benefits and limitation of multi-component seismic data processing.

With reliable multi-component data, we can access shear waves that will greatly widen the scope of information carried by the seismic data. Since the speed of S-waves is lower than that of P-waves, S-wave data offer higher resolution than P-wave data of equivalent frequencies. Because fluids cannot carry shear motion, S-wave data will not be influenced much by the presence of gas clouds or gas chimneys. More importantly, S-waves form a vector field, in contrast to P-waves which form a scalar field. The vector nature allows the use of S-wave data to detect vectorized or angular-dependent properties of the subsurface, such as the presence and orientation of fractures and velocity anisotropy. A joint use of P-wave and S-wave data enables us to reveal the elastic properties of the subsurface, such as detecting the presence and type of fluids, a core issue for the petroleum industry. In fact, as discussed in a previous section, studies of seismic anisotropy can be more effectively conducted using multi-component data.

In most cases, seismic imaging of subsurface structures, or structural imaging, can be done using single-component P-wave data. However, in many cases, imaging of subsurface stratigraphy requires joint use of P-wave and S-wave data. In other words, stratigraphic seismic imaging can be better achieved using multi-component rather than single-component data. Many studies have demonstrated the benefits of this. For instance, Figure 10.39 shows two horizontal slices at the top of the Gessoso Solfifera formation of the Emilio field in the Adriatic Sea (Gaiser *et al.*, 2002). The left slice displays the polarization azimuth of the PS_1-wave, a converted wave, and the right slice displays the percentage of shear-wave splitting coefficient above the formation. The trends of both types of data follow closely the trend of the anticlines and other structural features in the field. Both types of data contain information on the fractures and lithology, and were derived by processing multi-component seismic data.

Mode conversions between P- and SV-waves commonly occur across interfaces, producing converted waves that are useful in many studies. Even when using acoustic shots and receivers, such as in marine surveys, we can use converted waves propagating as shear waves in some solid portions of the raypath. Using multi-component data rather than single-component data will make it much easier to recognize and extract the converted waves. A special case is when the mode conversion takes place crossing a salt–sediment boundary where the V_S value of the salt (around 2.7 km/s) is comparable with the V_P values of the sediments surrounding the salt body. The raypath of such a converted wave, traveling at speed V_P outside the salt and V_S inside the salt, does not bend much across the top interfaces of salt bodies (see the PSSP raypaths in Figure 10.40) and carries high amplitude. Such converted waves are useful in imaging the bottom interfaces of salt bodies, as illustrated in Figure 10.40 where the energy of PSPP, PPSP, and PSSP waves will become artifacts when they are imaged using P-wave velocities. The lower panel of Figure 10.41 shows an

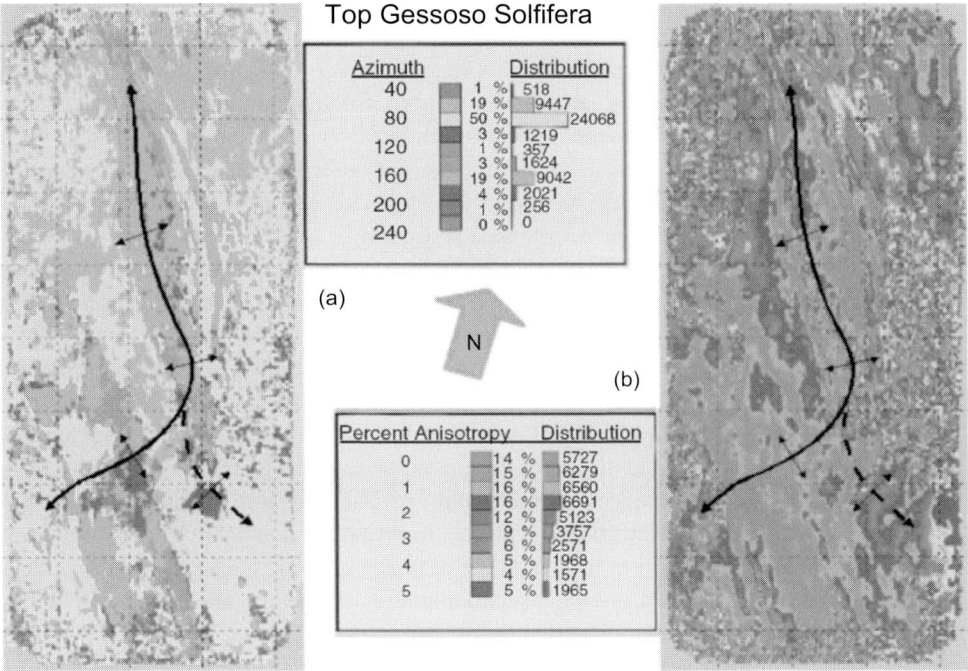

Figure 10.39 (a) Polarization azimuth of the PS$_1$-wave and (b) percentage of shear-wave splitting coefficient above the Gessoso Solfifera formation of the Emilio field in the Adriatic Sea (after Gaiser *et al.*, 2002). The PS$_1$-wave is polarized parallel to the crest of a doubly plunging anticline (thick black arrows), where anisotropy is generally higher. For color versions see plate section.

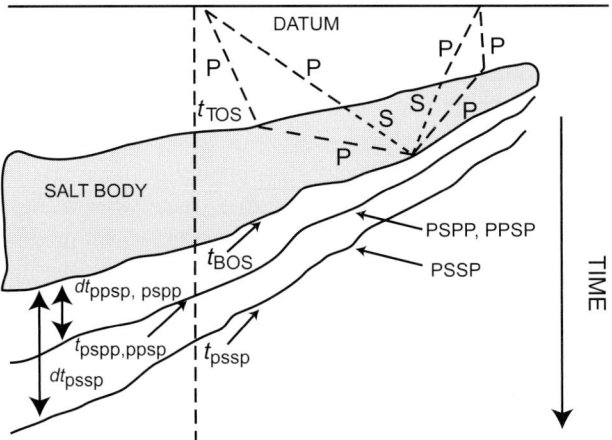

Figure 10.40 A cross-sectional diagram showing how the arrival times of the converted-wave reflections track the PPPP base-of-salt reflection after migration. Reflection raypaths of PPPP and PSSP waves are shown with dashed lines for P-wave and dotted lines for S-wave. (After Ogilvie & Purnell, 1996.)

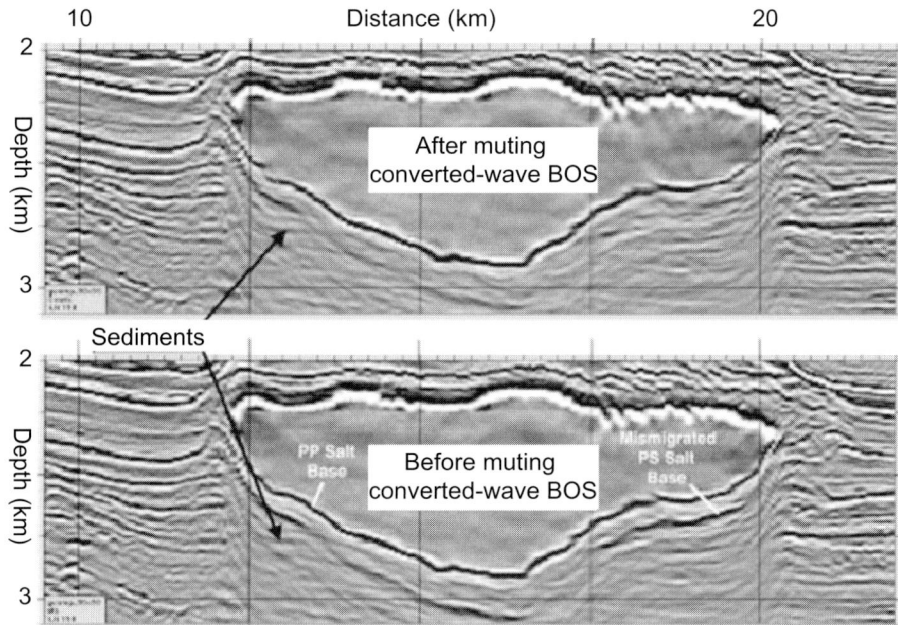

Figure 10.41 A cross-section of 3D pre-stack depth migrated cubes for a salt sheet after (upper) and before (lower) muting the PS, SP, and SS converted-wave reflection from the bottom of salt (BOS). (After Lu *et al.*, 2003.)

example of the imaging artifacts created by the converted waves. More detail on processing converted waves will be given in Section 10.4.3.

On the other hand, there are challenges for multi-component seismology in terms of high costs for equipment and field acquisitions, as well as a need for improving acquisition and processing technologies. From the perspective of data processing, the challenges of handling multi-component data may be highlighted by the following "how-to" questions:

- How should we estimate and account for non-omnidirectional sources, whose radiation varies with spatial angle?
- How should we assess and adjust the imbalance and inconsistency between different data components?
- How should we decompose or separate different wavefields, such as P-waves versus S-waves, upgoing waves versus downgoing waves?
- How should we adapt various data processing and imaging techniques to multi-component data and take advantage of the connection between different components?
- How should we process and analyze specialized multi-component acquisition setups, such as VSP and OBS?

Although the number of the challenges goes far beyond the above list, most issues in the processing of multi-component seismic data can be remedied by focusing on the fundamental principles of seismic wave propagation. Often a simple technique may produce significant improvement, such as the use of hodograms to identify the rotation angles of multi-component VSP data described in Section 2.1.5. In the following subsections, several of the issues are taken as examples of applying the fundamental principles.

10.4.2 Wavefield separation techniques

Most seismic imaging methods use a particular type of seismic wave, such as the surface-recorded primary reflections in exploration geophysics. In order to use a desired wave type, we need to extract it from data containing all wave types which may overlap with each other. A common practice is to construct decomposition filters via **plane-wave analysis**. For instance, decomposition of multi-component measurements into upgoing and downgoing P- and S-waves can be done using f–k filtering or median filtering, based on differential moveout between different wave types. An example of such separation for a 2D VSP dataset was shown in Figure 3.5 in Section 3.2.4.

Because multi-component data should be treated as a vector field in order to take advantage of the connection between different components, we may design the separation of P- and S-waves based on the first principles. If the particle motion history at every model position is available, we can apply divergence to obtain P-waves and curl to obtain S-waves. Recall that in seismic migration the observed waveforms at the surface are downward-continued to populate the entire model volume in the subsurface. Since such extrapolated wavefields contain the particle motion history of the wave propagation processes, we should be able to use the history to separate P- and S-waves at every model position.

Based on the above idea, Sun *et al.* (2004) demonstrated a separation of reflected P- and S-waves that are superimposed in 3D, three-component elastic seismograms. Using a finite-difference extrapolation algorithm, they compute the divergence of a single-component seismogram (P-waves) and curl of three-component seismograms (S-waves) of the extrapolated wavefield and record them independently at a fixed model depth. The P- and S-velocities in the elastic model are split into two independent models. Then the divergence seismogram, which contains P-waves only, is upward-extrapolated through the P velocity model to the original receiver locations at the surface to yield the separated P-waves. Next, the three-component seismograms of the curl, which contain S-waves only, are upward-extrapolated through the S velocity model to the surface receiver locations to yield the separated S-waves.

The separation process is illustrated using a simple 3D elastic model shown in Figure 10.42. The modeled multi-component seismograms along three profiles X, Y, and AB are shown as three rows of panels in Figure 10.43. Here the desired signals are reflections from the two planer interfaces of the model, hence the surface waves and direct waves were removed. Notice in this figure along profile Y the y-component has virtually zero amplitude. The reason is that profile Y is a dip profile, and the source did not generate oscillation in the offline or y-direction of this profile.

Figure 10.44 shows the results of the wavefield separation process. Because the divergence and curl involve spatial derivatives that generate a $\pi/2$ phase shift, the separated P- and S-waves need to be phase corrected, for instance by using the Hilbert transform discussed in Section 1.4 with respect to time. The first column of this figure shows the divergence or the retrieved P-waves along X, Y, and AB profiles. You can see the corresponding P-waves in the first column of Figure 10.43, although the amplitudes of the signal

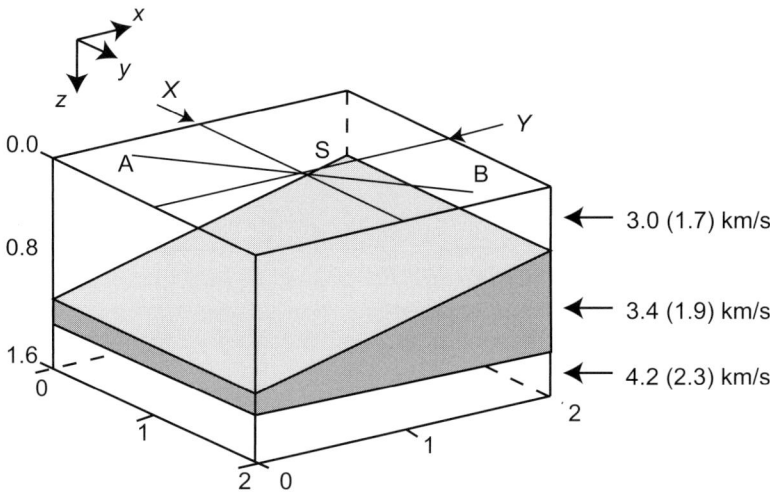

Figure 10.42 Perspective view of a 3D elastic model (Sun *et al.*, 2004). Distances are in kilometers. The velocities of three model layers are shown on the right, with V_P and V_s values outside and inside the parentheses. On the top surface of the model, S is the source, X, Y, and AB are three receiver profiles. The ends of AB are at A(0.4, 0.2, 0) and B(1.6, 1.8, 0).

in Figure 10.44 are apparently stronger owing to the amplitude normalization and the phase rotation of the separated P-waves.

The separated S-waves in the remaining three columns of Figure 10.44 are directly comparable with the corresponding panels in Figure 10.43. Along each profile, the three components of the S-waves in Figure 10.44 have similar traveltimes but differ in amplitude and phase angle. The z-component of the separated S-waves is weaker than the maximum amplitude of the two horizontal components. Along profile Y the x- and z-components of the curl are zero, because the S-waves here are generated from P-to-S mode conversion and the reflector dip is in the inline or x-direction of this profile.

10.4.3 Converted wave processing

10.4.3.1 Basics of converted wave and processing

In both earthquake seismology and exploration seismology the most common type of **converted waves** are generated by P-to-SV mode conversion at a subsurface interface such as the Moho discontinuity or the salt–sediment boundary. In exploration seismology the P–SV converted wave is commonly seen because manmade sources mostly emit P-waves, such as from the airguns offshore and dynamites onshore. There are three chief reasons to use converted waves. First, the raypaths and therefore seismic illumination of the converted waves differ from that of the pure-mode P-wave reflections. The extra illumination brought by the converted waves is extremely useful for imaging salt boundaries and other features of large velocity contrast. Second, the converted waves contain S-waves traversing through rock matrix, so the extracted S-waves can be used to image special targets such as gas clouds and chimneys where the P-wave images are severely blurred by the highly attenuating gas

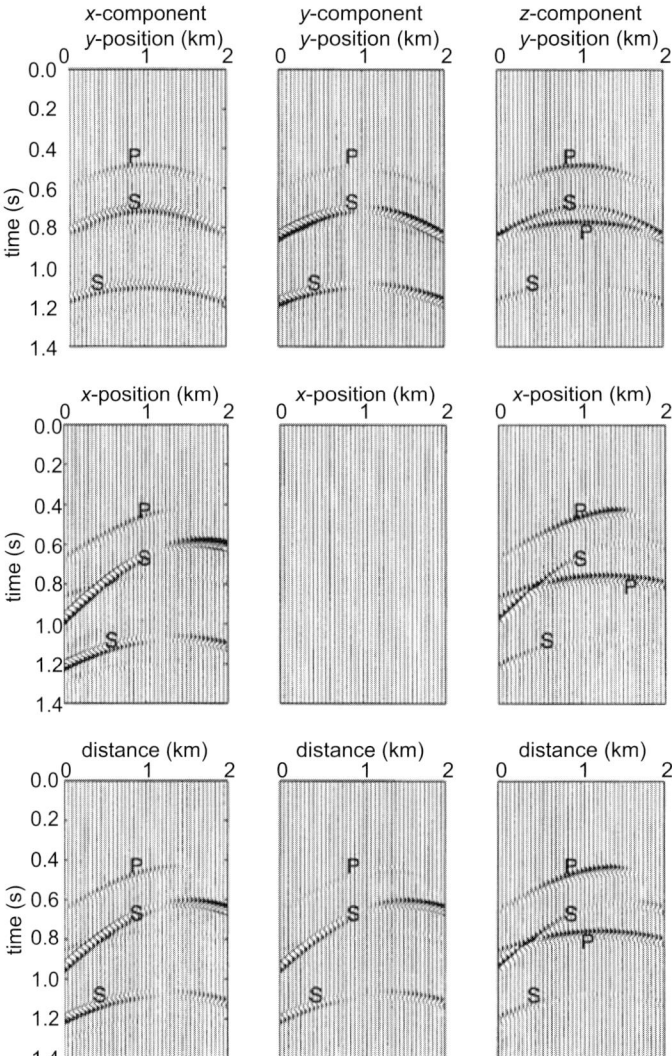

Figure 10.43 Synthetic multi-component seismograms for the model in the last figure after removing the surface waves and direct waves. The three columns are the x-, y-, and z-components, respectively. The three rows are along the X, Y, and AB profiles. The distance in the bottom row are from point A. The reflected P- and S-waves are marked. (After Sun *et al.*, 2004.)

pockets. Third, when strong converted waves exist and are untreated, they will generate artifacts in seismic imageries.

However, it is challenging to process converted waves owing to their overlapping with other wave types and the asymmetrical reflection raypaths of the converted waves, as shown in the schematic cross-sections in Figure 10.45. In laterally inhomogeneous media, the asymmetric raypath of converted-waves leads to the phenomenon of **diodic velocity**

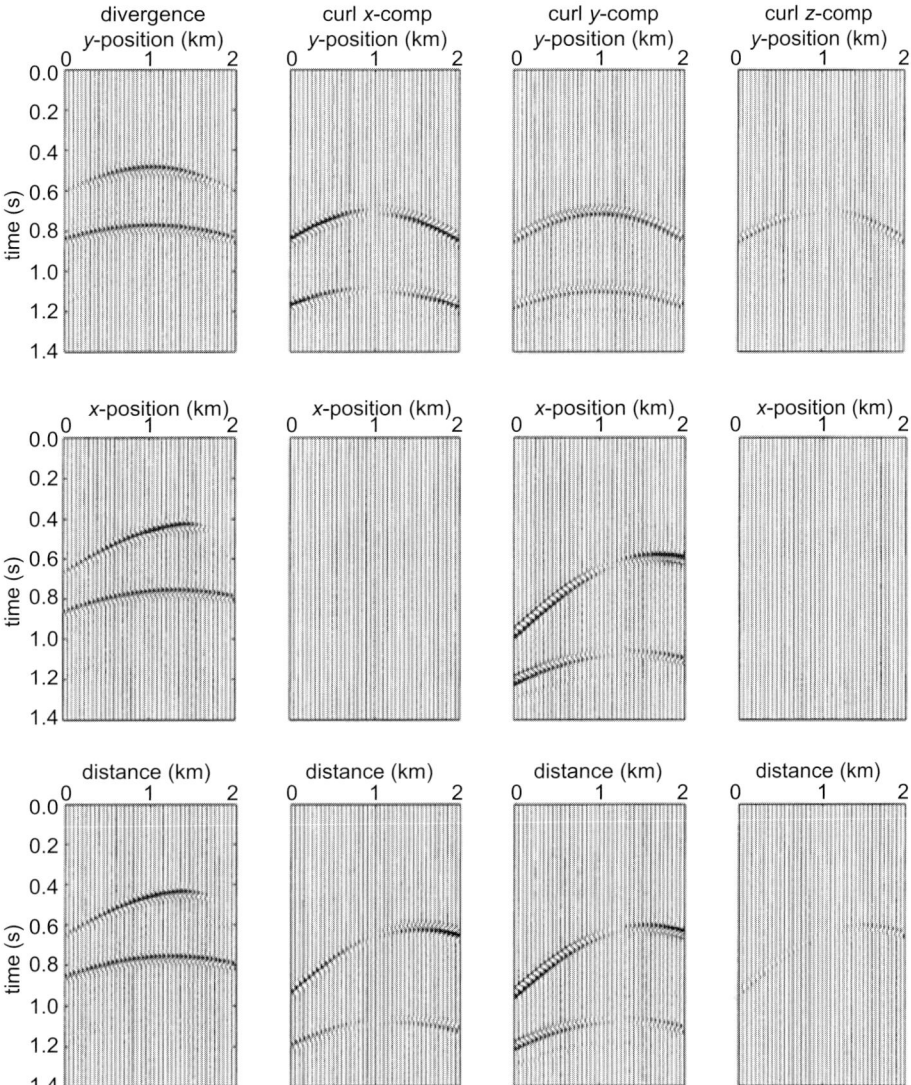

Figure 10.44 The P- and S-wave separated seismograms from the divergence and curl operations during finite-difference extrapolation plus a phase correction. The four columns are the divergence (P-waves) and the x-, y-, and z-components of the curl (S-waves), respective. The three rows from top down are along the X, Y, and AB profiles. The distance in the bottom row are from point A. (After Sun *et al.*, 2004.)

(Thomsen, 1999), which is an apparent violation of the reciprocity theorem. As shown in panel (b) of this figure, the traveltimes and ray coverage are no longer invariant when switching the source and receiver positions in a layer-cake medium. The name "diodic velocity" arises from the situation of an electronic diode, which operates differently in the forward and reverse cases.

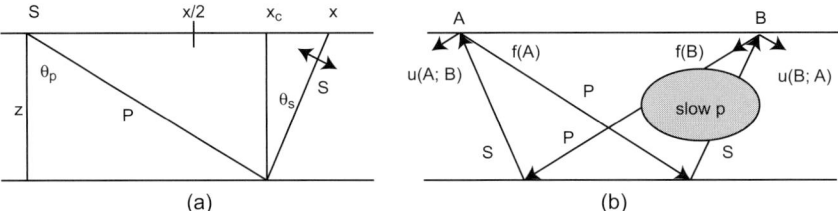

Figure 10.45 (a) Asymmetrical reflection raypath of a P-to-S converted wave from shot S to a receiver at distance x. The offset of the conversion point at x_c is much greater than that of the midpoint at $x/2$, and the reflection angle of the P-wave θ_p is greater than the reflection angle of S-wave θ_s. (b) Diodic velocity occurs for converted waves whenever there is a lateral variation in velocity. (After Thomsen, 1999.)

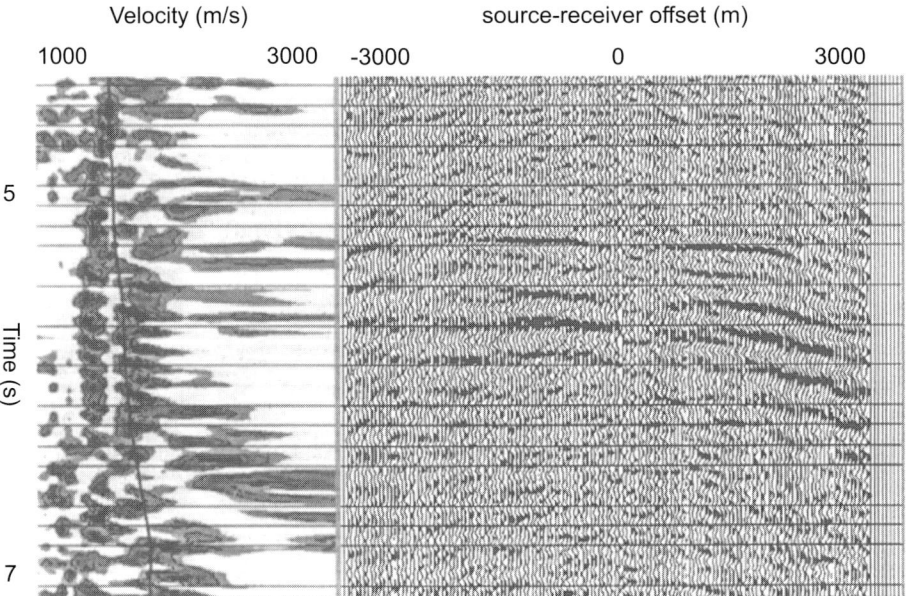

Figure 10.46 (Right) An asymmetric split-spread CMP super-gather of the inline horizontal-component data at the target level from the Valhall field in the North Sea (Thomsen, 1999). (Left) The velocity spectrum showing the diodic velocity character.

Figure 10.46 shows an asymmetric split-spread CMP super-gather of inline horizontal component data and the corresponding velocity spectrum from the Valhall field in the North Sea (Thomsen, 1999). The vertically aligned, double clusters in the velocity spectrum are diagnostic of the diodic velocity resulting from the split-spread of the acquisition setup. Such diagnostics are very valuable in processing converted waves where the first challenge is how to identify converted waves effectively.

Processing of converted wave data has been focused on identification, wavefield separation, and imaging with converted waves. Figure 10.47 shows an example flowchart of converted wave processing in six steps, focusing on S-waves since information on P-waves

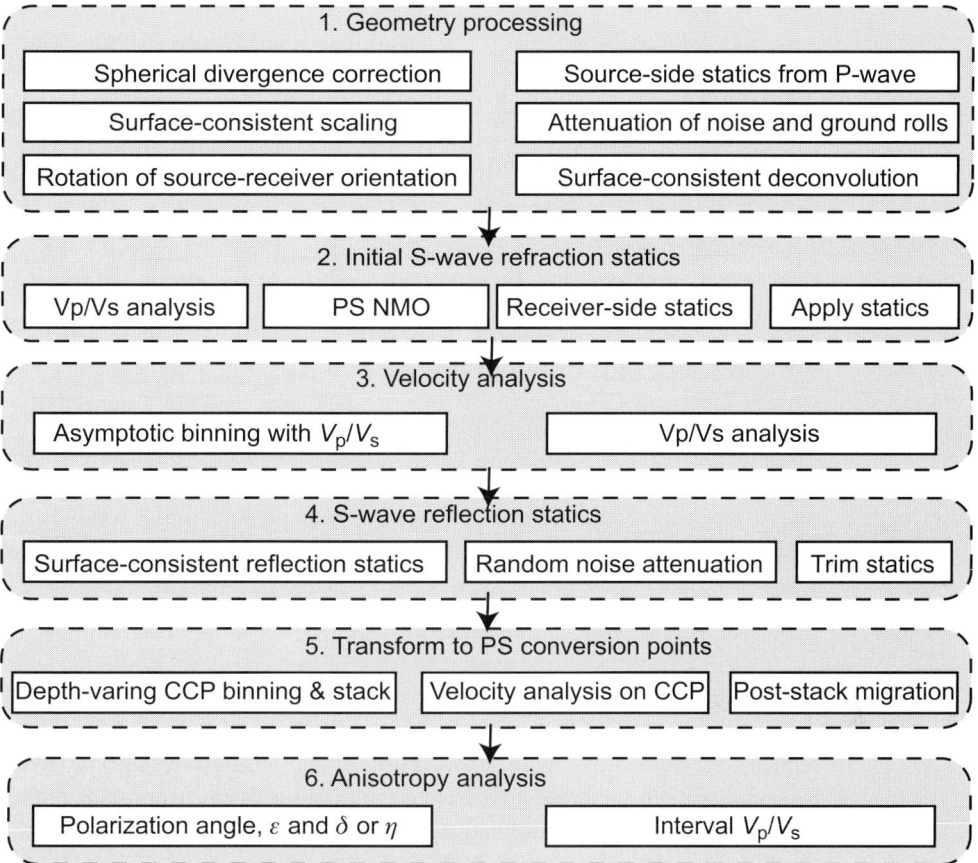

Figure 10.47 A flowchart of data processing for converted waves.

can be obtained from conventional processing. The first step of geometry processing, similar to that of processing conventional P-wave data, carries out the tasks of geometry QC, noise suppression, amplitude correction, source statics, and deconvolution to reduce the effects of source signature, narrow bandwidth, and multiples for marine data. The next three steps aim to correct for S-wave statics and velocity analysis. The long- and short-wavelength statics, respectively, are treated by the S-wave refraction statics and reflection plus trim statics. Embedded in the S-wave statics flow is the velocity analysis based on the V_P/V_S ratio. In Step 5 the **common-conversion point (CCP)** gathers are formed using the P-wave and S-wave stacking velocities from the previous steps. The last step is the analysis for layer interval anisotropic parameters and interval values of the V_P/V_S ratio.

10.4.3.2   Conversion point determination

It is not a trivial task to determine accurately the lateral position of the conversion point x_c and the reflection angles for P- and S-waves, α_P and α_S, respectively. Recently Yuan et al. (2006) derived exact solutions for calculating the P–SV conversion point in a dipping

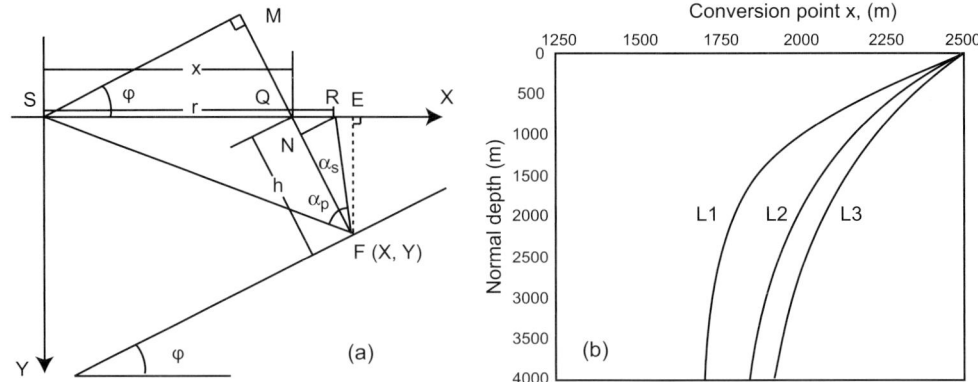

Figure 10.48 Computing P–SV conversion point for a dipping reflector below a constant-velocity layer (Yuan *et al.*, 2006). (a) Geometry for a reflector of dipping angle φ. S and R are the shot and receiver positions, respectively, and the origin is at S. The source–receiver offset is r. The P–SV wave conversion point is at F, which has a normal depth of QF = h. (b) Conversion-point curves as functions of normal depth. Curves labeled L1, L2, and L3 are computed for dip angle $\varphi = 0°$, 30°, and 60°, respectively, for the offset $r = 2.5$ km, $V_P = 2$ km/s, and $V_S = 1$ km/s.

reflector below a medium of constant velocity values. A schematic cross-section for their computation and the resultant conversion-point curves for three different dip angles of the reflector are shown in Figure 10.48. For a model with given interval velocities of P- and S-waves, we can generate a reference table of the conversion-point curves for different combination of the velocities and reflector geometry. Then we can use this reference table to estimate the positions of the P–S conversion points.

Based on the CCP gathers we can conduct semblance velocity analysis for V_P and V_S stacking velocities, and produce subsurface images using stack or post-stack migration. Figure 10.49 shows CCP stacks using VTI and isotropic velocity models. Here traveltimes of SS reflections were computed using a PP + PS = SS method. At the top of the Balder formation the VTI image is significantly improved with respect to the isotropic image. In multi-component data analysis, the issue of converted wave processing is linked closely with the issue of anisotropy, as shown in the last step of the data processing flowchart in Figure 10.47.

10.4.4   VSP data processing

One of the earliest applications of multi-component semiology in exploration geophysics is the **vertical seismic profile (VSP)** using a suite of borehole three-component receivers and surface shots. As introduced in Section 2.1.5, the records at different depths by VSP facilitate time-to-depth conversion for surface recorded seismic data and recognition of multiple reflections. Another important application of VSP data is to estimate seismic attenuation quantified by the Q factor (see Section 4.5.3, for example). VSP data assist the seismic–well tie process by taking advantage of their recording depth range and typically

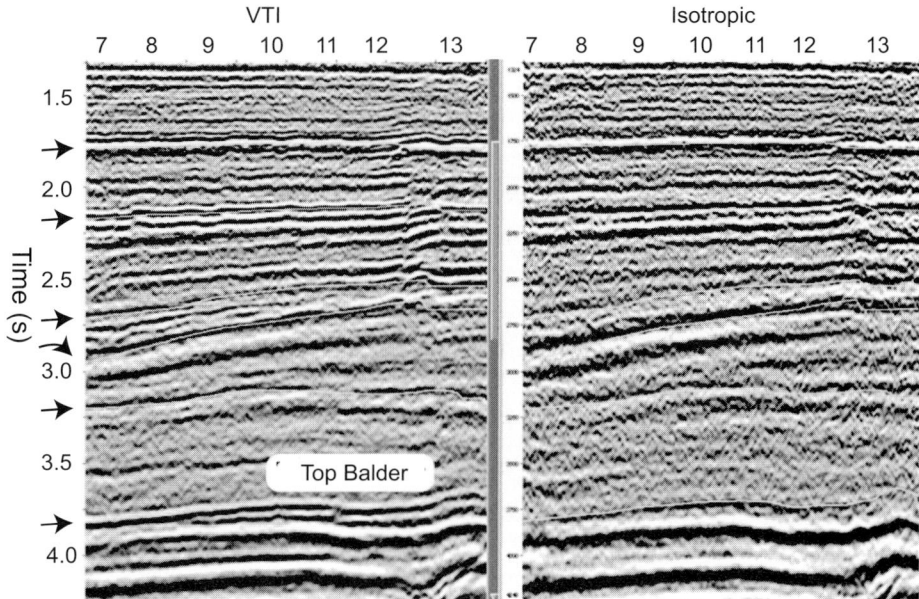

Figure 10.49 CCP stacks of P–SV waves in the Siri reservoir of the North Sea with VTI (left) and isotropic (right) models (Grechka *et al.*, 2002). Traveltimes of SS reflections were computed using a PP + PS = SS method. Resolution is improved in the VTI section, such as at the top of the Balder formation.

higher frequency content than the surface seismic data. The multi-component nature of VSP data provides information on shear waves and mode-converted waves. For different targets and acquisition conditions, a VSP survey can be designed using different arrangements of shots and receivers. Common setups include **offset VSP** using a fixed shot-to-receiver offset, **walkaway VSP** using shots of different offsets for each receiver, **azimuthal VSP** using different shot-to-receiver azimuths in 3D, and **reverse VSP** (**RVSP**) using surface receivers and wellbore sources. Sometimes RVSP are deployed to take the drilling head as the source, especially for so-called **look-ahead VSP** which tries to image rock strata below the drilling tool.

10.4.4.1 VSP imaging

The main limitation of VSP imaging is in the spatial range of seismic illumination, owing to its particular shot and receiver distributions in space. For a walkaway VSP using receivers deployed in a nearly vertical recording well, the illumination space provided by all available crisscrossing reflection waves form a spatial area called the **VSP corridor**. The images in the two right panels in Figure 7.27 are examples of the VSP corridor. Attempts to map targets far beyond the VSP corridor may result in severe smearing artifacts (e.g., Figure 7.30). Figure 10.50 shows an early comparison between migrated images based on a surface reflection dataset and a deviated-well VSP. Both images were produced using Kirchhoff migration of the vertical-component data. The two images are generally comparable.

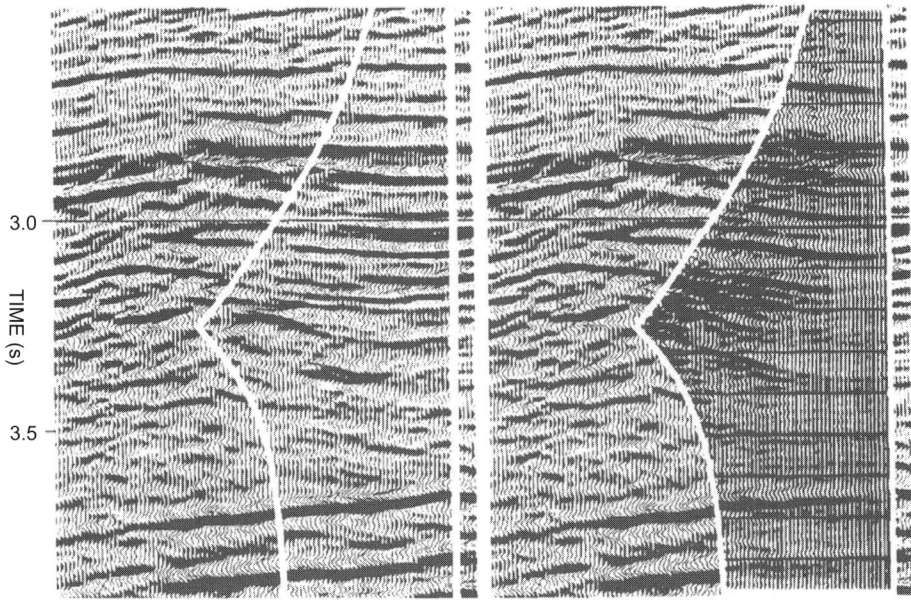

Figure 10.50 (Left) Migrated image of surface-seismic data. (Right) The surface-data image with an overlay of migrated image of a deviated-well VSP (Dillon, 1988). White lines outline the boundaries of the overlay.

As in the situation of processing surface seismic data, VSP data processing consists of signal enhancement such as extracting a particular wave mode, velocity model building, and seismic migration. Figure 10.51 shows a far-offset VSP common shot gather which recorded a source of vertical-motion vibrator in a study to image tight-gas sands in the East Texas Basin (O'Brien & Harris, 2006). We can tell the trends of downgoing and upgoing waves by first recognizing the trend of the first arrivals (labeled "P down" in this figure), downgoing in this case of a deep segment of receivers recording a surface source. The multi-component data were subjected to model-based rotation to optimize upgoing P–P reflections (labeled "P up" in panel (a) and, separately, upgoing P–S reflections (labeled "P-S down"' in (d)). Notice the difference between the alignments of the waves of panels (a) and (d). In this case the P–S mode conversion generated strong reflections. The rest of the panels in this figure show the results of further processing using a median filter to suppress waves of the opposite trends.

Figure 10.52 compares the Kirchhoff depth-migrated images using far-offset P–P versus P–S reflections. A successful separation of different wave modes as shown in the previous figure allows depth migration of the upgoing reflections of each wave mode. Because the S-wave velocities are lower than the P-wave velocities, the P–S imaging corridor is shifted toward the well relative to the P–P imaging corridor. Key geologic units, in reference to the gamma log inset in the upper left portion of Figure 10.52a, are highlighted in these seismic images. Patterns in the two types of reflection images are quite similar, especially in the shallow portion of the images. The resolution of the P–S image appears to be slightly higher than that of the P–P image, probably because of the shorter wavelengths of S-waves than

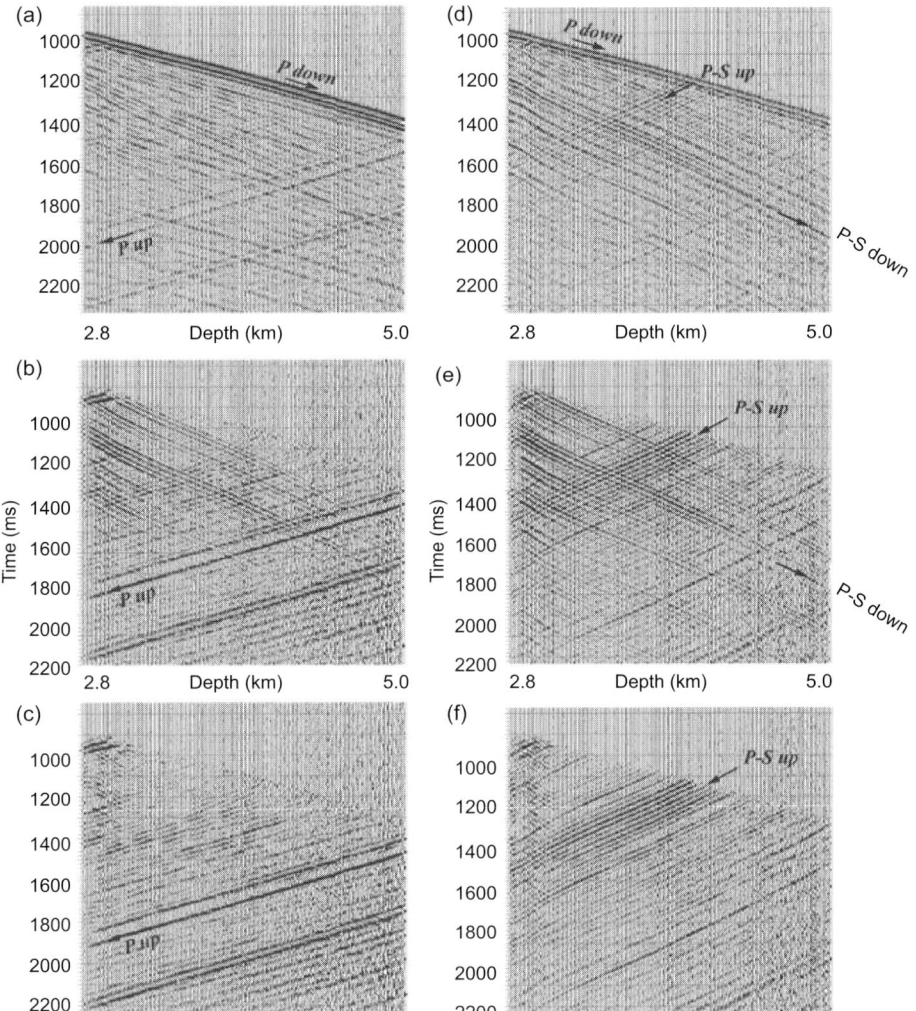

Figure 10.51 Far-offset VSP data recorded with a vertical-motion Vibroseis source after model-based rotation, with time ramp and trace balance applied (O'Brien & Harris, 2006). (a) Model-based rotation to optimize upgoing P–P reflections, followed by (b) median filter to suppress downgoing P-waves, and (c) additional median filter to suppress downgoing S-waves. Panels (d), (e), and (f) show the corresponding steps for model-based rotations to enhance upgoing P–S converted waves.

P-waves at the same frequencies. The amplitude of converted-wave reflectors is stronger than that of the P-wave reflectors above the Bossier Shale marker, but the amplitude trend is reversed below the marker. Before interpreting the geologic meaning of such amplitude variations, we need to make sure that the effect of AVO of different wave types is calibrated properly.

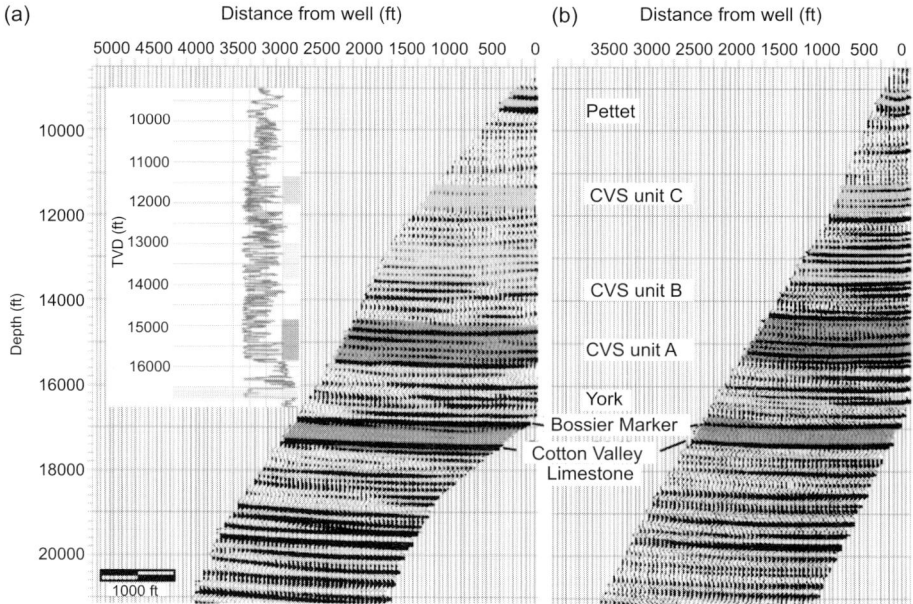

Figure 10.52 Far-offset VSP Kirchhoff depth-migration images (O'Brien & Harris, 2006). (a) P–P reflection image. (b) P–S mode conversion image. Insert shows the gamma log plotted at the same vertical scale. VSP and log data have been adjusted for datum differences. CVS denotes Cotton Valley Sand.

10.4.4.2   VSP versus surface data images

Figure 10.53 compares some P–P reflection images of an extracted line from the surface 3D data with the near-offset and far-offset VSP data. The near-offset VSP corridor stack matches very well with the surface imaged reflectors. With respect to the surface seismic data, the VSP data provide not only higher resolution, but also the proper depth-to-time conversion velocity. However, I suspect that some of the curved reflectors below 3 seconds on the far-offset VSP image may be contaminated by spearing artifacts due to the limited crisscrossing level of reflection waves available.

In order to take full advantage of multi-component seismic data, we should ideally treat each such dataset as a vector wavefield rather than a set of scalar fields. As an example, Hokstad (2000) developed an elastic multi-component Kirchhoff migration scheme that operates directly on vector traction and displacement or velocity data. The method is equivalent to a model-based separation of quasi P- and S-waves followed by a weighted diffraction stack. This way avoids the need for separating P- and S-waves in the pre-processing when the relative amplitudes of the various vector components are properly preserved or restored during pre-processing. In Figure 10.54, overlays of a VSP image solution from the elastic multi-component migration are embedded into surface seismic images. The VSP corridor of the P–S waves is much narrower than that of the P–P waves.

Although exploration seismology has advanced a long way since its birth, processing multi-component seismic data today still requires much more effort than processing conventional single-component data. In fact, most commercial seismic processing software

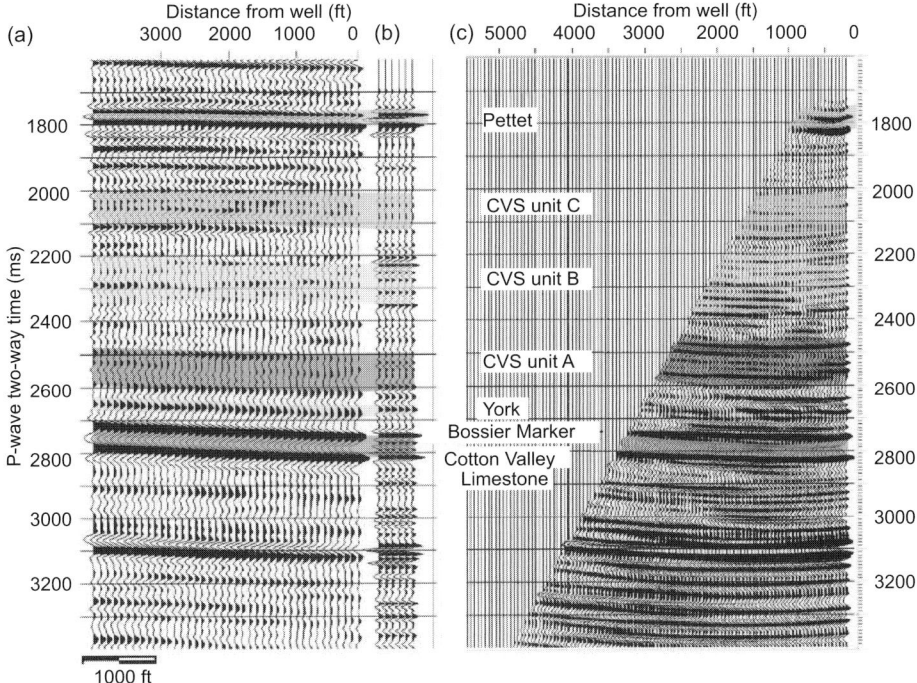

Figure 10.53 Comparison between P–P reflection images (O'Brien & Harris, 2006). (a) Section along the VSP line extracted from surface 3D cube. (b) Near-offset VSP corridor stack. (c) Far-offset VSP migrated image with trace spacing of 15 m.

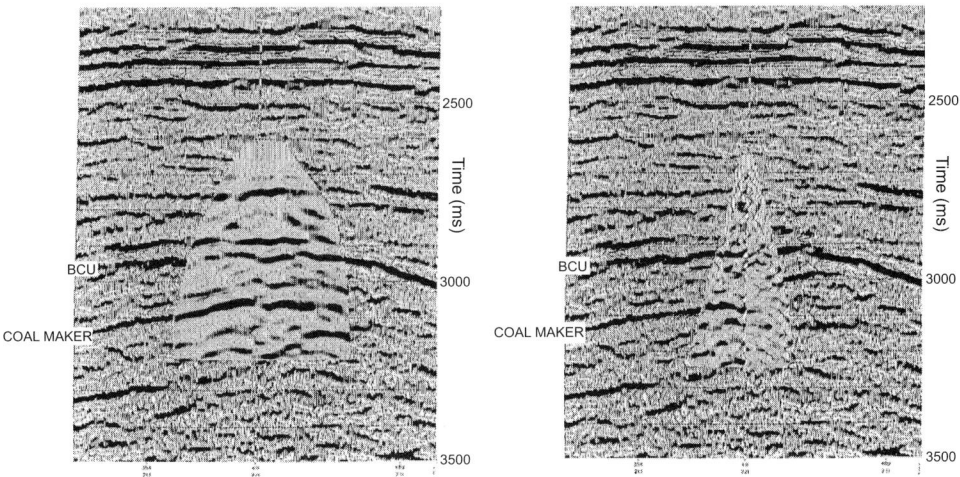

Figure 10.54 Anisotropic elastic multi-component Kirchhoff migration of a marine walkaway VSP, time-converted and embedded into surface seismics (Hokstad, 2000). (Left) P–P image and (right) P–SV image.

Box 10.3 Multi-component versus dense single-component surveys

Let us consider a situation encountered in a recent study using a mobile low-frequency geophone arrays. We have a fixed number (3N) of portable digital recorders that can be paired with either single- or three-component geophones. If we choose three-component geophones we will have N receivers, and if we choose single-component geophones we will have 3N receivers. Then which choice is better?

This question compares the benefits between the configuration of multi-component but fewer stations versus the configuration of vertical-component and more stations. In practice, we should first consider the scientific and/or business objectives. For instance, does the planned imaging target demand more multi-component data or denser station coverage? Next, we need to consider feasibility issues, such as whether the three-component data acquired can meet the required imaging quality. In onshore surveys, the horizontal components of seismic data can be of very poor quality in heavily weathered terrines. Another critical factor is the cost. For example, if it is very costly to access the station sites, such as in OBS as well as station sites in wellbores and caverns, we want to install as many recording components as possible. Since time is often worth more than money, we always prefer to use those acquisitions that can be done fast with sufficient quality.

packages do not offer a complete set of multi-component data processing modules. To make the multi-component seismic data much more useful, one needs to be familiar with the fundamentals of multi-component seismic acquisition, processing, and seismic wave theory, and ideally be able to code computer programs in order to test new ideas on processing multi-component data. Box 10.3 illustrates a situation that we have encountered.

Exercise 10.4

1. Compile in a table the criterion and diagnostics for recognizing a converted wave. Please specify the type of data, data domain, and principles of the recognition.

2. Conduct a literature search and summarize the idea and procedure of median filters that were used in Figure 10.51 to enhance upgoing reflections.

3. Section 8.4.2 described a simple case of tomographic inversion using first arrivals of VSP data. Prepare a data processing flowchart to build velocity model using first arrivals and traveltimes of different of reflection waves for a 2D walkaway VSP setup.

10.5   Processing for seismic attributes

A **seismic attribute** is a quantitative measure of a seismic characteristic of interest (Chopra & Marfurt, 2005). As the main products of seismic data processing, seismic images in the forms of sections and volumes are produced in order to detect subsurface targets for scientific or business interests. The targets may be the morphology of rock strata or faults, and the distribution of various types of fluids in rocks. However, it is challenging to interpret seismic images because they are merely responses to impedance variations of the subsurface, and these responses are band-limited, artifact-bearing, noisy, and related non-uniquely to the targeted properties. Such challenges are among the main reasons for developing seismic attributes, which are information about the targets drawn from computerized analysis of seismic imagery. Processing for seismic attributes involves various ways to make the seismic data more interpretable, or to reduce the negative impacts of the bandwidth, artifacts, noises, and non-uniqueness of seismic images.

Over the years, hundreds of seismic attributes have been generated to help understand and extract geologic information from reflection seismic data at a high level of objectivity and automation. Chopra and Marfurt (2005) advised that a good seismic attribute either is directly sensitive to the desired geologic feature or reservoir property of interest, or allows us to define the structural or depositional environment and thereby to infer some properties of interest. Thus, the highest priority in choosing a seismic attribute is to maximize its response to the targeted geologic or reservoir properties. Seismologically, we may define and quantify seismic attributes as particular quantities of geometric, kinematic, dynamic, or statistical features derived from seismic data (Liner *et al.*, 2004). Obviously, the quality of all seismic attributes relies on the quality of seismic data and their processing impacts. Interested readers can refer to a rich list of publications on this subject (e.g., Chopra & Marfurt, 2007).

10.5.1   Extraction of localized seismic attributes

10.5.1.1   Amplitude estimation

Since amplitude and timing are the two most fundamental elements of a time series, we may say that all seismic attributes are derived from amplitude and phase of the data, as introduced in Sections 1.3 and 1.4, respectively. In the early days of exploration geophysics, many bright spots of high reflection amplitude were identified at structural highs as promising prospects. The argument is that sands containing significant amount of hydrocarbons, especially natural gas, have anomalously lower values in density and velocity than the ambient rock strata, hence forming high-amplitude reflections. Figure 10.55 shows one of the examples of bright spots in which the darkest color denote events of the highest amplitude.

However, it is not a straightforward task to estimate amplitude directly because, as discussed in Section 1.3, the amplitude of seismic data is affected by many factors throughout the processes of wavefield generation, propagation, and acquisition, to processing. In fact, since seismographs record the voltage variations of the sensors, it is difficult to use absolute amplitude of seismic data at a fixed location alone. The practical value of seismic data lies in the variations of amplitude with respect to position, angle, frequency, and other variables. An example is a Section 10.5.1.3 on AVO, the variation of reflection amplitude with respect to the reflection angle at a fixed position, which is used to infer the presence and character of fluids at the position.

The two main tasks of data processing for amplitude estimation are noise suppression and corrections for all known systematic variations in amplitude. For the first task, we need to choose effective noise suppression methods based on the behavior of the given

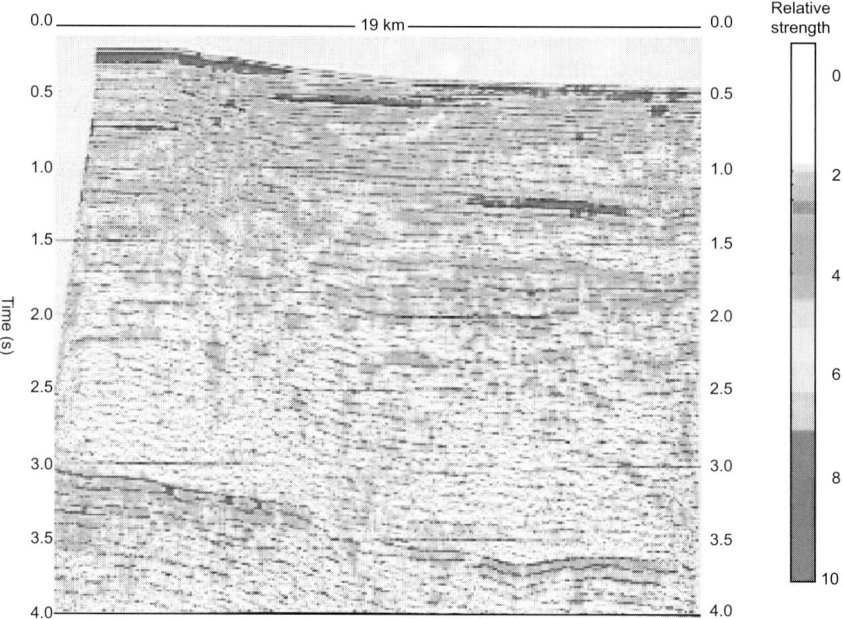

Figure 10.55 A section from the North Sea with bright spots (areas of high seismic reflectivity) identified. (After Anstey, 2005.) For color version see plate section.

data and expected signal behaviors. An emphasis here is on minimizing the impact of the processing on the original data amplitude. For the second task, as discussed in Section 1.3, we usually need to correct for source radiation pattern and geometrical spreading, and we may want to avoid the use of automatic gain control (AGC). Near-surface inhomogeneities often produce significant lateral variation in the amplitude of seismic data that has to be compensated using Q models or empirical relationships.

10.5.1.2   Instantaneous and local attributes

Instantaneous attributes are quantities defined at a time instance of seismic data, and thus are functions of local media properties. In Section 1.4.4 we have seen some original instantaneous attributes defined based on the analytical signal using the Hilbert transform, including instantaneous amplitude (envelope), instantaneous phase, and instantaneous frequency. Figure 10.56 displays these attributes for a seismic trace and a sketch of its complex trace. Envelope as a measure of local energy level of seismic traces has been related to the interval thickness of the corresponding geologic unit. Instantaneous phase is useful in stratigraphic interpretation to quantify unit boundaries of different scales. Instantaneous frequency is often related to seismic wave attenuators including hydrocarbon reservoirs.

A common practice in interpreting seismic data is to combine several attributes to form more specialized attributes to narrow down the non-uniqueness in interpretation. For instance, practitioners have created response attributes by specifying some instantaneous attributes at the peaks of the envelope function. As an example, **response phase** is the instantaneous phase at the point at which the envelope is at the maximum. One value of

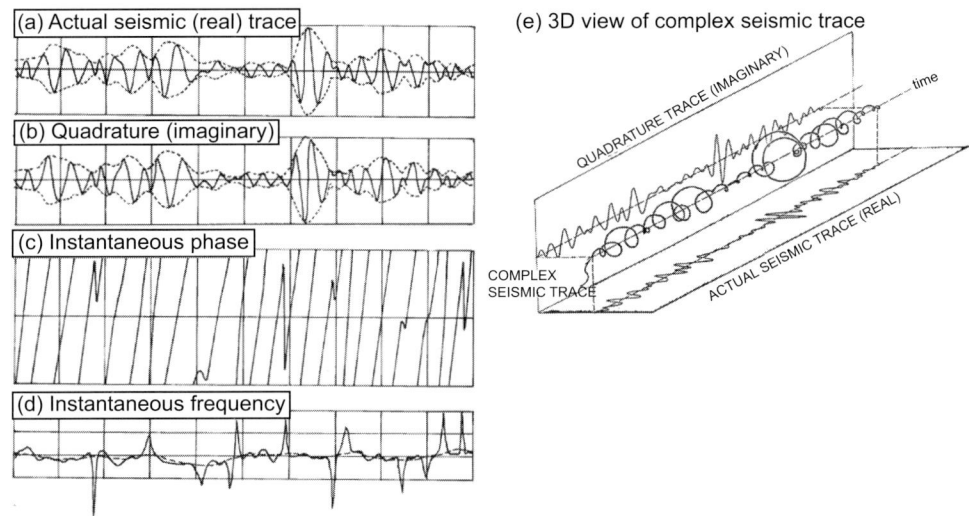

Figure 10.56 Displays of four basic instantaneous attributes in (a) to (d), and a 3D sketch of the complex trace in (e). Dotted curves in (a) and (b) are the envelopes and (e) is the radius of the complex trace from the time axis. Dashed curve in (d) is the weighted average frequency. (After Taner *et al.*, 1979.)

response phase is computed for each envelope peak and is assigned to the whole time interval covered by the envelope peak. Similarly, **response frequency** is the value of instantaneous frequency at each peak point of the envelope, and this single value is assigned to the time interval covered by the envelope peak.

From the perspective of seismic data processing, attributes generated from a single data point can be severely influenced by noise, while the commonality between neighboring points or traces may carry much higher level of signal. Stemmed from this notion, the concept of **local seismic attributes** has been introduced (Fomel, 2007) to specify local similarity and frequency. An example of the local frequency attribute is shown in Figure 10.57. The three panels on the left are test signals consisting of a synthetic chirp function, a synthetic trace from convolving a 40-Hz Ricker wavelet with a reflectivity profile, and a field data trace. Their local frequency traces are shown in the right panels of this figure. The frequency of the chirp signal in the top row is recovered. The dominant frequency of the synthetic signal in the middle row is correctly estimated at 40 Hz. The local frequency of the field data trace in the bottom row varies with time. We may regard such local seismic attributes as smoothed instantaneous attributes, which may be more tolerant to noise.

10.5.1.3   Amplitude versus offset (AVO)

AVO analysis quantifies seismic amplitude and phase across a NMO-corrected CMP gather from near trace to far trace. After the groundbreaking work of Shuey (1985) to simplify the Zoeppritz equations relating reflected P-wave AVO to elastic parameters of the reflector, AVO analysis has become a necessary step of seismic interpretation for fluid properties. Currently, a common objective of most AVO studies is to detect the presence and properties

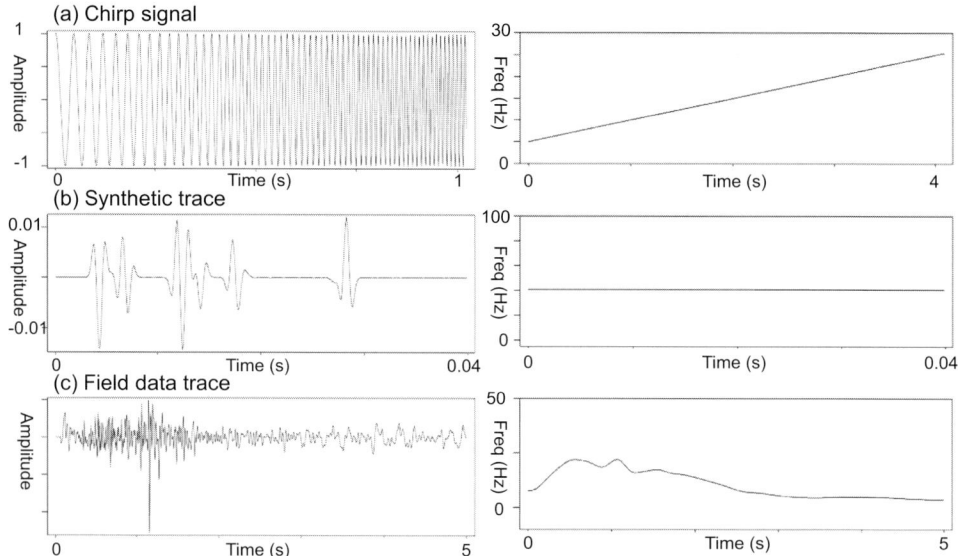

Figure 10.57 Three test signals (left panels) and the local frequency traces (right panels). (After Fomel, 2007.)

of various fluids including hydrocarbons in reservoir rock strata. Because the analysis focuses on a fixed reflector position, the effects from the overburden and source signature are minimized. This subsection follows a tutorial review of Ross (2000).

The objective of the AVO analysis is to identify those AVO responses that are characteristic of lithologies or fluid-filled reservoirs. For example, in high-porosity clastics, the anomalous AVO response is often associated with hydrocarbon saturation, whereas in lower porosity clastics and in some carbonates, the responses are more often associated with lithology and porosity. Figure 10.58 shows an example of measuring AVO attributes for a CMP record from Gulf of Mexico play. Shuey (1985) expressed the reflection coefficient as a function of average incidence angle θ:

$$R(\theta) = A + B \sin^2(\theta) \qquad (10\text{–}20)$$

where A and B terms are two AVO attributes, the intercept and gradient coefficients. The intercept is related to the normal incidence reflection coefficient, and the gradient is related to density and velocity contrasts around the position under examination. Other AVO attributes can be derived using different formulations and higher-order terms, or combining the basic attributes. The lower plot of this figure shows the best fitting AVO curve and the amplitude readings; the intercept is the zero-offset amplitude, and the gradient is the linear rate of change of the amplitude data.

One way to make joint use of two or more seismic attributes is to cross-plot them as the axes of the plotting coordinates. This is particularly useful for classification purposes. Figure 10.59 shows an example of cross-plotting two AVO parameters, the intercept and gradient, for an anomalous bright spot (high-amplitude event) of a gas sand in the Gulf of Mexico. The plot on the left has the AVO intercept along the abscissa and AVO gradient along the ordinate. Each point in the cross plot is taken from analyzing one location in the

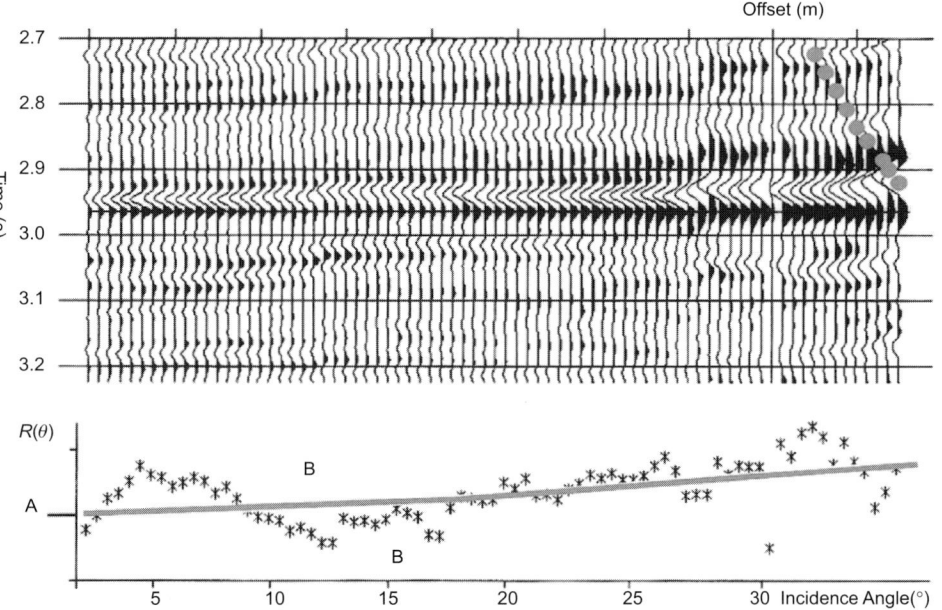

Figure 10.58 (Upper) A NMO-corrected CMP gather from Gulf of Mexico. The event at 2.97 s (peak) shows the AVO responses anticipated for 30 m of upper-Miocene gas deposits. (Lower) Amplitude values from the picked trough are plotted as a function of average incidence angle. The gray curve is the AVO trend, with the determined values of the intercept A and gradient B shown. (After Ross, 2000.)

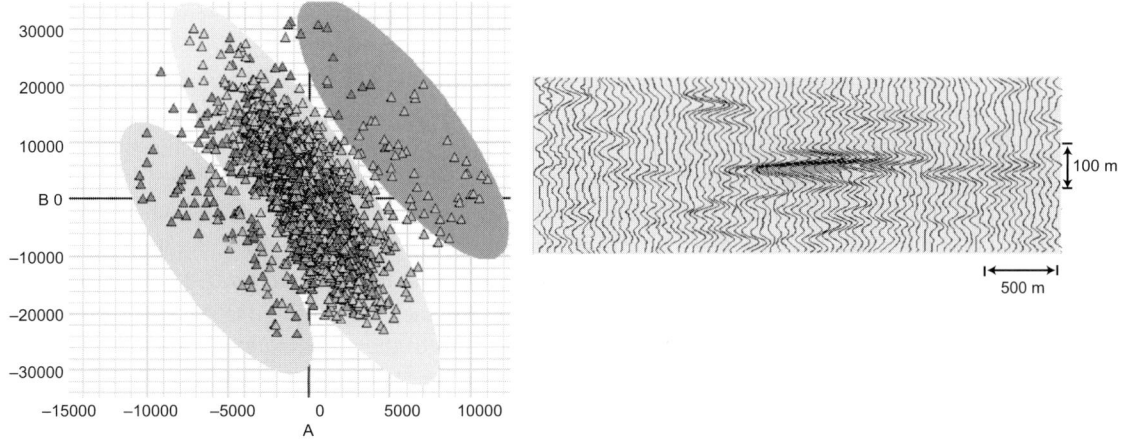

Figure 10.59 (a) Intercept-gradient cross plot using a 200 ms window centered on a bright spot anomaly from a gas sand in the Gulf of Mexico. (b) Color-coded seismic section using the cross-plotted zones in (a). (After Ross, 2000.) For color version see plate section.

relative amplitude processed seismic profile shown on the right, using the *A* and *B* values derived from processing the corresponding CMP gather as shown in the previous figure. The points in different trends or groups can be color-coded on the cross-plot as well as on the seismic profiles. The ability to cross-plot seismic attributes from a 3D data volume allows users to identify the most prosperous subsets that warrant detailed inspection.

Although there are noticeable trends in the cross-plot of Figure 10.59, their meaning is unclear before AVO modeling of synthetic seismograms of all incident angles based on well logs. The AVO modeling permits direct correlation of lithology and pore fluid measurements in wells with observed seismic data. AVO modeling is an important tool for assessing which AVO responses are indicative of hydrocarbon-charged pore fluids or pertinent lithologies. A typical AVO modeling flow has the following steps:

1. Edit and prepare the well logs for AVO modeling;
2. Create fluid/lithology replacement well logs;
3. Generate an *in situ* and a fluid/lithology replacement AVO model;
4. Generate the appropriate AVO attributes (such as AVO gradient and intercept) for both models of Step 3;
5. Cross-plot the attributes from each model simultaneously.

10.5.2   Extraction of geometric attributes

10.5.2.1   Seismic coherence

Seismic coherence is a measure of the lateral continuity between adjacent seismic traces. By mapping a normalized cross-correlation between neighboring traces, Bahorich and Farmer (1995) demonstrated the ability of seismic coherence to reveal fault surfaces in a 3D data volume without fault reflections. Since the cross-correlation process will suppress the influences from the source signature and the overburden, coherence images are able to capture stratigraphic features such as buried fluvial channels, deltas, and reefs that are not easily recognizable by interpreters. For instance, the coherence slice in Figure 10.60 reveals faults as alignments of incoherency, with a much higher level of clarity than the seismic time slice. A highly fractured region is also revealed in the upper right portion on the coherence slice. The data are from offshore East Coast of Canada where NW–SE trending faults and fractures are difficult to interpret before using the coherence attribute.

Although the computation of cross-correlation as defined in Section 2.3.2 is simple, the input data need to go through pre-processing to enhance the SNR and to recognize potential artifacts. We can alter the vertical and lateral scales of the processing windows to optimize the effectiveness of the coherence computation for the targets of interest. Figure 10.61 shows two time slices at 1.1 s and 1.2 s, respectively, through a coherence volume from the Fort Worth Basin. These slices reveal a complex system of lineaments and collapse features in the carbonate rocks, and a west–east oriented fault through the middle area. The circular collapse features are more pronounced at the deeper Ellenburger level than at the level of the Marbel Falls formation. The collapse features are aligned in conjugate with the NE–SW and NW–SE trends along the arrows in the figure, and some of the collapses are elongated rather than circular.

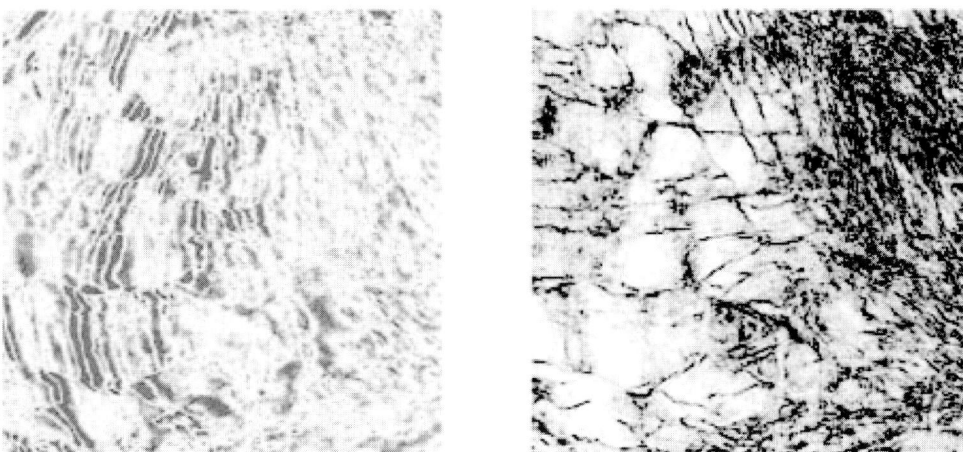

Figure 10.60 Time slices from (left) a seismic data volume and (right) the coherence volume. (After Chopra, 2002.)

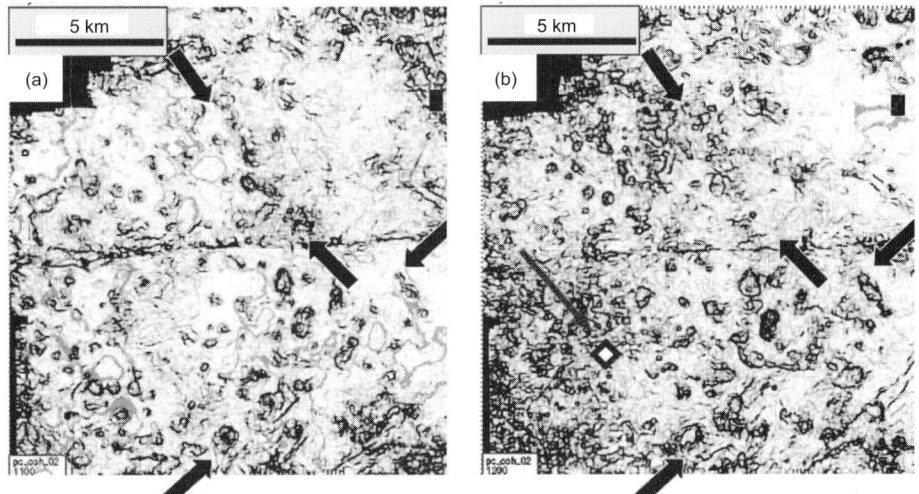

Figure 10.61 Collapse chimneys in carbonates of the Fort Worth Basin in two time slices of a coherence volume at: (a) 1.1 s at the Marble Falls; and (b) 1.2 s at the Ellenburger. Black arrows point to some of the collapse features. In the right panel an open diamond with an arrow denotes a cored Ellenburger well. (After Sullivan *et al.*, 2006.)

10.5.2.2   Curvature and lateral variations

Curvature is a 3D seismic attribute quantifying the degree to which the local reflector surface deviates from being planar. Curvature attributes have been used to map subtle features and predict fractures or small-scale faults. To extract curvature attributes, a quadratic surface is used to fit all plausible reflector trends at each position in the 3D seismic volume, and the best fitting quadratic surface is used to measure the curvature. In data processing the input seismic volume is subjected to a scanning operation; at each step a small cube of the

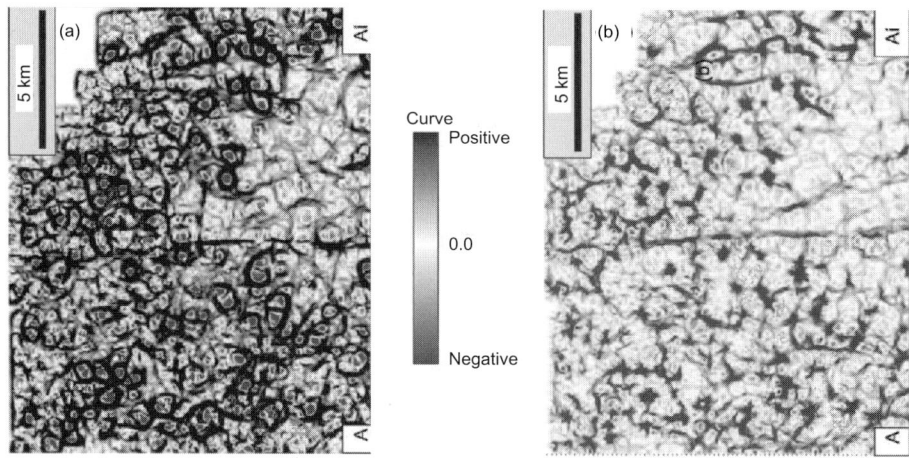

Figure 10.62 Time slices of curvature volumes at 1.2 s of the same data in Figure 10.61. (a) Most-positive curvature slice showing bowls with negative values. (b) Most-negative curvature slice showing domes as positive values. (After Sullivan *et al.*, 2006.)

input data surrounding the evaluation position is extracted to compute the optimal curvature values for the position. The above process is typical of **feature extraction** operators, which are data processing procedures to extract geologic features from seismic data. In Chapter 1 an example of feature extraction is given in Box 1.3 for detecting paleo-channels.

Figure 10.62 shows time slices of two types of curvature attributes for the same data shown in Figure 10.61 from the Fort Worth Basin (Sullivan *et al.*, 2006). The left-hand one is the most positive curvature, which depicts bowl-like features among the reflectors, and the righthand one is the most negative curvature, which depicts dome-like features among the reflectors. These slices are taken at 1.2 s in two-way time at the Ellenburger formation. Hence Figure 10.62 is compatible with the coherence slice shown in Figure 10.61b. In this case, the bowls in Figure 10.62a are interpreted as tectonic collapse features, linked by a complex system of faults and joints.

To verify the lateral variations revealed by different types of seismic attributes, it is often necessary to combine them into one display in comparative analyses. Figure 10.63 shows a map view of the fracture azimuths (ticks) estimated from the azimuthally varying AVO gradient for the top-chalk horizon, and some fault traces interpreted from 3D coherency analysis. Note the general alignment of fractures with large-scale faulting, especially near the faults trending from NW to SE. In the SE corner, fractures also appear to be perpendicular to the surface curvature defined by the time contours plotted at 20 ms intervals.

10.5.2.3   Texture attributes

Texture attributes are morphological features captured from seismic data mimicking the expertise of an experienced interpreter. In seismic stratigraphic interpretation, texture is often referred to fabrics of a stratigraphic facies such as a turbidite unit, or characteristic pattern of a geologic feature such as a salt body or a fractured siltstone layer. The processing for texture attributes is typically a feature extraction, like that for curvature attributes

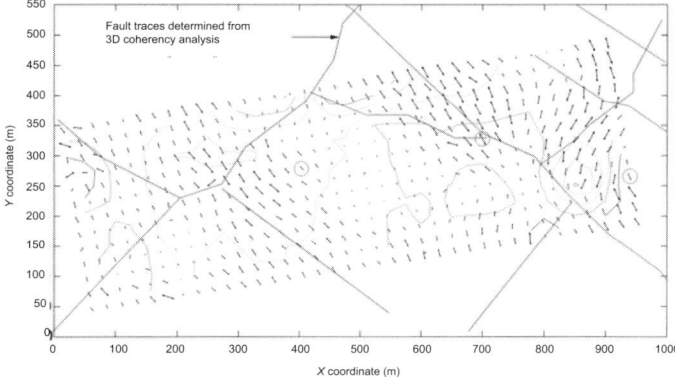

Figure 10.63 Determining faults and fractures at Valhall field via P-wave azimuthal AVO analysis (after Hall & Kendall, 2003). Ticks are fracture orientations derived from the near-offset AVO gradient anisotropy analysis. Lines are fault traces from 3D coherency analysis.

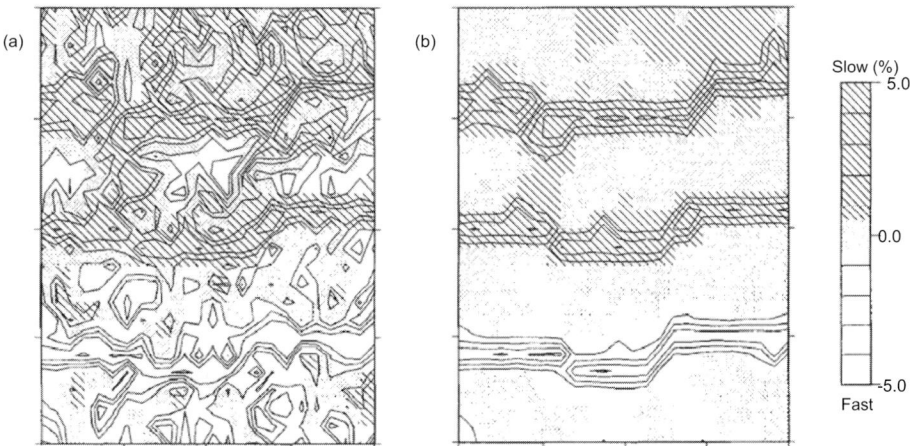

Figure 10.64 (a) Cross-section of synthetic velocity variation model with added random artifacts. (b) After spatial-coherency filtering. (After Zhou, 1993.)

discussed in the previous subsection. A filtering process is involved to scan, quantify, and extract the desired texture features. Obviously, we need to know some characteristics of the features in order to identify and extract them. Figure 10.64 shows a synthetic example in which three layers of distinguishable velocity variations are extracted from a synthetic input cross-section containing the layers and artifacts of random velocity anomalies. Here the feature extraction operator is a spatial coherency filter (Zhou, 1993) which searches through data for connectivity between neighboring pixels of similar values defined by connection matrices.

Gao (2003) gave the name of **texel** to a locally connected group of pixels or voxels for texture extraction. Like the idea of the connection matrices in the previous example,

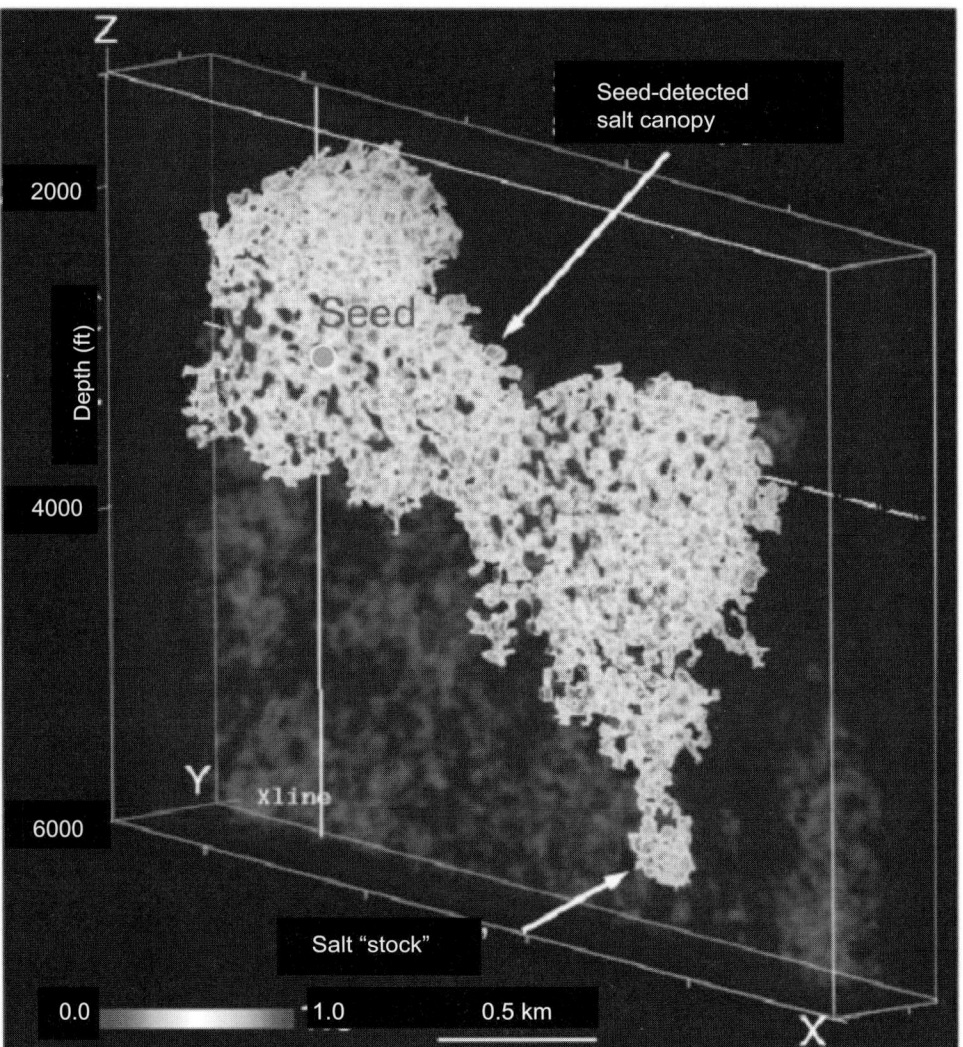

Figure 10.65 A salt body detected from a "seed" in a texture homogeneity cube (Gao, 2003). Since the texture homogeneity of salt is significantly higher than that in the surrounding areas, the whole salt body can be mapped by finding all "seeds", the attributes of the salt. Detecting and mapping such geological features are fundamental for constructing an accurate geologic model and for exploring hydrocarbons and other resources or subsurface features.

the texture extraction here was carried out based on evaluating the voxel co-occurrence matrix for each cubic texel in the data volume. The texture extraction method helped in detecting structural and stratigraphic features that are significant to seismic interpretation and hydrocarbon exploration. Figure 10.65 shows an example of extracting a salt body from a reflection amplitude volume. In an offshore depositional setting, the response of a salt body in seismic data volume has the characteristics of a dome-shaped feature with homogeneously low amplitude and low frequency, and its top surface has high reflection amplitude.

Two key factors for seismic attribute processing are the geologic plausibility of the extracted attributes and the associated computation cost. It is generally time-consuming to define the 3D geometry of stratigraphic features such as a salt body directly from the amplitude volume. Because amplitude samples within the salt body are similar to and connected with those in the surrounding areas, a seed-based propagation may cause "bleeding" across the salt boundary and thus is not effective for automatic salt detection. Similar problems exist for mapping geological features using amplitude data alone.

Figure 10.66 shows, based on the same seismic time slice, the texture contrasts along three orthogonal directions and the geologic interpretation (Gao, 2003). Here the texture contrasts in different directions help in illuminating faults and fractures along perpendicular orientations. Both high-angle normal and wrench faults as well as low-angle detachment and listric faults may be analyzed using multiple viewing directions. The texture attributes may better capture the internal facies variations than the coherence algorithm. We may see in Figure 10.66c that there are at least two stages of channel development. The channels to the east were developed prior to the fault displacement and were subsequently truncated and offset left laterally by the fault. The channels to the west were developed after the major fault displacement and ran across the fault. These interpretations are shown in Figure 10.66d based on observations from the texture data shown in the three texture panels and the regional geology of the study area.

10.5.3   Extraction of spectral attributes

Spectral attributes are frequency-dependent amplitude, phase, and other quantities extracted from reflection seismic data. Such attributes can be very useful in exploration geophysics, particularly for inferring relative thickness variation and other subtle geometric features from seismic data. For example, the amplitude and frequency of the spectral peak can be used to quantify thickness and even the presence of fluids in thin-bed reservoirs. By comparing reflection data with water and gas saturations, researchers have seen phase shifts and energy redistributions between different frequencies (e.g., Goloshubin *et al.*, 1996). Such studies indicated that reflections from a fluid-saturated layer have increased amplitude and delayed traveltime at low frequencies relative to reflections from a gas-saturated layer. Consequently, geophysicists have been searching for spectral attributes as possible direct hydrocarbon indicators.

10.5.3.1   Limitations of windowed Fourier decomposition

In terms of processing, extraction of some spectral attributes can be carried out using instantaneous and local attributes analysis (see Section 10.5.1.2). A more effective extraction can be facilitated via **spectral decomposition**, which was done traditionally via discrete Fourier transform (Section 3.2) using time windows suitable for the given data. However, the sample window is usually very short in order to detect local acoustic properties and layer thicknesses. A notion shown in Box 5.3 is that the spectrum of a seismic trace is nearly the same as the spectrum of the wavelet. However, this notion is no longer true for a short-window seismic trace. As illustrated in Figure 10.67 (Partyka *et al.*, 1999), the

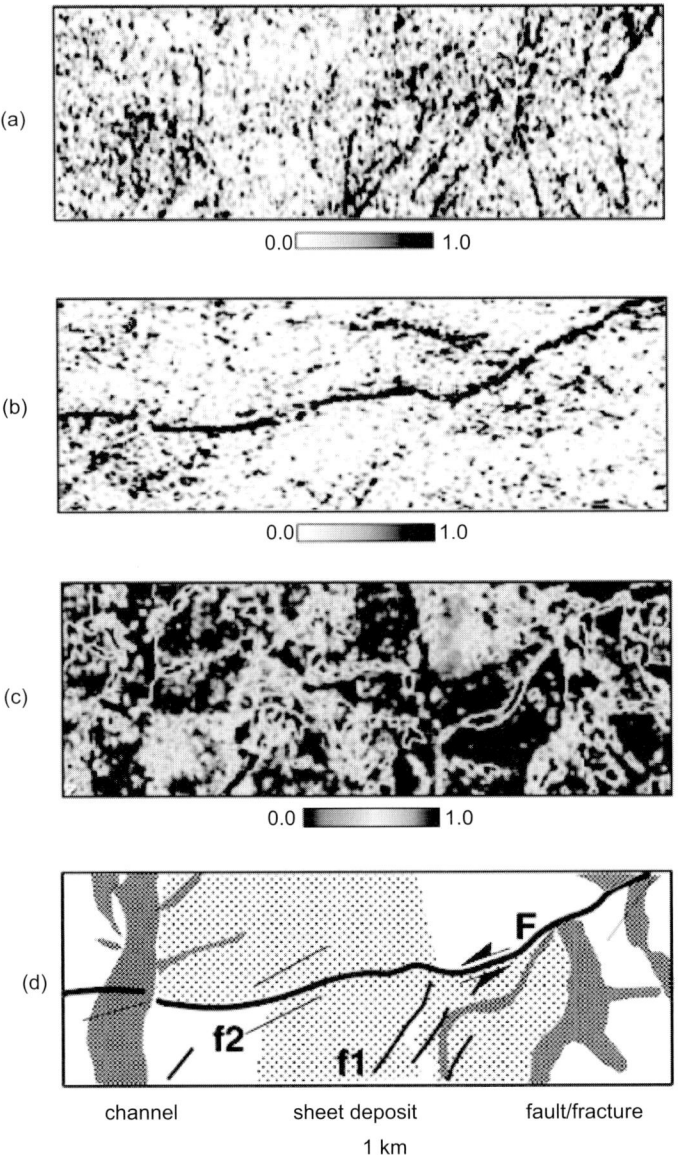

Figure 10.66 For a seismic time slice, (a) to (c) are texture contrasts along the x direction (E–W), y direction (N–S), and z direction (vertical), respectively. (d) Interpretation. The contrast along the x direction highlights the N–S trending fractures, whereas contrast evaluated along the y direction highlights the primary E–W trending fault and fractures, and contrast evaluated along the z direction (c) helps to identify depositional features such as channels. The geometric relationship between the major fault (F) and the two conjugate fractures suggests left-lateral displacement along the fault. Such an interpretation is consistent with the offset of the pre-fault depositional facies across the fault. (After Gao, 2003.)

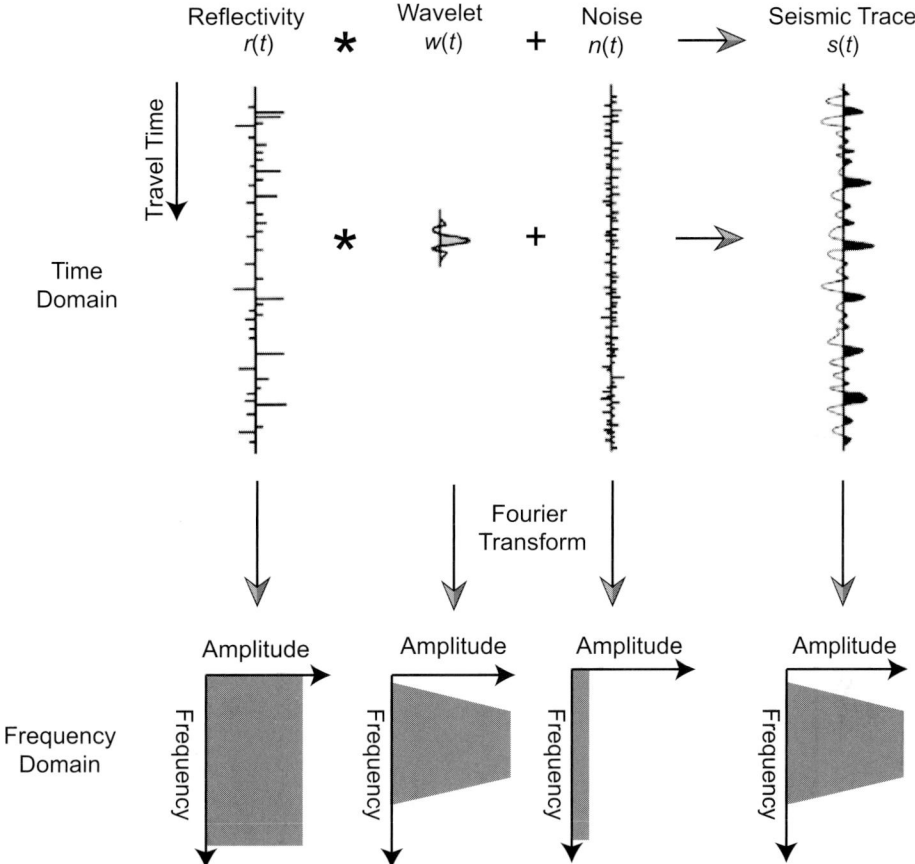

Figure 10.67 Short-window spectral decomposition and its relationship to the convolution model. The shaded traces are of the long-window spectral decomposition. The spectrum of the long-window reflectivity is white. In contrast, a short temporal window samples a small portion of the reflectivity, hence coloring the amplitude spectrum. (After Partyka *et al.*, 1999.)

spectrum of a short-window trace is colored or has notches, carrying information on the strata sampled by the window.

Deriving spectral attributes using windowed Fourier decomposition is also limited by the properties of the Fourier transform as a global decomposition using harmonics as base functions. Global decomposition is useful for extracting global properties such as amplitude spectrum over the whole space. In exploration geophysics, however, we are usually interested in the spectral attributes over targeted depth ranges that are of local rather than global scales.

10.5.3.2   Extracting spectral attributes via wavelet decomposition

Today, spectral decomposition is mostly done using **wavelet transform** to better describe local properties. We have seen in Sections 3.4.3 and 3.4.4 that wavelet transform is an effective way to decompose seismic traces into wavelet(s) of fixed shape but varying amplitude and lateral stretching scales. The idea here is to separate the seismic records at each given position into two parts: one is the wavelet, which represents properties independent of the given position such as the source signal, receiver's response, and wave propagation factors; and the other is the local scaling factors of the wavelet, which depend on physical properties of the given position, such as the spectral properties.

Several techniques are available in carrying out the wavelet decomposition. One intuitive way is the **matching-pursuit** technique, and an example flowchart is shown in Figure 10.68. In this case the input real seismic data trace is first paired up with its complex trace, which is taken as the data complex trace to be matched by the model complex traces of the chosen wavelet of different scales in an iterative loop. Each iteration step of the loop aims at determining the best-matching parameters of the wavelet at one scale, pursued from large to small scales sequentially. In the first iteration step the model complex wavelet of the largest scale is matched with the data complex trace to determine the best-fitting time positions and coefficients of the model complex trace. In each iteration step the residual trace between the data and model complex traces becomes the data complex trace of the next iteration step using a wavelet of the next scale. In this way the complex spectral parameters of all wavelet scales are determined sequentially.

10.5.3.3 Examples of applying spectral attributes

Figure 10.69 shows far-angle seismic stack sections from a deep-water West Africa reservoir at central frequencies of 15 Hz and 35 Hz. The vertical span of each section is 600 ms. We see some sand and shale layers up-dip against a growth fault on the right side. Comparing the two sections at positions above and below the oil–water contact (OWC), we see that the amplitude contrast between the oil sand and down-dip brine sand is higher at low frequencies in the left. A **low-frequency shadow** phenomenon like this is often explained as due to the dispersive effect of hydrocarbons. While the main factors for the spectral response of a reservoir include layer thickness, effective attenuation, reflectivity series and fluid type, the last two factors are of greatest importance in this case.

However, the spectral response of a hydrocarbon reservoir may not always be obvious at low frequencies, as often seen in shallow and unconsolidated sediments. Figure 10.70 shows an example of high-frequency anomalies for deep and tight carbonate reservoirs in the Central Tarim Basin, NW China (Li *et al.*, 2011). Here the dominant frequency of the seismic data is only about 18 Hz at the target level, which has a burial depth of nearly 6 km. In this case the researchers found that the 40 Hz iso-frequency volume shows good correspondence with the oil and gas production levels from eight wells in the area, as shown in the horizon slice along the production zone in Figure 10.70a. The three prolific oil and gas wells are located in the anomalous area in the attribute map, while the remaining five dry wells are located outside the anomalous area in the map.

The low- and high-frequency responses from the above two case studies demonstrate the non-uniqueness in using spectral attributes to infer the whereabouts of hydrocarbon reservoirs. In practice, detailed modeling and calibration between seismic and well observations are necessary to verify the value of different attributes as indicators.

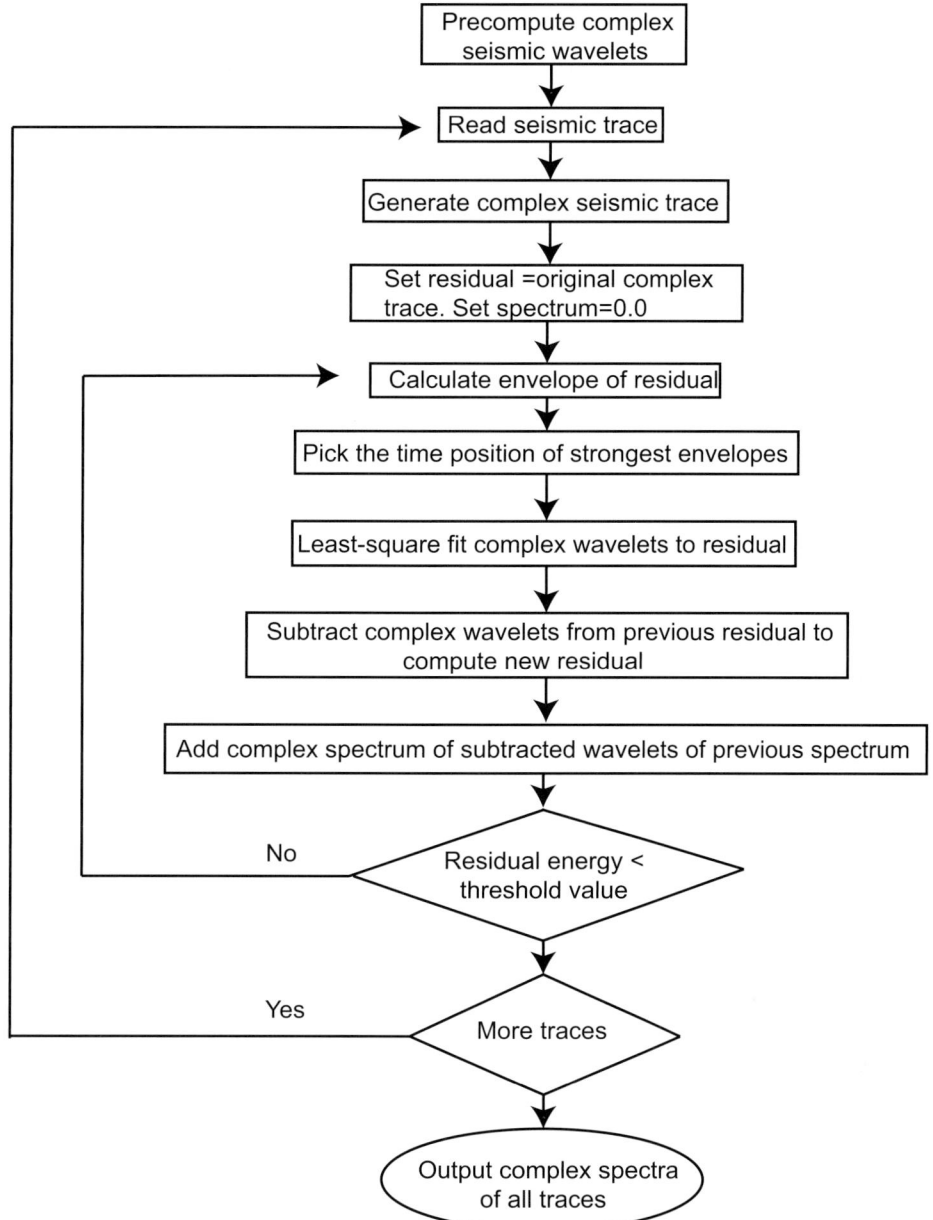

Figure 10.68 Flowchart of a wavelet-based spectral-decomposition algorithm using a matching-pursuit technique. The frequency interval is 0.5 Hz for the complex wavelet dictionary. (After Liu & Marfurt, 2007.)

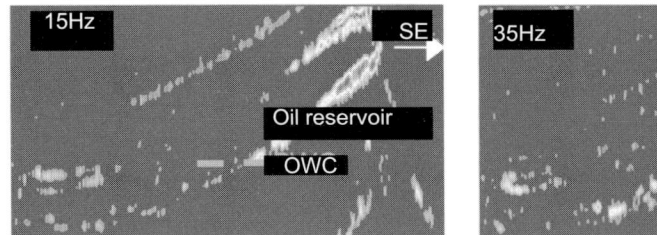
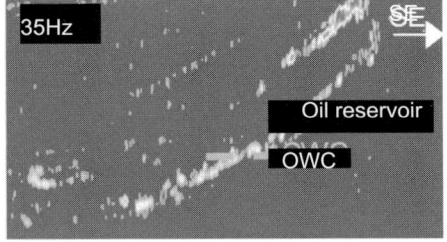

Figure 10.69 Two spectral decomposed sections showing interval average absolution amplitude, with central frequencies at 15 Hz and 35 Hz, from a deep-water West Africa reservoir. An oil sand shows as a brighter region on the 15 Hz section relative to the brine sand below the OWC. (After Chen *et al.*, 2008.)

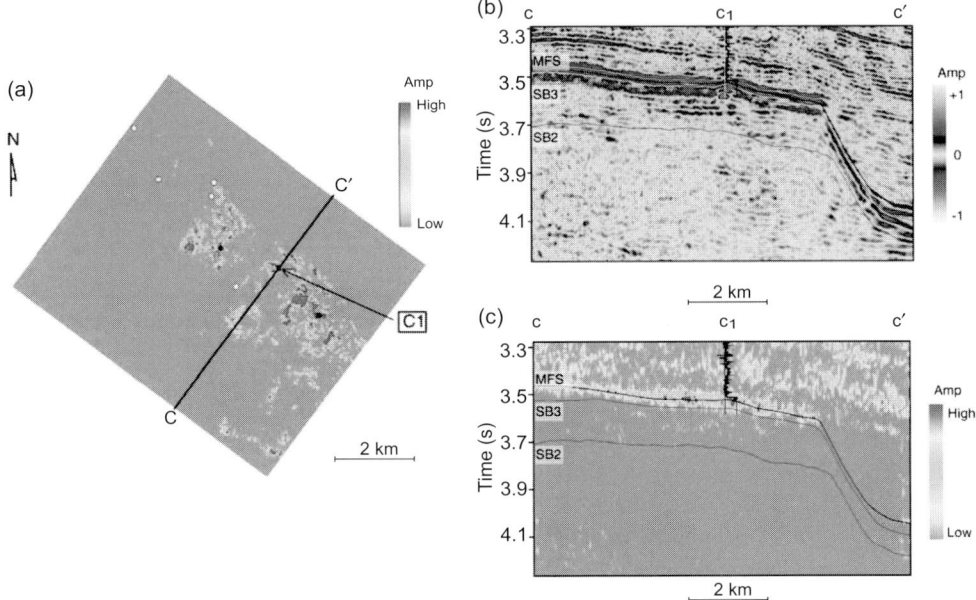

Figure 10.70 Two slices of a 40 Hz iso-frequency volume from the Central Tarim Basin in NW China as (a) horizon slice along the oil and gas zone between the maximum flooding surface (MFS) and sequence boundary 3 (SB3), and (b) vertical slice C–C′ whose location is shown in the mapview (a). In (a) the three solid black circles are oil- and gas-producing wells, and the five open circles are wells without a good reservoir. (c) Original seismic amplitude on vertical slice C–C′ across an oil and gas well C1. (After Li *et al.*, 2011.) For color versions see plate section.

10.5.4   Processing in seismic-to-well tie and impedance inversion

Sheriff (1991) describes **well tie** as running a seismic line by a well so that seismic events become correlated with subsurface (well-log) information. **Seismic-to-well tie**, or seismic–well tie and **well-to-seismic tie**, is a suite of techniques to match seismic data with wellbore measurements. It follows a major exploration geophysics strategy of combining two lines of complementary data: seismic imagery with broad spatial coverage but relatively low resolution and low fidelity, plus well logs with sparse spatial coverage but high resolution and high fidelity. Seismic-to-well tie is fundamental to seismic interpretation owing to its two key functions: (1) it enables horizons picked by an interpreter to be related directly to a well log; and (2) it provides a way to estimate the wavelet needed for inverting seismic data to impedance and other rock property indicators.

As its first function, the well tying process relates seismic imageries to wellbore measurements so that the rich and high-resolution information from all wellbore measurements can be extrapolated from the isolated 1D well(s) into the 2D and 3D seismic profiles and volumes. The well tying process is facilitated by matching the reflection seismic traces near each wellbore with log-based **synthetic seismograms**, which are produced by convolving a wavelet with calibrated reflectivity function based on well logs, as shown previously in Box 6.3.

For the second function of well tying, processed seismic reflection data are commonly regarded as band-limited versions of the subsurface reflectivity. This means that each seismic reflection trace is taken as the convolution of the reflectivity function at the location with a wavelet. Hence, a high-quality well tie provides a high-quality estimate of the wavelet that links the seismic trace with the wellbore-measured reflectivity at the same position.

10.5.4.1 Elements of seismic-to-well tie processing

With the given scientific objectives and data from seismic imaging and wellbore measurements, the processing for seismic-to-well tie is an iterative procedure to gradually improve the matching between processed seismic data and synthetic seismograms based on calibrated well logs. A typical processing flow includes the following elements:

1. Processing and QC of seismic data, choosing proper well-tie locations and intervals based on positions of the target zone, characteristics of the key horizons to tie, and properties of the available well logs; the output is a processed seismic section that can best represent the reflectivity in seismic bandwidth near the well.
2. Estimating wavelet using seismic data in conjunction with all available VSP and checkshot data, resulting in a suite of possible wavelets of different bandwidths and phase angles.
3. QC and calibrating the chosen sonic and density logs to produce the impedance trace, whose depth derivative is the broadband reflectivity; suppressing all noise unrelated to the *in situ* reflectivity, and "blocking" or smoothing the reflectivity to an optimal level for the given data.
4. Conducting depth-to-time conversion to produce a suite of calibrated reflectivity traces using velocity functions based on well-log, VSP, and checkshot data, with special attention to the timing, amplitude, and phase angle of key horizons.
5. Making synthetic seismograms by convolving each of the calibrated reflectivity traces with different blocking and velocity parameters from Step 4 with each of the wavelets from Step 2.
6. Quantifying well-tie by correlating the synthetic seismograms from Step 5 with the processed seismic section from Step 1, focusing on tying the key horizons.
7. Checking and fixing all possible causes of mis-ties, and reiterating the entire process.

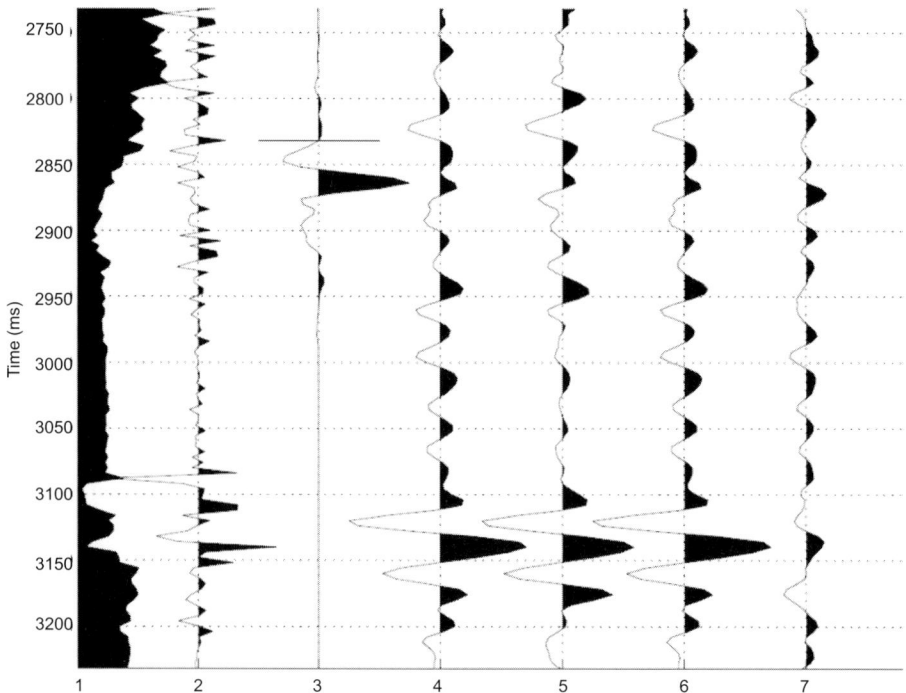

Figure 10.71 An example of seismic-to-well ties (White, 2003). These seven traces are functions along the same time scale. 1: Acoustic impedance based well-logs. 2: Reflectivity from 1. 3: Wavelet, with a horizontal bar denoting time zero. 5: Synthetic seismogram from convolving 2 with 3. 4 and 6: Seismic data trace at the wellbore location. 7: Residue or difference between Traces 4 and 5.

If we regard the geosciences as a combination of art and engineering, then the process of seismic-to-well tie may require more art than engineering. In the beginning stages of the process, people sometimes conduct "stretching and squeezing" of the wiggles of either the synthetic seismograms or the seismic data in order to match the key horizons. However, this practice is acceptable only after recognizing good correspondence between key horizons on the seismic section and the synthetic seismogram, and only a small amount of stretching and squeezing is acceptable. As much as possible, the stretching and squeezing should be constrained by some physical guidance, such as by varying the velocity functions used in the depth-to-time conversion in the above Step 4. Such constraints may help us identify the causes of the mis-tie.

Once we have identified possible causes of the mis-tie, we need to quantify the effects of each cause for the given situation, and hence find effective ways to improve the well-tie quality with the minimum amount of processing. For beginners in the subject, the best way to learn is to see examples and to put theory into practice. After understanding the principle and key steps, we need to look at more seismic-to-well tie examples from the literature and see what the data are telling us. One such example is shown in Figure 10.71.

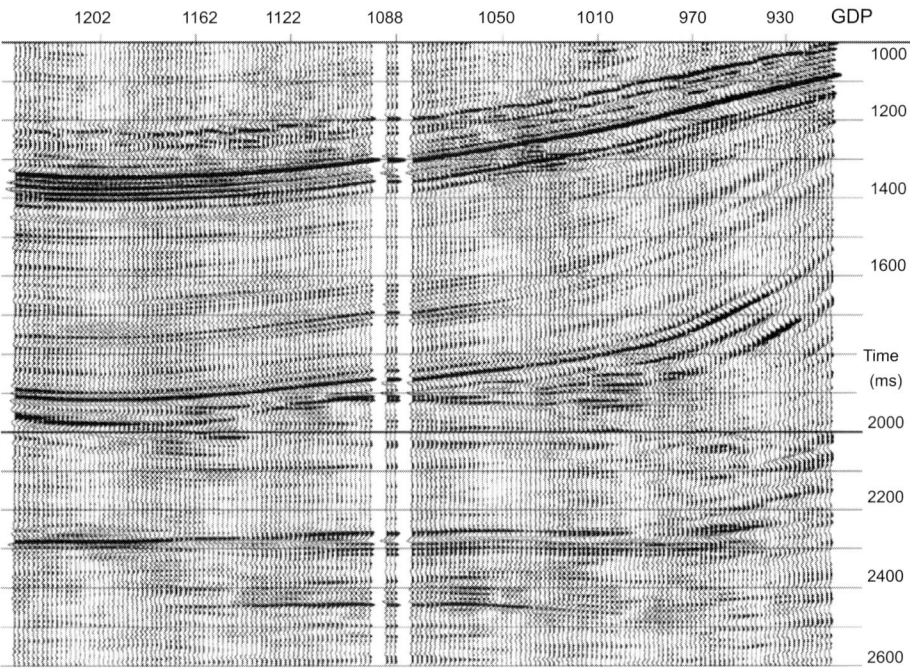

Figure 10.72 Three synthetic seismogram traces are spliced into a migrated seismic section after re-processing using whitened zero-phase (White & Hu, 1998).

10.5.4.2   Major challenges to seismic-to-well tie

Seismic–well tie is one of the most important processes in petroleum reservoir characterization. With a high level of well tie, such as the example shown in Figure 10.72, we suddenly gain a high level of confidence in all the matched seismic horizons with their geologic meanings and petrophysical information inferred from the well logs. In the case shown in this figure, the authors have carried out a careful re-processing with high-quality seismic and well-log data; the correlation coefficient for the well tie here is more than 95% over a long time window of 1600 ms. Unfortunately, such an excellent well tie is very rare. In most real applications, the correlation level is below 80% over a time window greater than 500 ms.

The common occurrence of mis-tie can be produced by a number of practical factors. Hence, at the heart of seismic-to-well tie is identifying the causes of mis-tie, and developing effective ways to improve the tie and quantify the uncertainty. Practically, searching for the causes of mis-tie needs to focus on the key factors for seismic-to-well tie. They include:

- Well-log QC
- Well tie location
- Timing and log calibration
- Accuracy of wavelet estimated

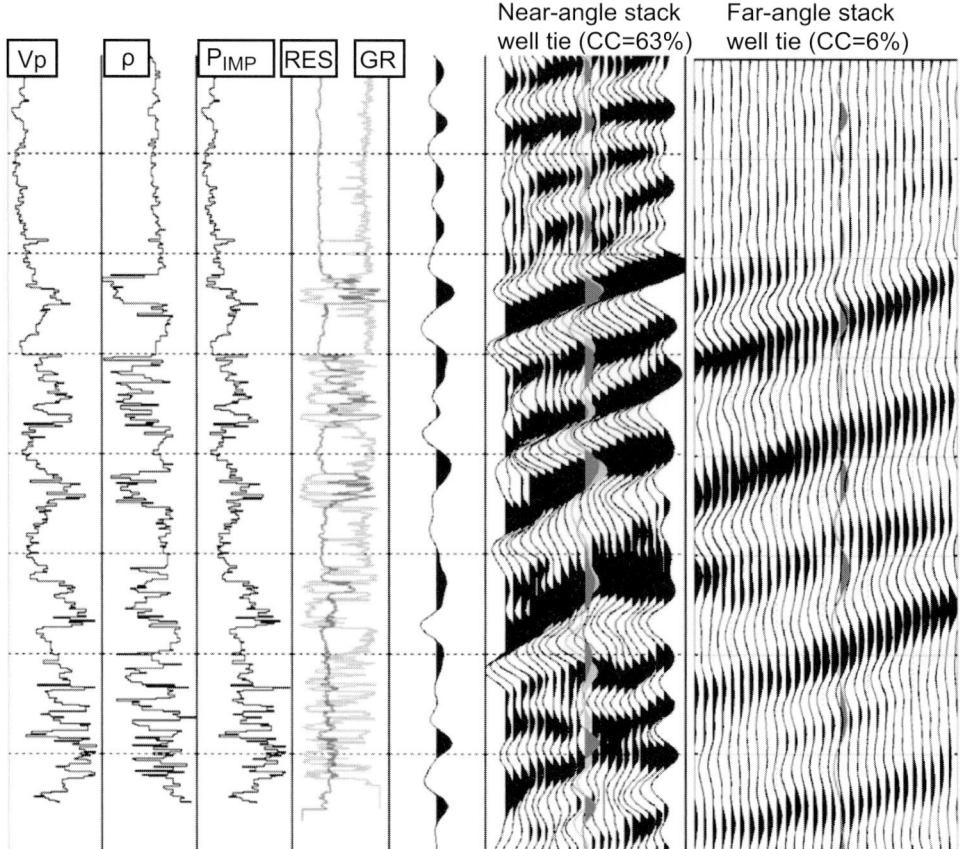

Figure 10.73 A case of adequate seismic-to-well tie at near offsets, but poor tie in the far offsets (after Gratwick & Finn, 2005). The left six traces are based on well logs: $1 = V_P$, $2 =$ density, $3 =$ impedance, $4 =$ resistivity, $5 =$ gamma ray, $6 =$ synthetic seismogram at near offset. The right two panels are seismic-to-well tie results for near and far offsets, with their synthetic seismograms shown as the middle traces. The cross-correlations are 63% and 6% for the near- and far-offset ties, respectively. The vertical axis is along traveltime, and the interval between the dashed lines is assumed to be 100 ms.

- Bandwidth of seismic data
- SNR of seismic data (e.g. multiples)
- Quality of velocity model (e.g. anisotropy)
- Methodology for complex imaging environments

As an example, Figure 10.73 demonstrates a case of seismic-to-well tie study of deep water reservoirs in West Africa, where the correlation coefficient of the well tie is 63% at near offsets (Panel 7) but only 6% in the far offsets (Panel 8). The seismic-to-well tie process is commonly applied to near-offset seismic data near the wellbore. Nevertheless, people have attempted to tie far-offset synthetic seismograms with far-offset seismic data in order to constrain certain reservoir properties that can be manifested in far-offset data. In this example case, Gratwick and Finn (2005) used far-offset seismic-to-well tie to prove the presence of velocity anisotropy in their mixed impedance reservoirs with unconsolidated clastics. By modeling with velocity anisotropy based on shear logs, they were able to improve the far-offset well tie to over 50% in the correlation coefficient.

There are situations in which, after many attempts, the quality of well tie is still poor. This can be for a number of possible reasons. For instance, in the case of a deviated wellbore or in the presence of steeply dipping rock layers, we cannot expect the convolution model to be appropriately applied. We may need to conduct elastic waveform modeling in order to generate the synthetic seismograms. Another physical limitation for seismic-to-well tie is that the seismic responses of many physical properties are not **scalable**; in other words, we cannot use their responses at one scale to predict the responses at a different scale. Examples of such properties include viscosity and permeability. The **scalability** of the seismic responses to different reservoir properties can often be evaluated by rock physics studies.

10.5.4.3   Inversion for seismic impedances

While an imaged seismic section is indicative of subsurface structures because seismic reflectors are most sensitive to geologic unconformities, the corresponding seismic impedance section is useful to infer petrophysical parameters and *in situ* conditions such as density, velocity, and pressure. The objective of seismic impedance inversion is to derive seismic impedance models based on seismic images and/or synthetic seismograms from seismic-to-well tie. Taking Figure 10.71 as an example, seismic impedance inversion takes the imaged seismic trace (Traces 4 or 6) to invert for the impedance trace (Trace 1). Since seismic impedance is a product of density and velocity, we have acoustic impedance for acoustic or P-wave velocities, and elastic impedance for both P- and S-wave velocities. In the presence of velocity anisotropy, the impedance becomes anisotropic as well. Following the general view of seismic imaging that is portrayed in Figure 8.1, seismic velocity models describe the long-wavelength and absolute values of the subsurface seismic responses, while the reflection imagery describes the short-wavelength and relative variation of seismic reflectivity. Therefore, we should jointly use seismic impedance and reflection profiles.

Figure 10.74 shows an example of using a velocity profile to assist in interpreting a crustal reflection profile in southern California. In large-scale studies there are often good correlations between P-wave velocity models and seismic impedance models. However, the data quality is often poor in crustal studies because of uneven and insufficient shot and receiver coverage, and, in the case of this figure, the limitation of 2D seismic lines in the presence of strong 3D inhomogeneities. Here, on the depth-migrated reflection profile in panel (c), it is difficult to tell the positions of the Moho discontinuity and the bottoms of sedimentary basins from the large number of reflectors distributed throughout the image. In panel (b), some velocity contours from a tomographic velocity model shown in panel (a) are projected as dashed curves on the reflection profile. These velocity contours help in identifying the position of the Moho, which by definition is along the top of the deepest velocity layer with 8 km/s in P-wave velocity. The basement boundary of the sedimentary basins along this profile is interpreted as along the velocity layer with 5.5 km/s in P-wave

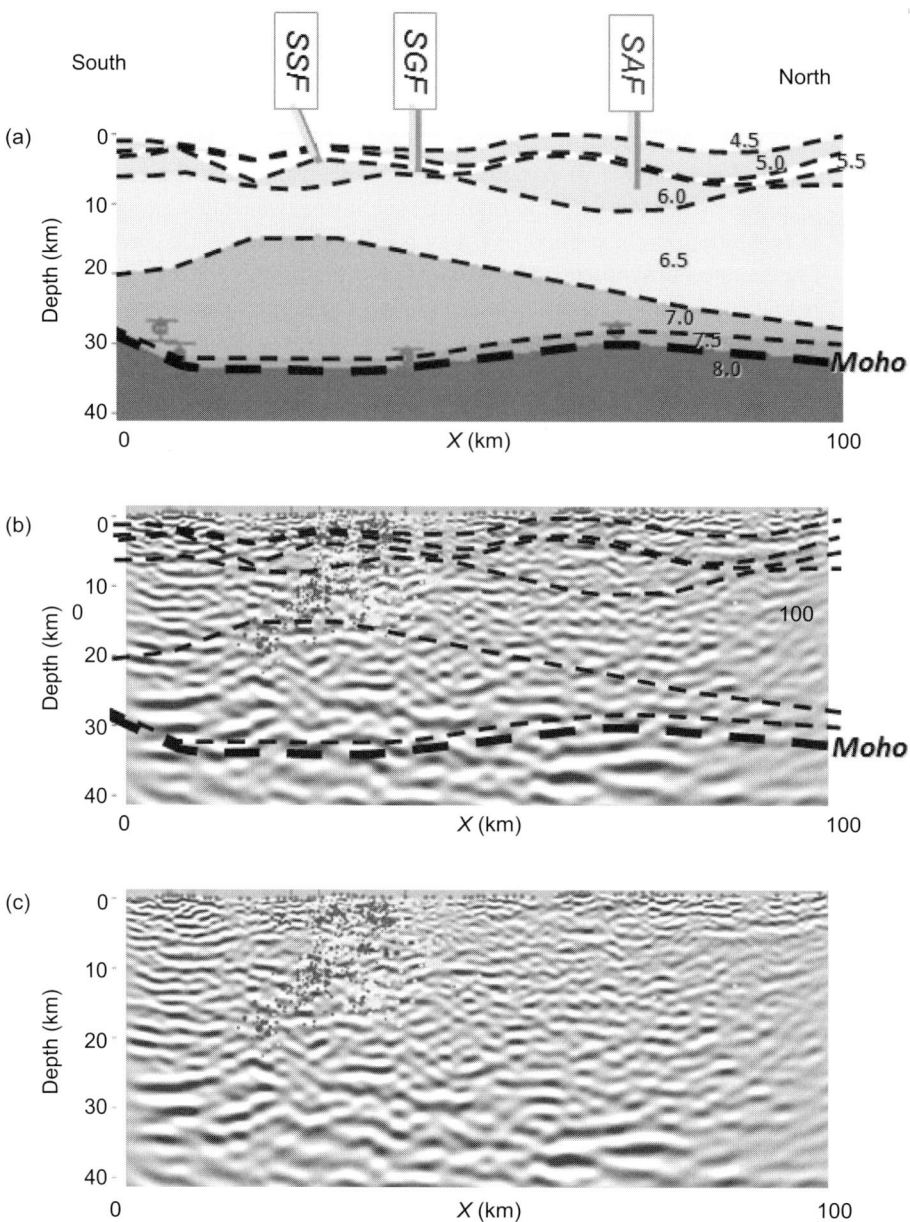

Figure 10.74 Use of absolute velocities to interpret a crustal reflection seismic profile in southern California. (a) A portion of the tomographic velocity profile shown in Figure 4.14b. Three faults are Santa Susana (SSF), San Gabriel (SGF), and San Andreas (SAF). The numbers are layer velocities in km/s. Panels (b) and (c), respectively, are depth-migrated reflection profile shown in Figure 4.15b with and without velocity contours from (a). The gray dots denote earthquake foci.

velocity. Considering that the velocity model is just a low-resolution approximation of the real velocity structure, we can interpret these major crustal boundaries based on the velocity model, seismic reflection profile, and geologic information for the area.

As we have seen in the introduction to Section 10.5.4, seismic sections based on reflection data are commonly regarded as band-limited versions of the reflectivity structure. A depth-varying reflectivity function is just the vertical derivatives of the corresponding seismic impedance function, as demonstrated in the first two traces of Figure 10.71. Hence in principle we should be able to produce a seismic impedance section by integrating each of the seismic traces along the vertical axis. Indeed, we can carry out seismic impedance inversion following an integration approach. This is based on the normal-incidence assumption that the reflectors in seismic section define the positions and reflection coefficients of a set of layers of constant seismic impedance $Z_n = \rho_n V_n$, where ρ_n and V_n, respectively, are the density and velocity of the nth layer. This approach carries a layer-stripping process using a recursive formula from the nth layer to the $(n+1)$th layer:

$$Z_{n+1} = Z_n \frac{1 + R_n}{1 - R_n} \tag{10-21}$$

where R_n is the normal-incidence reflection coefficient between the nth and $(n+1)$th layers. As shown in Figure 10.71, a good estimate of the reflection coefficient requires a good estimation of the wavelet used for deconvolving the seismic data.

The advantage of the aforementioned integration approach is its simplicity and applicability with reflection seismic profiles alone. In practice, however, this approach faces a number of challenges that must be dealt with properly, as summarized in the following.

- Lack of low frequencies in seismic data will pose a major problem if there are no wellbore data, because the low-frequency signal cannot be estimated properly by integrating band-limited seismic data.
- Another challenge is the presence of noise, particularly colored noise such as multiple reflections and artifacts due to poor seismic illumination; integrating data of low SNR will produce messy and unreliable results.
- Detections of wavelet and data polarity are two non-trivial tasks without well-logs and seismic-to-well tie. Another associated challenge is in determining the phase of the wavelet.
- Since the recursive formula (10-21) is based on the normal incidence assumption, we have to make sure that all non-zero-offset effects, such as the AVO effect, are minimized satisfactorily. These effects often demand corresponding calibrations and assessments on the uncertainties.

If we have well-log data and are able to achieve a reasonable level of seismic-to-well tie, then we are in an advantageous position to derive high-quality seismic impedance profiles using the well tie and wavelet. In fact when we have well logs available, we should first attempt to achieve an optimal level of well tie, before inverting for seismic impedance profiles using both the reflection seismic data and well-tie solutions. There are different methods to carry out the inversion, such as the deterministic deconvolution discussed in Section 6.5. As an illustration, Figure 10.75 compares the characteristics of the interpreted and true impedance profiles. Even with the best data quality, an inverted impedance curve

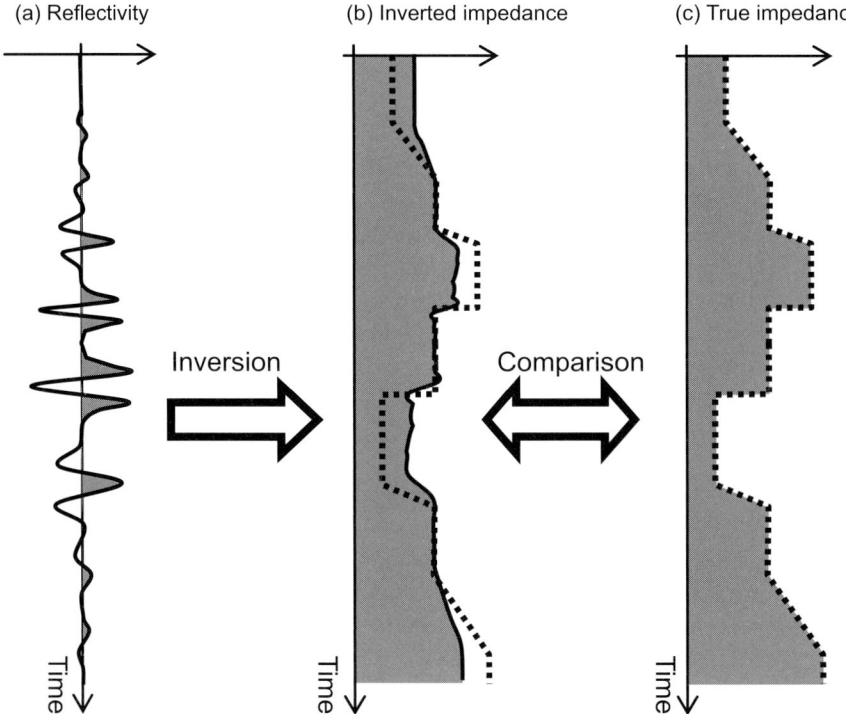

Figure 10.75 Sketch of seismic impedance inversion. (a) Migrated seismic trace or synthetic seismogram based on well logs. (b) Inverted seismic impedance, with the true impedance as a dashed curve. (c) True impedance.

tends to follow the average values of the true impedance curve; many details of the true impedance curve, such as first-order discontinuities or steps, will be smoothed out and produce artifacts in the inverted solution. There are also many non-linear and statistical inversion methods which are beyond the scope of this book.

Exercise 10.5

1. Discuss the sensitivity of the AVO attribute with respect to noise. List the measures that may help reduce the influences of various types of noise.

2. The amplitude of two time series S1 and S2 can be matched to a similar level by multiplying one of them by a constant a. How would you determine the value of a in order to optimize the misfit between the two time series?

3. Describe two ways to quantify the seismic-to-well tie level. Write computer code to quantify the "similarity" between Trace 5 and Trace 6 in Figure 10.71 (Hint: this can be done in Excel.) Would you apply a variable weighting for different events?

10.6   Summary

- In practice, we use the basic skills of seismic data processing to solve most problems, but we may need special tools to address certain issues. It is useful to see how basic data processing skills can be used in some special data processing topics.

- Seismic data acquisition plays a key role in maintaining seismic data quality to better serve the intended objectives, as shown in monitoring different types of sources such as background noise and microseismicity induced by fracking.

- Seismic illumination analysis is a valuable tool for assessing seismic resolution by quantifying the spatial coverage of the traversing seismic waves through the subsurface targets. Since data resolution and fidelity depend on the data bandwidth, it is critical to preserve low-frequency signals.

- Because many seismic imaging methods use only primary reflection data, suppressing multiple reflections is necessary before interpreting the images. Demultiple via Radon transform relies on the differential moveout between primaries and multiples. Demultiple via deconvolution exploits the periodicity of the multiples. SRME suppresses all surface-related multiple energy via pre-stack inversion.

- Seismic anisotropy is a medium property which causes a variation of seismic velocities as a function of the traversing angle. Information in seismic anisotropy helps us in improving the fidelity of seismic images, fault imaging, and detecting the dominant orientations of fractures. Processing of P-wave data often uses the transverse isotropy (TI) models described by the axial velocity V_{P0} and two Thomsen parameters ε and δ.

- Although it is costly in time and money, multi-component seismic data offers information on elastic properties that will be necessary for certain studies. Processing of multi-component data is illustrated using wavefield separation methods, converted wave processing, and VSP data processing and imaging.

- A seismic attribute is a quantitative measure of a seismic characteristic of interest. Interpretations of seismic data use various seismic attributes based on trace properties such as AVO, geometric properties such as coherence and curvature, and morphological features that can be extracted in either the spatial or spectral domains. Processing of seismic attributes relies largely on how much we know about their geologic characteristics and seismic expressions.

- Seismic-to-well tie follows a major exploration geophysics strategy by combining seismic imaging with well logs. It enables reflecting horizons picked by an interpreter to be related directly to a well log, and provides a way to estimate the wavelet needed for inverting seismic data to impedance and other indicators of rock and fluid properties.

- Seismic impedance inversion derives models of seismic velocity and density based on seismic images and/or synthetic seismograms from seismic-to-well tie. While seismic sections are indicative of reflection structures such as unconformities, seismic impedance models are useful to inferring petrophysical parameters and *in situ* conditions of the subsurface.

FURTHER READING

Chopra, S. and K. J. Marfurt, 2005, Seismic attributes – A historical perspective, *Geophysics*, 70, 3SO–28SO.

Sun, R., G. A. McMechan, H. Hsiao, and J. Chow, 2004, Separating P- and S-waves in pre-stack 3D elastic seismograms using divergence and curl, *Geophysics*, 69, 286–297.

Tsvankin, I., J. Gaiser, V. Grechka, M. van der Baan, and L. Thomsen, 2010, Seismic anisotropy in exploration and reservoir characterization: An overview, *Geophysics*, 75, 75A15–75A29.

White, R. E., 2003, Tying well-log synthetic seismograms to seismic data: the key factors, *SEG Expanded Abstracts*, 22, 2449–2452.

REFERENCES

Aki, K. and P. G. Richards, 1980, Quantitative Seismology, Freeman and Co.

Alkhalifah, T. and I. Tsvankin, 1995, Velocity analysis for transversely isotropic media, Geophysics, 60, 1550–1566.

Alkhalifah, T., I. Tsvankin, K. Larner, and J. Toldi, 1996, Velocity analysis and imaging in transversely isotropic media: Methodology and a case study, The Leading Edge, 15, 371–378.

Alvarez, G. and K. Larner, 2004, Relative performance of moveout-based multiple-suppression methods for amplitude variation with offset (AVO) analysis and common midpoint (CMP) stacking, Geophysics, 69, 275–285.

Al-Yahya, K., 1989, Velocity analysis by iterative profile migration, Geophysics, 54, 718–729.

Anderson, D. I., 1989, Theory of the Earth, Blackwell Scientific Publications.

Anstey, N., 2005, Attributes in color: the early years, Canadian Society of Exploration Geophysicists Recorder, 30, 12–15.

Atkinson, F. V., 1964, Discrete and Continuous Boundary Problems, Academic Press.

Backus, G. E. and J. F. Gilbert, 1967, Numerical application of a formalism for geophysical inverse problems, Geophys. J. Roy. Astron. Soc., 13, 247–276.

Backus, G. E. and J. F. Gilbert, 1968, The resolving power of gross earth data, Geophys. J. Roy. Astron. Soc., 16, 169–205.

Backus, G. E. and J. F. Gilbert, 1970, Uniqueness in the inversion of gross Earth data, Phil. Trans. Roy. Soc. London Ser. A 266, 123–192.

Bahorich, M. and S. L. Farmer, 1995, 3D seismic discontinuity for faults and stratigraphic features: The coherence cube, 65th Annual International Meeting, SEG Exp. Abst., 93–96.

Barnes, A. E., 1998, The complex seismic trace made simple, The Leading Edge, 17, 473–476.

Barry, K. M., D. A. Cavers and C. W. Kneale, 1975, Recommended standards for digital tape formats, Geophysics, 40, 344–352.

Baysal, E., D. D. Kosloff and J. W. C. Sherwood, 1983, Reverse time migration, Geophysics, 48, 1514–1524.

Bednar, J. B., 2005, A brief history of seismic migration time, Geophysics, 70, 3MJ-20MJ.

Berkhout, A. J. and D. J. Verschuur, 2006, Imaging of multiple reflections, Geophysics, 71, SI209–SI220.

Best, A. I., C. McCann and J. Sothcott, 1994, The relationships between the velocities, attenuations and petrophysical properties of reservoir sedimentary rocks, Geophys. Prosp., 42, 151–178.

Bickel, S. H, 1990, Velocity–depth ambiguity of reflection traveltimes, Geophysics, 55, 266–276.

Bijwaard, H. and W. Spakman, 1999, Tomographic evidence for a narrow whole mantle plume below Iceland, Earth & Planet. Sci. Lett., 166, 121–126.

Bijwaard, H., W. Spakman and E. R. Engdahl, 1998, Closing the gap between regional and global travel time tomography, J. Geophys. Res., 103, 30055–30078.

Biondi, B. L., 2004, 3-D Seismic Imaging, SEG.

Blackman, R. B. and J. W. Tukey, 1959, The measurement of power spectra from the point of view of communications engineering, Dover.

Bullen, K. E., 1963, An Introduction to the Theory of Seismology 3rd edn, Cambridge Univ. Press.

Burg, J. P., 1975, Maximum entropy spectral analysis, Unpublished PhD thesis, Stanford University.

Caldwell, J. and W. Dragoset, 2000, A brief overview of seismic air-gun arrays, The Leading Edge, 19, 898–902.

Cao, H., J. Xie, Y. Kim and H. Zhou, 2008, Multi-scale migration tomography for constraining the artifacts in depth imaging, Geophysics, 73, VE217–VE222.

Celis, V. and K. L. Larner, 2002, Selective-correlation velocity analysis, SEG Exp. Abst., 21, 2289–2292.

Chang, W. and G. A. McMechan, 1986, Reverse-time migration of offset vertical seismic profiling data using the excitation-time imaging condition, Geophysics, 51, 67–84.

Chang, W. and G. A. McMechan, 1987, Elastic reverse-time migration, Geophysics, 52, 1365–1375.

Chang, X., Y. Liu, H. Wang, F. Li and J. Chen, 2002, 3-D tomographic static correction, Geophysics, 67, 1275–1285.

Chen, G., G. Matteucci, B. Fahmy and C. Finn, 2008, Spectral-decomposition response to reservoir fluids from a deepwater West Africa reservoir, Geophysics, 73, C23–C30.

Chon, Y. T. and T. J. Dillon, 1986, Tomographic mapping of the weathered layer, 56th Annual International Meeting, SEG Exp. Abst., 593–595.

Chopra, S., 2002, Coherence cube and beyond, First Break, 20/1, 27–33.

Chopra, S. and K. Marfurt, 2005, Seismic attributes – A historical perspective, Geophysics, 70, 3SO–28SO.

Chopra, S. and K. Marfurt, 2007, Seismic attributes for prospect identification and reservoir characterization, SEG.

Choy, G. L. and P. G. Richards, 1975, Phase distortion and Hilbert transformation in multiply reflected and refracted body waves, Bull. Seism. Soc. Am., 65, 55–70.

Chun, J. H. and C. Jacewitz, 1981, Fundamentals of frequency-domain migration, Geophysics, 46, 717–732.

Claerbout, J. F., 1971, Toward a unified theory of reflector mapping, Geophysics, 36, 467–481.

Claerbout, J. F., 1985a, Fundamentals of Geophysical Data Processing, Blackwell Science.

Claerbout, J. F., 1985b, Imaging the Earth's Interior, Blackwell Science.

Claerbout, J. F., 1986, Simultaneous pre-normal moveout and post-normal moveout deconvolution, Geophysics, 51, 1341–1354.

Claerbout, J. F., 1992, Earth Sounding Analysis, Blackwell Science.

Clayton, R. W. and B. Engquist, 1980, Absorbing side boundary conditions for wave-equation migration, Geophysics, 45, 895–904.

Clayton, R. W., B. McClary and R. A. Wiggins, 1976, Comments on the paper: Phase distortion and Hilbert transform in multiply reflected and refracted body waves, Bull. Seism. Soc. Am., 66, 325–326.

Constance, P. E., M. B. Holland, S. L. Roche et al., 1999, Simultaneous acquisition of 3-D surface seismic data and 3-C, 3-D VSP data, 69th Annual International Meeting, SEG Exp. Abst., 104–107.

Cooley, J. W. and J. W. Tukey, 1965, An algorithm for the machine calculations of complex Fourier series, Math. Comp., 19, 297–301.

Cox, M., 1999, Static Corrections for Seismic Reflection Surveys, SEG.

Daubechies, I., 1988, Orthonormal bases of compactly supported wavelets, Commun. Pure Appl. Math., 41, 909–996.

Davis, G., S. G. Mallat and Z. Zhang, 1994, Adaptive time-frequency approximations with matching pursuits. In Wavelets: Theory, Algorithms, and Applications, C. K. Chui, L. Montefusco and L. Puccio (eds.), Academic Press, 271–293.

Deal, M. M., G. Matteucci and Y. C. Kim, 2002, Turning ray amplitude inversion: mitigating amplitude attenuation due to shallow gas, 72nd Annual International Meeting, SEG Exp. Abst.

De Amorim, W. N., P. Hubrail and M. Tygel, 1987, Computing field statics with the help of seismic tomography, Geophys. Prosp., 35, 907–919.

Deighan, A. J. and D. R. Watts, 1997, Ground-roll suppression using the wavelet transform, Geophysics, 62, 1896–1903.

Dillon, P. B., 1988, Vertical seismic profile migration using the Kirchhoff integral, Geophysics, 53, 786–799.

Dines, K. A. and R. J. Lytle, 1979, Computerized geophysical tomography, Proc. IEEE, 67, 1065–1073.

Dix, C. H., 1955, seismic velocities from surface measurements, Geophysics, 20, 68–86.

Docherty, P., 1992, Solving for the thickness and velocity of the weathering layer using 2-D refraction tomography, Geophysics, 57, 1307–1318.

Dragoset, W. H. and Ž. Jeričević, 1998. Some remarks on surface multiple attenuation, Geophysics, 63, 772–789.

Dreger, D. S., S. R. Ford and W. R. Walter, 2008, Source analysis of the Crandall Canyon, Utah, mine collapse, Science, 321, 217.

Duncan, W., 2005, Integrated geophysical study of Vinton Dome, Louisiana, Unpublished Ph.D. dissertation, University of Houston.

Duncan, G. and G. Beresford, 1995, Some analyses of 2-D median $f–k$ filters, Geophysics, 60, 1157–1168.

Duncan, P. M. and L. Eisner, 2010, Reservoir characterization using surface microseismic monitoring, Geophysics, 75, A139–A146.

Dziewonski, A. M., 1984, Mapping the lower mantle: Determination of lateral heterogeneity in P velocity up to degree and order 6, J. Geophys. Res., 89, 5929–5952.

Dziewonski, A. M. and D. L. Anderson, 1981, Preliminary reference Earth model, Phys. Earth Planet Inter., 25, 297–356.

Dziewonski, A. M. and J. H. Woodhouse, 1987, Global images of the Earth's interior, Science, 236, 37–48.

Eisner, L., S. Williams-Stroud, A. Hill, P. M. Duncan and M. Thornton, 2010, Beyond the dots in the box: Microseismicity-constrained fracture models for reservoir simulation, The Leading Edge, 29, 326–333.

Embree, P., J. P. Burg and M. M. Backus, 1963, Wide-band velocity filtering – the pie-slice process, Geophysics, 28, 948–974.

Estill, R. and K. Wrolstad, 1993, Interpretive aspects of AVO application to offshore Gulf Coast bright-spot analysis. In Offset-dependent Reflectivity Theory and Practice of AVO Analysis, J. P., Castagna and M. M. Backus (eds.), SEG, 267–284.

Fagin, S., 1996, The fault shadow problem. Its nature and elimination, The Leading Edge, 15, 1005–1013.

Fink, M., 1999, Time-reversed acoustics, Sci. Am, 281, 91–97.

Fomel, S., 2001, Three-dimensional seismic data regularization, Unpublished PhD dissertation, Stanford University.

Fomel, S., 2007, Local seismic attributes, Geophysics, 72, A29–A33.

Foster, D. J. and C. C. Mosher, 1992, Suppression of multiple reflections using the Radon transform, Geophysics, 57, 386–395.

Fuis, G. S., R. W. Clayton, P. M. Davis *et al.*, 2003, Fault systems of the 1971 San Fernando and 1994 Northridge earthquakes, southern California, Relocated aftershocks and seismic images from LARSE II, Geology, 31, 171–174.

Fuis, G. S., M. D. Kohler, M. Scherwath *et al.*, 2007, A comparison between the transpressional plate boundaries of the South Island, New Zealand, and southern California, USA, the Alpine and San Andreas Fault systems. In A Continental Plate Boundary, Tectonics at South Island, New Zealand, D. Okaya, T. Stern and F. Davey (eds.), AGU Monograph 175, 307–327.

Fukoa, Y., M. Obayashi, H. Inoue and M. Nenbai, 1992, Subducting slabs stagnant in the mantle transition zone, J. Geophys. Res., 97, 4809–4822.

Gaiser., J. E., E. Loinger, H. Lynn and L. Vetri, 2002, Birefringence analysis at Emilio field for fracture characterization, First Break, 20, 505–514.

Gao, D., 2003, Volume texture extraction for 3D seismic visualization and interpretation, Geophysics, 68, 1294–1302.

Gao, D., 2009, 3D seismic volume visualization and interpretation: An integrated workflow with case studies, Geophysics, 74, W1–W12.

Gardner, G. H. F., L. W. Gardner and A. R. Gregory, 1974, Formation velocity and density – the diagnostic basics for stratigraphic traps, Geophysics, 39, 770–780.

Gazdag, J., 1978, Wave equation migration with the phase-shift method, Geophysics, 43, 1342–1351.

Gazdag, J. and P. Squazzero, 1984, Migration of seismic data by phase shift plus interpolation, Geophysics, 49, 124–131.

Gilbert, P., 1972, Iterative methods for the reconstruction of three-dimensional objects from projections, J. Theor. Biol., 36, 105–117.

Goloshubin, G. M., A. M. Verkhovsky and V. V. Maurov, 1996, Laboratory experiments of seismic monitoring, Expanded Abst., 58th EAEG meeting, P074.

Gordon, R., R. Bender and G. T. Herman, 1970, Algebraic reconstruction techniques for three-dimensional electron microscopy and X-ray photography, J. Theor. Biol., 29, 471–481.

Grant, F. S. and G. F. West, 1965, Interpretation Theory in Applied Geophysics, McGraw-Hill.

Gratwick, D. and C. Finn, 2005, What's important in making far-stack well-to-seismic ties in West Africa? The Leading Edge, 24, 739–745.

Gray, S. H., 1986, Efficient traveltime calculation for Kirchhoff migration, Geophysics, 51, 1685–1688.

Gray, S. H., J. Etgen, J. Dellinger and D. Whitmore, 2001, Seismic migration problems and solutions, Geophysics, 66, 1622–1640.

Grechka, V., I. Tsvankin, A. Bakulin, C. Signer and J. O. Hansen, 2002, Anisotropic inversion and imaging of PP and PS reflection data in the North Sea, The Leading Edge, 21, 90–97.

Green, W. R., 1975, Inversion of gravity profiles by use of a Backus-Gilbert approach, Geophysics, 40, 763–772.

Grenander, U. and G., Szego, 1958, Toeplitz Forms and their Applications, University of California Press, Berkeley and Los Angeles.

Guerra, R. and S. Leaney, 2006, $Q(z)$ model building using walkaway VSP data, Geophysics, 71, V127–V131.

Guitton, A., 2004, Amplitude and kinematic corrections of migrated images for nonunitary imaging operators, Geophysics, 69, 1017–1024.

Gurevich, B., V. B. Zyrianov and S. L. Lopatnikov, 1997, Seismic attenuation in finely layered porous rocks: Effects of fluid flow and scattering, Geophysics, 62, 319–324.

Hall, S. and J. M. Kendall, 2003, Fracture characterization at Valhall: Application of P-wave amplitude variation with offset and azimuth (AVOA) analysis to a 3-D ocean-bottom data, Geophysics, 68, 1150–1160.

Hampson, D., 1986, Inverse velocity stacking for multiple elimination, J. Can. Soc. Explor. Geophys., 22, 44–55.

Hatton, L., M. H. Worthington and J. Makin, 1986, Seismic Data Processing: Theory and Practice, Blackwell Science.

Hauksson, E. and J. S. Haase, 1997, Three-dimensional Vp and Vp/Vs velocity models of the Los Angeles basin and central Transverse Ranges, California, J. Geophys. Res., 102, 5423–5453.

Helbig, K., 1981, Systematic classification of layer-induced transverse isotropy, Geophys. Prosp., 29, 550–577.

Herrin, E., 1968, Seismological tables for P-phases, Bull. Seism. Soc. Am., 60, 461–489.

Hestenes, M. R. and E. Stiefel, 1952, Methods of conjugate gradients for solving linear systems, J. Res. Natl Bureau Standards, 49, 409–436.

Hilterman, F., 1990, Is AVO the seismic signature of lithology? A case history of Ship Shoal-South Addition, The Leading Edge, 9, 15–22.

Hokstad, K., 2000, Multicomponent Kirchhoff migration, Geophysics, 65, 861–873.

Hornby, B. E., L. M. Schwartz and J. A. Hudson, 1994, Anisotropic effective-medium modeling of the elastic properties of shales, Geophysics, 59, 1570–1583.

Hornbostel, S. C., 1999, A noise-optimized objective method for predictive deconvolution, Geophysics, 64, 552–563.

Hsu , C. and M. Schoenberg, 1993, Elastic waves through a simulated fractured medium, Geophysics, 58, 964–977.

Hudson, J. A., 1981, Wave speeds and attenuation of elastic waves in material containing cracks, Geophys. J. Royal Astron. Soc., 64, 133–150.

Humphreys, E., R. W. Clayton and B. H. Hager, 1984, A tomographic image of mantle structure beneath southern California, Geophys. Res. Lett., 11, 625–627.

Iwasaki, T., H. Shiobara, A. Nishizawa et al., 1989, A detailed subduction structure in the Kuril trench deduced from ocean bottom seismographic refraction studies, Tectonophysics, 165, 315–336.

Jacob, K. H., 1970, Three-dimensional seismic ray tracing in a laterally heterogeneous spherical earth, J. Geophys. Res., 75, 6675–6689.

Jeffreys, H. and K. E. Bullen, 1940, Seismological Tables, British Association for the Advancement of Science.

Jenkins, G. M. and D. G. Watts, 1968, Spectral Analysis and its Applications, Holden-Day.

Jiang, F. and H. Zhou, 2011, Traveltime inversion and error analysis for layered anisotropy, J. Appl. Geophys., 73, 101–110.

Jones, I. F., 2008, A modeling study of preprocessing considerations for reverse-time migration, Geophysics, 73, T99–T106.

Julian, B. R., 1970, Ray tracing in arbitrarily heterogeneous media, Tech. Note 1970–45, Lincoln Laboratory, Massachsuetts Institute of Technology.

Julian, B. R. and D. Gubbins, 1977, Three-dimensional seismic ray tracing, J. Geophys., 43, 95–113.

Kallweit, R. S. and L. C. Wood, 1982, The limits of resolution of zero-phase wavelets, Geophysics, 47, 1035–1046.

Kearey, P. and M. Brooks, 1984, An Introduction to Geophysical Exploration. Blackwell Science.

Kennett, B. L. N. and E. R. Engdahl, 1991, Traveltimes for global earthquake location and phase identification, Geophys. J. Int., 105, 429–465.

Kim, W.-Y., L. R. Sykes, J. H. Armitage *et al.*, 2001, Seismic waves generated by aircraft impacts and building collapses at World Trade Center, New York City, Eos Trans. AGU, 82(47), 565, doi:10.1029/01EO00330.

Knapp, R. W., 1990, Vertical resolution of thick beds, thin beds, and thin-bed cyclothems, Geophysics, 55, 1183–1190.

Kohler, M. D., H. Magistrale and R. W. Clayton, 2003, Mantle heterogeneities and the SCEC reference three-dimensional seismic velocity model version 3, Bull. Seismol. Soc. Am., 93, 757–774.

Koren, Z., I. Ravve, G., Gonzalez and D. Kosloff, 2008, Anisotropic local tomography, Geophysics, 73, VE75–VE92.

Kosloff, D. and E. Baysal, 1982, Forward modeling by a Fourier method, Geophysics, 47, 1402–1422.

Lacombe, C., S. Butt, G. MacKenzie, M. Schons and R. Bornard, 2009, Correcting for water-column variations, The Leading Edge, 28, 198–201.

Lacoss, R. T., 1971, Data adaptive spectral analysis methods, Geophysics, 36, 661–675.

Lanczos, C., 1961, Linear Differential Operators, Chapter 1, Van Nostrand.

Latimer, R. B., R. Davidson and P. van Riel, 2000, An interpreter's guide to understanding and working with seismic-derived acoustic impedance data, The Leading Edge, 19, 242–256.

Lawson, C. L. and D. J. Hanson, 1974, Solving Least Squares Problems, Prentice-Hall.

Lazarevic, I., 2004, Impact of Kirchhoff and wave equation pre-stack depth migrations in improving lateral resolution in a land data environment, Unpublished Master's thesis, University of Houston.

Levenburg, K., 1944, A method for the solution of certain non-linear problems in least squares, Q. J. Appl. Math., 2, 164–168.

Levinson, N., 1947, The Wiener RMS error criterion in filter design and prediction. In Extrapolation, Interpolation and Smoothing of Stationary Time Series, ed. N. Wiener, MIT Press, Appendix B.

Li, Y., X. Zheng and Y. Zhang, 2011, High-frequency anomalies in carbonate reservoir characterization using spectral decomposition, Geophysics, 76, V47–V57.

Lindseth, R. O., 1979, Synthetic sonic logs: A process for stratigraphic interpretations, Geophysics, 44, 3–26.

Liner, C. L., 1999, Concepts of normal and dip moveout, Geophysics, 64, 1637–1647.

Liner, C., C. Li, A. Gersztenkorn and J. Smythe, 2004, SPICE: A new general seismic attribute, 72nd Ann. Intl Meeting, SEG Exp. Abst., 433–436.

Lines, L., 1993, Ambiguity in analysis of velocity and depth, Geophysics, 58, 596–597.

Lines, L. R., W. Wu, H. Lu, A. Burton and J. Zhu, 1996, Migration from topography: Experience with an Alberta Foothills data set, Can. J. Expl. Geophys., 32, 24–30.

Liu, J. and K. J. Marfurt, 2007, Instantaneous spectral attributes to detect channels, Geophysics, 72, P23–P31.

Liu, H., H. Zhou, W. Liu, P. Li and Z. Zou, 2010, Tomographic velocity model building of near surface with reversed-velocity interfaces: A test using the Yilmaz model, Geophysics, 75, U39–U47.

Liu, W., B. Zhao, H. Zhou *et al.*, 2011, Wave-equation global datuming based on the double square root operator, Geophysics, 76, B1–B9.

Loewenthal, D. and I. R.Mufti, 1983, Reversed time migration in spatial frequency domain, Geophysics, 48, 627–635.

Lu, R. S., D. E. Willen and I. A. Watson, 2003, Identifying, removing, and imaging P–S conversions at salt–sediment interfaces, Geophysics, 68, 1052–1059.

Luo, Y., S. Al-Dossary, M. Marhoon and M. Alfaraj, 2003, Generalized Hilbert transform and its applications in geophysics, The Leading Edge, 21, 198–202.

Maeland, E., 2004, Sampling, aliasing, and inverting the linear Radon transform, Geophysics, 69, 859–861.

Mallat, S., 1999, A Wavelet Tour of Signal Processing, 2nd edn, Academic Press.

Mares, S. 1984, Introduction to Applied Geophysics, Reidel.

Margrave, G. F., 1998, Theory of non-stationary linear filtering in the Fourier domain with application to time-variant filtering, Geophysics, 63, 244–259.

Mayne, W. H., 1962, Common reflection point horizontal data stacking techniques, Geophysics, 27, 927–938.

McMechan, G. A., 1983, Migration by extrapolation of time-dependent boundary values, Geophys. Prosp., 31, 413–420.

McNamara, D. E. and R. P. Buland, 2004, Ambient noise levels in the continental United States, Bull. Seismol. Soc. Am., 94, 1517–1527.

Menke, W., 1989, Geophysical Data Analysis, Discrete Inverse Theory, Academic Press.

Michelena, R. J., 1993, Singular value decomposition for cross-well tomography, Geophysics, 58, 1655–1661.

Morley, L. and J. Claerbout, 1983, Predictive deconvolution in shot-receiver space, Geophysics, 48, 515–531.

Moser, T. J., 1991, Shortest path calculation of seismic rays, Geophysics, 56, 59–67.

Mufti, I. R., J. A. Pita and R. W. Huntley, 1996, Finite-difference depth migration of exploration-scale 3-D seismic data, Geophysics, 61, 776–794.

Mulder, W. A. and A. P. E. ten Kroode, 2002, Automatic velocity analysis by differential semblance optimization, Geophysics, 67, 1184–1191.

Neidell, N. S., and M. T. Taner, 1971, Semblance and other coherency measures for multichannel data, Geophysics, 36, 482–497.

Newman, P., 1973, Divergence effects in a layered Earth, Geophysics, 38, 481–488.

Norris, M. W. and A. K. Faichney (eds.), 2001, SEG Y rev1 Data Exchange Format, SEG Technical Standards Committee.

Nur, A. and G. Simmons, 1969, Stress-induced velocity anisotropy in rocks: An experimental study, Journal of Geophysical Research, 74, 6667.

O'Brien, J. and R. Harris, 2006, Multicomponent VSP imaging of tight-gas sands, Geophysics, 71, E83–E90.

Ogilvie, J. S. and G. W. Purnell, 1996, Effects of salt-related mode conversions on subsalt prospecting, Geophysics, 61, 331–348.

Okaya, D. A., 1995, Spectral properties of the Earth's contribution to seismic resolution, Geophysics, 60, 241–251.

Oldenburg, D. W., S. Levy and K. P. Whittall, 1981, Wavelet estimation and deconvolution, Geophysics, 46, 1528–1542.

Olson, A. H., 1987, A Chebyshev condition for accelerating convergence of iterative tomographic methods – solving large least squares problems, Phys. Earth Planet. Inter., 47, 333–345.

Pai, D. M., 1985, A new solution method for wave equations in inhomogeneous media, Geophysics, 50, 1541–1547.

Paige, C. C. and M. A. Saunders, 1982, LSQR: An algorithm for sparse linear equations and sparse least squares, ACM Trans. Math. Software, 8, 43–71.

Partyka, G., J. Gridley and J. Lopez, 1999, Interpretational applications of spectral decomposition in reservoir characterization, The Leading Edge, 18, 353–360.

Peacock, S., C. McCann, J. Sothcott and T. R. Astin, 1994, Experimental measurements of seismic attenuation in microfractured sedimentary rock, Geophysics, 59, 1342–1351.

Pereyra, V., W. H. K. Lee and H. B. Keller 1980, Solving two-point seismic ray tracing problems in a heterogeneous medium, Bull. Seismol. Soc. Am., 70, 79–99.

Peterson, J., 1993, Observation and modeling of seismic background noise, US Geol. Surv. Tech. Rep., 93–322, 1–95.

Pralica, N., 2005, A revisit of tomostatics by deformable layer tomography, Unpublished MS thesis, University of Houston.

Press, W. H, S. A. Teukolsky, W. T. Vetterling and B. P. Flannery, 1992, Numerical Recipes in C, 2nd edn, Cambridge University Press.

Pulliam, R. J., D. W. Vasco and L. R. Johnson, 1993, Tomographic inversions for mantle P wave velocity structure based on the minimization of l2 and l1 norms of International Seismological Centre travel time residuals, J. Geophys. Res., 98, 699–734.

Qin, F., Y. Luo, K. B. Olsen, W. Cai and G. T. Shuster, 1992, Finite-difference solution of the eikonal equation along expanding wavefronts, Geophysics, 57, 478–487.

Rajasekaran, S. and G. A. McMechan, 1995, Pre-stack processing of land data with complex topography, Geophysics, 60, 1875–1886.

Rajasekaran, S. and G. A. McMechan, 1996, Tomographic estimation of the spatial distribution of statics, Geophysics, 61, 1198–1208.

Rathor, B. S., 1997, Velocity–depth ambiguity in the dipping reflector case, Geophysics, 62, 1583–1585.

Rathore, J. S., E. Ejaer, R. M. Holt and L. Renlie, 1995, P- and S-wave anisotropy of a synthetic sandstone with controlled crack geometry, Geophys. Prosp., 43, 711–728.

Ricker, N., 1940, The form and nature of seismic waves and the structure of seismograms, Geophysics, 5, 348–366.

Ricker, N., 1953a, The form and laws of propagation of seismic wavelets, Geophysics, 18, 10–40.

Ricker, N., 1953b, Wavelet contraction, wavelet expansion, and the control of seismic resolution, Geophysics, 18, 769–792.

Rickett, E. R., 2003, Illumination-based normalization for wave-equation depth migration, Geophysics, 68, 1371–1379.

Robinson, E. A., 1998, Model-driven predictive deconvolution, Geophysics, 63, 713–722.

Robinson, E. A. and S. Treitel, 1980, Geophysical Signal Analysis, Prentice-Hall.

Ross, C. P., 2000, Effective AVO crossplot modeling: A tutorial, Geophysics, 65, 700–711.

Rutledge, J. T., W. S. Phillips and B. K. Schuessler, 1998, Reservoir characterization using oil-production-induced microseismicity, Clinton County, Kentucky, Tectonophysics, 289, 129–152.

Ryan, H., 1994, Ricker, Ormsby, Klauder, Butterworth – A choice of wavelets, CSEG Recorder, 19, 7, 8–9.

Sacchi, M. D. and T. J. Ulrych, 1996, Estimation of the discrete Fourier transform, a linear inversion approach, Geophysics, 61, 1128–1136.

Sarkar, D. and I. Tsvankin, 2006, Anisotropic migration velocity analysis: Application to a dataset from West Africa, Geophys. Prosp., 54, 575–587.

Schneider, W. A., 1971, Developments in seismic data processing and analysis (1968–1970), Geophysics, 36, 1043–1073.

Schoenberger, M., 1974, Resolution comparison of minimum-phase and zero-phase signals, Geophysics, 39, 826–833.

Schoenberger, M. and L. M. Houston, 1998, Stationarity transformation of multiples to improve the performance of predictive deconvolution, Geophysics, 63, 723–737.

Schoenberger, M. and F. K. Levin, 1974, Apparent attenuation due to intrabed multiples, Geophysics, 39, 278–291.

Shapiro, N. M., M. Campillo, L. Stehly and M. H. Ritzwoller, 2005, High-resolution surface-wave tomography from ambient seismic noise, Science, 307, 1615–1618.

Sharma, P. V., 1976, Geophysical Methods in Geology, Elsevier.

Sheriff, R. E., 1991, Encyclopedic Dictionary of Exploration Geophysics, 3rd edn, SEG.

Sheriff, R. E. and L. P. Geldart, 1995, Exploration Seismology, 2nd edn, Cambridge University Press.

Shuey, R. T., 1985, A simplification of Zoeppritz equations, Geophysics, 50, 609–614.

Soroka, W. L., T. J. Fitch, K. H. Van Sickle and P. D. North, 2002, Successful production application of 3-D amplitude variation with offset: The lessons learned, Geophysics, 67, 379–390.

Stewart, R. R., 1985, Median filtering: Review and a new f/k analogue design, J. Can. Soc. Expl. Geophys., 21, 54–63.

Stoffa, P. L., J. T. Fokkema, R. M. D. Freire and W. P. Kessinger, 1990, Split-step Fourier migration, Geophysics, 55, 410–421.

Stolt, R. H., 1978, Migration by Fourier transform, Geophysics, 43, 23–48.

Storchak, D. A., J. Schweitzer and P. Bormann, 2003, The IASPEI standard seismic phase list, Seismol. Res. Lett., 74, 6, 766–772.

Strobbia, C., A. Gluschcheko, A. Laake, P. Vermeer, T. J. Papworth and Y. Ji, 2009, Arctic near-surface challenges: the point receiver solution to coherent noise and statics, First Break, 27/2, 69–76.

Sullivan, E. C., K. J. Marfurt, A. Lacazette and M. Ammerman, 2006, Application of new seismic attributes to collapse chimneys in the Fort Worth Basin, Geophysics, 71, B111–B119.

Sun, R., G. A. McMechan, H. Hsiao and J. Chow, 2004, Separating P- and S-waves in pre-stack 3D elastic seismograms using divergence and curl, Geophysics, 69, 286–297.

Symes, W. W. and J. J. Carazzone, 1991, Velocity inversion by differential semblance optimization, Geophysics, 56, 654–663.

Taner, M. T., F. Koehler and K. A. Alhilali, 1974, Estimation and correction of near-surface time anomalies, Geophysics, 39, 441–463.

Taner, M. T., F. Koehler and R. E. Sheriff, 1979, Complex seismic trace analysis, Geophysics, 44, 1041–1063.

Taner, M. T., D. E. Wagner, E. Baysal and L. Lu, 1998, A unified method for 2-D and 3-D refraction statics, Geophysics, 63, 260–274.

Tatham, R. H. and M. D. McCormack, 1991, Multicomponent Seismology in Petroleum Exploration, SEG.

Thomsen, L., 1986, Weak elastic anisotropy, Geophysics, 51, 1954–1966.

Thomsen, L., 1999, Converted-wave reflection seismology over inhomogeneous, anisotropic media, Geophysics, 64, 678–690.

Thornton, M. P. and H. Zhou, 2008, Crustal-scale pre-stack depth imaging for 1994 and 1999 LARSE surveys, Geophys. Prosp., 56, 577–585.

Thorson, J. R. and J. F. Claerbout, 1985, Velocity-stack and slant-stack stochastic inversion, Geophysics, 50, 2727–2741.

Thurber, C. H. and W. L. Ellsworth, 1980, Rapid solution of ray tracing problems in heterogeneous media, Bull. Seismol. Soc. Am., 70, 1137–1148.

Tichelaar, B. W. and L. J. Ruff, 1989, How good are our best models? Jackknifing, bootstrapping, and earthquake depth, Eos, May 16, 593–606.

Tieman, H. J., 1994, Investigating the velocity–depth ambiguity of reflection traveltimes, Geophysics, 59, 1763–1773.

Timur, A., 1968, Velocity of compressional waves in porous media at permafrost temperatures, Geophys. J. R. Astr. Soc., 71, 1–36.

Treitel, S. and L. R. Lines, 1982, Linear inverse theory and deconvolution, Geophysics, 47, 1153–1159.

Treitel, S., J. L. Shanks and C. W. Frasier, 1967, Some aspects of fan filtering, Geophysics, 32, 789–800.

Tsvankin, I., J. Gaiser, V. Grechka, M. van der Baan and L. Thomsen, 2010, Seismic anisotropy in exploration and reservoir characterization: An overview, Geophysics, 75, 75A15–75A29.

Ulrych, T. J., M. D. Sacchi and A. Woodbury, 2001, A Bayes tour of inversion: A tutorial, Geophysics 66, 55–69.

van der Baan, M., 2004, Processing of anisotropic data in the τ–p domain: I – Geometric spreading and moveout corrections, Geophysics, 69, 719–730.

van der Baan, M. and J. M. Kendall, 2002, Estimating anisotropy parameters and traveltimes in the τ–p domain, Geophysics, 67, 1076–1086.

Vermeer, G. J. O., 1998, 3-D symmetric sampling, Geophysics, 63, 1629–1647.

Verschuur, D. J., A. J. Berkhout and C. P. A. Wapenaar, 1992, Adaptive surface-related multiple elimination, Geophysics, 57, 1166–1177.

Versteeg, R. and G. Grau, 1991, The Marmousi experience: Proceedings of the 1990 EAEG Workshop on Practical Aspects of Seismic Data Inversion, 52nd EAEG Meeting, EAEG.

Ver West, B., 2002, Suppressing peg-leg multiples with parabolic Radon demultiple, EAGE 64th Conference and Exhibition, Exp. Abst., Z-99.

Vidale, J. 1990, Finite-difference calculation of traveltimes in three dimensions, Geophysics, 55, 521–526.

Waldhauser F. and W. Ellsworth, 2000, A double-difference earthquake location algorithm: Method and application to the northern Hayward Fault, California, Bull. Seismol. Soc. Am., 90, 1353–1368.

Wang, H., and H. Zhou, 1993, A comparison between spherical harmonic inversion and block inversion in whole mantle tomography, Eos Trans. AGU, 74(43), 418.

Waters, K. H., 1987, Reflection Seismology, 3rd edn, Wiley.

Wei, W. S. S., 1990, Time Series Analysis, Addison-Wesley.

Wesson, R. L., 1971, Travel-time inversion for laterally inhomogeneous crustal velocity models, Bull. Seismol. Soc. Am., 61, 729–746.

White, R. E., 1973, The estimation of signal spectra and related quantities by means of the multiple coherence function, Geophys. Prosp., 21, 660–703.

White, R. E., 1992, The accuracy of estimating Q from seismic data, Geophysics, 57, 1508–1511.

White, R. E., 2003, Tying well-log synthetic seismograms to seismic data: the key factors, SEG Exp. Abst., 22, 2449–2452.

White, R. E. and T. Hu, 1998, How accurate can a well tie be? The Leading Edge, 17, 1065–1071.

Whitmore, N. D. and L. R. Lines, 1986, Vertical seismic profiling depth migration of a salt dome flank, Geophysics, 51, 1087–1109.

Widess, M. B., 1973, How thin is a thin bed? Geophysics, 38, 1176–1180.

Widess, M. B., 1982, Quantifying resolving power of seismic systems, Geophysics, 47, 1160–1173.

Wielandt, E. 1987, On the validity of the ray approximation for interpreting delay times. In Seismic Tomography, G. Nolet (ed.), Reidel, 85–98.

Wiener, N., 1947, Extrapolation, Interpolation and Smoothing of Stationary Time Series, N. Wiener (ed.), MIT Press.

Wiggins, R. A., 1966, Omega-k filter design, Geophys. Prosp., 14, 427–440.

Wiggins, R. A., 1977, Minimum entropy deconvolution, Proc. Int. Symp. Computer Aided Seismic Analysis and Discrimination, Falmouth, MA, IEEE Comp. Soc., 7–14.

Winkler, K. W. and A. Nur, 1982, Seismic attenuation: Effects of pore fluids and frictional sliding, Geophysics, 47, 1–15.

Wold, H., 1938, Stationary Time Series, Almquist & Wiksell, Stockholm.

Wyllie, M. R. J., A. R. Gregory and G. H. F. Gardner, 1958, An experimental investigation of factors affecting elastic wave velocities in porous media, Geophysics, 23, 459–493.

Xie, X., S. Jin and R. Wu, 2006, Wave-equation-based seismic illumination analysis, Geophysics, 71, S169–S177.

Yan, Z., R. W. Clayton and J. Saleeby, 2005, Seismic refraction evidence for steep faults cutting highly attenuated continental basement in the central Transverse ranges, California, Geophys. J. Int., 160, 651–666.

Yilmaz, Ö., 1987, Seismic Data Processing, SEG Series on Investigations in Geophysics, Vol. 2, SEG.

Yilmaz, Ö., 2001, Seismic Data Analysis: Processing, Inversion, and Interpretation of Seismic Data, SEG.

Youn, O. K. and H. Zhou, 2001, Depth imaging with multiples, Geophysics, 66, 246–255.

Young, R. P. and J. J. Hill, 1986, Seismic attenuation spectra in rock mass characterization; a case study in open-pit mining, Geophysics, 51, 302–323.

Yuan, C., S. Peng and C. Li., 2006, An exact solution of the conversion point for the converted waves from a dipping reflector with homogeneous (isotropic) overburden, Geophysics 71, 7–11.

Zhang, J. and M. N. Toksöz, 1998, Nonlinear refraction traveltime tomography, Geophysics, 63, 1726–1737.

Zhang, Y., J. C. Sun and S. H. Gray, 2003, Aliasing in wavefield extrapolation pre-stack migration, Geophysics, 68, 629–633.

Zhou, B. and S. Greenhalgh, 1994, Wave-equation extrapolation-based multiple attenuation: 2-D filtering in the f–k domain, Geophysics, 59, 1377–1391.

Zhou, H., 1988, How well can we resolve the deep seismic slab with seismic tomography? Geophys. Res. Lett., 15, 1425–1428.

Zhou, H., 1989, Travel time tomographic studies of seismic structures around subducted lithospheric slabs, Unpublished PhD dissertation, Caltech.

Zhou, H., 1990, Mapping of P wave slab anomalies beneath the Tonga, Kermadec and New Hebrides arcs, Phys. Earth Planet. Inter., 61, 199–229.

Zhou, H., 1993, Traveltime tomography with a spatial-coherency filter, Geophysics, 58, 720–726.

Zhou, H., 1994, Rapid 3-D hypocentral determination using a master station method, J. Geophys., Res., 99, 15439–15455.

Zhou, H., 1996, A high-resolution P wave model for top 1200 km of the mantle, J. Geophys. Res., 101, 27791–27810.

Zhou, H., 2003, Multiscale traveltime tomography, Geophysics, 68, 1639–1649.

Zhou, H., 2004a, Multi-scale tomography for crustal P and S velocities in southern California, Pure Appl. Geophys, 161, 283–302.

Zhou, H., 2004b, Direct inversion of velocity interfaces, Geophys. Res. Lett., 31, 447–450.

Zhou, H., 2006, Multiscale deformable-layer tomography, Geophysics, 71, R11–R19.

Zhou, H., K. Al-Rufaii, J. Byun and S. L. Roche, 2001, Retrieval of high-resolution components by deterministic deconvolution: A field example, 71st Annual International Meeting, SEG Exp. Abst., 20, 1827–1830.

Zhou, H. and G. Chen, 1995, Waveform response to the morphology of 2D subducting slabs, Geophys. J. Int., 121, 511–522.

Zhou, H., P. Li, Z. Yan and H. Liu, 2009, Constrained deformable layer tomostatics, Geophysics, 74, WCB15–WCB26.

Zhou, H., J. A. Mendoza, C. A. Link, J. Jech and J. A. McDonald, 1993, Crosswell imaging in a shallow unconsolidated reservoir, The Leading Edge, 12(1) 32–36.

Zhou, H. B., D. Pham, S. Gray and B. Wang, 2004, Tomographic velocity analysis in strongly anisotropic TTI media, SEG Exp. Abst., 23, 2347–2350.

Zhu, L., 2002, Deformation in the lower crust and downward extent of the San Andreas Fault as revealed by teleseismic waveforms, Earth Plan. Space, 54, 1005–1010.

Zhu, L. and H. Kanamori, 2000, Moho depth variation in southern California from teleseismic receiver functions, J. Geophys. Res., 105, 2969–2980.

Zhu, J. and L. R. Lines, 1998, Comparison of Kirchhoff and reverse-time migration methods with applications to pre-stack depth imaging of complex structures, Geophysics, 63, 1166–1176.

Zhu, X., B. G. Angstman, D. P. Sixta and B. G. Angstman, 1992, Tomostatics: Turning-ray tomography + static corrections, The Leading Edge, 11, 15–23.

Ziolkowski, A., 1984, *Deconvolution*, International Human Resources Development Corporation.

Ziolkowski, A., P. Hanssen, R. Gatliff *et al.*, 2003, Use of low frequencies for sub-basalt imaging, Geophys. Prosp., 51, 169–182.

INDEX